When do fishes become juveniles?

Developments in environmental biology of fishes 19

Series Editor
EUGENE K. BALON

When do fishes become juveniles?

Guest Editors:

Gordon H. Copp, Vladimír Kováč & Karol Hensel

Reprinted from *Environmental biology of fishes*, Volume 56 (1–2), 1999
with addition of species and subject index

SPRINGER-SCIENCE+BUSINESS MEDIA, B.V.

A C.I.P. Catalogue record for this book is available from the library of Congress

ISBN 978-90-481-5305-3 ISBN 978-94-017-3678-7 (eBook)
DOI 10.1007/978-94-017-3678-7

Cover design by Eugene and Janusz Balon
using tails of a roach larva and juvenile from the 1956 paper

The logo
designed by Elżbieta Sierakowska-Copp

Printed on acid-free paper

CONTENTS

Prelude: looking at early development in fishes, by V. Kováč & G.H. Copp 7–14

Keynote presentation

Alternative ways to become a juvenile or a definitive phenotype (and on some persisting linguistic offenses), by E.K. Balon 17–38

Part 1. Reflections on early ontogeny and metamorphosis

Features of transition from larva to juvenile in fishes with different types of early ontogeny, by D.A. Pavlov 41–52

Thyroxine as a mediator of metamorphosis of Atlantic halibut, *Hippoglossus hippoglossus*, by J.S. Solbakken, B. Norberg, K. Watanabe & K. Pittman 53–65

Early development of the sofie, *Chondrostoma toxostoma*, by R.E. Gozlan, G.H. Copp & J.-N. Tourenq 67–77

The reproductive biology and early ontogeny of the mouthbrooding Banggai cardinalfish, *Pterapogon kauderni* (Perciformes, Apogonidae), by A. Vagelli 79–92

The onset of the juvenile period in carp, *Cyprinus carpio*: a literature survey, by L. Vilizzi & K.F. Walker 93–102

Part 2. Organism-environment relationships

Morphometry of the stone loach, *Barbatula barbatula*: do mensural characters reflect the species' life history thresholds? by V. Kováč, G.H. Copp & M.P. Francis 105–115

Correspondence between ontogenetic shifts in morphology and habitat use in minnow *Phoxinus phoxinus*, by P.D. Simonović, P. Garner, E.A. Eastwood, V. Kováč & G.H. Copp 117–128

Fish, flows and flood plains: links between freshwater fishes and their environment in the Murray-Darling River system, Australia, by P. Humphries, A.J. King & J.D. Koehn 129–151

Comparison of growth plasticity in the laboratory and field, and implications for the onset of juvenile development in sofie, *Chondrostoma toxostoma*, by R.E. Gozlan, G.H. Copp & J.-N. Tourenq 153–165

Part 3. Ontogeny of predator-prey interactions

A review of predation impact by 0+ fish on zooplankton in fresh and brackish waters of the temperate northern hemisphere, by T. Mehner & R. Thiel 169–181

Seasonal and diel utilisation of inshore microhabitats by larvae and juveniles of *Leuciscus cephalus* and *Leuciscus leuciscus*, by E. Baras & J. Nindaba 183–197

Seasonal shifts in day-time resource use of 0+ barbel, *Barbus barbus*, by A. Bischoff & J. Freyhof 199–212

Vertical distribution and feeding activity of metamorphosing sole, *Solea solea*, before immigration to the Bay of Vilaine nursery (northern Bay of Biscay, France), by F. Lagardère, R. Amara & L. Jossard 213–228

Part 4. Behaviour and ontogeny
Ontogeny of aggressive behaviour in schools of yellowtail, *Seriola quinqueradiata*, by Y. Sakakura & K. Tsukamoto 231–242
School formation and concurrent developmental changes in carangid fish with reference to dietary conditions, by R. Masuda & K. Tsukamoto 243–252
Ontogeny of diel pattern of stream-margin habitat use by emerging brown trout, *Salmo trutta*, in experimental channels: influence of food and predator presence, by J.-M. Roussel & A. Bardonnet 253–262
Size-based variation in somatic energy reserves and parental expenditure by male small-mouth bass, *Micropterus dolomieu*, by R.W. Mackereth, D.L.G. Noakes & M.S. Ridgway 263–275

To be a juvenile and not to be a larva: an attempt to synthesize, by K. Hensel 277–280

Species and subject index 281–289

Environmental Biology of Fishes **56:** 7–14, 1999.

Prelude: looking at early development in fishes

How the parturition of our workshop began

The seed that gave fruit to the workshop from which this volume was derived[1] was planted sometime in July 1992 during Vlado's first exchange visit to Gordon's laboratory in England. The laboratory was not big, it was not particularly tidy and it was rather modest with respect to equipment. It was there that the first measurements of some small specimens of roach were taken, ignoring the question whether they were larvae or juveniles. A couple of weeks later, we started to discuss the possible correlations between changes in mensural characters and habitat use that fish undergo during their early ontogeny and the possible influence of human activities (fragmentation of river systems) on the phenotype. We found some answers (e.g. Kováč & Copp 1996), however, as usual, each answer triggered a proliferation of new questions, some of which were later not expressed exactly in the way we intended (Copp & Kováč 1996).

So, following the saying 'two heads are better than one', and some three years and more heads later (not to mention some more reading but also some more litres of beer and wine), we came up with the idea to organize a workshop on this topic. Gordon was driving us home after a long day at the field laboratory when he suggested (following similar suggestion by Milan Peňáz in 1995) that we should organize a workshop to bring together colleagues to discuss the problem of early ontogeny, especially of larva-to-juvenile transition from various points of view. It should be a rather informal event, we thought, organised as the first workshop of the Fish Ontogeny Network of Europe (FONE), a forum we and twenty-something other colleagues had tried to get funded by the EC.

The workshop

Although the central theme of the workshop was indeed its title 'When do fish become juveniles – looking beyond metamorphosis to juvenile development', many other important questions of early ontogeny were discussed. For example, besides the questions on metamorphosis and the beginning of juvenile development, two other problems frequently came up: (1) is fish ontogeny saltatory or non-saltatory, and (2) does the larva period begin with hatching or with the onset of exogenous feeding? No matter whether asked explicitly or implicitly, these questions usually promoted the most emotional debates of the workshop. There have already been countless, both oral and written, arguments on various aspects of these questions (e.g. Richards 1976, Balon 1976, Hempel 1979, Houde 1981, Blaxter 1988), and the story seems never ending. We have taken part in many such discussions, and have always tried to listen carefully to the arguments of people standing on opposite sides. One main conclusion can be derived from these debates – that the major problem is a *lack of understanding*! Thus, what is actually the conflict between these two views of ontogenetic processes, which seem to be contradictory and in total opposition?

What do 'gradualists' say?

Do fish (and other creatures) develop gradually, as time passes, or is their development non-gradual, i.e. saltatory, with distinguishable natural intervals separated by thresholds? (e.g. Balon 1986a, 1989a,b, 1990, 1999 this volume).

[1] Some articles in this volume were not presented orally at the workshop, but submitted independently and subsequently invited to be included due to their relevance to the Workshop's theme.

Some fishery biologists (who deal only indirectly with the developmental biology of fishes) say that the theory of saltatory ontogeny is nonsense. In general, 'gradualists'[2] believe that ontogeny is a gradual process during which small and inconspicuous changes in form and structure accumulate continuously. These slow changes then result somehow in the development and growth of an individual, and are associated with changes in the individual's ecological role and position. Interestingly, it appears that 'gradualists' are grouped mainly from scientists specialising in fisheries ecology or behavioural ecology, which was also apparent from the debates we had during our workshop. However, this is not to say that all so-called ecologists ignore the theory of saltatory ontogeny. Some (e.g. Bruton 1990, Copp 1990, 1992, Holden & Bruton 1994, Garner 1996, Roussel & Bardonnet 1999 this volume) have found the saltatory life-history model an extremely useful framework within which to interpret ecological data because of its relevance to the environmental biology of fishes. However, its relevance to ecology and ethology is even wider, mirroring a similar perception of geomorphological and ecological succession processes (Amoros et al. 1987, Copp 1989).

What do 'saltatorists' say?

The non-gradual view of ontogeny has been developing for at least half a century (Vasnetsov 1953). A 'theory of saltatory ontogeny' has been finally formulated in terms of new emerging concepts (e.g. dualism of Tao and the nonequilibrium thermodynamics) by Balon (e.g. 1986a, 1990). The 'theory of saltatory ontogeny' views development as a sequence of longer stabilised states (steps) alternating with shorter less stable intervals (thresholds). During the stabilised state, cells, tissues and organs develop and grow *gradually* and *continuously* until they are ready for new or additional functions. As soon as they reach this level of development, the stabilised state is replaced, for a short period of time, by a less stable interval during which new organ-to-organ or organism-to-environment interactions are achieved (Balon 1986a, 1989b). For example, if a fish wants to catch a prey for the first time in its life, it must have completed structures that enable it to do so, otherwise it fails. It must be able to move, detect a prey item, capture it, swallow and digest it. Therefore, a certain level of development is necessary with respect to the muscles, the senses, the nervous, circulatory, digestive, and respiratory systems, etc., to coincide in functional readiness. Although these organ systems may develop gradually, all such attempts to catch prey by an individual that has not achieved the final readiness of development in the given stabilized interval (step) must fail, such as in the yellow tail, *Seriola quinqueradiata* (Sakakura & Tsukamoto 1999 this volume). However, once the completion of each particular system within the step coincided, the individual is suddenly successful and starts feeding exogenously. This, however, can be achieved only owing to the rapid change of capability of the organism to carry out new additional activities – to catch and digest the prey. This is the threshold in its ontogeny. Similar failures are also apparent in subsequent dietary shifts, such as the suffocation that occurs in precocious attempts at piscivory (see Sakakura & Tsukamoto op. cit.).

Gradual or saltatory? Asking the wrong question!

Here we go! We have two points of view, which seem to be opposite to each other. Gradual or saltatory? So is it now just to make a choice, which is right and which is wrong? Unfortunately (or perhaps fortunately . . .), it is not so simple, and any attempt to make such a choice would be meaningless because the two points of view are not mutually exclusive, each incorporating elements of the other: on the one hand, the gradualist perspective must recognize at least the major saltatory thresholds that separate periods of development otherwise they are unable to distinguish embryos, larvae, juveniles, and adults; on the other hand, and as stated above for the theory of saltatory ontogeny, gradual and continuous development are inherent to the 'stabilised' intervals (steps) between the thresholds (Balon 1989a,b). So, to address this ongoing misunderstanding, we had better try to answer another question: what are the major conflicts in this debate?

[2] Though an imprecise term, 'gradualist' is used here to designate those espousing the gradual process of development.

Misunderstanding 1: saltatory versus gradual

Why do 'gradualists' refuse the theory of saltatory ontogeny? An anonymous referee, who reviewed one of Vlado's articles on early development (Kováč 1995), wrote: 'Have any "jumps" been found in the development of the mudminnow (which I doubt very much)? From my experience with early development of fish the concept of any "saltatory" behaviour of ontogeny is nonsense and not confirmed by any empirical data. Also the concept of a "free embryo" cannot be accepted since hatching might not be relevant in terms of morphology but it certainly is from a physiological and ecological point of view. Therefore I think the larval period starts with hatching.' Lauri Urho (unpublished manuscript) goes even further stating that 'we need to break away from the conventional way of considering larva life *only from the ontogenetic point of view...*', and that 'the onset of feeding may well be an important step in the life of fish larvae, but it is not as *practical* a boundary as hatching in studies of fish survival and ecology' (Italics ours).

These quotations illustrate a line of reasoning often used by 'gradualists' and many fishery ecologists. However, the theory of saltatory ontogeny *does not* deny the gradual pattern in development completely. It only states that this occurs within the intervals designated as stabilised states, i.e. steps. Furthermore, the approach of the anonymous referee is apparently based on the following misconceptions: (1) if the saltatory concept says there are 'jumps' (read 'thresholds') in the development, then we should see them easily; (2) the theory of saltatory ontogeny views the development of fish mainly from a morphological (and/or so called 'ontogenetic') point of view; and (3) we cannot believe in things we cannot prove empirically.

Countless examples from the history of science exist to dispute this latter argument. Balon (1989a) refers to the story of 'quantum jumps' by Niels Bohr, which have been observed in atoms over half a century later after their existence was predicted and accepted! To be more trivial, did people not believe that electricity had existed before they recognized how to use it? And eventually, even the onset and the end of ontogeny of each individual represent thresholds: an ovum is mature and ready to be activated and/or to accept a sperm well before it really happens, whereas the ontogeny of a zygote starts suddenly; death is sudden and non-gradual, though dying can be often a longer process. Eventually, some progress towards empirical proof of saltatory thresholds is being made through more comprehensive laboratory and field studies of ontogeny and ecology (Gozlan 1998, Solbakken et al. 1999 this volume), but this will be discussed later.

Misunderstanding 2: 'ontogenetic' point of view

Although Urho's (workshop communication) call to break away from considering larva life '*only from the ontogenetic point of view*', has not been formulated ideally, it indeed corroborates the opinion of many workshop participants that a holistic approach in the study of ontogeny is necessary. This has been expressed either explicitly or implicitly for example in Balon, Gozlan et al., Kováč et al., Vilizzi & Walker (all 1999 in this volume). So what is misunderstanding two? Even the earliest papers on the theory of saltatory ontogeny (e.g. Balon 1975, 1979, 1986a) addressed implicitly that there is no extra 'ontogenetic point of view' in the theory of saltatory ontogeny. This theory apparently involves all aspects of biology, e.g. morphology, ecology, physiology, behaviour, genetics, etc. A major theme of the 'theory of saltatory ontogeny' is that developmental thresholds are nothing but a new cell-to-cell, tissue-to-tissue, organ-to-organ and/or organism-to-*environment* interactions. Are we not using different terms to express the same idea?

Misunderstanding 3: the onset of larva period

The third debate is about whether the larva period begins with hatching or with the onset of exogenous feeding. Discussions on the advantages and disadvantages of these 'interpretations' represent a kind of utilitarianism, which should be avoided in scientific investigations. Biological phenomena do exist regardless of what humans think about them. Gradual ontogeny may be easier to understand and more comfortable and practical to use, but the arbitrary selection of 'normal stages' does not provide a realistic representation of development (Balon 1981, 1986a, 1989a,b). It is undoubtedly easier to distinguish between creatures 'imprisoned' in their egg envelopes and free swimming, designating the former as eggs and the latter as larvae. However, such an approach has little to do with the biological nature of the process called ontogeny. It is just a technique that makes the separation and indentification of specimens simpler (see e.g. Snyder 1976, Moser 1981, 1984, Simon & Vondruska 1991). From a scientific point of view, however, it may lead to incorrect conclusions (depending on the objectives of each particular study), because it ignores general biological principles common for most living organisms in this world (Balon 1976, 1999 this voume). If a fishery biologist or ecologist does not wish to undertake detailed studies in which distinction between free embryos and larvae is required, then that is a personal choice, which will nonetheless influence the scientific conclusions that can be drawn from that study (e.g. Mackereth et al. 1999 this volume). Practicality does not justify criticism of studies that address the biological and ecological nature of ontogeny just because they are less 'practical'.

So, when do fish become juveniles?

We believe it is fair to say that the workshop ended with as many new questions as those answered. However, progress was made at least in emphasizing that the ontogeny of fishes is as much about ecological issues as morphological or physiological (see Humphries et al. 1999, Solbakken et al. 1999, both this volume). The studies that led us directly to the workshop (Kováč & Copp 1996, Copp & Kováč 1996) highlighted a poorly defined aspect of early fish development. However, this was not Gordon's first grapple with defining juvenile transition. During his doctoral studies (Copp & Peňáz 1988), he encountered discrepancies between Snyder's (1976) 'practical' definition of the onset of juvenile development and shifts in the logarithmic relationship between length and weight. This led to a complete reprocessing of thousands of specimens using the more ecologically, comprehensive-based saltatory life-history model (Balon 1975, 1986b), which has provided a useful, more developmentally-based framework within which subsequent ecological investigations have been designed and interpreted (e.g. Copp 1990, 1992, Copp & Mann 1993, Garner 1996, Garner & Copp 1997, various articles in this volume).

Nonetheless, a major flaw of some of our studies (Copp & Kováč 1996, Kováč et al. 1999 this volume, Simonović et al. 1999 this volume) was that a post priori hypothesis was being tested with preserved specimens not necessarily collected for the purpose of these investigations. What was needed was a more comprehensive study of a species' ontogeny (such as that taken by e.g. McElman & Balon 1979, 1980, Crawford & Balon 1994a,b,c), involving as many aspects of its morphology, physiology, niche and behaviour as possible, undertaken within the context of the species' entire life history and combining field and laboratory studies to highlight adaptations to ambient environmental conditions during early ontogeny. Gordon was fortunate enough to receive funding and to be able to recruit an enthusiastic enough candidate to attempt such an ambitiously intensive study. Gozlan's (1998) doctoral studies revealed marked differences between wild and laboratory-reared sofie, *Chondrostoma toxostoma*, with respect to the rate of ontogenetic development, somatic growth, relative growth, i.e. morphology, as well as in swimming capacity. Laboratory reared sofie took twice as many degree days to achieve larva step 8 as did wild specimens, but at half the size (Gozlan et al. 1999a,b this volume), suggesting that laboratory-reared fish are dwarf 'altricial' forms compared to the 'precocial' form of wild specimens (see Balon 1989b, 1990). Variations in ontogenetic rates of development thus could partly explain the discrepancy in sizes between the roach, *Rutilus rutilus*, described by Balon (1956) in the laboratory and by Copp & Kováč (1996) in the field, as well as those seen in the minnow, *Phoxinus phoxinus*, in different rivers (Simonović et al. 1999 this volume).

Nevertheless, definitions of the transition to juvenile development remain controversial, as the completion of several adult features (e.g. olfactory pits divided by membranous septum, scale cover, mucus, inferior mouth, etc.) does not coincide with 'stabilised' relative growth (i.e. the end of metamorphosis) in all species with indirect development (e.g. Gozlan et al. 1999b). Because the end of metamorphosis may coincide with shifts in habitat and swimming ability (Gozlan 1998, Gozlan et al. 1999b, Kováč et al. 1999), we suggest that this interval between the acquisition/completion of adult structures and the stabilisation of relative growth is almost entirely metamorphic and warrants consideration as a separate developmental interval, perhaps a new 'metabiotic' phase of the larva period in some species (see Wald 1981, Balon 1999 this volume) or a step in others (e.g. the sofie), depending on species-specific ontogeny. This proposed interval could encompass the 'second metamorphosis' described by Youson (1988) for lampreys, and more recently for stripped jack, *Pseudocaranx dentex*, by Masuda & Tsukamoto

(1999 this volume). Justification for a separate interval is based on clearly definable start and end points as well as an extended temporal interval (e.g. from late summer until the following spring in the sofie, or the morphological shifts, scale cover and changes in schooling behaviour of 'second metamorphosis' in stripped jack). However, the hormonally-driven basis of metamorphosis in Atlantic halibut, *Hippoglossus hippoglossus* (Solbakken et al. 1999 this volume), with a 'window of opportunity' in which eyes migrate (adaptation to a benthic habitat) suggests that similar physiological processes may be the driving force in the metamorphosis of other species such as the sofie, playing a role in over-winter recruitment success and the adaptation to a new habitat at the end of the first year of life.

Looking beyond the first workshop on when fish become juveniles

Whilst the workshop that yielded this volume was viewed unanimously to have been a success, the major conclusion agreed by the participants was that early ontogeny in general and metabiosis/metamorphosis in particular deserve much more attention and comparative study (e.g. Balon 1980, 1985, Crawford & Balon 1994a,b,c, Kováč 1994, Gozlan 1998) than given and done so far. It is safe to say that greater progress can be made if ichthyologists coming from ecological and developmental biology orientations work together and avoid terminological arguments through clear understanding of the organismal biology. As convenors of the workshop, we come respectively from developmental biology and ecological orientations but have found collaboration extremely fruitful both scientifically and culturally, because we both recognise that ontogeny is not about morphology only, and it is not about ecology only; it is about the behaviour, morphology, physiology and anything else of an organism during its entire life history. The integrations of these various disciplines within fishes we study is unavoidable, and our approach to their study should be equally integrated.

Acknowledgements

The EC is thanked for funding the TEMPUS exchange programme that brought the two authors together, more by accident than design, eventually leading to a fruitful research collaboration and, more importantly, a nice friendship. We also thank staff and students of the Institute of Ecology, Faculty of Natural Sciences, Comenius University, Bratislava, for their support and help, with special thanks directed to J. Halgoš, as well as staff of the University of Hertfordshire, Hatfield. We also wish to thank the European Commission for its financial support of Central and Eastern European scientists to attend the Workshop, and finally to the Workshop participants themselves for their contributions and friendship.

References cited

Amoros, C., J.-C. Rostan, G. Pautou & J.-P. Bravard. 1987. The reversible process concept applied to the environmental management of large river systems. Env. Manag. 11: 607–618.

Balon, E. 1956. Neres a postembryonálny vývoj plotice (*Rutilus rutilus* ssp.) (Spawning and postembryonic development of the roach). Biologické práce 2: 7–60 (in Slovak).

Balon, E.K. 1975. Terminology of intervals in fish development. J. Fish. Res. Board Can. 32: 1663–1670.

Balon, E.K. 1976. A note concerning Dr Richards' comments. J. Fish. Res. Board Can. 33: 1254–1256.

Balon, E.K. 1979. The theory of saltation and its application in the ontogeny of fishes: steps and thresholds. Env. Biol. Fish. 4: 97–101.

Balon E.K. (ed.) 1980. Charrs: salmonid fishes of the genus *Salvelinus*. Perspectives in Vertebrate Science 1, Dr W. Junk Publishers, The Hague. 928 pp.

Balon, E.K. 1981. Saltatory processes and altricial to precocial forms in the ontogeny of fishes. Amer. Zool. 21: 573–596.

Balon, E.K. (ed.) 1985. Early life history of fishes: new developmental, ecological and evolutionary perspectives. Developments in Environmental Biology of Fishes 5, Dr W. Junk Publishers, Dordrecht. 280 pp.

Balon, E.K. 1986a. Saltatory ontogeny and evolution. Rivista di Biologia/Biology Forum 79: 151–190 (in English and Italian).

Balon, E.K. 1986b. Types of feeding in the ontogeny of fishes and the life-history model. Env. Biol. Fish. 16: 11–24.

Balon, E.K. 1989a. The Tao of life: from the dynamic unity of polar opposites to self-organization. pp. 7–40. *In*: M.N. Bruton (ed.) Alternative Life-History Styles of Animals, Perspectives in Vertebrate Science 6, Kluwer Academic Publishers, Dordrecht.

Balon, E.K. 1989b. The epigenetic mechanisms of bifurcation and alternative life-history styles. pp. 467–501. *In*: M.N. Bruton (ed.) Alternative Life-History Styles of Animals, Perspectives in Vertebrate Science 6, Kluwer Academic Publishers, Dordrecht.

Balon, E.K. 1990. Epigenesis of an epigeneticist: the development of some alternative concepts on the early ontogeny and evolution of fishes. Guelph Ichthyol. Rev. 1: 1–42.

Balon, E.K. 1999. Alternative ways to become a juvenile or a definitive phenotype (and on some persisting linguistic offenses). Env. Biol. Fish. 56: 17–38 (this volume).

Blaxter, J.H.S. 1988. Pattern and variety in development. pp. 1–58. *In*: W.S. Hoar & D.J. Randall (ed.) Fish Physiology, Vol. 11A, Eggs and Larvae, Academic Press, San Diego.

Bruton, M.N. (ed.) 1990. Alternative life-history styles of fishes. Developments in Environmental Biology of Fishes 10, Kluwer Academic Publishers, Dordrecht. 327 pp.

Copp, G.H. 1989. The habitat diversity and fish reproductive function of floodplain ecosystems. Env. Biol. Fish. 26: 1–26.

Copp, G.H. 1990. Recognition of cohorts and growth of larval and juvenile roach *Rutilus rutilus* (L.), using size class ordination of developmental steps. J. Fish Biol. 36: 803–819.

Copp, G.H. 1992. Comparative microhabitat use of cyprinid larvae and juveniles in a lotic floodplain channel. Env. Biol. Fish. 33: 181–193.

Copp, G.H. & V. Kováč. 1996. When do fish with indirect development become juveniles? Can. J. Aquat. Sci. 53: 746–752.

Copp, G.H. & R.H.K. Mann. 1993. Comparative growth and diet of tench *Tinca tinca* (L.) larvae and juveniles in river floodplain biotopes in France and England. Ecol. Freshwat. Fish. 2: 58–66.

Copp, G.H. & M. Peňáz. 1988. Ecology of fish spawning and nursery zones in the flood plain, using a new sampling approach. Hydrobiologia 169: 209–224.

Crawford, S.S. & E.K. Balon. 1994a. Alternative life histories of the genus *Lucania*: 1. Early ontogeny of *L. parva*, the rainwater killifish. Env. Biol. Fish. 40: 349–389.

Crawford, S.S. & E.K. Balon. 1994b. Alternative life histories of the genus *Lucania*: 2. Early ontogeny of *L. goodei*, the bluefin killifish. Env. Biol. Fish. 41: 331–368.

Crawford, S.S. & E.K. Balon. 1994c. Alternative life histories of the genus *Lucania*: 3. An ecomorphological explanation of altricial (*L. parva*) and precocial (*L. goodei*) species. Env. Biol. Fish. 41: 369–402.

Garner, P. 1996. Microhabitat use and diet of 0+ cyprinid fishes in a lentic, regulated reach of the River Great Ouse, England. J. Fish Biol. 48: 367–382.

Garner, P. & G.H. Copp. 1997. Variation in the length-weight relationships in 0+ cyprinid fishes in the River Great Ouse, UK. Folia Zool. 46: 273–278.

Gozlan, R.E. 1998. Environmental biology of the sofie *Chondrostoma toxostoma* (Cypinidae), with emphasis on early development. Ph.D. Thesis, University of Hertfordshire, Hatfield. 210 pp.

Gozlan, R.E., G.H. Copp & J.-N. Tourenq. 1999a. Early development of the sofie, *Chondrostoma toxostoma*. Env. Biol. Fish. 56: 67–77 (this volume).

Gozlan, R.E., G.H. Copp & J.-N. Tourenq. 1999b. Comparison of growth plasticity in the laboratory and field, and implications for the onset of juvenile development in sofie, *Chondrostoma toxostoma*. Env. Biol. Fish. 56: 153–165 (this volume).

Hempel, G. 1979. Early life history of marine fish. The egg stage. University of Washington Press, Seattle. 70 pp.

Holden, K.H. & M.N. Bruton. 1994. The early ontogeny of the southern mouthbrooder, *Pseudocrenilabrus philander* (Pisces, Cichlidae). Env. Biol. Fish. 41: 311–329.

Houde, E.D. 1981. Book critique of 'G. Hempel, 1979, Early life history of marine fish. The egg stage.' Env. Biol. Fish. 6: 391–392.

Humphries, P., A.J. King & J.D. Koehn. 1999. Fish, flows and floodplains: links between freshwater fishes and their environment in the Murray–Darling Riter system, Australia. Env. Biol. Fish. 56: 129–151 (this volume).

Kováč, V. 1994. Early development of three *Gymnocephalus* species: reflections on the evolution of the genus. Env. Biol. Fish. 40: 241–253.

Kováč, V. 1995. Reproductive behaviour and early development of the European mudminnow, *Umbra krameri*. Folia Zool. 44: 57–80.

Kováč, V. & G. H. Copp 1996. Ontogenetic patterns of relative growth in young roach *Rutilus rutilus*: within river-basin comparison. Ecography 19: 153–161.

Kováč, V., G. H. Copp & M. P. Francis. 1999. Morphometry of the stone loach, *Barbatula barbatula*: do mensural characters reflect the species' life-history thresholds? Env. Biol. Fish. 56: 105–115 (this volume).

Mackereth, R.W., D.L.G. Noakes & M.S. Ridgway. 1999. Size-based variation in somatic energy reserves and parental expenditure by male smallmouth bass, *Micropterus dolomieu*. Env. Biol. Fish. 56: 263–275 (this volume).

Masuda, R. & K. Tsukamoto. 1999. School formation and concurrent developmental changes in carangid fish with reference to dietary conditions. Env. Biol. Fish. 56: 243–252 (this volume).

McElman, J.F. & E.K. Balon. 1979. Early ontogeny of walleye, *Stizostedion vitreum*, with steps of saltatory development. Env. Biol. Fish. 4: 309–348.

McElman, J.F. & E.K. Balon. 1980. Early ontogeny of white sucker, *Catostomus commersoni*, with steps of saltatory development. Env. Biol. Fish. 5: 191–224.

Moser, H.G. 1981. Morphological and functional aspects of marine fish larvae. pp. 89–131. *In*: R. Lasker (ed.) Marine Fish Larvae, Washington Sea Grant Program, Seattle.
Moser, H.G. (ed.) 1984. Ontogeny and systematics of fishes. Amer. Soc. Ichthyol. Herpet. Special Publ. 1. 760 pp.
Pavlov, D.A. 1999. Features of transition from larva to juvenile in fishes with different types of early ontogeny. Env. Biol. Fish. 56: 41–52 (this volume).
Richards, W.J. 1976. Some comments on Balon's terminology of fish developmental intervals. J. Fish. Res. Board Can. 33: 1253–1254.
Roussel, J.-M. & A. Bardonnet. 1999. Ontogeny of diel pattern of stream-margin habitat use by emerging brown trout, *Salmo trutta*, in experimental channels: influence of food and predator presence. Env. Biol. Fish. 56: 253–262 (this volume).
Sakakura, Y. & K. Tsukamoto. 1999. Ontogeny of aggressive behaviour in schools of yellowtail, *Seriola quinqueradiata*. Env. Biol. Fish. 56: 231–242 (this volume).
Simon P.S. & J.T. Vondruska. 1991. Larval identification of the ruffe, *Gymnocephalus cernuus* (Linnaeus) (Percidae: Percini), in the St. Louis River Estuary, Lake Superior drainage basin, Minnesota. Can. J. Zool. 69: 436–442.
Simonović, P.D., P. Garner, E.A. Eastwood, V. Kováč & G.H. Copp. 1999. Correspondence between ontogenetic shifts in morphology and habitat use in minnow *Phoxinus phoxinus*. Env. Biol. Fish. 56: 117–128 (this volume).
Snyder, D.E. 1976. Terminologies for intervals of larval fish development. pp. 41–58. *In*: J. Boreman (ed.) Great Lakes Fish Egg and Larvae Identification: Proceedings of a Workshop, National Poer Plant Team, Power Plant Project FWS/OBS 76/23, Ann Arbor.
Solbakken, J.S., B. Norberg, K. Watanabe & K. Pittman. 1999. Thyroxine as a mediator of metamorphosis of Atlantic halibut, *Hippoglossus hippoglossus*. Env. Biol. Fish. 56: 53–65 (this volume).
Vasnetsov, V.V. 1953. Etaps in the development of bony fishes. pp. 207–217. *In*: E.N. Pavlovsky (ed.) Otcherky po Obshtch. Vopr. Ichthyol., AN SSSR Press, Moscow-Leningrad (in Russian).
Vilizzi, L. & K.F. Walker. 1999. The onset of the juvenile period in carp, *Cyprinus carpio*: a literature survey. Env. Biol. Fish. 56: 93–102 (this volume).
Wald, G. 1981. Metamorphosis: an overview. pp. 1–39. *In*: L.I. Gilbert & E. Frieden (ed.) Metamorphosis, A Problem in Developmental Biology, Plenum Press, New York.
Youson, J.H. 1988. First metamorphosis. pp. 135–196. *In*: W.S. Hoar & D.J. Randall (ed.) Fish Physiology, Vol. 11B, Academic Press, New York.

Bratislava and Hatfield
22 December 1998

Vladimír Kováč
Gordon H. Copp

Juvenile workshop images: a – Gordon Copp (chairing) and Vlado Kováč (presenting) a paper at the workshop; b – Karin Pittman, Gordon Copp and Vlado Kováč as the most serious listeners at one of the workshop sessions; c – a group of German, Canadian, Norwegian and Japanese representatives of the workshop participants during the excursion to the Danube's River Gabčíkovo water works; d – the exhausted conveners from Slovakia and UK at the Vojka restaurant before the workshop dinner; e – a group photograph of most participants after the workshop dinner; f – Schöner Naci silver statue, a landmark at the Bratislava promenade, and Karin Pittman; g – Karol Hensel examines a sturgeon embeded in the sidewalk. All photographs by E.K. Balon.

Keynote presentation

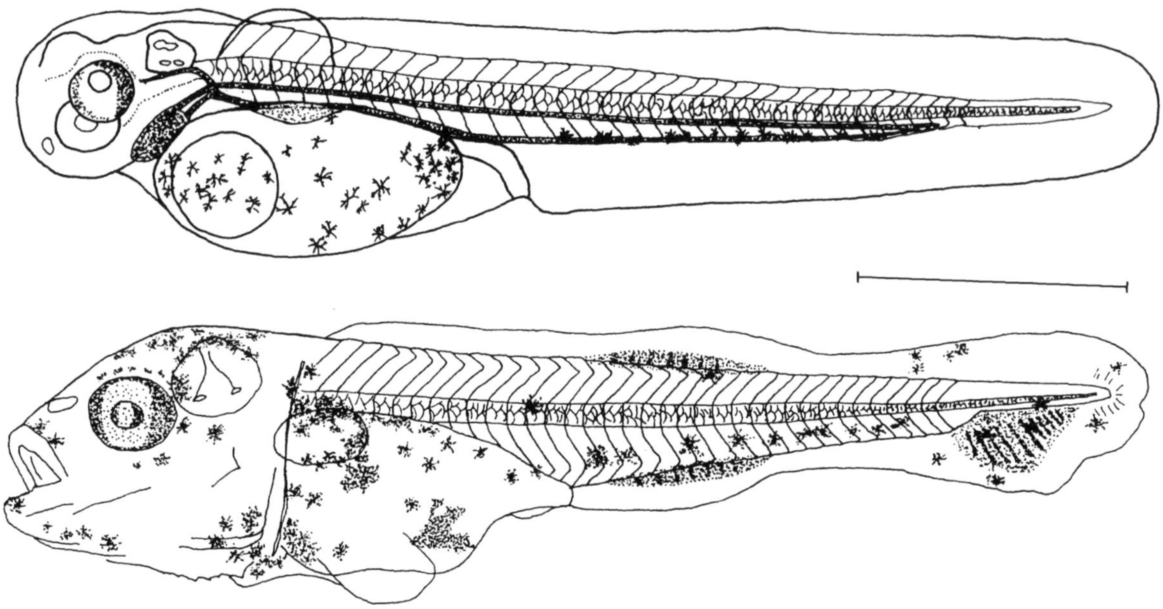

Environmental Biology of Fishes **56**: 17–38, 1999.

Alternative ways to become a juvenile or a definitive phenotype (and on some persisting linguistic offenses)

Eugene K. Balon
Axelrod Institute of Ichthyology, University of Guelph, Guelph, Ontario N1G 2W1, Canada (e-mail: ebalon@uoguelph.ca)

Received 4 September 1997 Accepted 24 November 1998

Key words: embryo, larva, metamorphosis, indirect or direct development, life-history model, altricial, precocial, allometry, fish biologists vs. fishery biologists, alprehost, ontogeny, evolution

Synopsis

Lack of knowledge of early and juvenile development often makes it difficult to decide when a fish becomes a juvenile or, for that matter, a definitive phenotype. According to the established life-history model, a fish develops naturally in a saltatory manner, its entire life consisting of a sequence of stabilized self-organizing steps, separated by distinct less stabilized thresholds. Changes are usually introduced during thresholds. In principle, there are two ways to reach the juvenile period: by indirect or by direct development. Indirectly developing fishes have a distinct larva period that ends in a cataclysmic or mild remodeling process, called metamorphosis, from which the fishes emerge as juveniles. During metamorphosis, most temporary organs and structures of the embryos and larvae are replaced by definitive organs and structures that are also possessed by the adults. In contrast, directly developing fishes have no larvae. Their embryos develop directly into juveniles and do not need major remodeling. Consequently, the beginning of their juvenile period is morphologically and functionally less distinct than in indirect development. The life-history model helps to find criteria that identify the natural boundaries between the different periods in the life of a fish, among them, the beginning of the juvenile period. Looking at it from a different angle, when ontogeny progresses from small eggs with little yolk, larvae are required as the necessary providers of additional nutrients ('feeding entities' similar to amphibian tadpoles or butterfly caterpillars) in order to accumulate materials for the metamorphosis into the definitive phenotypes. Directly developing fishes start with large demersal eggs provided with an adequate volume of high density yolk and so require no or little external nutrients to develop into the definitive phenotype. These large eggs are released and develop in concentrated clutches. It therefore becomes possible and highly effective to guard them in nests or bear them in external pouches, gill chambers or the buccal cavity. Viviparity is the next natural step. Now the maternal investment into large yolks can be supplemented or replaced by direct food supply to the developing embryos like, for example, the secretion of uterine histotrophe or nutrient transfer via placental analogues. When the young of guarders and bearers start exogenous feeding, they are much larger or better developed than larvae of nonguarders and the larva period in the former is reduced to a vestige or eliminated entirely. In the latter case, the juvenile period begins with the first exogenous feeding. Such precocial fishes are more specialized and able to survive better in competitive environments. In contrast, altricial forms retain or revert to a life-history style with indirect development and high fecundity when dispersal is advantageous or essential. Fishes become juveniles when the definitive phenotype is formed in most structures, either indirectly from a larva via metamorphosis or directly from the embryo.

Introduction

> *Although some of the terminology of biology is of the functional workaday kind (...) too much of it is a mixture of grandiloquence and linguistic offenses.*
>
> Peter Medawar & Jean Medawar (1983) in 'Aristotle to Zoos'

We usually write articles and books in a temporal order that parallels the development of our ideas. We expect them to be read in the same order so that the developments of the ideas and the differences between them are understood. More important, we expect that the ideas presented in the latest article of such sequence will be referred to as representing the most mature and therefore valid version of these ideas. Rarely do these expectations come true; ideas are often misunderstood or applied incorrectly (see also Balon 1985). It is useful, therefore, to review ideas from time to time (Balon 1990) and even more useful if others do it for us (e.g. Bruton 1989, 1990, Copp & Kováč 1996, Kováč & Copp 1999, Vilizzi & Walker 1999). However, sometimes even the newest reviews fail miserably to advance or clarify ideas (e.g. Webb 1999). Caveat emptor!

In my initial studies of ontogeny, I followed Vasnetsov (1948, 1953) in using 'etaps' to describe non-gradual development. While the theory underlying this concept appeared wrong to me when applied to socioeconomics, and was later severely discredited (e.g. Medvedev 1969), I felt that the concept, when used in ontogenetic studies, supported an appropriate life-history model (Balon 1956a,b, 1958, 1959a,b,c, 1960, and until 1977). In search of a better explanation, I became later aware of the processes of selforganization, the Taoist idea of harmony and dichotomy, and non-equilibrium thermodynamics (not known at the time of Vasnetsov and Kryzhanovsky et al. 1953). It took some effort to understand their implication for the development of a useful life-history model, in spite of my ideological predisposition (Balon 1989c).

I remember reading Prigogine (1980) over and again and failing to comprehend his ideas. But rereading the book after the interpretations by Capra (1975) and Jantsch (1980) I began to understand the relationships; this understanding was later enhanced by the writings of Eigen & Schuster (1979), Maturana & Varela (1988) and Gottlieb (1992). In the meantime, all this conceptualization ended in the inception of a theory that views ontogeny as saltatory patterns caused by specific processes of selforganization. Named the 'theory of saltatory ontogeny', it was first formulated in its most complete form by Balon (1986a), after earlier less successful attempts (e.g., Balon 1979a,b, 1980a, 1981a, 1983, 1984, 1985). These publications read separately and in non-chronological order caused much misunderstanding, and forced the publication of even more background material (Balon 1989a,b) and reviews (Crawford & Balon 1996). In spite of all this repetition, some authors have mistaken my 'theory of saltatory ontogeny' for the 'etap concept' by Vasnetsov (1948, 1953), and at best considered it merely an English translation (e.g. Peňáz 1981, 1983). I am disappointed when being ignored (e.g. Picard & Voßwinkel 1996) and even more so when finding my ideas misinterpreted in the literature (e.g. Fuiman et al. 1998), but pleased by Makeyeva's (1988), Copp & Kováč's (1996), Moyle & Cech's (1996), Bond's (1996), and Vilizzi & Walker's (1999) comprehension, and even by the late Greenwood's (1989) sophistry.

My life-history model is supported and explained by the theory of saltatory ontogeny (e.g. Balon 1986a,b). Before the formulation of this model, the various intervals of ontogeny, selected arbitrarily, were given common sense names that were rarely defined by **natural** boundaries. Some names were taken from lay usage referring only to fish and ignoring other animal groups. It was, therefore, rarely realized that a fish larva is comparable to a butterfly's caterpillar or a frog's tadpole. Nor was it realized that there are fishes with direct development and no larvae as there are, for example, leptodactyline and telmatobiine frogs of the genera *Adenomera* and *Eleutherodactylus* without tadpoles (e.g. Duellman 1989) or 'lecithotrophic' bivalves, sea stars, sea urchins, brittle stars, and sea cucumbers (e.g. Kasyanov et al. 1998). These are not 'atypical' creatures whose 'life histories may deviate significantly from the norm' (Houde 1994, p. 92) but common alternatives if we care to look beyond the prevailing orthodoxy.

Further biases have been caused by 'fishery biologists' who, in the necessary attempt at cooperation, adjust their 'terminology' to the jargon of their non-biologist counterparts in management, trade and politics. The result is a ridiculous life-history sequence: *egg–larva–fry–fingerling–subadult–spawner–* (and ultimately) *trophy*, where 'embryo', a term from the ivory-tower discipline of developmental biology has no place.

In contrast, 'ichthyologists' or 'fish biologists' usually study fishes as objects of scientific curiosity, to understand the living world around us, the function of its ecosystems, the evolution of its biodiversity, and, of course, as the essential background for the fishery related disciplines. As between applied and basic science, the requirements on precision differ, and most communications by the two kinds of biologists are aimed at different recipients. While some might never have given this a thought, 'fishery biologists' often use lay terms to make their communications easier to understand, whereas 'fish biologists' may be less concerned as their communications are aimed mainly at well-informed peers[1].

As Maurice Kottelat (1997, p. 7) wrote, 'Any research on fish, due to their economic and food significance, has too often been appropriated by fisheries or agriculture agencies (I really mean *appropriated*; for many agencies, to [pose] as "the" specialists does not involve a moral or ethical responsibility and commitment to produce sound research but merely tap[p]ing at the money source to keep the administration running and, not seldom, to prevent other institutions or agencies from entering the field). Fisheries biologists might certainly be competent and efficient for some kind of fisheries management, but few only have been successful as taxonomists'. And as students of fish ontogenies or early life histories they are not doing much better (e.g. Rass 1946, 1948, Peňáz 1975, 1981, 1983, Snyder 1976, Moser 1981). However, there is not always a clear-cut separation between the two groups of 'biologists' and the practitioners become sometimes confused (e.g. Kratt & Smith 1977, Blaxter 1988, Fuiman & Higgs 1997), special technical nomenclature for the 'larval fish taxonomy' notwithstanding (e.g. Rass 1953, Leis & Rennis 1983, Moser 1984, Okiyama 1988, Leis & Trnski 1989). The ichthyoplankton systematics would certainly gain in resolution and explanatory power adopting the life-history model based on a wider understanding of ontogeny (see footnote 7), and that would certainly prevent claiming ichthyoplankton where none exist (e.g. Makeyeva & Pavlov 1998).

[1] The third category – fishery scientists – was created for the mathematicians helping biology. Since their emergence, they were tracking the world's fish catches or landings through constant adjustment of various models (which suited developing agencies and politicians). Even after it was clearly documented that catches are driven by 'overcapitalization' (see Clark 1977) the fishery scientists continued the 'statistical' self deceptions. Will the new field of bioeconomics (see also Landa 1998) contribute to solving the problem (see Ghiselin, M.T. & J.T. Landa. 1996. The International Code of Zoological Nomenclature: a law and bioeconomics approach. Paper presented at the 'Bioeconomics' panel, Western Economics Association International Conference, 28 June to 2 July 1996, San Francisco, and the new 'Journal of Bioeconomics' launched recently)?

Parts of the life-history model explained a little more

There are certain truth that cannot be said so as to be understood without being believed, . . .

Mary Catherine Bateson (1984)
in 'With a Daughter's Eye'

Fishes are part of all organisms inhabiting our planet, part of the common evolutionary history and the result of the same formative processes. Their ontogenies are comparable especially among vertebrates. Yet most often their life histories are studied as if entirely unrelated. This way the explanatory values of comparison are lost (e.g. placental mammals and fishes, Wourms 1981, Wourms et al. 1988).

Embryos are, of course, equally important in the life-history of fishes. Since the onset of exogenous feeding marks the beginning of the larva or juvenile period, respectively, the embryo is characterized by the use of mainly endogenous nutrient supplies (but remember also the contribution of absorptive nutrient delivery, Balon 1986b). Therefore, the traditional manner to consider as embryos only those developing fish still inside the egg envelopes (e.g. Moser 1981, 1984, Blaxter 1988, Williamson 1992, Kamler 1992) can no longer be upheld whatever the rationalization or excuse; after all, mammalian embryos also persist after hatching until parturition.

This brings us to hatching as a process. Before hatching, special gland cells (e.g. Mabee et al. 1998) on the anterior part of most fish embryos produce proteolytic enzymes that dissolve or weaken the egg envelopes in order for the embryo to break free, or as we most frequently call it – hatch. Why then do so many scientists and managers say that an 'egg hatched' or write about a 'freshly hatched larva' when during the process of hatching the embryo leaves the egg envelope and only becomes a larva sometime later when it begins acquiring food externally? This is not a matter of terminology – and I would not like to argue on that level – but a matter of understanding what we are talking about.

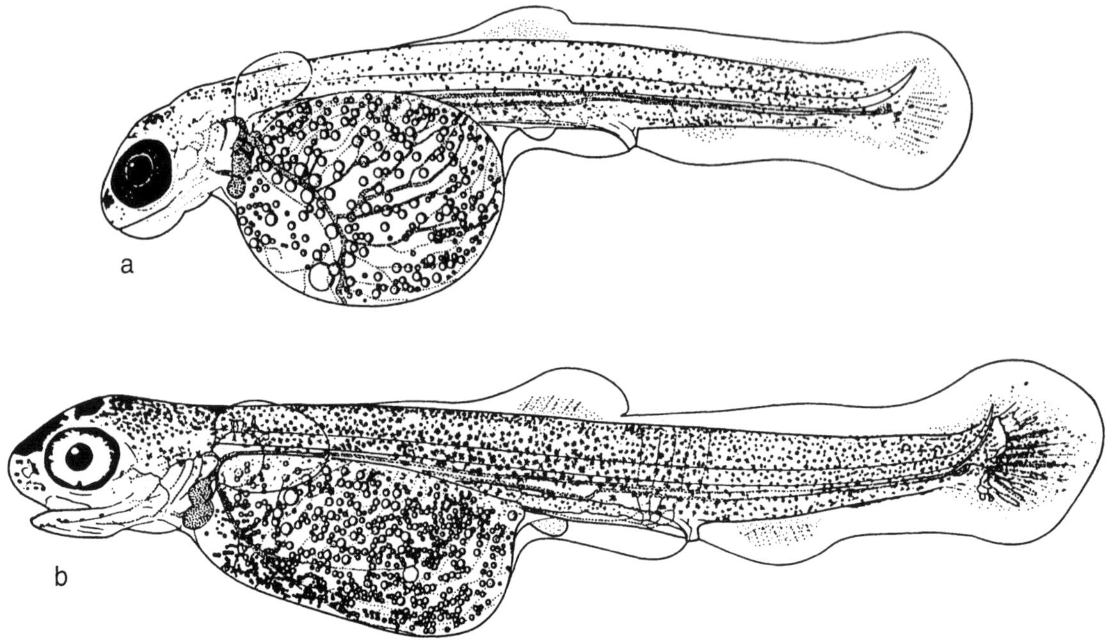

Figure 1. Sunapee charr, *Salvelinus alpinus oquassa*, embryos incubated at 9.5°C (above) and 4.4°C: a – hatched early at 15.9 mm TL, b – hatched late at 19.8 mm TL (from Balon 1980b).

In all vertebrates, including livebearers, an embryo first leaves the egg envelope or envelopes[2] in the process of hatching. Hatching – breaking free from the egg envelope – however, is not a fixed threshold, but is triggered by environmental cues (e.g. low oxygen tension, light intensity, release of hatching enzymes as a cue to adjacent eggs) at different times during the embryo period (Figure 1). If the free embryo is retained in the mother's body for further development or even nutrition after hatching, then it leaves the body during the process of parturition. Release or deposition of eggs in oviparous fishes is not parturition because it precedes the formation of an embryo, as hatching is not synonymous with birth because it always precedes parturition. Birth, however, is in a broad sense a synonym of parturition, but both are meaningless for the establishment of the biological age: parturition also not only varies in time according to environmental stimuli but occurs long after the life (ontogeny) of an individual started. Every individual can hatch or be born at very different states of development! Therefore, birth certificates are biological nonsense and I am surprised that the 'prolifers' did not question them long ago. The true beginning of ontogeny – activation – was explained in detail elsewhere (Balon 1985, 1990).

What is a 'biological concept of the larva', for example (e.g. Leis & Rennis 1983)? Is it a phrase similar in its ambiguity to the even more often quoted 'life-history theory' (e.g. Stearns 1992). I failed to find the true meaning of these phrases. – As already mentioned, a fish larva and its interval in the ontogeny of a fish are comparable to that of a tadpole in frogs and – to some extent – a caterpillar in butterflies.

[2] The terminology of egg envelopes in fishes (and other organisms) is inconsistent and often illogical (see Balon 1977). Called 'case', 'shell', 'capsule', 'chorion' or 'membrane', it often makes cell biologists despair. It seems best to retain the term *membrane* for the three-layered outer boundaries of eucaryotic cells (plasma membrane) of a normal, somatic cell. The terms case, capsule or shell should then be used for the tertiary protective structures formed after the egg leaves the ovary, and chorion for the secondary envelope synthesized and secreted by the follicle cells. The *primary* egg envelope 'secreted by the superficial protoplasm of the oocyte' (Yamamoto 1975, p. 33), often called vitelline membrane, if radiated becomes the *zona radiata*; the secondary egg envelope which is formed by the follicle cells, when present, e.g. as the outer adhesive coat, may be synonymous with the *chorion* (but see, e.g. Anderson 1974). While much of the origin of primary, secondary, tertiary and compound envelopes (Wourms 1987) is still under debate (Laale 1980), the use of *envelope* is the best compromise as a general term, and possibly as plural if known that both the primary, secondary or compound envelopes are present (modified from footnote 7 in Balon 1990).

Caterpillars, larvae and tadpoles are sometimes called 'feeding machines' (Wassersug 1975) or, let me invent a more fitting term – 'feeding entities' – which acquire orally ingested and intestinally digested food when not enough endogenous nutrients (yolk) are available to complete the definitive phenotype. Larvae metamorphose into juveniles. Clearly then, the start of external feeding is characteristic of larvae. A yolksac larva is, therefore, a misnomer maintained at best by tradition only (e.g. Blaxter 1988) as it cannot be applied even to the step of mixed feeding because, for example, in *Cyphotilapia frontosa* such mixed feeding occurs already in free embryos and continues in early juveniles (see later).

Most larvae, especially in invertebrates, are 'those which drift freely for a time in the upper waters, as members of the plankton, before they finally settle down to be transformed into creatures of sessile habit' (Hardy 1962, p. 3). This statement, not true even for all invertebrates, marine fishes and especially not for freshwater fishes, may be the reason why so many biologists have such a hard time to abandon their 'larval' misinterpretation (e.g. Moser 1984, Williamson 1992). Lately, the conventional assumption that all marine fishes have planktonic stages was completely refuted by the discovery of clear cases of direct development, for example, in the damselfish *Acanthochromis polyacanthus* (Thresher 1984, 1985), in the Banggai cardinalfish, *Pterapogon kauderni* (Allen & Steene 1995, Marini 1996, Vagelli & Kind[3], Vagelli 1999 this volume), and the spotted handfish, *Brachionichthys hirsutus* (Bruce et al. 1998). In addition, there are fishes which disperse via planktonic stages but have part of the early stages demersal, recruiting into the parent stock like, for example, the sanddivers *Trichonotus* spp. (Balon et al.[4]). Such fishes buried in the substrate at first as embryos and free embryos are able then as planktonic larvae to maintain position in spite of sea currents (Stobutzki & Bellwood 1997). There is no reason to accept anymore the prevailing opinion about the ubiquitousness of planktonic dispersal in marine fishes as eggs, embryos and larvae. Local populations can maintain themselves in spite of currents and planktonic transport (Lobel 1997).

The life history of organisms consists of distinct intervals, the longest of which are named **egg**, **embryo**, **larva**, **juvenile**, **adult** and **senescence** not only in fishes, but also in other animals. However, while some fishes have larvae, a fish is hardly 'larval' in the true meaning of the word. Did anyone ever go to a 'tadpoled amphibian conference' as some attend the 'larval fish conferences'? Linguists would certainly agree – as those I asked did – that the use of a noun for a life-history interval as an adjective should be discouraged (see Balon 1984) or at most restricted to parts of the organism only, like 'larval finfold' or 'larval notochord'.

The overuse of the patois of lay fishers – like **fry** – for every small fish instead of its correct designation as embryo, larva, juvenile or small adult is even more annoying and confusing (see footnote 2 explaining its etymology and figure 0, both in Balon 1990). In no other field of biology do we have an equivalent term for the state of early ontogeny used so often.

The egg size fallacy

> *... fish eggs measuring 1 mm in diameter are more than 23 000 times larger, by volume, than a human egg, and the eggs of a coelocanth [sic] (...) are more than a million-fold larger than a human egg.*
>
> Brooks et al. (1997, p. 389)
> in 'Egg quality in fish: what makes a good egg?'

Most life-history studies of fishes provide as information on the eggs only the outer diameters, the egg size[5] (e.g. Marshall 1953, Bagenal 1971, Ware 1975, Hempel 1979, Policansky 1981, Thresher 1984, 1988, Pepin 1991, Pepin & Myers 1991). This parameter, however, is not at all sufficient for a meaningful interpretation because 'egg size' alone does not reflect the endowment of an egg and embryo with endogenous nutrients. Volume and density of yolk are more significant for many aspects of life history, at least in oviparous fishes (e.g. Balon 1978, 1985, Flegler-Balon 1989), and can greatly affect the ontogeny. This knowledge of the amount and density of the yolk allows us to predict not

[3] Vagelli, A. & M. Kind. 1998. The reproductive biology of the mouthbrooding cardinalfish *Pterapogon kaudermi* (Perciformes, Apogonidae). 78th Annual Meeting of American Society of Ichthyologists and Herpetologists, University of Guelph, Guelph (abstract 612).

[4] Balon, E.K., E. Clark, C. Flegler-Balon & L.M. Benveniste. 1998. Reproduction and early ontogeny of the sand diver, *Trichonotus setiger*, from Papua New Guinea. 78th Annual Meeting of the American Society of Ichthyologists and Herpetologists, University of Guelph, Guelph (abstract 215).

[5] Egg size should be measured only after activation, and at a time when the 'swelling' of the envelope(s) is complete. If size of ovarian eggs (oocytes) is used, then it will not be comparable with the size after activation (e.g. Balon 1985, 1990).

only egg quality but also the type of ontogeny (Balon 1990), beyond the egg quality (Kjørsvik et al. 1990).

Some eggs owe their large size primarily to a wide perivitelline space filled with low density fluid for buoyancy, but contain very little yolk of low density. Other eggs are large because of a thick, gelatinous, adhesive and buoyancy assisting chorion, or because of one or more oil globule(s) embedded in the yolk (Figure 2). In spite of the large size of such eggs, they represent species that have all indirect ontogenies with small planktonic larvae and high mortalities (Balon 1975, 1981a).

In contrast, many smaller eggs have little or no perivitelline space, no chorion surrounding the zona radiata (the primary egg envelope, see Laale 1980) but relatively large, high density yolk that supports direct ontogeny with no larvae (see Balon 1985, 1990 figure 3, Flegler-Balon 1989). Herewith lays the catch: even for a simple distinction of 'planktonic and demersal eggs', size alone is of limited value for correct interpretations of reproductive investment and should be substituted by yolk measurements (e.g. Houde 1994). But yolk size and density may be more meaningful if presented together with the volume of cytoplasm. The ratio of yolk volume to cytoplasm volume reveals much about the type of ontogeny – indirect or direct – or predicts it, and about the life-history style, be it altricial or precocial (Bruton 1989).

In most developing teleostean eggs, the cytoplasm initially surrounds the yolk and forms a peripheral cytoplasmic layer. Upon activation much of it starts flowing toward the animal pole, thus forming a distinct blastodisc, a process called 'bipolar differentiation' (e.g. Balinsky & Fabian 1981, Makeyeva 1992, Depêche & Billard 1994). This process varies depending on species, but the blastodisc typically becomes largest and best separated from the yolk just prior to epiboly (see also Devillers 1961). Whereas bipolar differentiation (Figure 3) concentrates the cytoplasm at the animal pole below the micropyle, where the pronuclei of the sperm cell and oocyte will join in the actual fertilization process (Balon 1985), the cytoplasm at this earlier time is not yet as distinct from the yolk as it is later. After many cleavages, as a small-cell morula, the blastodisc is clearly separated from the yolk and provides a good estimate of cytoplasm volume (Balon 1977, McElman & Balon 1979, 1980, Cunningham & Balon 1985, Paine & Balon 1986). Depending on the species, the measurements of the yolk and cytoplasm components of an egg are best made between the beginning of

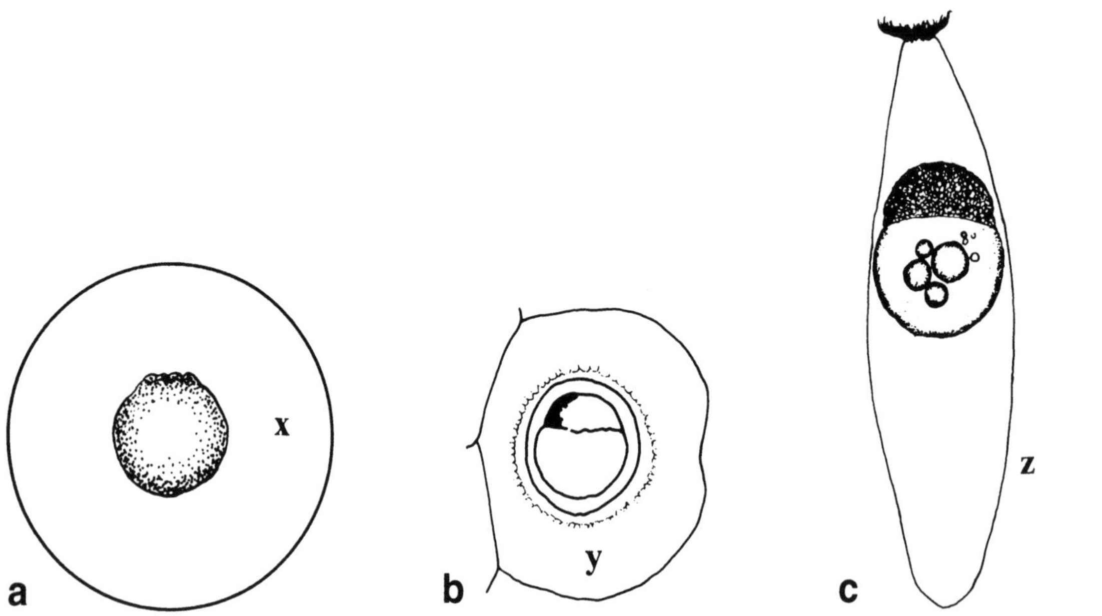

Figure 2. The size of an activated egg depends on external structures more than on the all important amounts of cytoplasm and yolk: a – an egg of *Sarcocheilichthys sinensis*; its size of 5.3 mm in diameter is caused mainly by a large perivitelline space (x); b – egg of *Pseudogobio rivularis* with gelatinous chorion (y); c – egg of *Rhinogobius simulans* with elongated envelopes (z) but spherical yolk and blastodisc.

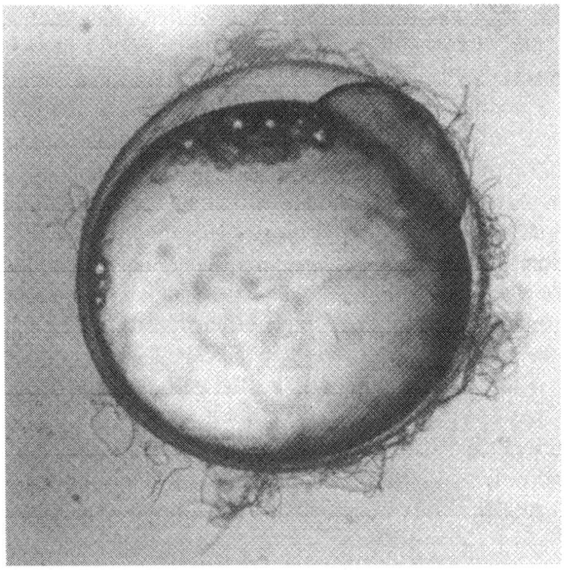

Figure 3. Lucania goodei egg in lateral view. The small cup on the upper right represents the single-cell blastodisc, which is situated on top of the larger, spherical yolk mass (photomicrograph by S.S. Crawford).

cleavage and the beginning of epiboly (Trinkaus 1984). In most fishes, the ooplasmic and the deuteroplasmic components of an egg separate clearly into cytoplasm and yolk and together assume an almost spherical shape. In life-history studies, one should therefore disregard the perivitelline space and the egg envelope(s), which often assume a nonspherical shape, and concentrate rather on cytoplasm and yolk as the variables of importance (see Crawford et al. 1999).

The proportion of yolk, its density and overall quality, are determined during vitellogenesis and may vary even within the same species, depending on the female's age, on feeding conditions during the maturation of the ovaries, and on the annual climate changes (Bromage et al. 1990, Hickley 1990, DeMartini 1991, Brooks et al. 1997). The ratio of yolk to cytoplasm becomes an epigenetic problem: individuals with less yolky eggs may produce **altricial** phenotypes, whereas others within the same species have more yolk and will produce **precocial** phenotypes (Balon 1985), comparable to the 'maintenance' and 'dispersal' phenotypes of Geist (1971). Over time, this epigenetic process which Balon (1989b) named 'alprehost', may lead to genetic evolutionary divergence (see Balon 1986a, 1988b, 1990).

Though 'egg size', in a conventional sense, occasionally yields very general correlations (e.g. Barlow 1981, Thresher 1988), a more meaningful interpretation is rarely possible. Even in more sophisticated studies (e.g. Houde 1994), the 'weight at hatch' and the duration of the larva period were correlated with fewer meaningful attributes than the decisive nutrient deliveries (Balon 1986b) would provide. Concentrating on the density and volume of yolk, on absorptive structures and placental analogues, fish ontogenies can be understood in a much more significant way.

Indirect and direct ontogenies

Not only is every living organism, even the simplest, unique, different from every other, even of its own kind; it also differs from its former self from moment to moment. In a sense, as every Hindu or Buddhist is taught, one dies each moment to be reborn to live the next moment, yet not altogether the same since one was changed by having lived the moment before.

George Wald (1981)
in 'Metamorphosis: an overview'

The life-history model is based on the concept of nongradual development and is supported by the theory of saltatory ontogeny in which natural boundaries – such as the thresholds – are considered between steps. Some decisive events like hatching, first oral feeding, metamorphosis, or physiological transformation from parr to smolt, may not be a part of a saltatory threshold, but an interval undergoing heterochronous shift according to environmental stimuli.[6] 'Stages' in common usage refer to intervals of varying, often undefined, lengths. Here these intervals are organized in a hierarchical system. In the process the term – **stage** – is confined again to its original meaning in embryology **as an instantaneous state**, not an interval. The true saltatory intervals, from the shortest to the longest akin to second, minute and hour, are called: **step**, **phase** and **period** (Table 1).

If one now applies this same model to the development of different species, then the modifications of each ontogeny and their significance become immediately evident, and so do the facts that there are fishes with direct development. When one abandons the confusing term 'yolksac larva' or 'lecithotrophic larva' for free embryo it becomes quite clear that direct ontogenies are

[6] In this respect, the process of metamorphosis is similar to the process of hatching, and may, therefore, not be a part of a saltatory threshold nor of a specific step. Both processes are not fixed in ontogeny but flexible in time according to environmental stimuli.

Table 1. Examples of the indirect, intermediate and direct ontogenies within the **life-history model**, and the hierarchy of intervals based on natural boundaries of the saltatory development.

Periods and *phases* in the following three types of ontogenies:

Indirect	Transitory	Direct
Embryo	**Embryo**	**Embryo**
cleavage egg	*cleavage egg*	*cleavage egg*
embryo	*embryo*	*embryo*
free embryo	*free embryo*	*free embryo*
Larva	**Alevin**	
finfold larva	(a vestige of larva)	
finformed larva	*alevin*	**Juvenile**
Juvenile	**Juvenile**	**Adult**
transition juvenile	*parr*	
Adult	*smolt*	
Senescence	*juvenile*	
	Adult	
	Senescence	**Senescence**

Period = the longest interval of ontogeny separated by the most decisive thresholds.
Phase = the next longest interval into which periods are divided as morphological units for identification purposes but of lesser saltatory significance – the phases of embryo period are: cleavage, embryo and free embryo (or eleutheroembryo); of larva period (if present): finfold larva (or apterolarva), and finformed larva (pterolarva); juvenile period, e.g. in some coral reef fishes can have as first phase 'transition juvenile', in anadromous salmonins can be divided into: parr, smolt and juvenile (or grilse for paedomorphic males).
Step = the shortest natural interval of ontogeny separated by thresholds; most important as an epigenetic stabilized state in the saltatory ontogeny.
Stage = an instantaneous state of ontogeny; should not be used to denote interval as common in lay jargon.

life histories without larvae, comparable to such ontogenies in Chaetognatha and many other invertebrates (Kasyanov et al. 1998) and vertebrates (e.g. Jägersten 1972, Wassersug & Duellman 1984, Duellman 1989, Flegler-Balon 1989, Williamson 1992). The recognition of transitory life histories with vestiges of a larva period enables us to interpret the evolution of ontogenies, and finally to understand the processes responsible for such ontogenies (Balon 1983, 1985, 1988a,b, 1990, Pavlov 1999). Attempts at syntheses without understanding this life-history model usually fail miserably even in their quasi 'practical application' (e.g. Houde 1994).

Let me expand further on the life-history model (Figure 4). Simplifying, we can briefly say that indirect development is mostly a consequence of poor vitellogenesis, which leaves the embryo with insufficient nutrients to develop directly into a definitive phenotype. Very small ova usually produced in large quantities with little, low-density yolk require the external acquisition of additional nutrients. The larva period becomes a natural prolongation of the life history after or shortly before embryos have exhausted their endogenous food supply (Balon 1986b). Larvae, with their special temporary organs that enable them to use concentrations of prey in other than adult habitats, are efficient 'feeding entities' capable not only to continue absorptive acquisition of external dissolved organic matter (e.g. Koller 1930, Stephens 1982, Pfeiler 1986) but also the oral ingestion of food particles, which are digested in the intestine (Figure 5). Most larvae, therefore, thrive in the sea within the planktonic soup. I suspect, however, that dispersal is their primary function (see also Hardy 1962), their ability to maintain position or travel actively, however, not to be overlooked (e.g. Stobutzki & Bellwood 1997). Direct development is otherwise far safer, without the cost of cataclysmic metamorphosis, and the necessity and cost of forming temporary organs.

Behind the mask of a larva (Flegler-Balon 1989), the 'feeding entity' acquires some tissues and structures very different to those in the definitive organism, and so these have to be remodeled through the process of metamorphosis. Some fishes even shrink (like in elopomorphs) during this process and so lose the survival advantage of larger size (Figure 6). With the completion of metamorphosis, the fish acquire the definitive phenotype and look like a small adult; but, as their gonads are not mature, they are juveniles. Most often, the end of metamorphosis and the beginning of the juvenile period coincide with a change of the habitat. On coral reefs, for example, the freshly metamorphosed juveniles depart from the plankton[7] and settle

[7] In most publications concerned with the settlement of planktonic fish larvae into the coral reef habitats no evidence is presented that what really settles are freshly metamorphosed juveniles and not larvae (e.g. Kaufman et al. 1992). On the basis of comparative ontogeny [cf. eel leptocephalus to elver (Tesch 1977, Williamson et al. 1993), or, an amphidromous goby's marine larva to a freshwater ascending juvenile (e.g. Iguchi & Mizuno 1991, Balon & Bruton 1994, Mariyama et al. 1998)] and the above life-history model, larvae should metamorphose still in the plankton and freshly metamorphosed juveniles settle (e.g. Amarullah et al. 1991, Subiyanto et al. 1993). 'Presettlement juveniles' (Leis 1993) may not exist and 'transition juveniles' (Kaufman et al. 1992) may be the first step of the juvenile period, and so claims that it is 'larvae' that settle (e.g. Bell et al. 1987, Sweatman 1988, Breitburg 1989, Sale 1991, Booth 1992, Risk 1998) should not be accepted without further corroboration. However, until more

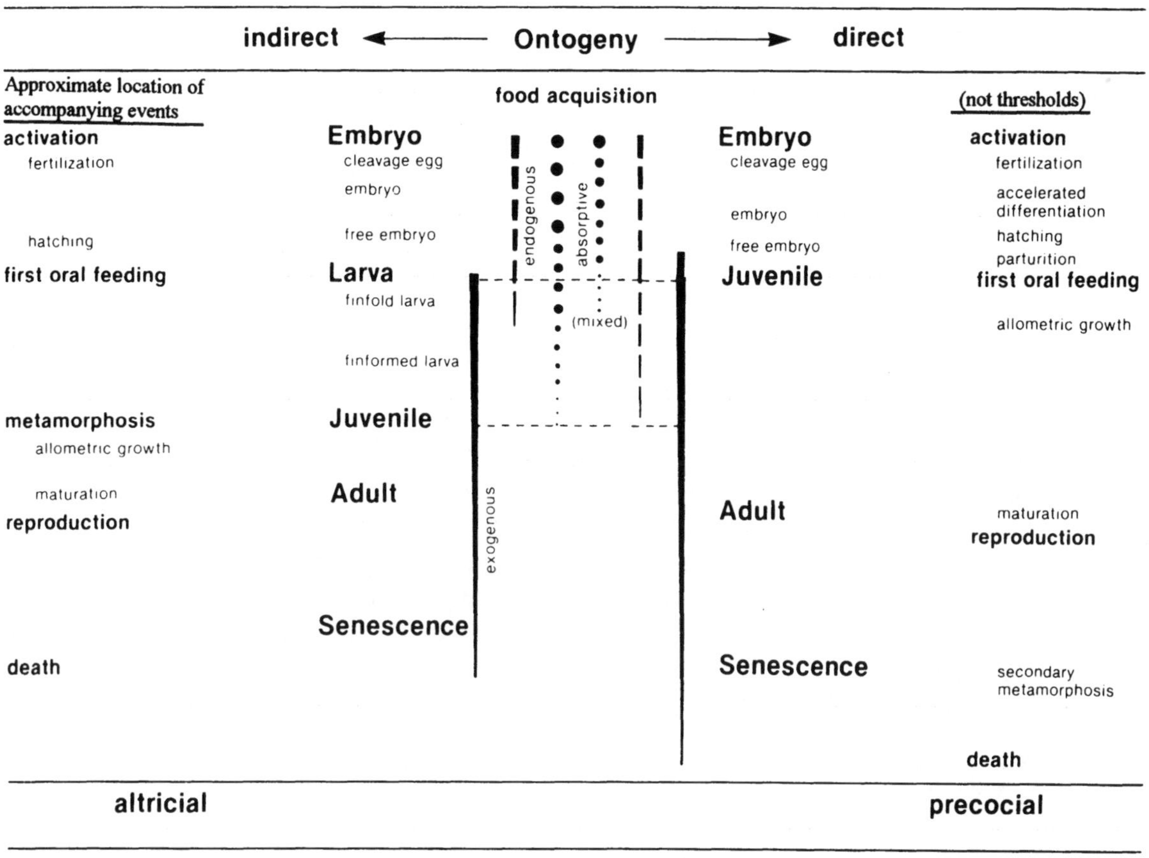

Figure 4. Comparison of the types of nutrient acquisition within the indirect (left) and direct development of the life-history model [intermediate (e.g. salmonids) and extreme ontogenies (e.g. *Cyphotilapia frontosa* and *Latimeria chalumnae*) are ignored]. The decisive (bold) and some accompanying events in either type of ontogeny are listed in the columns on the extreme left and right, respectively. Solid vertical line = orally ingested and intestine digested (exogenous), dashed line = endogenous, and dotted line = absorptive nutrient uptake (from Balon 1986b).

in their permanent territories on the reef (e.g. Victor 1986a, Harmelin-Vivien 1989, Danilowicz 1997) or, as in amphidromous gobies (e.g. Balon & Bruton 1994) or catadromous eels (Tesch 1977), ascend the rivers from the sea (Figure 7). Metamorphosis, therefore, is more than a change in form, as the name may suggest. The consequences of metamorphosis are changes of anatomy, physiology, behavior and ecology, 'altogether a profound change of life, a *metabiosis*' (Wald 1981, p. 1).

When the dispersal of buoyant zygotes is not needed, direct development becomes a much more attractive life-history style if only for its survival advantages. During maternal deposition of more and richer yolk into larger eggs, enough nutrients are available to produce larger juveniles than in indirect life styles. In livebearers, yolk can be supplemented or replaced by embryonic absorption of various histotrophes (uterine milk), by nutrient transfer through placental analogues or even by uterine cannibalism (Wourms 1981, Balon 1981b, 1991).

insight is available, the opinion of Leis (in a letter of April 1998) may prevail: 'One problem we have faced in reef-fish research is the confusion of morphological terms like larva, juvenile, with the ecological concepts of presettlement and postsettlement. Often, there is correspondence, but just as often, there is not. Thus, some reef fish species settle onto the reef already as juveniles by anyone's definition, while other species settle as larvae. There are many which fit in between. For the ecologist, the important thing may be the habitat transition between the pelagic and the benthic environments. For the morphologist, the larval–juvenile transition is the important one. In my ecological work, I naturally emphasise the ecological transition, and am not too concerned if any individual species is a larva or a juvenile at settlement. In my taxonomic work, I emphasise the morphological transition.'

Voluminous and dense yolk reduces buoyancy and only a few such eggs can be produced; no additional nutrient acquisition by the larvae is needed nor is a costly cataclysmic metamorphosis required. Embryos develop for longer but directly into larger juveniles that are able to compete in nursery or even in adult habitats. Their small numbers are fully compensated for by better survival. Eggs not scattered individually in the plankton but deposited in clutches are, of course, more vulnerable to invertebrate and vertebrate predators. On the other hand, aggregations of these large tasty morsels are easier to guard by their parents than the scattered small eggs. Consequently, direct development occurs more frequently (Figure 8) in the reproductive guilds of guarders and bearers (Balon 1975, 1990).

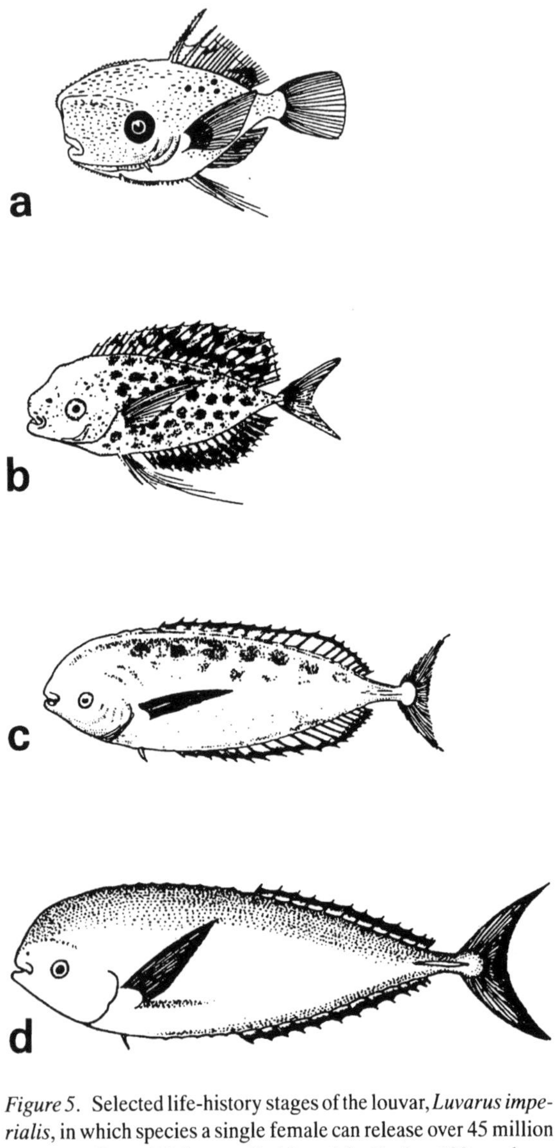

Figure 5. Selected life-history stages of the louvar, *Luvarus imperialis*, in which species a single female can release over 45 million eggs: a – 19 mm TL larva called hystricinella, b – 100 mm larva called astrodermella, c – 600 mm juvenile called luvarella, and d – a 2 m adult (after Roule 1924).

Altricial and precocial fishes

The lenses have not changed because they are prescribed and ground within the ideological prosthetic device which dictates how we receive and apprehend the nature of reality.

John A. Livingstone (1994)
in 'Rogue primate'

As summarized so well by Flegler-Balon (1989, p. 71) 'Fish larvae – like any larva – are characterized by temporary organs and sometimes strikingly different body proportions; some are so different in appearance from the adults that they were initially considered different species. While growing into the definitive phenotype, these larvae have to undergo a more or less drastic metamorphosis. This indirect development is typical for fish with many eggs, little yolk and, in most cases, no parental care; it is especially common in pelagic marine species. With increasing parental care – from egg scatterers to brood hiders to external and internal bearers – the eggs become yolkier and less numerous [Balon 1990, Crawford & Balon 1996]. A higher amount and density of yolk enable the young to grow to a larger size and to further differentiate before feeding actively and also allow them to develop more directly into the definitive phenotype. Non-guarding egg-scattering fishes share some characteristics with altricial birds: small eggs, little yolk, and smaller and less developed young (at hatching in birds, at the onset of exogenous feeding in fishes). In contrast, most guarders and bearers have in common with precocial birds large eggs with a large amount of dense yolk and larger, more developed young. Because of these parallels, one might also distinguish between altricial and precocial life-history styles in fishes'.

Some have tried to identify a similar dichotomy in life histories, like the r- and K-selection concept (e.g. Pianka 1978) using arbitrary population variables such as size and age at maturity and fecundity. The main reason, however, against using the r- and K-selection concept instead of altricial and precocial states is that

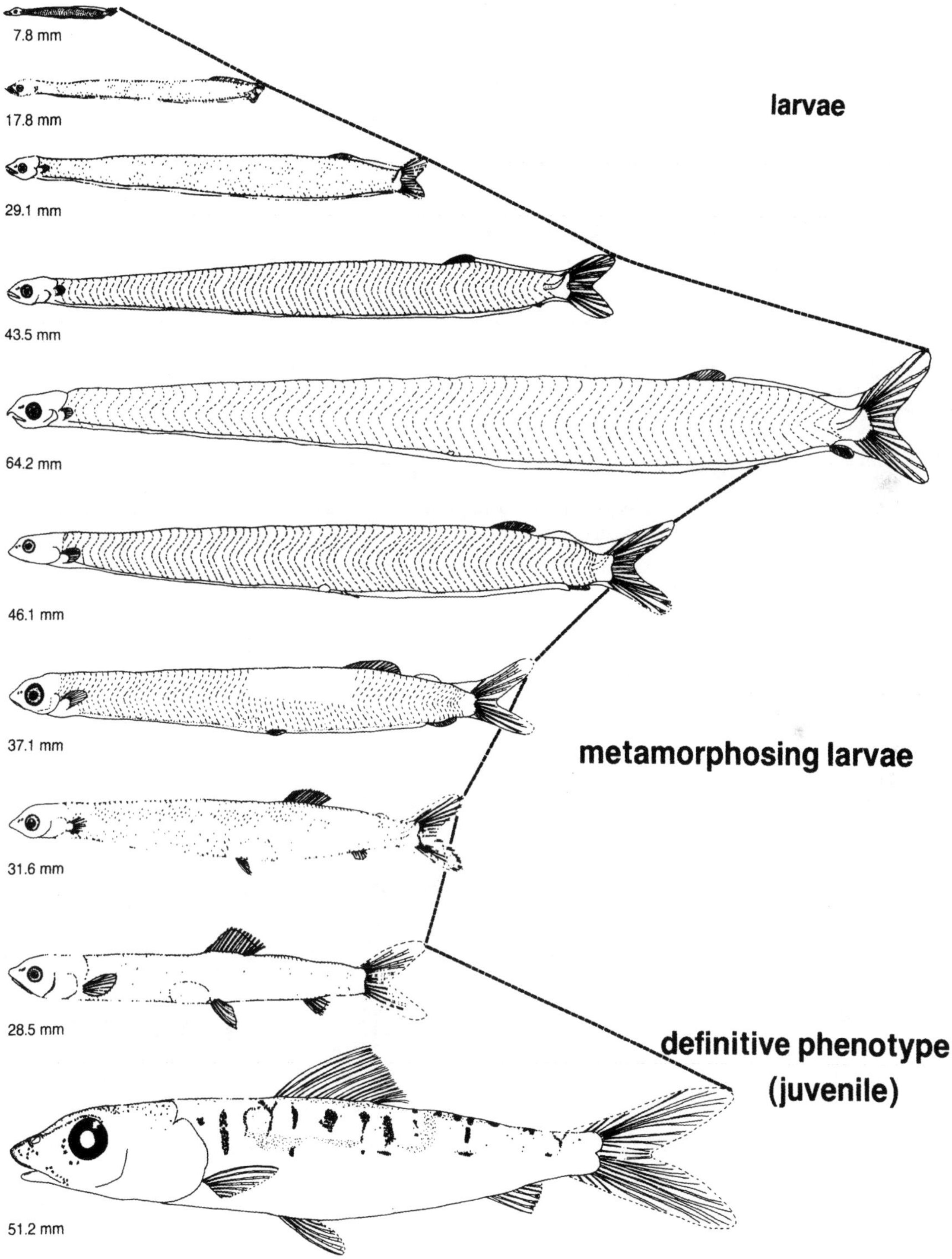

Figure 6. Development of the bonefish, *Albula vulpes*, from the smallest 7.8 mm long to the largest 64.2 mm long orally feeding larvae, until metamorphosis causes shrinkage and remodeling into a 51.2 mm long juvenile (from Flegler-Balon 1989).

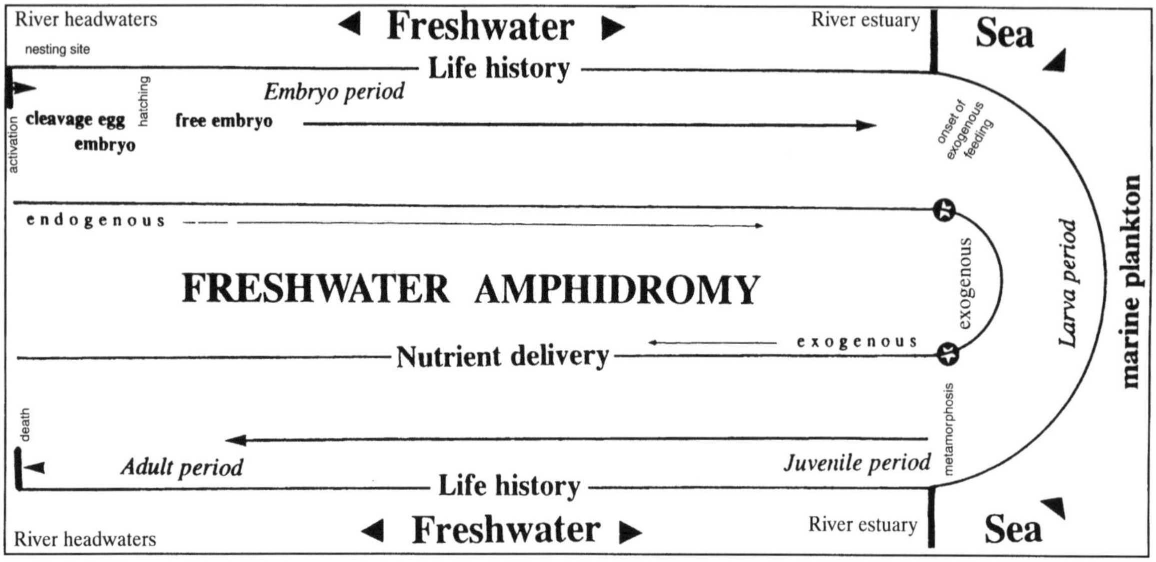

Figure 7. Schematic representation of the freshwater amphidromy according to the life-history model for the goby *Sicyopterus lagocephalus* (from Balon & Bruton 1994).

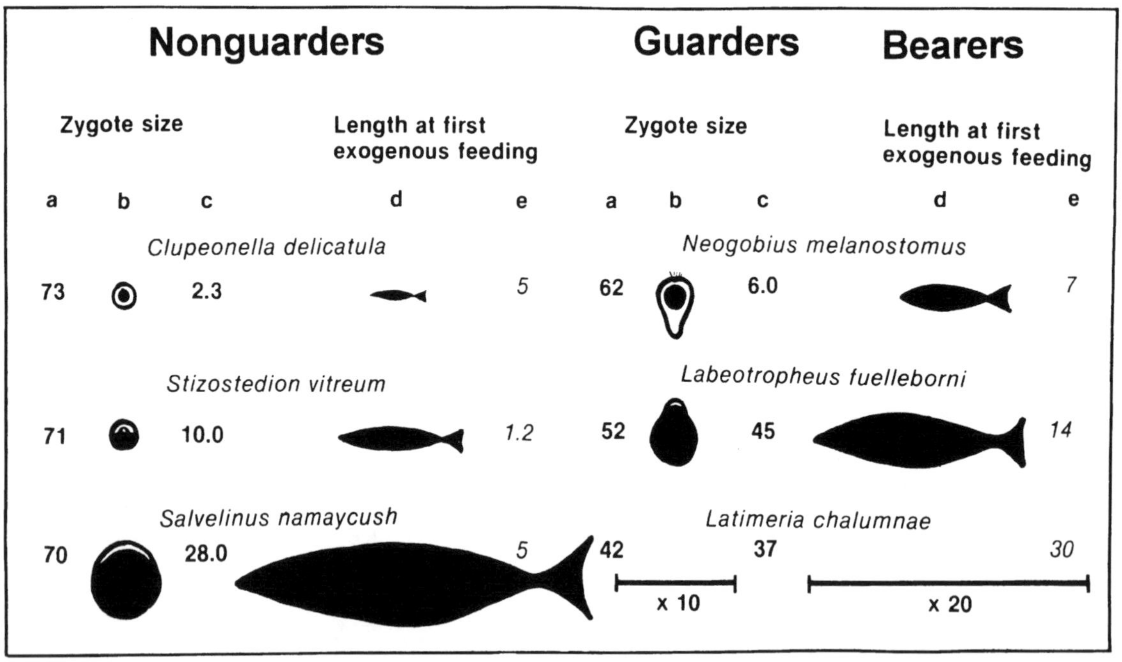

Figure 8. Zygote sizes, densities and lengths at first exogenous feeding in selected examples of nonguarders, guarders and bearers: a = % of moisture content, b = relative size of yolk and envelopes at activation, c = % of lipid content, d = relative size at first exogenous feeding, e = size at first exogenous feeding in % of the length of an average adult (from Balon 1990).

the latter reflect the epigenetic processes responsible for their formation and the character of the entire life history. In contrast, r- and K-selection indicate life-history variables of merely arbitrary or unknown origin, and I question the suitability of the term 'selection' in this context (e.g. as in Constantz 1979, Felsenstein 1979, Luckinbill 1979, Stearns 1980).

'Two sets of terms are found in the literature [writes Nice 1962, p. 18]: precocial and altricial (. . .) *Praecox* means ripened beforehand; *altrix* means a nurse, from *alere*, to nourish. The first word gives a generalized picture of the state at hatching, while the second refers to the necessity for parental feeding (. . .)' When the 'life-history model' is also applied to animals other than birds and developmental attributes besides parental care are allowed for, such as yolk volume, state of development at a certain time, an altricial bird hatchling and a fish larva are comparable states. Both require exogenous food to transform them from a less developed state into a definitive phenotype, much more so than the directly developing precocial bird or fish, which are both capable of independent existence at the end of 'incubation' thanks to a larger endogenous food supply. Parental care like guarding differs, therefore, from care aimed also at feeding (Crawford & Balon 1996), and a frog tadpole or a fish larva are both independent 'feeding entities' but fully comparable to an altricial bird nestling or a young marsupial despite the latter dependence on parental food delivery (Balon 1990).

The altricial life history of an indirectly developing fish may have a tendency to change in a succession of generations (Figure 9), and given some environmental stability (others would call 'selective pressures'), toward more precocial life histories (Figure 10). Larger and more yolky eggs coincide with a decrease in fecundity, a shorter larva period and a longer embryo period, accompanied by shorter adult and longer senescent periods and also changes from iteroparity to semelparity. Ever more specialized life histories along these generation trajectories would ultimately lead to overspecialization and extinction if it where not for occasional paedomorphosis, which returns a precocial life history back to an altricial one. The processes whereby these occur were termed 'altricial-precocial homeorhetic states' ('alprehost' for short) and explained in detail elsewhere (Balon 1985, 1988b, 1990, Bruton 1989).

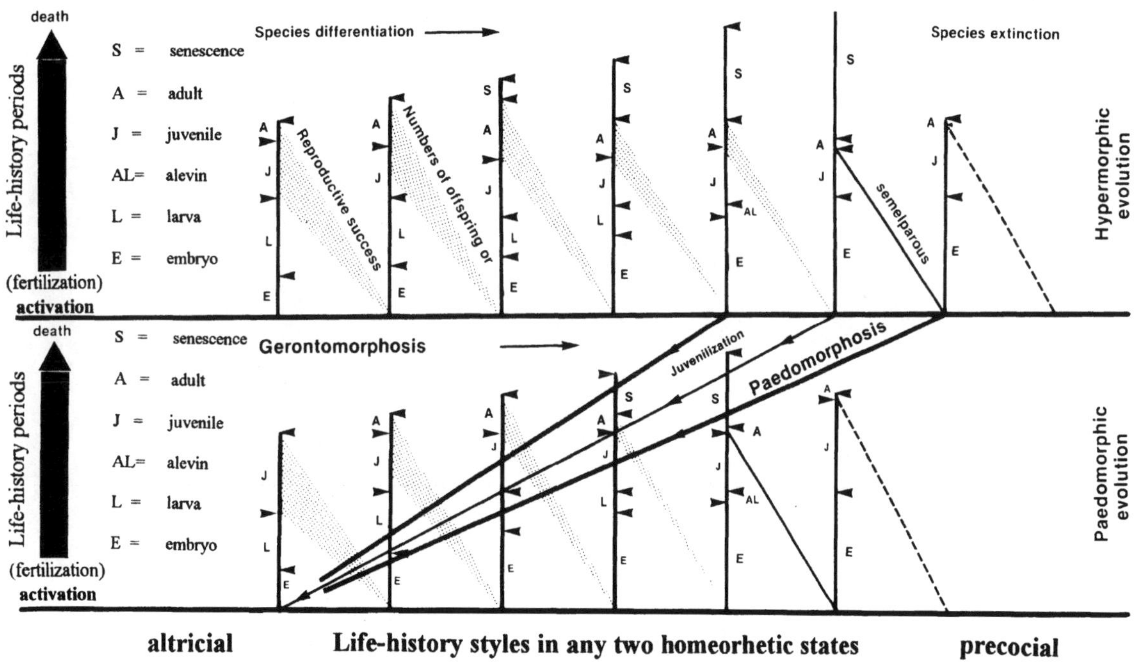

Figure 9. Scheme illustrating possible ontogenies (generations, life-history styles) in sequence of increased specialization (from left to right) and, under certain conditions, despecialization by paedomorphosis (from upper right to lower left). The relative duration of developmental periods (arrowheads) changes during specialization from altricial (left) towards precocial (right), concomitantly the reproductive success (number of offspring) is reduced and paralleled by truncation of adult, elimination of larva and prolongation of senescence periods (from Balon 1985).

When does the juvenile period start?

No drastic metamorphosis! – each youngster keeps her skin: Her larval frills are not thrown off, but eaten from within.
Walter Garstang (1962)
in 'Mülleria and the Ctenophore'

The juvenile is not a subadult, as many fishery biologists would like us to believe, the same way as an embryo is not a sublarva. A juvenile is the beginning of a definitive phenotype in which most of the embryonic and larval temporary structures degenerated and most vital permanent adult organs or structures have formed. However, 'the larva frequently possesses rudiments of adult structures, the adult vestiges of larval structures' (Wald 1981, p. 5), be it the type of hemoglobin or visual pigment, a finfold or capillary plexus. Also, the shape of a juvenile may still be somewhat different from the shape of an adult and some less vital structures like scales or skeletal calcifications may still be missing. The shape changes into the adult form by allometric growth. Gonads of a juvenile contain only developing and immature gametes. Maturation of the gonads marks the beginning of the adult period.

The precise start of the juvenile period is another matter. In indirectly developing altricial fish a very different looking larva becomes a juvenile after its cataclysmic metamorphosis is completed, after the weird ribbon-like or willow leaf-like looking fish (e.g. leptocephalus) is remodelled into a more adult-like looking creature (Figures 6, 11). The behavior of the juvenile and its habitat often change dramatically from those of a larva as in coral reef fishes. On the other hand, larvae can postpone metamorphosis for some time if settlement territory is not available (see footnote 7) (e.g. Victor 1986b, Kaufman et al. 1992). More difficult to recognize is the exact transition to juvenile in fishes with not so different larvae and a less dramatic metamorphosis (e.g. Fostner et al. 1983), and in intermediate or directly developing fishes in which metamorphosis in a strict sense is eliminated. In such cases, only detailed observations can reveal the sequence of the loss of temporary organs and structures and the gain of

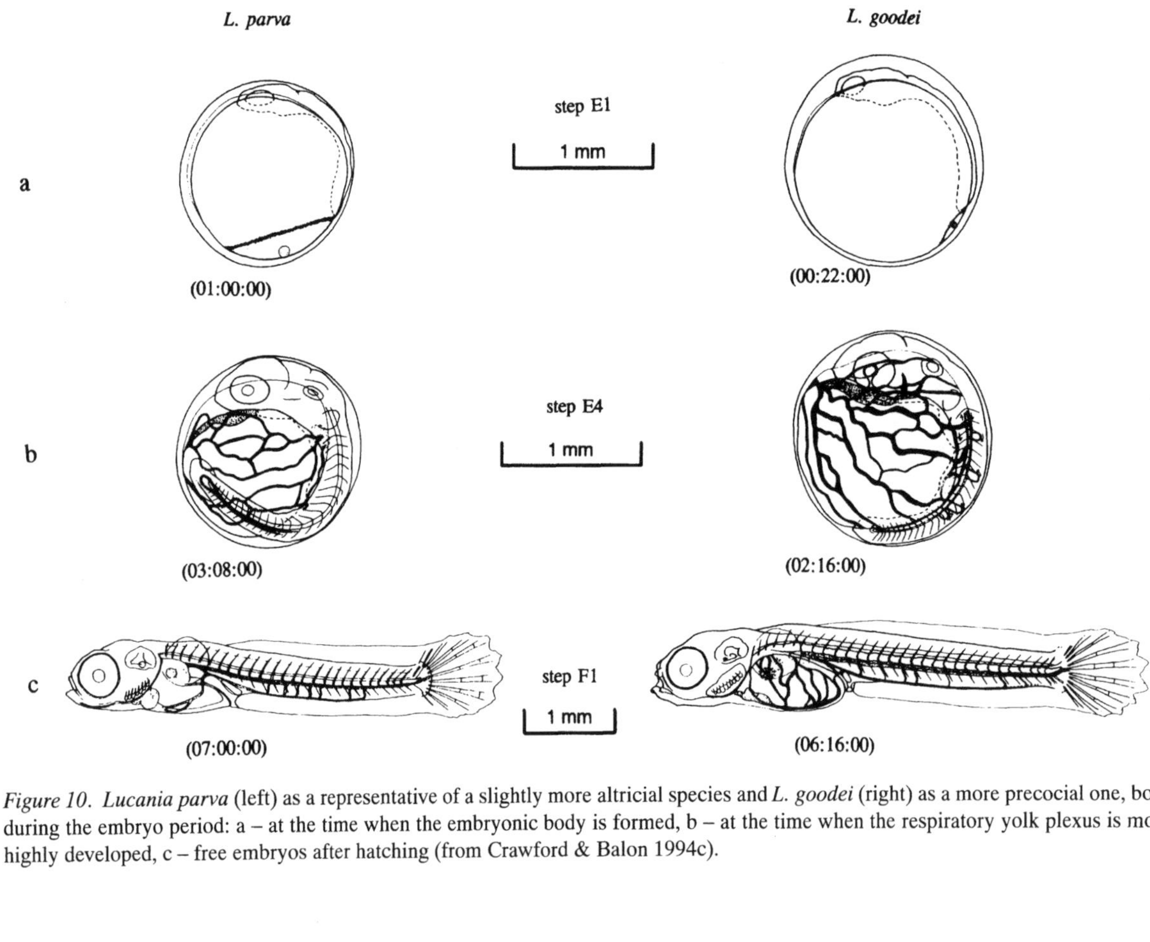

Figure 10. Lucania parva (left) as a representative of a slightly more altricial species and *L. goodei* (right) as a more precocial one, both during the embryo period: a – at the time when the embryonic body is formed, b – at the time when the respiratory yolk plexus is most highly developed, c – free embryos after hatching (from Crawford & Balon 1994c).

permanent ones (see Paine & Balon 1984, Cunningham & Balon 1985, 1986a,b, Crawford & Balon 1994a,b,c). This is best achieved through a 'composite description' (Balon & Flegler-Balon 1985). In the absence of detailed knowledge of a specific ontogeny, only rough approximations of the beginning of the juvenile period may be suggested, but it helps to have a clear definition of a juvenile in mind.

Figure 11. Three stages demonstrating the differences between an 8 mm long larva (above), 56 mm metamorphosing larva and 90 cm long adult of *Trachipterus trachypterus* (from Flegler-Balon 1989).

The case of a 'yolksac juvenile' as an example of further specialization, a conclusion

Extinction is a necessary half of the evolutionary process, as we know. But the other half must always be the appearance of new forms to occupy new and presumably changing environments.
John A. Livingston (1994)
in 'Rogue primate'

Evolution never stops its tireless introduction of novelties; and 'Natural systems are not orchestras replaying a 180-year old score by Beethoven but groups of talented musicians improvising to a restless audience' (Bruton 1989, p. 4). In order to escape uncertainty and avoid competition in ever more complex communities, extremely 'specialized' life histories sometimes evolve if conditions are right. Such conditions presented themselves in the stable African Rift Valley lakes in which explosive speciation, both allopatric and sympatric, has formed a high diversity of cichlid taxa (e.g. Greenwood 1974, Ribbink et al. 1983, Witte 1984, Poll 1986). Along with the evolution of species and their feeding specializations (Fryer 1959, 1996, Fryer & Iles 1972, Lowe-McConnell 1987, 1996, Coulter 1991, Seehausen 1996) modifications to their life histories also appeared.

The altricial cichlid *Tilapia rendalli*, for example, deposits 1.8 mm large eggs already into a specially constructed depression in the substrate (nest) and, in addition to fanning, transfers the embryos by mouth several times into fresh clean nests (unpublished observations). *Oreochromis niloticus* picks up its 2.8 mm eggs soon after deposition and retains them in the female's mouth, whereas *Labeotropheus trewavasae* picks up its 4.4 mm eggs one by one immediately upon release into the water column (Balon 1977). This sequence demonstrates the evolution of mouthbrooding from substrate tending and nest guarding (e.g. Balon 1981c). When the young are released from the protection of the nest or the buccal cavity of the female and start feeding exogenously, their sizes and ages vary accordingly: at the onset of feeding, the larvae of tending species *T. rendalli* are 5 mm in total length (TL) and 6 days old (from activation), whereas the juveniles of the continuous mouthbrooders are 8 mm and 11 days old (*O. niloticus*) or 15 mm long and 24 days old (*L. trewavasae*). Although the tenders may still have vestiges of a larva period, the precocial mouthbrooders have direct development without larvae and without metamorphosis. In some species, the released juveniles return to the mother's buccal cavity to hide at the slightest sign of danger. In most species, the tender as well as the brooder juveniles soon leave the adult habitat and live in shallow nurseries that are inaccessible to larger predators.

Is it possible to specialize even further by producing even larger eggs with more yolk? *Cyphotilapia frontosa* is a mouthbrooding cichlid that lives and reproduces in the deeper waters of Lake Tanganyika. This fish produces eggs of (on average) 5.6 mm in diameter but fewer than the other three species mentioned before. These few eggs, therefore, are relatively loose in the buccal pouch of the female. The embryos hatch after 5 days of incubation at 6.5 mm length and have already a cartilaginous axial skeleton (Balon 1985). Fourteen days after activation, and while still inside the buccal cavity, the embryos start exogenous feeding on detritus and small plankton inhaled by the mother. In addition to a still large and dense yolk, these 11 mm long free embryos already have advanced differentiation of their fins as well as advanced skeletal calcification. They feed endogenously both from the large yolk and exogenously on the inhaled food, as if mixed feeding could spill over in both directions at the embryo/juvenile boundary. At 27 days and 17 mm TL the young have still half of their yolk left (Figure 12) but have otherwise already completed the formation of the definitive phenotype, including juvenile coloration and calcification of the skeleton. Hence, while becoming 'yolksac juveniles' these young stay in the protection of the maternal buccal pouch until released another 27 days later at 23 mm TL. At that size they can already survive in the adult habitat and need not undertake the dangerous journey to the nursery grounds.

After these observations were first made in the laboratory (Balon 1985), studies have been carried out confirming them in the field (Yanagisawa & Ochi 1991). Other cichlids with similar advanced life-history specializations have also been subsequently found in Lake Tanganyika (e.g. Yanagisawa & Sato 1990). The development of the marine ariid catfish, *Galeichthys feliceps*, studied by Tilney & Hecht (1993) demonstrates that the most precocial mouthbrooding with buccal feeding of embryos and juveniles may not be restricted to cichlids only.[8] Clearly, an increase in

[8] Similarly, more directly developing fishes were found in the marine environment once this case has been reported, thus proving the previous generalization wrong.

yolk density and volume combined with an earlier start of exogenous feeding and a consequently extended duration of mixed (endo- and exogenous) feeding, will render a juvenile that is rather large, but more important, better developed when it becomes independent, larger than it would have been had it had one food source only. The living coelacanth, *Latimeria chalumnae*, was found to benefit from a combination of a large and dense yolk (eggs 90 mm in diameter), histotrophy and the placenta-like juxtaposition of maternal and fetal tissues (and possibly ingestion of sibling debris) in order to deliver an advanced juvenile 420 mm long at parturition (Balon 1991). Some elasmobranchs are known to produce even larger young (Compagno 1990).

I hope that I have shown with the cases presented here that even the ultimate life-history style (precocial, direct development) may still be improved upon until overspecialization spells the death knell to the species. Even then the end may not be imminent if

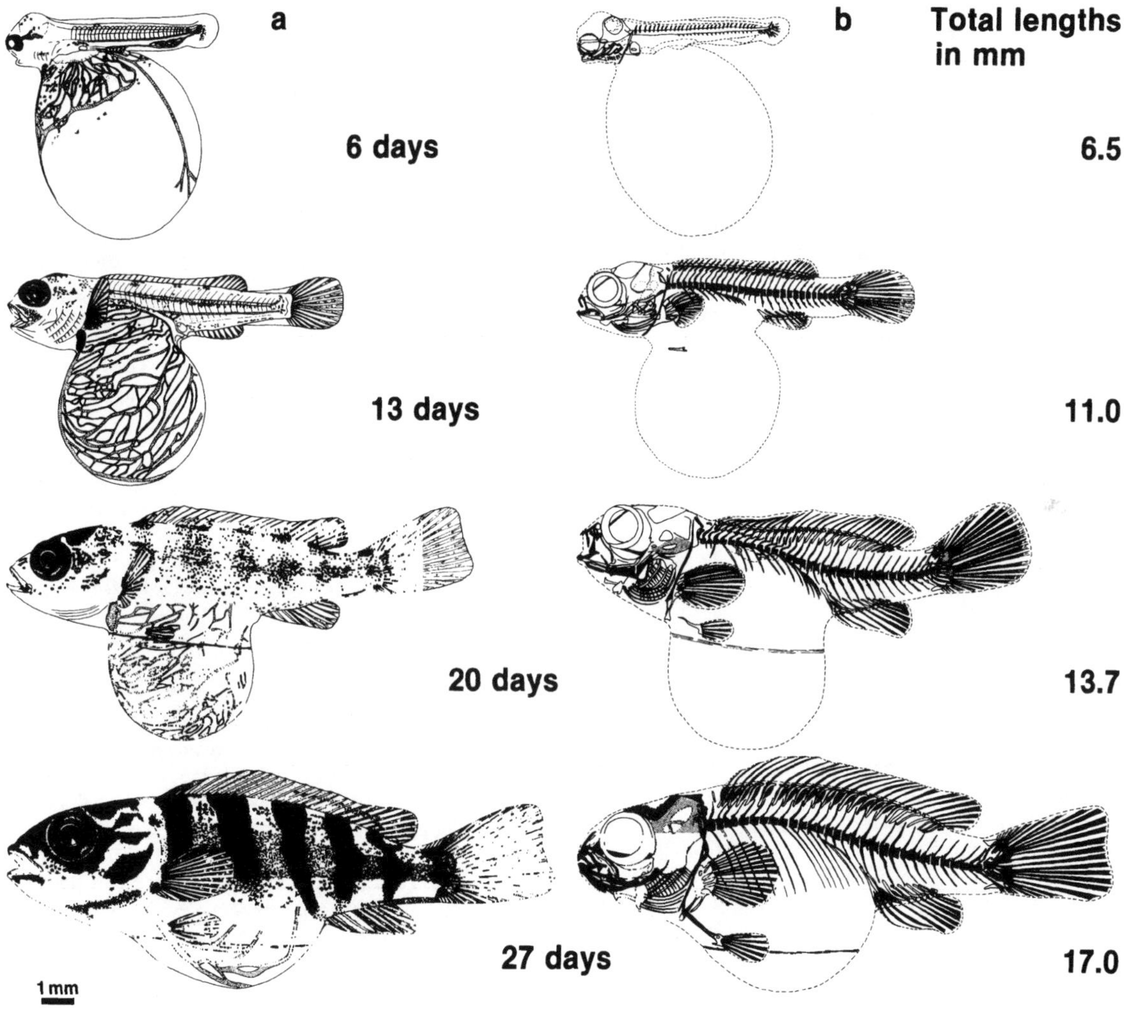

Figure 12. Four decisive stages of *Cyphotilapia frontosa* based on live (a) and cleared and stained (b) specimens removed from the buccal pouch of a brooding female. At the time of mating, inseminated eggs (5.6 × 4.0 mm in diameter) are picked up by the female and mouthbrooded until a few large juveniles (23 mm TL) are released 54 days later. In the buccal pouch of the female the embryos hatch on the 5th day with a large yolksac still, but already with the first cartilaginous elements of the axial skeleton. They start feeding orally when 13 days old. By this time, they have developed differentiated fins and many calcified skeletal elements (black in b). When 27 days old, they become fully formed juveniles that retain about half of the yolk but have a completely calcified skeleton. At that time, they are only half through the brooding interval and feed on particles inhaled by the mother (from Balon 1985).

paedomorphosis enables return from Hades' gates to a parallel path for a while (Balon 1988a,b). More important, however, I hope that a heuristic handling of the confusion of names and concepts that proliferate in the 'life-history literature' has been achieved in place of various biological nonsenses, fisheries conformities and terminological chaos (paraphrased from Kottelat 1997, p. 18).

What all this means is that I can send you a message only if – if what? – if you know it already.

Giuseppe Sermonti (1997)
in 'Resonant messages'

Acknowledgements

I am most grateful to Gordon Copp and Vladimír Kováč for the invitation to present this keynote lecture. Because most of it was already published in various forms before, it would never have been compiled again without this invitation. As usual, Christine Flegler-Balon edited with care the various drafts and I have substituted in many places her words for mine. Many thanks for comments and suggestions for improvements made by Mike Bruton, Gordon Copp, Vlado Kováč, and Karin Pittman.

References cited

Allen, G.R. & R.C. Steene. 1995. Notes on the ecology and behaviour of the Indonesian cardinalfish (Apogonidae) *Pteragon kauderni* Koumans. Rev. Franc. Aquariol. 22(1–2): 7–9.

Amarullah, M.H., Subiyanto, T. Noichi, K. Shigemitsu, Y. Tamamoto & T. Senta. 1991. Settlement of larval Japanese flounder (*Paralichthys olivaceus*) along Yanagihama Beach, Nagasaki Prefecture. Bull. Fac. Fish. Nagasaki Univ. 70: 7–12.

Bagenal, T.B. 1971. The interrelation of the size of fish eggs, the date of spawning and the production cycle. J. Fish Biol. 3: 207–219.

Balinsky, B.I. & B.C. Fabian. 1981. An introduction to embryology, 5th edition. Saunders College Publishing, Philadelphia. 768 pp.

Balon, E. 1956a. Vývoj hlavátky [*Hucho hucho* (L)] počas endogénneho spôsobu výživy po vyliahnutí (Postembryonic development of the huchen during the time of endogenous feeding). Pol'nohospodárstvo 3: 433–455 (in Slovak).

Balon, E. 1956b. Neres a postembryonálny vývoj plotice (*Rutilus rutilus* ssp.) (Spawning and postembryonic development of the roach). Biologické práce 2: 7–60 (in Slovak).

Balon, E. 1958. Vývoj dunajského kapra (*Cyprinus carpio carpio* L.) v priebehu predlarválnej fázy a larválnej periódy (Development of the Danubian carp during the prelarval phase and larval period). Biologické práce 4: 5–54 (in Slovak).

Balon, E.K. 1959a. Die Entwicklung des akklimatisierten *Lepomis gibbosus* (Linné 1748) während der embryonalen Periode in den Donauseitenwassern. Zeitschrift für Fischerei 8: 1–27.

Balon, E.K. 1959b. Die Entwicklung der Texas-Cichlide (*Herichthys cyanoguttatus* Baird et Girard) nach dem Schlüpfen. Zoologischer Anzeiger 162: 339–355.

Balon, E.K. 1959c. Die embryonale und larvale Entwicklung der Donauzope (*Abramis ballerus* subsp.). Biologické práce 5: 1–87.

Balon, E.K. 1975. Reproductive guilds of fishes: a proposal and definition. J. Fish. Res. Board Can. 32: 821–864.

Balon, E.K. 1977. Early ontogeny of *Labeotropheus* Ahl, 1927 (Mbuna, Cichlidae, Lake Malawi), with a discussion on advanced protective styles in fish reproduction and development. Env. Biol. Fish. 2: 147–176.

Balon, E.K. 1978. Reproductive guilds and the ultimate structure of fish taxocenes: amended contribution to the discussion presented at the mini-symposium. Env. Biol. Fish. 3: 149-152. (reprinted in Dev. Env. Biol. Fish. 5: 83–86, 1985).

Balon, E.K. 1979a. The theory of saltation and its application in the ontogeny of fishes: steps and thresholds. Env. Biol. Fish. 4: 97–101.

Balon, E.K. 1979b. The juvenilization process in phylogeny and the altricial to precocial forms in the ontogeny of fishes. Env. Biol. Fish. 4: 193–198.

Balon, E.K. 1980a. Early ontogeny of the lake charr, *Salvelinus* (*Cristivomer*) *namaycush*. pp. 485–562. *In*: E.K. Balon (ed.) Charrs: Salmonid Fishes of the Genus *Salvelinus*, Perspectives in Vertebrate Science 1, Dr W. Junk Publishers, The Hague.

Balon, E.K. 1980b. Early ontogeny of the North American landlocked arctic charr – sunapee *Salvelinus* (*Salvelinus*) *alpinus oquassa*. pp. 568–606. *In*: E.K. Balon (ed.) Charrs: Salmonid Fishes of the Genus *Salvelinus*, Perspectives in Vertebrate Science 1, Dr W. Junk Publishers, The Hague.

Balon, E.K. 1981a. Saltatory processes and altricial to precocial forms in the ontogeny of fishes. Amer. Zool. 21: 573–596.

Balon, E.K. 1981b. Additions and amendments to the classification of reproductive styles in fishes. Env. Biol. Fish. 6: 377–390.

Balon, E.K. 1981c. About processes which cause the evolution of guilds and species. Env. Biol. Fish. 6: 129–138.

Balon, E.K. 1983. Epigenetic mechanisms: reflections on evolutionary processes. Can. J. Fish. Aquat. Sci. 40: 2045–2058.

Balon, E.K. 1984. Reflections on some decisive events in the early life of fishes. Trans. Amer. Fish. Soc. 113: 178–185.

Balon, E.K. (ed.) 1985. Early life histories of fishes: new developmental, ecological and evolutionary perspectives. Dev. in Env. Biol. Fish. 5, Dr W. Junk Publishers, Dordrecht. 280 pp.

Balon, E.K. 1986a. Saltatory ontogeny and evolution. Rivista di Biologia/Biology Forum 79: 151–190 (in English and Italian).

Balon, E.K. 1986b. Types of feeding in the ontogeny of fishes and the life-history model. Env. Biol. Fish. 16: 11–24.

Balon, E.K. 1988a. Tao of life: universality of dichotomy in biology. 1. The mystic awareness. Rivista di Biologia/Biology Forum 81: 185–230 (in English and Italian).

Balon, E.K. 1988b. Tao of life: universality of dichotomy in biology. 2. The epigenetic mechanisms. Rivista di Biologia/Biology Forum 81: 339–380 (in English and Italian).

Balon, E.K. 1989a. The Tao of life: from the dynamic unity of polar opposites to self-organization. pp. 7–40. *In*: M.N. Bruton (ed.) Alternative Life-History Styles of Animals, Perspectives in Vertebrate Science 6, Kluwer Academic Publishers, Dordrecht.

Balon, E.K. 1989b. The epigenetic mechanisms of bifurcation and alternative life-history styles. pp. 467–501. *In*: M.N. Bruton (ed.) Alternative Life-History Styles of Animals, Perspectives in Vertebrate Science 6, Kluwer Academic Publishers, Dordrecht.

Balon, E.K. 1989c. The confessions of s structuralist. Rivista di Biologia/Biology Forum 82: 135–136.

Balon, E.K. 1990. Epigenesis of an epigeneticist: the development of some alternative concepts on the early ontogeny and evolution of fishes. Guelph Ichthyol. Rev. 1: 1–42.

Balon, E.K. 1991. Probable evolution of the coelacanth's reproductive style: lecithotrophy and orally feeding embryos in cichlid fishes and in *Latimeria chalumnae*. Env. Biol. Fish. 32: 249–265.

Balon, E.K. & M.N. Bruton. 1994. Fishes of the Tatinga River, Comoros, with comments on freshwater amphidromy in the goby *Sicyopterus lagocephalus*. Ichthyol. Explor. Freshwat. 5: 25–40.

Balon, E.K. & C. Flegler-Balon. 1985. Microscopic techniques for studies of early ontogeny in fishes: problems and methods of composite descriptions. pp. 33–56. *In*: E.K. Balon (ed.) Early Life Histories of Fishes: New Developmental, Ecological and Evolutionary Perspectives, Dev. in Env. Biol. Fish. 5, Dr W. Junk Publishers, Dordrecht.

Barlow, G.W. 1981. Patterns of parental investment, dispersal and size among coral-reef fishes. Env. Biol. Fish. 6: 65–85.

Bateson, M.C. 1984. With a daughter's eye. William Morrow and Comp., New York. 242 pp.

Bell, J.D., M. Westoby & A.S. Steffe. 1987. Fish larvae settling in seagrass: do they discriminate between beds of different leaf density. J. exp. Mar. Biol. Ecol. 111: 133–144.

Blaxter, J.H.S. 1988. Pattern and variety in development. pp. 1–58. *In*: W.S. Hoar & D.J. Randall (ed.) Fish Physiology, volume 11A Eggs and Larvae, Academic Press, San Diego.

Bond, C.E. 1996. Biology of fishes, 2nd ed. Saunders College Publishing, Philadelphia. 750 pp.

Booth, D.J. 1992. Larval settlement patterns and preferences by domino damselfish *Dascyllus albisella* Gill. J. exp. Mar. Biol. Ecol. 155: 85–104.

Breitburg, D. 1989. Demersal schooling prior to settlement by larvae of the naked goby. Env. Biol. Fish. 26: 97–103.

Bromage, N.R., P. Hardiman, J. Jones, J. Springate & V. Bye. 1990. Fecundity, egg size and total egg volume differences in 12 stocks of rainbow trout, *Oncorhynchus mykiss*. Aquaculture and Fisheries Management 21: 269–284.

Brooks, S., C.R. Tyler & J.P. Sumpter. 1997. Egg quality in fish: what makes a good egg? Reviews in Fish Biology and Fisheries 7: 387–416.

Bruce, B.D., M.A. Green & P.R. Last. 1998. Threatened fishes of the world: *Brachionichthys hirsutus* (Lacépède, 1804) (Brachionichthyidae). Env. Biol. Fish. 52: 418.

Bruton, M.N. (ed.) 1989. Alternative life-history styles of animals. Perspectives in Vertebrate Science 6, Kluwer Academic Publishers, Dordrecht. 617 pp.

Bruton, M.N. (ed.) 1990. Alternative life-history styles of fishes. Developments in Environmental Biology of Fishes 10, Kluwer Academic Publishers, Dordrecht. 327 pp.

Capra, F. 1975. The Tao of physics. An exploration of the parallels between modern physics and Eastern mysticism. Shambhala, Boulder. 330 pp.

Clark, C.W. 1977. Overcapitalization in commercial fisheries: symptoms, causes, and cures. Env. Biol. Fish. 2: 3–5.

Compagno, L.J.V. 1990. Alternative life-history styles of cartilaginous fishes in time and space. Env. Biol. Fish. 28: 33–75.

Constantz, G.D. 1979. Life history patterns of a livebearing fish in contrasting environments. Oecologia (Berlin) 40: 189–201.

Copp, G.H. & V. Kováč. 1996. When do fish with indirect development become juveniles? Can. J. Fish. Aquat. Sci. 53: 746–752.

Coulter, G.W. (ed.) 1991. Lake Tanganyika and its life. Nat. Hist. Museum Publ. and OUP, Oxford. 354 pp.

Crawford, S.S. & E.K. Balon. 1994a. Alternative life histories of the genus *Lucania*: 1. Early ontogeny of *L. parva*, the rainwater killifish. Env. Biol. Fish. 40: 349–389.

Crawford, S.S. & E.K. Balon. 1994b. Alternative life histories of the genus *Lucania*: 2. Early ontogeny of *L. goodei*, the bluefin killifish. Env. Biol. Fish. 41: 331–368.

Crawford, S.S. & E.K. Balon. 1994c. Alternative life histories of the genus *Lucania*: 3. An ecomorphological explanation of altricial (*L. parva*) and precocial (*L. goodei*) species. Env. Biol. Fish. 41: 369–402.

Crawford, S.S. & E.K. Balon. 1996. Cause and effect of parental care in fishes. An epigenetic perspective. pp. 53–107. *In*: J.S. Rosenblatt & C.T. Snowdon (ed.) Parental Care: Evolution, Mechanisms, and Adaptive Significance, Advances in the Study of Behavior 25, Academic Press, San Diego.

Crawford, S.S., E.K. Balon & K.S. McCann. 1999. A mathematical technique for estimating blastodisc:yolk volume ratios instead of egg sizes. Env. Biol. Fish. 54: 227–232.

Cunningham, J.E.R. & E.K. Balon. 1985. Early ontogeny of *Adinia xenica* (Pisces, Cyprinodontiformes): 1. The development of embryos in hiding. Env. Biol. Fish. 14: 115–166.

Cunningham, J.E.R. & E.K. Balon. 1986a. Early ontogeny of *Adinia xenica* (Pisces, Cyprinodontiformes): 2. Implications of embryonic resting interval for larval development. Env. Biol. Fish. 15: 15–45.

Cunningham, J.E.R. & E.K. Balon. 1986b. Early ontogeny of *Adinia xenica* (Pisces, Cyprinodontiformes): 3. Comparison and evolutionary significance of some patterns in epigenesis of egg-scattering, hiding and bearing cyprinodontiforms. Env. Biol. Fish. 15: 91–105.

Danilowicz, B.S. 1997. The effects of age and size on habitat selection during settlement of a damselfish. Env. Biol. Fish. 50: 257–265.

DeMartini, E.E. 1991. Annual variations in fecundity, egg size, and the gonadal and somatic conditions of queenfish *Seriphus politus* (Sciaenidae). U.S. Fish. Bull. 89: 9–18.

Depêche J. & R. Billard. 1994. Embryology in fish, a review. Société Française d'Ichtyologie, Paris. 123 pp.

Devillers, C. 1961. Structural and dynamic aspects of the development of the teleostean egg. pp. 379–428. *In*: M. Abercrombrie

& J. Brachet (ed.) Adavances in Morphogenesis, Academic Press, New York.
Duellman, W.E. 1989. Alternative life-history styles in anuran amphibians: evolutionary and ecological implications. pp. 101–126. *In*: M.N. Bruton (ed.) Alternative Life-History Styles of Animals, Perspectives in Vertebrate Science 6, Kluwer Academic Publishers, Dordrecht.
Eigen, M. & P. Schuster. 1979. The hypercycle. A principle of natural self-organization. Springer-Verlag, Berlin. 92 pp.
Felsenstein, J. 1979. r- and K-selection in a completely chaotic population model. Amer. Natur. 113: 499–510.
Flegler-Balon, C. 1989. Direct and indirect development in fishes – examples of alternative life-history styles. pp. 71–100. *In*: M.N. Bruton (ed.) Alternative Life-History Styles of Animals, Perspectives in Vertebrate Science 6, Kluwer Academic Publishers, Dordrecht.
Fostner, H., S. Hinterleitner, K. Mahr & W. Wieser. 1983. Towards a better definition of 'metamorphosis' in *Coregonus* sp.: biochemical, histological, and physiological data. Can. J. Fish. Aquat. Sci. 40: 1224–1232.
Fryer, G. 1959. The trophic interrelationships and ecology of some littoral communities of Lake Nyasa, with special reference to the fishes, and discussion of the evolution of a group of rock frequenting Cichlidae. Proc. Zool. Soc. Lond. 132: 153–281.
Fryer, G. 1996. Endemism, speciation and adaptive radiation in great lakes. Env. Biol. Fish. 45: 109–131.
Fryer, G. & T.D. Iles. 1972. The cichlid fishes of the Great Lakes of Africa: their biology and evolution. Oliver & Boyd, Edinburgh. 641 pp.
Fuiman, L.A. & D.M. Higgs. 1997. Ontogeny, growth and the recruitment process. pp. 225–249. *In*: R.C. Chambers & E.A. Trippel (ed.) Early Life History and Recruitment in Fish Populations, Chapman & Hall, London.
Fuiman, L.A., K.R. Poling & D.M. Higgs. 1998. Quantifying developmental progress for comparative studies of larval fishes. Copeia 1998: 602–611.
Garstang, W. 1962. Larval forms with other zoological verses. Basil Blackwell, Oxford. 77 pp.
Geist, V. 1971. Mountain sheep. A study in behavior and evolution. University of Chicago Press, Chicago. 383 pp.
Gottlieb, G. 1992. Individual development and evolution. Oxford University Press, New York. 231 pp.
Greenwood, P.H. 1974. The cichlid fishes of Lake Victoria, East Africa: the biology and evolution of a species flock. Bull. Br. Mus. nat. Hist. (Zool.) Suppl. 6: 1–134.
Greenwood, P.H. 1989. Ontogeny and evolution: saltatory or otherwise? pp. 245–259. *In*: M.N. Bruton (ed.) Alternative Life-History Styles of Animals, Perspectives in Vertebrate Science 6, Kluwer Academic Publishers, Dordrecht.
Hardy, A. 1962. Introduction. pp. 1–21. *In*: W. Garstang, Larval Forms with Other Zoological Verses, Basil Blackwell, Oxford.
Harmelin-Vivien, M.L. 1989. Implications of feeding specialization on the recruitment processes and community structure of butterflyfishes. Env. Biol. Fish. 25: 101–110.
Hempel, G. 1979. Early life history of marine fish: the egg stage. University Press, Seattle. 70 pp.
Hinckley, S. 1990. Variation of egg size of walleye pollack *Theragra chalcogramma* with a preliminary examination of the effect of egg size on larval size. U.S. Fish. Bull. 88: 471–483.
Houde, E.D. 1994. Differences between marine and freshwater fish larvae: implications for recruitment. ICES J. Mar. Sci. 51: 91–97.
Iguchi, K. & N. Mizuno. 1991. Mechanisms of embryonic drift in the amphidromous goby, *Rhinogobius brunneus*. Env. Biol. Fish. 31: 295–300.
Jägersten, G. 1972. Evolution of the metazoan life cycle. Academic Press, London. 276 pp.
Jantsch, E. 1980. The self-organizing universe. Scientific and human implications of the emerging paradigm of evolution. Pergamon Press, Oxford. 343 pp.
Kamler, E. 1992. Early life history of fish. An energetics approach. Chapman & Hall, London. 267 pp.
Kasyanov, V.L., G.A. Kryuchkova, V.A. Kulikova & L.A. Medvedeva. 1998. Larvae of marine bivalves and echinoderms. Science Publishers, Enfield. 288 pp.
Kaufman, L., J. Ebersole, J. Beets & C.C. McIvor. 1992. A key phase in the recruitment dynamics of coral reef fishes: post-settlement transition. Env. Biol. Fish. 34: 109–118.
Kjørsvik, E., A. Mangor-Jensen & I. Holmefjord. 1990. Egg quality in fishes. Adv. Mar. Biol. 26: 71–113.
Koller, G. 1930. Versuche an marinen Wirbellosen über die Aufnahme gelöster Nahrstoffe. Zeitschr. vergl. Physiol. 11: 437–447.
Kottelat, M. 1997. European freshwater fishes. An heuristic checklist of the freshwater fishes of Europe (exclusive of former USSR), with an introduction for non-systematists and comments on nomenclature and conservation. Biologia (Bratislava) 52 (supplement 5): 1–271.
Kováč, V. & G.H. Copp. 1999. Prelude: looking at early development in fishes. Env. Biol. Fish. 56: 7–14 (this volume).
Kratt, L.F. & R.J.F. Smith. 1977. A post-hatching sub-gravel stage in the life history of the Arctic grayling, *Thymallus arcticus* Pallas. Trans. Amer. Fish. Soc. 106: 241–243.
Kryzhanovsky, S.G., N.N. Disler & E.N. Smirnova. 1953. Ecomorphological principles of development in percids. Trudy Inst. Morph. Zhiv. Severtsova 10: 3–138 (in Russian).
Laale, H.W. 1980. The perivitelline space and egg envelopes of bony fishes: a review. Copeia 1980: 210–226.
Landa, J.T. 1998. Bioeconomics of schooling fish: selfish fish, quasi-free riders, and other fishy tales. Env. Biol. Fish. 53: 353–364.
Leis, J.M. 1993. Minimum requirements for published descriptions of larval fish development. Japan. J. Ichthyol. 40: 393–394.
Leis, J.M. & D.S. Rennis. 1983. The larvae of Indo-Pacific coral reef fishes. New South Wales University Press, Kensington and University of Hawaii Press, Honolulu. 269 pp.
Leis, J.M. & T. Trnski. 1989. The larvae of Indo-Pacific shorefishes. University of Hawaii Press, Honolulu. 371 pp.
Livingston, J.A. 1994. Rogue primate. An exploration of human domestication. Key Porter Books, Toronto. 229 pp.
Lobel, P.S. 1997. Comparative settlement age of damselfish larvae (*Plectroglyphidon imparipennis*, Pomacentridae) from Hawaii and Johnston Atoll. Biol. Bull. 193: 281–283.
Lowe-McConnell, R.H. 1987. Ecological studies in tropical fish communities. Cambridge University Press, Cambridge. 382 pp.
Lowe-McConnell, R.H. 1996. Fish communities in the African Great Lakes. Env. Biol. Fish. 45: 219–235.

Luckinbill, L.S. 1979. Selection and the r/K continuum in experimental populations of protozoa. Amer. Natur. 113: 427–437.

Mabee, P.M., D.S. Cua, S.B. Barlow & J.V. Helvik. 1998. Morphology of the hatching glands in *Betta splendens* (Teleostei: Perciformes). Copeia 1998: 1021–1026.

Makeyeva, A.P. 1988. Review of 'Early life histories of fishes: new developmental, ecological and evolutionary perspectives (ed. by E.K. Balon)'. Voprosy ichtiologii 28: 697–700 (in Russian).

Makeyeva, A.P. 1992. Embryology of fishes. Izd. Moskovskogo Universiteta, Moscow. 216 pp. (in Russian).

Makeyeva, A.P. & D.S. Pavlov. 1998. Freshwater ichthyoplankton of Russia (an atlas). Moscow University Press, Moscow. 215 pp. (in Russian).

Marini, F.C. 1996. My notes and observations on raising and breeding the banggai cardinalfish. The Journal of MaquaCulture 4: 1–4.

Marshall, N.B. 1953. Egg size in Arctic, Antarctic and deep-sea fishes. Evolution 7: 328–341.

Maturana, H.R. & F.J. Varela. 1988. The tree of knowledge. The biological roots of human understanding. Shambhala, Boston. 263 pp.

McElman, J.F. & E.K. Balon. 1979. Early ontogeny of walleye, *Stizostedion vitreum*, with steps of saltatory development. Env. Biol. Fish. 4: 309–348 (reprinted in Dev. Env. Biol. Fish. 5: 92–131, 1985).

McElman, J.F. & E.K. Balon. 1980. Early ontogeny of white sucker, *Catostomus commersoni*, with steps of saltatory development. Env. Biol. Fish. 5: 191–224.

Medawar, P.B. & J.S. Medawar. 1983. Aristotle to zoos. A philosophical dictionary of biology. Harvard University Press, Cambridge. 305 pp.

Medvedev, Z.A. 1969. Rise and fall of T.D. Lysenko. Columbia University Press, New York. 284 pp.

Moriyama, A., Y. Yanagisawa, N. Mizuno & K. Omori. 1998. Starvation of drifting goby larvae due to retention of free embryos in upstream reaches. Env. Biol. Fish. 52: 321–329.

Moser, H.G. 1981. Morphological and functional aspects of marine fish larvae. pp. 89–131. *In*: R. Lasker (ed.) Marine Fish Larvae, Washington Sea Grant Program, Seattle.

Moser, H.G. (ed.) 1984. Ontogeny and systematics of fishes. Amer. Soc. Ichthyol. Herpet. Special Publ. 1. 760 pp.

Moyle, P.B. & J.J. Cech, Jr. 1996. Fishes. An introduction to ichthyology, 3rd. ed. Prentice Hall, Upper Saddle River. 590 pp.

Nice, M.M. 1962. Development of behavior in precocial birds. Trans. Linn. Soc. New York, volume 8. 211 pp.

Okiyama, M. (ed.) 1988. An atlas of the early stage fishes in Japan. Tokai University Press, Tokyo. 1157 pp. (in Japanese).

Paine, M.D. & E.K. Balon. 1984. Early development of the rainbow darter, *Etheostoma caeruleum*, according to the theory of saltatory ontogeny. Env. Biol. Fish. 11: 277–299.

Paine, M.D. & E.K. Balon. 1986. Early development of johny darter, *Etheostoma nigrum*, and fantail darter, *E. flabellare*, with a discussion of its ecological and evolutionary aspects. Env. Biol. Fish. 15: 191–220.

Pavlov, D.A. 1999. Features of transition from larva to juvenile in fishes with different types of early ontogeny. Env. Biol. Fish. 56: 41–52 (this volume).

Peňáz, M. 1975. Early development of the grayling *Thymallus thymallus* (Linnaeus, 1758). Acta Sc. Nat. Brno 9 (11): 1–35.

Peňáz, M. 1981. Ecomorphological principles and saltation in the early ontogeny of salmonid fishes. pp. 95–100. *In*: M. Peňáz & M. Prokeš (ed.) Topical Problems in Ichthyology, Czech. Acad. Sci., Brno.

Peňáz, M. 1983. Ecomorphological laws and saltation in the early ontogeny of Salmonoidei. Folia Zool. (Brno) 32: 365–373.

Pepin, P. 1991. The effect of temperature and size on development, mortality and survival rates of the pelagic early life history stages of marine fish. Can. J. Fish. Aquat. Sci. 48: 503–518.

Pepin, P. & R.A.Myers. 1991. Significance of egg and larval size to recruitment variability of temperate marine fish. Can. J. Fish. Aquat. Sci. 48: 1820–1828.

Pfeiler, E. 1986. Towards an explanation of the developmental strategy in leptocephalous larvae of marine teleost fishes. Env. Biol. Fish. 15: 3–13.

Pianka, E.R. 1978. Evolutionary ecology. Harper & Row, New York. 397 pp.

Picard, F. & R. Voßwinkel. 1996. Normal developmental stages and levels of organogenesis of the rainbow trout *Oncorhynchus mykiss* (Walbaum, 1792). Z. Fischk. (Solingen) 3: 127–204.

Policansky, D. 1981. Egg size and the timing of reproductive effort in fishes. Amer. Zool. 21: 987.

Poll, M. 1986. Classification des Cichlidae du lac Tanganika, tribus, genres et espèces. Acad. Roy. Belg. Mem. Cl. Sci. Coll. 8°, 2e series 45, Fasc. 2: 1–163.

Prigogine, I. 1980. From being to becoming. Time and complexity in the physical sciences. W.H. Freeman, San Francisco. 272 pp.

Rass, T.S. 1946. Stages of ontogeny in teleost fishes. Zool. Zhurnal 25: 137–148 (in Russian).

Rass, T.S. 1948. On the life periods and regularities of development and growth of fishes. Izvestiya Akad. Nauk SSSR, Ser. Biologiya 3: 295–305 (in Russian).

Rass, T.S. 1953. Importance of eggs and larval structure for fish systematics. pp. 181–198. *In*: E.N. Pavlovsky (ed.) Otcherki po Obshchim Voprosam Ikhtiologii, Akad. Nauk SSSR Press, Moscow (in Russian).

Ribbink, A.J., B.A. Marsh, A.C. Marsh, A.C. Ribbink & B.J. Sharp. 1983. A preliminary survey of the cichlid fishes of rocky habitats in Lake Malawi. S. Afr. J. Zool. 18: 147–310.

Risk, A. 1998. The effects of interactions with reef residents on the settlement and subsequent persistence of ocean surgeonfish, *Acanthurus bahianus*. Env. Biol. Fish. 51: 377–389.

Roule, L. 1924. L'ontogénèse et la croissance avec hypermétamorphose de *Luvarus imperialis* Raf. Ann. Inst. Océan., Paris (n.s.) 1: 121–157.

Sale, P.F. (ed.) 1991. The ecology of fishes on coral reefs. Academic Press, San Diego. 754 pp.

Seehausen, O. 1996. Lake Victoria rock cichlids. Verduyn Cichlids, Zevenhuizen. 304 pp.

Sermonti, G. 1997. Resonant messages. Rivista di Biologia/ Biology Forum 90: 187–189.

Snyder, D.E. 1976. Terminologies for intervals of larval fish development. pp. 41–58. *In*: J. Boreman (ed.) Great Lakes Fish Egg and Larvae Identification: Proceedings of a Workshop, National Power Plant Team, Power Plant Project FWS/OBS 76/23, Ann Arbor.

Stearns, S.C. 1980. A new view of life-history evolution. Oikos 35: 266–281.
Stearns, S.C. 1992. The evolution of life histories. Oxford University Press, Oxford. 249 pp.
Stobutzki, I.C. & D.R. Bellwood. 1997. Sustained swimming abilities of the late pelagic stages of coral reef fishes. Mar. Ecol. Prog. Ser. 149: 35–41.
Stephens, G.C. 1982. Recent progress in the study of 'Die Ernährung der Wassertiere und der Stoffhaushalt der Gewässer'. Amer. Zool. 22: 611–619.
Subiyanto, I. Hirata & T. Senta. 1993. Larval settlement of the Japanese flounder on sandy beaches of the Yatsushito Sea, Japan. Nippon Suisan Gakkaishi 59: 1121–1128.
Sweatman, H.P.A. 1988. Field evidence that settling coral reef fish larvae detect resident fishes using dissolved chemical cues. J. Exp. Mar. Biol. Ecol. 124: 163–174.
Tesch, F.-W. 1977. The eel. Biology and management of anguillid eels. Chapman & Hall, London. 434 pp.
Tilney, R.L. & T. Hecht. 1993. Early ontogeny of *Galeichthys feliceps* Valenciennes from the south east coast of South Africa. J. Fish Biol. 43: 183–212.
Thresher, R.E. 1984. Reproduction in reef fishes. T.F.H. Publications, Neptune City. 399 pp.
Thresher, R.E. 1985. Distribution, abundance, and reproductive success in the coral reef fish *Acanthochromis polyacanthus*. Ecology 66: 1139–1150.
Thresher, R.E. 1988. Latitudinal variation in egg sizes of tropical and sub-tropical North Atlantic shore fishes. Env. Biol. Fish. 21: 17–25.
Trinkaus, J.P. 1984. Cells into organs. The forces that shape the embryo. Prentice-Hall, Englewood Cliffs. 543 pp.
Vagelli, A. 1999. The reproductive biology and early ontogeny of the mouthbrooding Banggai cardinalfish, *Pterapogon kaudermi* (Perciformes, Apogonidae). Env. Biol. Fish. 56: 79–92 (this volume).
Vasnetsov, V.V. 1948. 'Etaps' of development in organ systems connected with feeding in roach, bream and carp. pp. 233–253. *In*: V.V. Vasnetsov (ed.) Morphological Features Determining Feeding of Roach, Bream and Carp at All Stages of Development, Akad. Nauk Press, Moscow-Leningrad (in Russian).
Vasnetsov, V.V. 1953. Etaps in the development of bony fishes. pp. 207–217. *In*: E.N. Pavlovsky (ed.) Otcherky po Obshtch. Vopr. Ikhtiol., AN SSSR Press, Moscow-Leningrad (in Russian).
Victor, B.C. 1986a. Duration of the planktonic larval stage of one hundred species of Pacific and Atlantic wrasses (family Labridae). Mar. Biol. 90: 317–326.
Victor, B.C. 1986b. Delayed metamorphosis with reduced larval growth in a coral reef fish (*Thalassoma bifasciatum*). Can. J. Fish. Aquat. Sci. 43: 1208–1213.
Vilizzi, L. & K.F. Walker. 1999. The onset of the juvenile period in carp, *Cyprinus carpio*: a literature survey. Env. Biol. Fish. 56: 93–102 (this volume).
Wald, G. 1981. Metamorphosis: an overview. pp. 1–39. *In*: L.I. Gilbert & E. Frieden (ed.) Metamorphosis, A Problem in Developmental Biology, Plenum Press, New York.
Ware, D.M. 1975. Relation between egg size, growth and natural mortality of larval fish. J. Fish. Res. Board Can. 32: 2503–2512.
Wassersug, R.J. 1975. The adaptive significance of the tadpole stage with comments on the maintenance of complex life cycles in anurans. Amer. Zool. 15: 405–417.
Wassersug, R.J. & W.E. Duellman. 1984. Oral structures and their development in egg-brooding hylid frog embryos and larvae: evolutionary and ecological implications. J. Morphol. 182: 1–37.
Webb, J.F. 1999. Larvae in fish development and evolution. pp. 109–158. *In*: B.K. Hall & M.H. Wake (ed.) The Origin and Evolution of Larval Forms, Academic Press, San Diego.
Williamson, D.I. 1992. Larvae and evolution. Toward a new zoology. Chapman & Hall, London. 223 pp.
Williamson, G.R., C.L. Deedler & F. de Graaf. 1993. One-day metamorphosis of *Anguilla anguilla* leptocephali into glass eels. Japan. J. Ichthyol. 40: 282–283.
Witte, F. 1984. Ecological differentiation in Lake Victoria haplochromines: comparison of cichlid species flocks in African lakes. pp. 157–167. *In*: A.A. Echelle & I. Kornfield (ed.) Evolution of Fish Species Flocks, University of Maine Press, Orono.
Wourms, J.P. 1981. Viviparity: the maternal-fetal relationship in fishes. Amer. Zool. 21: 473–515.
Wourms, J.P., B.D. Grove & J. Lombardi. 1988. The maternal-embryonic relationship in viviparous fishes. pp. 1–134. *In*: W.S. Hoar & D.J. Randall (ed.) Fish Physiology, vol. 11b, Academic Press, San Diego.
Yanagisawa, Y. & H. Ochi. 1991. Food intake by mouthbrooding females of *Cyphotilapia frontosa* (Cichlidae) to feed both themselves and their young. Env. Biol. Fish. 30: 353–358.
Yanagisawa, Y. & T. Sato. 1990. Active browsing by mouthbrooding females of *Tropheus duboisi* and *Tropheus moorii* (Cichlidae) to feed the young and/or themselves. Env. Biol. Fish. 27: 43–50.

Part 1. Reflections on early ontogeny and metamorphosis

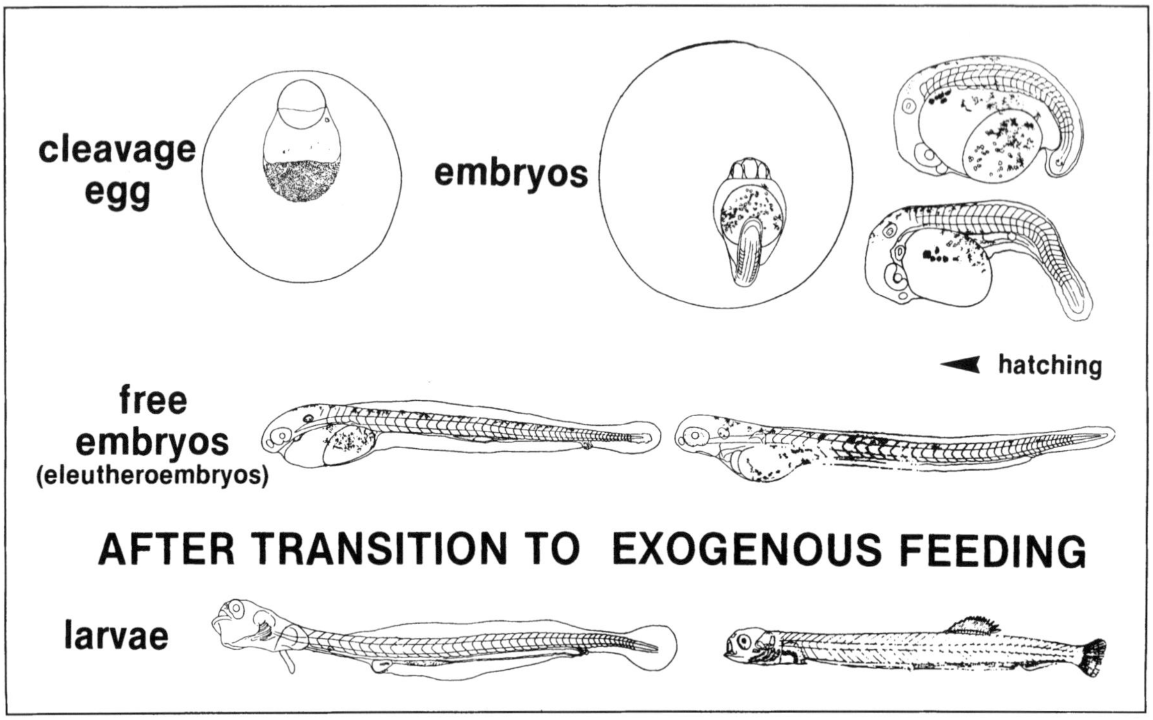

Clupeonella cultriventris

Environmental Biology of Fishes **56:** 41–52, 1999.

Features of transition from larva to juvenile in fishes with different types of early ontogeny

Dimitri A. Pavlov
Department of Ichthyology, Faculty of Biology, Moscow State University, 119899 Moscow, Russia
(e-mail: pavlov@dpavlov.home.bio.msu.ru)

Received 14 May 1997 Accepted 1 December 1998

Key words: early development, herring, wolffish, eelpout, heterochrony

Synopsis

The transition from the larva state to the juvenile state (i.e. morphological condition characterised by mainly adult characters) was examined in three marine fish species: herring *Clupea pallasi marisalbi*, wolffish *Anarhichas lupus*, and eelpout *Zoarces viviparus*, based on external morphology and skeletal development. In spite of the different reproductive styles (oviparity, facultative viviparity, and obligate viviparity, respectively) and different types of early ontogeny (indirect, transitory, and direct, respectively), the beginning of the juvenile state occurred at similar total lengths (TL), which were approximated as 35 mm TL in herring and eelpout, and 32 mm TL in wolffish. Features of ontogeny were compared, assuming that the beginning of the juvenile state represented an uniform characteristic of morphological development for these species. It was proposed that the beginning of the larva or juvenile periods (sensu Balon) could not coincide with the beginning of the larva and juvenile states in the ontogeny of some species.

Introduction

The concept of non-gradual ontogeny of fishes, initially proposed by Vasnetsov (1953) and Kryzhanovsky et al. (1953), stimulated investigations to identify objective boundaries for distinguishing different intervals of fish development. A new understanding of non-gradual ontogeny was developed in the theory of saltatory ontogeny (Balon 1986, 1989, 1990). This theory, based on Taoist dichotomy and self-organization, explained the processes of a succession of stabilized states bordered by formative thresholds.

According to the theory of saltatory ontogeny, periods are the longest intervals of ontogeny, separated by the strongest thresholds. However, in many fish species the beginning of the juvenile period (sensu Balon) has not been determined precisely, and, in some cases, the transitional status of an organism (neither larva, nor juvenile) has been described (Altukhov 1975, Fukuhara & Fushimi 1986).

The theory of saltatory ontogeny considers development as a sequence of stabilised states (steps) and rapid changes (thresholds). Morphological transitions are often associated with physiological and behavioural shifts and all these changes are usually used to distinguish developmental intervals (Balon 1990). In the present study, I use morphological criteria only to determine the developmental status of marine teleosts from different systematic and ecological groups (irrespective to their relations with the environment). I believe that such an approach will allow better understanding of the nature of diversity of life styles in fishes. For the definition of the morphological status of the organism, the term 'state', which has no saltatory significance, is used in the present study. The beginning of the larva state is the morphological condition when the organism has the ability to feed orally. The beginning of the juvenile state is the morphological condition when the organism no longer has larval characters and most adult characters have appeared. The criteria applied in the present study to determine the beginning of the juvenile state are the formation of main definitive organs and skeletal elements, and the disappearance of larval characters. The morphologically-based 'states' differ from 'periods' (sensu Balon), which designate the main natural (homeorhetic) intervals of ontogeny as a saltatory process.

Following general opinion, a juvenile is a small adult with mainly definitive characters (Vasnetsov 1953, Balon 1990, Copp & Kováč 1996). Various characteristics (external morphology, skeletal development, gonad condition, physiological parameters, relations with environment, and behaviour) have been used to determine the juvenile state (Vasnetsov 1953, Kryzhanovsky et al. 1953, Ryzhkov 1976, Balon 1980, 1985, Peňáz 1983, Zhiteneva & Seitov 1986, Pavlov 1989, Copp & Kováč 1996). At the same time, in some species, larval (or even embryonic) organs (like yolksac, and preanal finfold) can be still present in fish assumed to be juvenile (Balon 1985, 1990, Pavlov 1989). Owing to heterochronies in morpho-functional events, which are induced by environmental fluctuations, it is difficult to determine the exact onset of juvenile development, and the beginning of the juvenile state cannot be determined based on only one character. For example, the onset of scale formation is often used as an important character to distinguish the beginning of juvenile development (Balon 1980, 1985). However, as is known in the Atlantic salmon, *Salmo salar*, and brown trout, *Salmo trutta* (Jensen & Johnsen 1982), that the formation of scales is delayed, and in cold years they may not appear at all in some juveniles.

Because of the very large diversity of life styles, even within one group of teleost fishes, serious problems arise during comparison of the transition from embryo to larva, and from larva to juvenile in different species. According to the definition of Balon (1989, p. 475), 'The transition to exogenous feeding, i.e. to the orally ingested and intestinally digested acquisition of nutrients, marks the beginning of the next period of life history, be it larva in case of indirect, or juvenile in case of direct ontogeny'. The larva period terminates the process of remodelling, or metamorphosis (Balon 1989). A great variation in morphology of the organisms at the beginning of larva period is described in fishes with different types of early ontogeny (Balon 1985, 1990, Flegler-Balon 1989). At the beginning of the juvenile period, interspecific variation in morphology apparently should not be so large, but still exists and is caused partly by interactions between the organism and its environment.

For example, according to Balon (1985, 1990), the cichlid fish *Cyphotilapia frontosa* of Lake Tanganyika begins to feed orally inside the buccal pouch of its mother, long before being released from the pouch at an advanced state of ontogeny, as a 'yolksac juvenile'. In round goby *Neogobius melanostomus*, another fish with direct development, oral feeding is observed even inside the egg envelope (Moskal'kova 1985): the particles, consisting of mainly proteins and carbohydrates, appeared as a result of digestion of the yolk, are released from the intestine and sucked into the oral cavity along with perivitelline fluid. This form of feeding allows better digestion of proteins. In both cases, the beginning of oral feeding is apparently accompanied by morphological change at least in the structure of digestive tract. At the same time, the organism continues to stay at the previous stabilized state until a dramatic change of its environment coincides with the releasing from the mother's buccal cavity in *C. frontosa*, or from the egg shell in *N. melanostomus*. Thus, not only a new organ-to-organ interaction, but also a new organism-to-environment interaction is associated with the beginning of juvenile period (sensu Balon).

The species examined in the present study are common to the White Sea: the spring spawning herring, *Clupea pallasi marisalbi* Berg, wolffish, *Anarhichas lupus* L., and eelpout, *Zoarces viviparus* (L.). The two latter species belong to the same suborder Zoarcoidei, but represent two families (Anarhichadidae and Zoarcidae). The aim of the present study was to identify the beginning of the juvenile state in herring, wolffish, and eelpout through the description of features in transitory stages, and to determine whether the morphologically-defined onset of the juvenile 'state' corresponds to that of the 'juvenile period' (sensu Balon).

Material and methods

Herring

Spawners of herring at maturity stage V were caught in Palkina Creek situated in Kandalaksha Bay near Kandalaksha, the White Sea. The average total length (TL) of the fish was 21.7 cm (range 19–24 cm), and average age (determined by scales) was 6 years (range 4–8 years) (n = 16). The gonads were removed, placed in plastic bags, and transported to the White Sea Biological Station of Moscow State University (WBS MSU) within 4 h. Then, the eggs from three pairs of spawners were artificially inseminated on

27 April 1996. The eggs, attached to the nets and to glass plates, were incubated in the aquarium (100 l) with a closed system of water circulation. A stable water temperature of 10.9°C (SD = 0.1) and natural photoperiod (with the light from a window) were used during the incubation. Mass hatching was observed on 8 May, at age 11 d after insemination.

The free embryos were transferred into a small aquarium (40 l) with low water level (10 cm), and as larvae (30 days from hatching), to larger aquarium (150 l) with a water level of 20–30 cm. The mean water temperature for the 60 days after hatching was 12.3°C (SD = 1.7). A natural light cycle (with the light from a window) and additional continuous light from a lamp (100 W) were used throughout rearing. Larvae were fed dry pellets produced for tropical aquarium fish and natural zooplankton on day 3 after hatching, and only zooplankton was offered after a week from hatching. Egg mortality was low, and that of larvae and juveniles up to 60 days from hatching was approximately 80% of the initial number of larvae. At this age, TL ranged from 23.1 to 38.9 mm, with a mean of 31.0 mm (SD = 3.3) (n = 50). On each fifth day of development, ten specimens were anaesthetised by MS-222, and their lengths (SL and TL) were measured under the binocular microscope. Specimens were fixed in 4% formaldehyde buffered with Ca_2CO_3. Specimens were cleared and stained five months later according to the method described by Balon (1985), with blue structures being cartilaginous and red parts being calcified (i.e. ossification has begun). The terminology of skeletal structures used in the present study is given according to Cunningham & Balon (1986) and Potthoff & Tellock (1993).

Wolffish

Larvae and juveniles of wolffish were collected in May and June 1990 and 1991 near the WBS MSU and preserved in 4% formaldehyde. In addition, fertilized eggs of the same species were obtained in 1995 from broodstock kept at the Flødevigen Marine Research Station, Institute of Marine Research (IMR), Norway. This broodstock was established in 1986–1988 from wolffish larvae. The methods of artificial insemination and incubation of eggs were described before (Moksness & Pavlov 1996). The eggs from several females were incubated at constant temperature 7.0°C (SD = 0.2). After hatching (within 133 days from insemination), the young were kept in the tank (280 l) at 5.5–6.0°C and fed natural zooplankton and dry pellets. Specimens were anaesthetised by MS-222 on each 6–8 day of development, measured for TL, fixed in 4% formaldehyde, then cleared and stained within a week after fixing; those from the White Sea were cleared and stained in January 1997. Wolffish young of the same size groups, reared in the IMR and collected in the White Sea, had similar external morphology and skeleton structure.

The description of the ontogenetic stages in herring and wolffish was undertaken on typical specimens, used both for the analysis of their external morphology (on live anaesthetised specimens) and skeletal elements.

Eelpout

About 30 live embryos of eelpout were removed from the ovarian cavity of an adult female, which was caught near the WBS MSU in October 1987. Ten embryos were anaesthetised by MS-222, measured and fixed in 4% formaldehyde. They ranged from 35.2 to 43.0 mm SL, with a mean of 41.1 mm (SD = 2.2). The specimens were cleared and stained in January 1997.

Results

Herring

At 39 days after hatching, the herring were 21.1 mm SL, 22.6 mm TL (Figures 1a, 2a,b). A prolonged preanal finfold extended from the anterior part of the body venter to the anus. In the pectoral and pelvic fins, rays were absent. The numbers of rays in the dorsal and anal fins (18–19 and 14–17, respectively) were close to the definitive numbers (18–20 and 16–19) for the White Sea herring, respectively (Svetovidov 1952). The number of caudal fin rays was 21–24; these were mainly principal rays; several secondary caudal rays were not developed. Melanophores were distributed proportionally on the flanks, and the largest pigment cells were situated on the ventral part of the trunk. In the head skeleton, the dorsal part of the skull, auditory capsule, palatine–pterygoid–quadrate complex, ventral part of hyomandibular and symplectic were cartilaginous. From the bones of opercular series, only operculum appeared. The pectoral girdle consisted of three ossified bones (cleithrum, supracleithrum, and posttemporal) and coraco-scapular cartilage. The ossification process had begun in the vertebral column. In the axial precaudal skeleton, anlagen of neural arches were cartilaginous, and their distal margins were not fused in the pairs. However, in the caudal skeleton, the neural arches were ossifying, and their distal margins were fused in all pairs. Of the first preural centra, 2–3 were still not separated from each other. Epurals as well as proximal and distal parts of several caudal neural arches, haemal arches, and hypurals were cartilaginous.

At 46 days after hatching, the herring were 26.3 mm SL, 28.4 mm TL (Figures 1b, 2c,d). The length and width of the preanal finfold decreased: remnants of the preanal finfold existed between the pelvic fins and anus. The inflated swimbladder was seen above the pelvic fins and intestine. The body venter was covered with guanine and with large melanophores. The rays in the pectoral and pelvic fins had appeared. The number of rays in the caudal fin had reached 29–35. In the head skeleton, the nasal, frontal, and parietal bones were ossifying, and the suboperculum had appeared. Posterior process developed from the caudal margin of maxilla. Teeth had appeared on the dentary. The quadrate was still not separated from the palatine–pterygoid–quadrate complex. This complex and auditory capsule remained mainly cartilaginous. In the axial skeleton, the distal margins of all neural arches were joined, and several cartilaginous proximal radials appeared in the dorsal part of the body from the head to the dorsal fin. These structures remained during subsequent development, despite the absence of fins in this area. In the caudal skeleton, all preural centra were separated from each other.

At 60 days after hatching, the herring were 30.0 mm SL, 35.0 mm TL (Figures 1c, 2e,f). The preanal finfold had disappeared. The average number of caudal fin rays was 32.5 (SD = 1.7) (n = 50). In the head skeleton, the quadrate, preoperculum and interoperculum had formed. Almost all elements were undergoing ossification, with the exception of nasal area and that of auditory capsule. In the coraco-scapular cartilage of the pectoral girdle, proximal pectoral radials had formed. In the caudal peduncle, the third epural appeared in

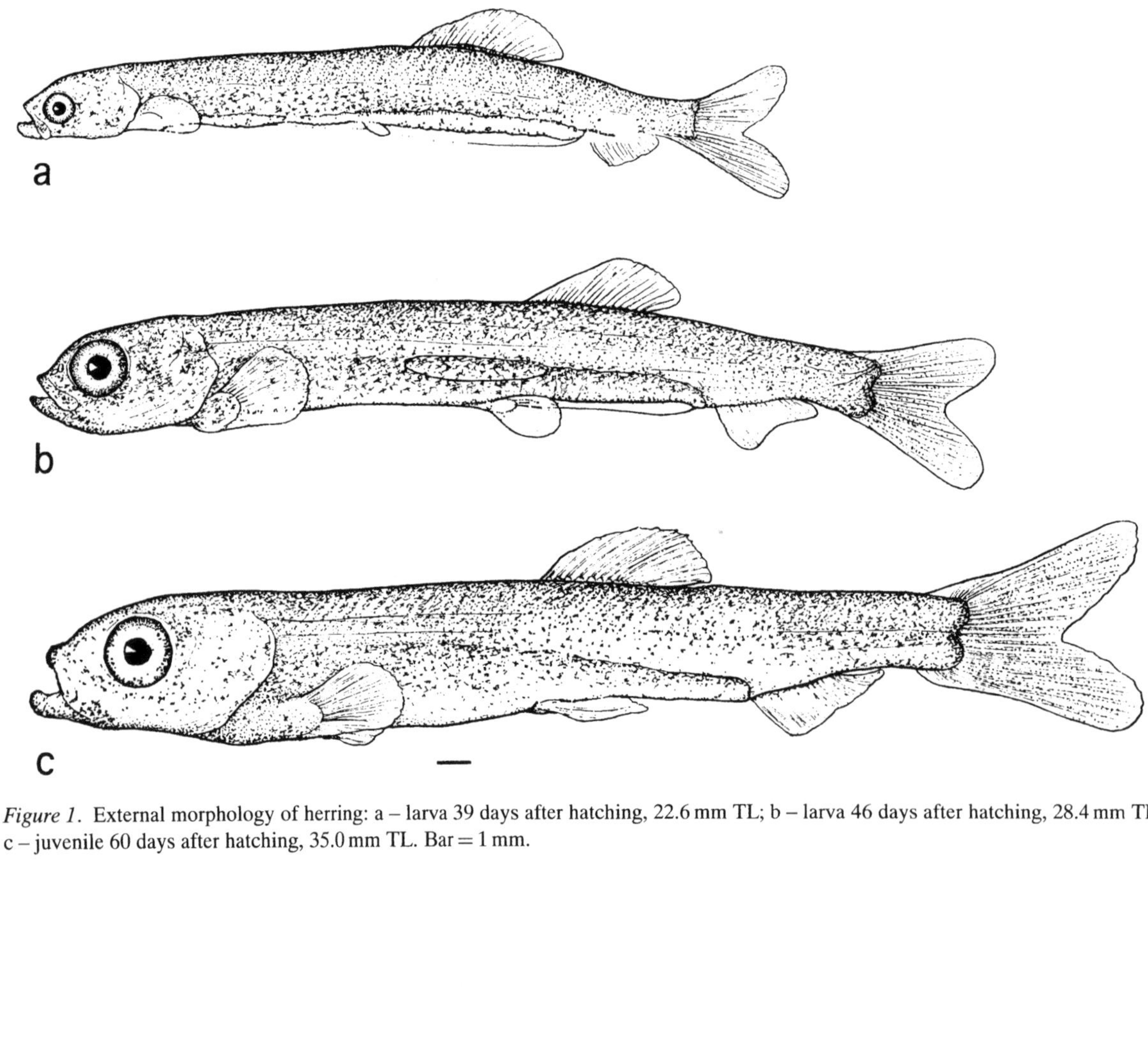

Figure 1. External morphology of herring: a – larva 39 days after hatching, 22.6 mm TL; b – larva 46 days after hatching, 28.4 mm TL; c – juvenile 60 days after hatching, 35.0 mm TL. Bar = 1 mm.

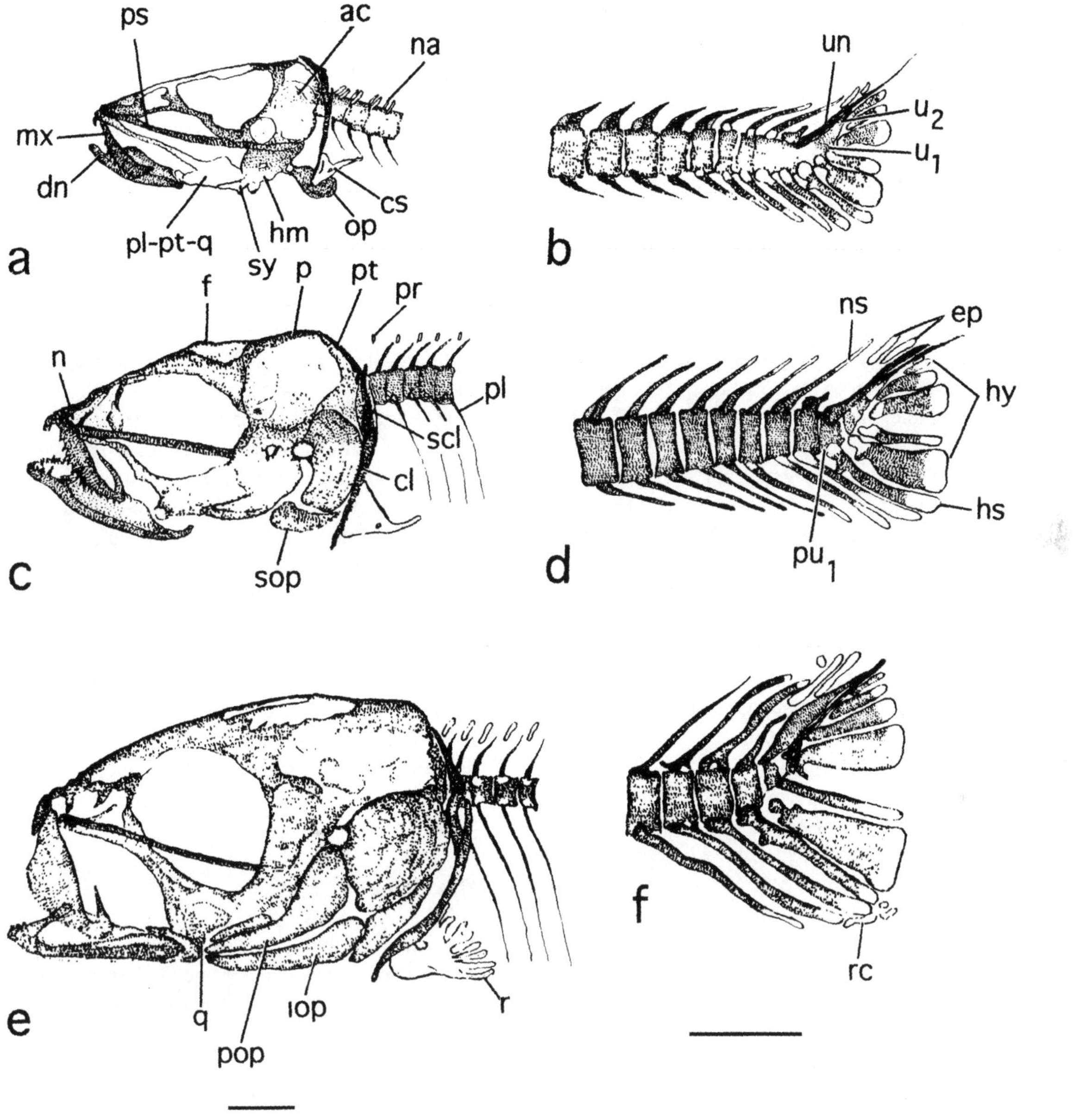

Figure 2. Development of the head skeleton and caudal skeleton of herring: a, b – larva 39 days after hatching, 22.6 mm TL; c, d – larva 46 days after hatching, 28.4 mm TL; e, f – juvenile 60 days after hatching, 35.0 mm TL. ac = auditory capsule; cl = cleithrum; cs = coraco-scapular cartilage; dn = dentary; ep = epural; f = frontal; hm = hyomandibular; hs = haemal spine; hy = hypural; iop = interoperculum; mx = maxilla; n = nasal; na = neural arch; ns = neural spine; op = operculum; p = parietal; pl = pleural rib; pl–pt–q = palatine–pterygoid–quadrate; pop = preoperculum; pr = proximal radial; ps = parasphenoid; pt = posttemporal; pu = preural centrum; q = quadrate; r = proximal pectoral radial; rc = radial cartilage; scl = supracleithrum; sop = suboperculum; sy = symplectic; u = ural centrum; un = uroneural. Cartilage white; bone stippled. Bar = 1 mm.

its dorsal part, and three radial cartilages formed in its ventral part. The distal margins of hypurals and of last neural and haemal spines were still cartilaginous. The behaviour of the fish changed, forming a school with very strong reactions to external stimuli. At earlier intervals of development, schooling behaviour was not so well developed, and the school often divided into separated groups of individuals.

Wolffish

At hatching, wolffish embryos were 18.7 mm SL, 21.0 mm TL (Figures 3a,c). Free embryos had a remnant of the yolksac with an oil droplet (about 0.8 mm diameter). Eye diameter was about two times lower than head depth. The finfold almost completely differentiated into dorsal, anal, and caudal fins. Principal caudal fin rays had five segments each. The mean number of rays in the dorsal, anal, and caudal fins were 74.5 (SD = 0.8), 46.4 (SD = 0.8), and 23.8 (SD = 1.0), respectively (n = 40). The numbers were close to those in adults (Barsukov 1959). Melanophores were evenly distributed on the trunk. At the end of the caudal peduncle, pigmentation was absent. The largest melanophores were located at the base of the pectoral fins and on the body venter. In addition, the ventral part of the trunk was covered with guanine. The neurocranium was cartilaginous. All elements of the splanchnocranium had formed. Among them, palatine, pterygoid, hyomandibular, symplectic, articular, hyoid, and branchial arches were cartilaginous. Teeth located on the dentary, premaxilla, palatine, and vomer. In the pectoral girdle, the cleithrum, supracleithrum and posttemporal ossification had begun. The scapula and coracoid had not separated from each other and represented one cartilage. In the cartilaginous blade located posterior from coraco-scapular cartilage, three apertures had formed. Two anlagen of pelvic girdles were cartilaginous. The vertebral column had undergone ossification, but clear borders between centra had not formed. The young were pelagic. They had a positive reaction to light and moved towards the water surface by means of continuous undulations of the trunk. The young began to feed on zooplankton just after hatching.

At 45 days after hatching, the wolffish were 27.0 mm SL, 32.0 mm TL (Figures 3b,d). The yolksac had been resorbed. Body depth was approximately two times higher than that in an embryo at hatching. Eye diameter was about 2.7 times lower than head depth. Several (8–11) pigment bands had appeared on the flanks. Separated pigment spots distributed on the dorsal and anal fins. Almost all elements of the skull had begun ossification. The branchial and hyoid arches were partly cartilaginous. In the pectoral girdle, the proximal radials of the cartilaginous blade were still not completely separated from each other. In the axial skeleton, ossification of the vertebral column appeared to be complete, and the centra were separated from each other. In the dorsal and anal fins, proximal and distal radials remained cartilaginous. The positive photoreaction had been lost, and the juveniles preferred to spend all the time near the bottom. The young moved to the water surface only when they were hungry, seeking the food, which was offered from the upper side of the tank.

Eelpout

At approximately 1.5 months after hatching, eelpout were 43.0 mm SL, 44.4 mm TL (Figures 4a,b,c). A specimen removed from the female's oviduct had an adult body shape, but still possessed a large yolksac with an oil droplet (about 1.9 mm in diameter) in its anterior part. Several pigment spots were present, distributed mainly on the dorsal part of the trunk and on the dorsal fin. Almost all skeletal elements of the head had begun ossification. In the pectoral girdle, cartilaginous elements were represented by coraco-scapular cartilage and cartilaginous blade. As in wolffish, the proximal radials of the cartilaginous blade were not completely separated from each other. In the pelvic girdle, the basipterygium had begun ossification, but the proximal radials were cartilaginous. In the axial skeleton, the distal ends of all proximal radials remained cartilaginous. In the caudal skeleton, the epural, distal part of hypural plate, and the distal end of the last haemal spine were cartilaginous. The fishes were able to swim and feed just after removing from the female's body.

Discussion

Spawning of herring in Kandalaksha Bay of the White Sea occurs mainly under the ice, from the end of April to the beginning of May at −0.2 to +0.3°C. The embryos hatch in June, 45–50 days after egg deposition. Water temperature at hatching reaches 10–12°C. The beginning of the juvenile state is observed in August–September (Altukhov 1975, Ivanchenko 1975). According to the classification system of fish reproductive styles (Balon 1975, 1981, 1990), herring belong to the ethological section of nonguarders, and to the ecological guild of nonobligatory plant spawners. In herring, the disappearance of larval organs, as well as the formation of all skeletal elements and their ossification, showed that the organism reached the juvenile state by approximately 35 mm TL. Similarly, Girsa & Lapin (1978) reported for herring from the same stock that by 35 mm TL, formation of the stomach was completed, the swimbladder had an anterior duct,

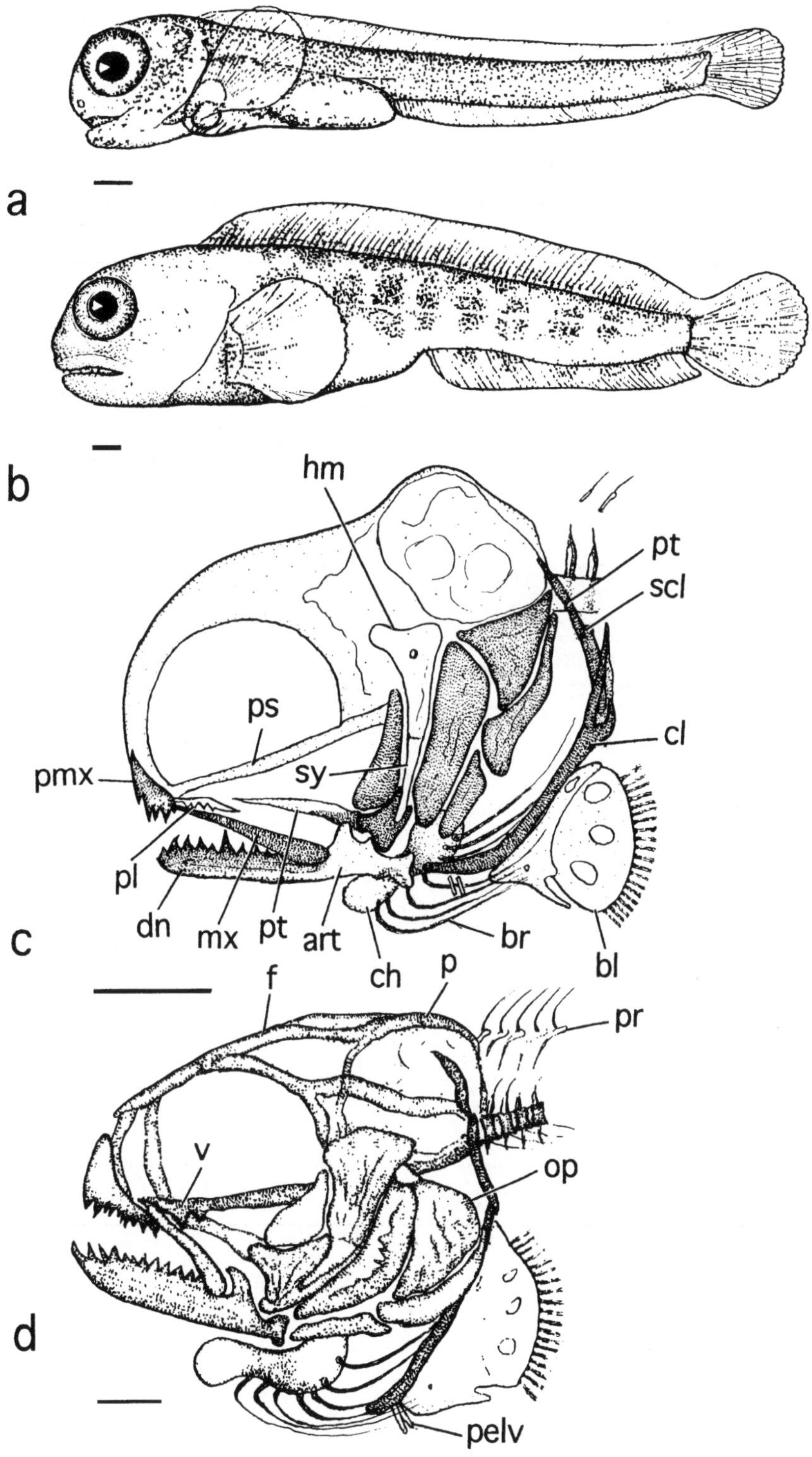

Figure 3. External morphology and head skeleton of wolffish: a, c – embryo at hatching, 21.0 mm TL; b, d – juvenile 45 days after hatching, 32.0 mm TL. art = articular; bl = blade; br = branchiostegal rays; ch = ceratohyal; pelv = pelvic girdle; pl = palatine; pmx = premaxilla; pt = pterygoid; v = vomer. For other abbreviations, see Figure 2. Cartilage white; bone stippled. Bar = 1 mm.

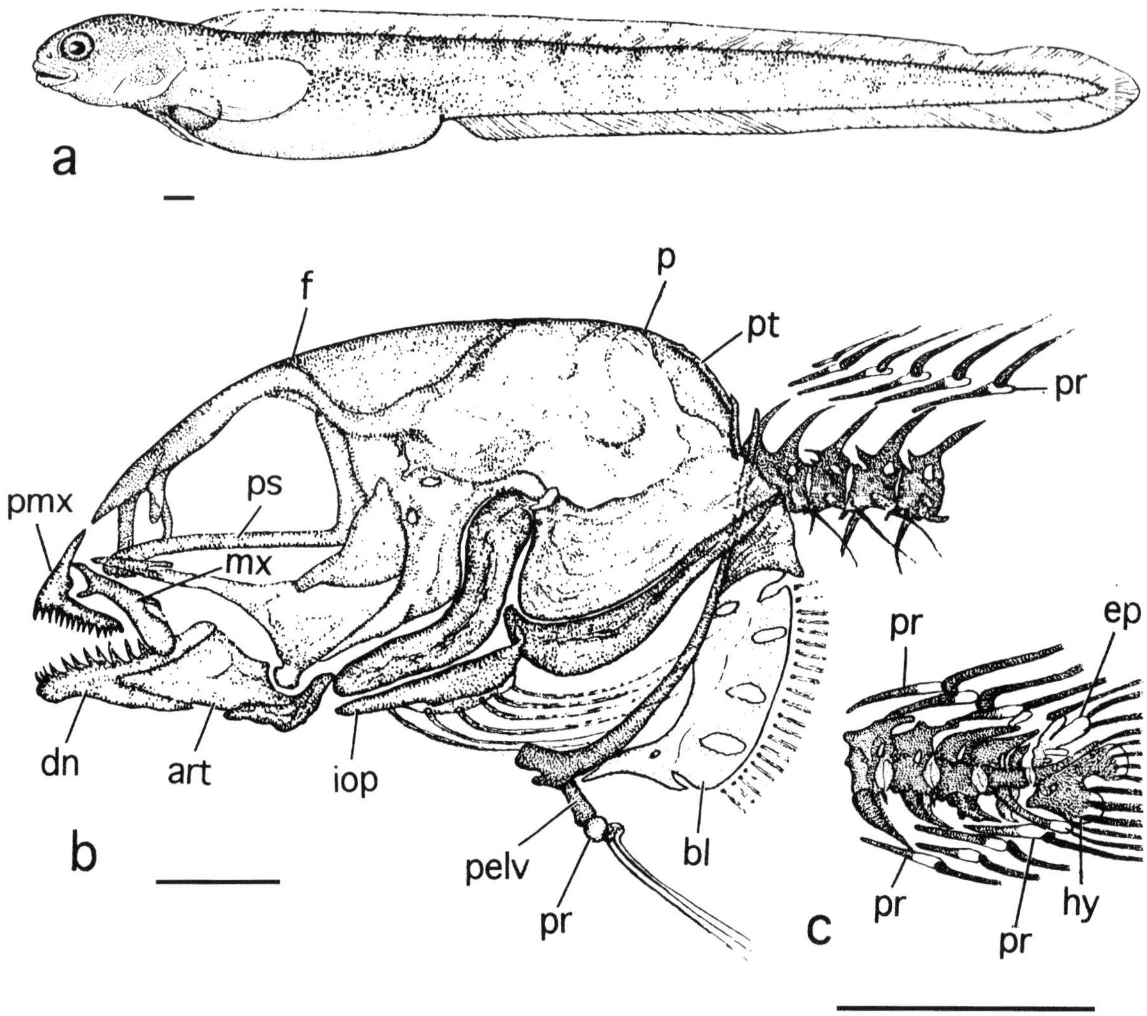

Figure 4. External morphology (a), head skeleton (b), and caudal skeleton (c) in embryo eelpout, 44.4 mm TL. For abbreviations, see Figures 2 and 3. Cartilage white; bone stippled. Bar = 1 mm.

separating into two ducts before the auditory capsules, and a prolonged posterior duct. In the wild, herring of this size, and in the juvenile state, were found in schools, and migrated from the shallow water to more exposed areas.

Spawning of wolffish in the White Sea is observed mainly at the end of summer, with fertilized eggs released 8–15 h after ovulation (Johannessen et al. 1993, Pavlov 1994). Within this time, copulation between spawners and internal insemination of eggs occurs. After deposition, the egg mass is protected by the male, apparently until hatching, for about 9.5 months. Hatching occurs in spring and after a comparatively short period (1–1.5 months) of pelagic life the larvae shift to benthic life (Pavlov & Novikov 1993).

Owing to internal insemination, this species can be considered ethologically as bearers, and ecologically as facultative lecithotrophic live bearers (Balon 1990). According to Balon (1985), a short retention of fertilized eggs within the oviduct should not be considered as viviparity. However, based on the opinion of Wourms (1981), wolffish can be placed into the group of fishes with facultative viviparity. In wolffish, a substantial change of external morphology and almost completed skeleton ossification (assumed from the retention of alizarin red-S) suggested that the beginning of the juvenile state occurred by about 32 mm TL. At this same length, the appearance of borders separating centra in wolffish caudal skeleton was registered in a previous study (Pavlov & Moksness 1997). The beginning of the juvenile state was associated with the transition from pelagic to benthic life and a change in behaviour. As was observed before in nature, wolffish over 30 mm TL are rarely caught in the pelagic zone (Pavlov & Novikov 1993). In the laboratory, wolffish of 50–60 mm TL occupied shelters and no longer demonstrated a positive reaction to light (Pavlov 1995).

In the White Sea, the spawning period of eelpout is prolonged; females with embryos inside were registered both in summer and in winter (Andriashev 1954). Eggs leave the ovarian follicle before internal insemination and then develop in the ovarian cavity during gestation of 3.5–4 months. The embryos hatch by an age of about two months and at a length about 17 mm, reaching 33–46 mm at parturition as a result of the absorption of histotrophe (Andriashev 1954, Soin 1968, Kristofferson et al. 1973). Eelpout can be considered ethologically as bearers, and ecologically as histotrophic live bearers (Balon 1990). All specimens of eelpout (36–44 mm TL) examined in the present study appeared to achieve the juvenile state. Soin (1968) reported eelpout progeny to be born at 33–45 mm TL, but the yolksacs were still present. Therefore, the eelpout specimens used in the present study were close to parturition. Based on the degree of skeleton development and ossification in my specimens, eelpout may be assumed to reach the juvenile state by about 35 mm TL, despite the presence of a large yolksac. The presence of yolksacs in apparently juvenile individuals has also been reported in some other species with direct development, e.g. *Labeotropheus* (Balon 1977), *Cyphotilapia frontosa* (Balon 1985), and *Latimeria chalumnae* (Balon 1991).

Thus, the three species have different sequences of the following events of their life cycles: ovulation (o), egg deposition (d) or parturition (p), insemination (i), and hatching (h). Herring: o–d–i–h. Wolffish: o–i–d–h. Eelpout: o–i–h–p. The three species possess different types of early ontogeny (Balon 1985, 1990). Herring undergo indirect ontogeny with metamorphosis and a long period of larva development. In herring, the threshold (sensu Balon) separating larvae and juveniles has not been determined precisely. Altukhov (1975) described a transitional 'stage' (at about 30 mm TL) in spring spawning herring of the Kandalaksha Bay, the White Sea, characterised by appearance of silvery pigment on flanks, and increase of body depth. According to Ivanchenko (1975), herring larva of the same stock becomes juvenile at 29–33 mm TL, but the morphological criteria at the beginning of the juvenile period have not been described. Wolffish hatch at an advanced state of development with many juvenile characters, almost resorbed yolksac, and begin to feed externally just after hatching. According to the first observations, the onset of the juvenile state coincides with the total resorption of the yolksac (Pavlov 1986). However, subsequent study of wolffish development has shown that some skeletal elements develop and ossify at later stages, and its ontogeny can be referred to as transitory (Pavlov 1997). Eelpout at parturition have an adult morphology (Soin 1968). Therefore, the larva period is eliminated, and the species' ontogeny can be regarded as direct. The exact morphological condition at the beginning of juvenile period in eelpout remains unknown.

In the three species, at the beginning of the juvenile state, parts of the pectoral girdle and several small elements of the axial skeleton remained cartilaginous; ossification of these structures occurred much later than that in the majority of other bones and apparently cannot be used as a criterion for juvenile status. The anlagen of scales were not observed under the microscope; in adult wolffish and eelpout the scales are very small (Andriashev 1954). In spring spawning herring of Kandalaksha Bay, the anlagen of first scales were found in specimens ranging from 29 to 52 mm TL, and their formation probably depended on environmental conditions (Ivanchenko 1983).

In the three species, fecundity and egg size values were different: herring had the highest fecundity and the smallest eggs, wolffish had the largest eggs, and eelpout the lowest fecundity (Table 1). However, these species reached the juvenile state at similar lengths. Wolffish at the beginning of the juvenile state had the smallest TL in % of the average TL of spawning females. Amongst the three species, herring had the longest interval of metamorphosis, and probably had the benefits of being a transparent larva for a long time, living in the pelagic environment. In wolffish, a comparatively short period of metamorphosis from larva to juvenile was connected with the pelagic phase of its ontogeny during which larvae were distributed by water currents. Such a distribution is an important event of its life cycle, as the occurrence of juveniles and adults seems to be limited by the numbers of suitable shelters (Pavlov & Novikov 1993). In eelpout, the size of embryo increased substantially as a result of the absorption of histotrophe (Kristofferson et al. 1973). After parturition, the young were found at the bottom, and their morphology and behaviour were similar to those observed in adult fish (Andriashev 1954). Thus, the

Table 1. Absolute fecundity, diameter of mature oocyte from (1) Svetovidov (1952), (2) Barsukov (1959), (3) Andriashev (1954), and Soin (1968), and total length (TL) at the beginning of the larva and juvenile states in herring, wolffish, and eelpout.

Species	Absolute fecundity	Diameter of mature oocyte (mm)		TL at the beginning of the larva state		TL at the beginning of the juvenile state	
		min–max	mean	mean (mm)	% of adult TL	mean (mm)	% of adult TL
Herring	4000–13100[(1)]	1.3–1.6[(1)]	1.4	10.0	4.6	35.0	16.1
Wolffish	650–7000[(2)]	3.7–6.4[(2)]	5.5	22.0	4.9	32.0	7.1
Eelpout	9–132[(3)]	2.8–3.2[(3)]	3.0	35.0	17.1	35.0	17.1

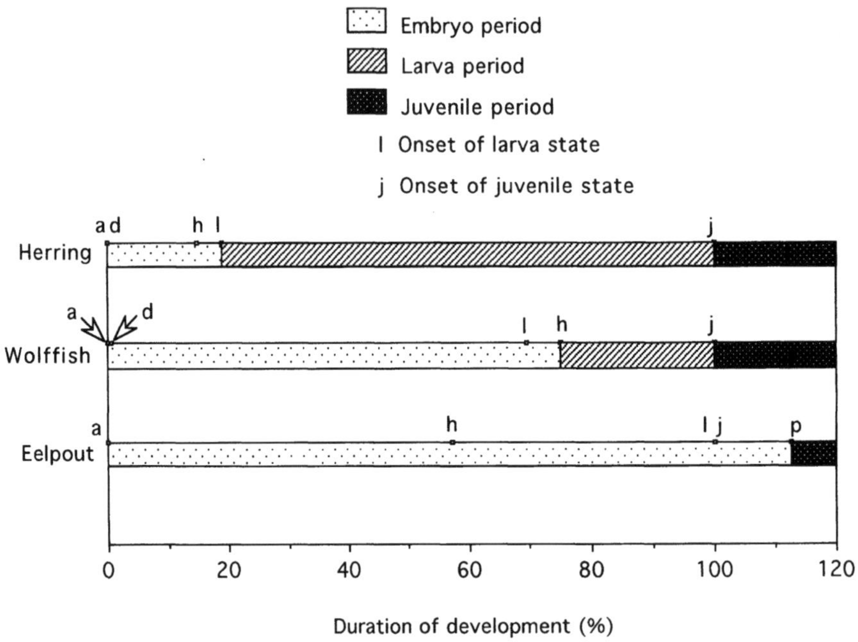

Figure 5. Scheme representing comparative features of early ontogeny in herring, wolffish, and eelpout. Duration of development from egg activation to onset of juvenile state = 100% (approximately 74, 173, and 120 days, respectively, with the latter value according to Soin (1968). a = activation; d = egg deposition; h = hatching; p = parturition.

results of the present study can support a general trend of increasing size at first exogenous feeding (expressed as % average TL at spawning) in the evolutionary process of transformation from indirect to direct type of early ontogeny (Flegler-Balon 1989). However, the size of an individual at the beginning of the juvenile state is apparently determined mainly by ecological adaptations of the species.

To compare features of early ontogeny in the three species, the following scheme is suggested (Figure 5). For each species, the duration of development from egg activation to the beginning of the juvenile state was taken as 100%. In herring, egg release from the female's body coincided with egg activation. The onset of the larva (i.e. ability to oral feed) and juvenile states coincided with the beginning of the larva and juvenile periods (sensu Balon). In wolffish, egg release occurred several hours after egg insemination and activation. Wolffish larvae (19–20 mm TL) were able to feed exogenously immediately after premature hatching, which can be caused by comparatively high incubation temperature, destruction of the zona radiata, or mechanical pressure (Pavlov & Moksness 1995). Normal hatching occurred at approximately 21–24 mm TL. Therefore, in wolffish, normal hatching, which coincides with first feeding (i.e. the onset of the larva period, sensu Balon), was observed later than the beginning of larva state, whereas, both the juvenile 'state' and the juvenile 'period' (sensu Balon) started at the same time. In eelpout, individuals of 36–44 mm TL were able to feed externally immediately after being removed from the female's oviduct. In eelpout, achievement of the larva and juvenile 'states' probably occurs at the same time, but both events took place before normal parturition. The larva period (sensu Balon) is absent, and only parturition could be regarded as a threshold leading to a life in the new environment, designating the onset of the juvenile period (sensu Balon). Therefore, in eelpout, the juvenile 'period' seems to start after the beginning of the juvenile 'state'.

This scheme could be applied to the development of some species with unusual life styles. For example, *Cyphotilapia frontosa* began to feed orally and then reached the juvenile state inside the buccal pouch of its mother (Balon 1985, 1990). In this species, the larva state began inside the buccal pouch, long before the release of progeny from the mother's mouth. This species also apparently lacks a larva period (sensu Balon).

Some terminological problems arose to designate developmental events in fish species with different types of early ontogeny. Organisms could be referred to based on their developmental periods (embryo, larva, or juvenile), irrespective of whether they have reached the larva or juvenile states. According to Balon (1985), the ontogeny of some fish species is intermediate between indirect and direct development; the transition to external feeding occurs at advanced states of ontogeny, but some temporary organs represent a vestige of larvae. In some salmonids, the period of development between the onset of exogenous feeding and the beginning of the juvenile period has been called 'alevin', which may be present in wolffish. However, it seems to be difficult to determine the difference between alevin and larva.

Thus, according to the theory of saltatory ontogeny, we can more or less clearly distinguish the larva period (free-swimming organism with mixed or only external feeding and undeveloped adult characters), and the juvenile period (free-swimming organism with external feeding, adult characters, and eliminated larva characters) in fishes with various types of early ontogeny. The degree of morphological development at the beginning of the larva period can be very different in various fish species. At the same time, similar morphological states in fishes from different ecological groups can be determined and used for inter- and intraspecific comparison. In particular, the beginning of the juvenile state can be easily distinguished in many different fish species by the disappearance of larva characters and appearance of majority adult characters. As was shown in the present study, the beginning of the juvenile 'state' does not necessarily coincide with the beginning of the juvenile 'period' (sensu Balon).

Acknowledgements

I would like to thank several people for help in preparation and revision of the manuscript. Erlend Moksness took part in the collection of material for this paper. Eugene Balon, Gordon H. Copp, Vladimír Kováč and an anonymous referee made constructive comments on the structure of the manuscript. The technical assistance of many members of the White Sea Biological Station, Moscow State University (Russia) and of the Flødevigen Marine Research Station, Institute of Marine Research (Norway) is highly appreciated. This study was supported in part by the Council of Grants of the President of Russian Federation and Governmental Foundation for Leading Scientific Schools (project no. 96-15-98020) and by the Norwegian Research Council, Department NFFR.

References cited

Altukhov, K.A. 1975. Biology of early developmental stages in herring of the Kandalaksha Gulf. pp. 185–226. *In*: B.E. Bykhovskii (ed.) Biology of the White Sea Herring, Investigations of Fauna of the Seas 16, Zoologicheskii Institut Akad. Nauk SSSR, Leningrad (in Russian).

Andriashev, A.P. 1954. Fishes of the northern seas of the USSR. Akad. Nauk SSSR, Leningrad. 566 pp. (in Russian).

Balon, E.K. 1975. Reproductive guilds of fishes: a proposal and definition. J. Fish. Res. Board Can. 32: 821–864.

Balon, E.K. 1977. Early ontogeny of *Labeotropheus* Ahl, 1927 (Mbuna, Cichlidae, Lake Malawi), with a discussion on advanced protective styles in fish reproduction and development. Env. Biol. Fish. 4: 97–101.

Balon, E.K. 1980. Comparative ontogeny of charrs. pp. 703–720. *In*: E.K. Balon (ed.) Charrs: Salmonid Fishes of the Genus *Salvelinus*, Perspectives in Vertebrate Science 1, Dr W. Junk Publishers, The Hague.

Balon, E.K. 1981. Additions and amendments to the classification of reproductive styles in fishes. Env. Biol. Fish. 6: 377–389.

Balon, E.K. (ed.) 1985. Early life histories of fishes: new developmental, ecological and evolutionary perspectives. Developments in Env. Biol. Fish. 5, Dr W. Junk Publishers, Dordrecht. 280 pp.

Balon, E.K. 1986. Saltatory ontogeny and evolution. Rivista di Biologia/Biology Forum 79: 151–190.
Balon, E.K. 1989. The epigenetic mechanisms of bifurcation and alternative life-history styles. pp. 7–40. *In*: M.N. Bruton (ed.) Alternative Life-History Styles of Animals, Kluwer Academic Publishers, Dordrecht.
Balon, E.K. 1990. Epigenesis of an epigeneticist: the development of some alternative concepts on the early ontogeny and evolution of fishes. Guelph Ichthyol. Rev. 1: 1–42.
Balon, E.K. 1991. Probable evolution of the coelacanth's reproductive style: lecithotrophy and orally feeding embryos in cichlid fishes and in *Latimeria chalumnae*. Env. Biol. Fish. 32: 249–265.
Barsukov, V.V. 1959. The wolffish (Anarhichadidae). Fauna SSSR, Moscow N.S. 73. (Translation by Smithsonian Inst. 1972, NTIS: Springfield). 292 pp.
Copp, G.H. & V. Kováč. 1996. When do fish with indirect development become juveniles? Can. J. Fish. Aquat. Sci. 53: 746–752.
Cunningham, J.E.R. & E.K. Balon. 1986. Early ontogeny of *Adinia xenica* (Pisces, Cyprinodontiformes): 2. Implications of embryonic resting interval for larval development. Env. Biol. Fish. 15: 15–45.
Flegler–Balon, C. 1989. Direct and indirect development in fishes – examples of alternative life-history styles. pp. 71–100. *In*: M.N. Bruton (ed.) Alternative Life-History Styles of Animals, Kluwer Academic Publishers, Dordrecht.
Fukuhara, O. & T. Fushimi. 1986. Development and early life history of ayu reared in the laboratory. Bull. Japan. Soc. Sci. Fish. 52: 75–80.
Girsa, I.I. & Yu.E Lapin. 1978. Features of morphology and behaviour of the young of the White Sea herring. pp. 117–136. *In*: Yu.E. Lapin (ed.) The Ecology of Fishes of the White Sea, Nauka Press, Moscow (in Russian).
Ivanchenko, O.F. 1975. Some data about occurrence of young herring *Clupea harengus* Pallasi n. *maris-albi* Berg in the Kandalaksha Bay of the White Sea. pp. 255–259. *In*: B.E. Bykhovskii (ed.) Biology of the White Sea Herring, Investigations of Fauna of the Seas 16, Zoologicheskii Institut Akad. Nauk SSSR, Leningrad (in Russian).
Ivanchenko, O.F. 1983. The bases of the herring mariculture at the White Sea. Nauka Press, Leningrad. 39 pp. (in Russian).
Jensen, A.J. & B.O. Johnsen. 1982. Difficulties in aging Atlantic salmon (*Salmo salar*) and brown trout (*Salmo trutta*) from cold rivers due to lack of scales as yearlings. Can. J. Fish. Aquat. Sci. 39: 321–325.
Johannessen, T., J. Gjøsæter & E. Moksness. 1993. Reproduction, spawning behaviour and captive breeding of the common wolffish *Anarhichas lupus* L. Aquaculture 115: 41–51.
Kristofferson, R., S. Broberg & M. Pekkarinen. 1973. Histology and physiology of embryotrophe formation, embryonic nutrition and growth in the eel pout, *Zoarces viviparus* (L.). Ann. Zool. Fennici 10: 467–477.
Kryzhanovsky, S.G., N.N. Disler & E.N. Smirnova. 1953. Ecomorphological principles of development in percids. Trudy Inst. Morph. Zhiv. Severtsova 10: 3–138 (in Russian).
Moksness, E. & D.A. Pavlov. 1996. Management by life cycle of wolffish, *Anarhichas lupus* L., a new species for cold-water aquaculture: a technical paper. Aquaculture Res. 27: 865–883.
Moskal'kova, K.I. 1985. 'Coprophagous' embryos of the teleost fish *Neogobius melanostomus* (Pallas) (Pisces, Gobiidae). Dokl. Akad. Nauk SSSR 282(5): 1251–1254 (in Russian).
Pavlov, D.A. 1986. Developing the biotechnology culturing White Sea wolffish, *Anarhichas lupus marisalbi*. II. Ecomorphological peculiarities of early ontogeny. J. Ichthyol. 26(6): 156–169.
Pavlov, D.A. 1989. Salmonid fishes (developmental biology and breeding). Moscow University Press, Moscow. 216 pp. (in Russian).
Pavlov, D.A. 1994. Fertilization in the wolffish, *Anarhichas lupus*: external or internal? J. Ichthyol. 34 (1): 140–151.
Pavlov, D.A. 1995. Growth of juveniles of White Sea common wolffish, *Anarhichas lupus* L., in captivity. Aquaculture Res. 26: 195–203.
Pavlov, D.A. 1997. Development of head skeleton and paired fins girdles in wolffish, *Anarhichas lupus*, at different temperature regimes. J. Ichthyol. 37 (4): 294–303.
Pavlov, D.A. & G.G. Novikov. 1993. Life history and peculiarities of common wolffish (*Anarhichas lupus*) in the White Sea. ICES J. Mar. Sci. 50: 271–277.
Pavlov, D.A. & E. Moksness. 1995. Development of wolffish eggs at different temperature regimes. Aquaculture Int. 3: 315–335.
Pavlov, D.A. & E. Moksness. 1997. Development of axial skeleton in wolffish, *Anarhichas lupus* L. (Pisces, Anarhichadidae), at different temperatures. Env. Biol. Fish. 49: 401–416.
Peňáz, M. 1983. Ecomorphological laws and saltation in the early ontogeny of salmonoidei. Folia Zool. 32: 365–373.
Potthoff, T. & J.A. Tellock. 1993. Osteological development of the snook, *Centropomus undecimalis* (Teleostei, Centropomidae). Bull. Mar. Sci. 52: 669–716.
Ryzhkov, L.P. 1976. Morphophysiological features and transformation of substances and energy in the early ontogeny of freshwater salmonid fishes. Karelia Press, Petrozavodsk. 250 pp. (in Russian).
Soin, S.G. 1968. Adaptive principles of fish development. Univ. Press, Moscow. 90 pp. (in Russian).
Svetovidov, A.N. 1952. Clupeidae. Fauna SSSR, fishes. 2 (1). Israel Progr. for Sci. Translat., Jerusalem (1963). 428 pp.
Vasnetsov, V.V. 1953. Etaps in the development of bony fishes. pp. 207–217. *In*: E.N. Pavlovskii (ed.) Otcherki po Obshch. Vopr. Ikhtiol., Akad. Nauk SSSR Press, Moscow-Leningrad (in Russian).
Wourms, J.P. 1981. Viviparity: the maternal-fetal relationship in fishes. Amer. Zool. 21: 473–515.
Zhiteneva, T.S. & V.P. Seitov. 1986. On the method for designation of developmental etaps in the late ontogeny of bream *Abramis brama* (L.) (Cyprinidae) from Rybinskoe water reservoir. Trudy Inst. Biol. Vnutr. Vod Akad. Nauk SSSR 53/56: 173–184 (in Russian).

Environmental Biology of Fishes **56:** 53–65, 1999.

Thyroxine as a mediator of metamorphosis of Atlantic halibut, *Hippoglossus hippoglossus*

Jostein Sigurd Solbakken, Birgitta Norberg[a], Kuninori Watanabe & Karin Pittman
Department of Fisheries and Marine Biology, University of Bergen, Bergen High Technology Center, N-5020 Bergen, Norway (e-mail: jostein.solbakken@imr.no)
[a]*Institute of Marine Research, Austevoll Aquaculture Research Station, N-5392 Storebø, Norway*

Received 10 September 1998   Accepted 11 January 1999

Key words: endocrinology, histology, cortisol, thyroid, environmental factors, window of opportunity, larva, juvenile, aquaculture

Synopsis

The response of morphological, histological and endocrinological development to exogenous l-thyroxine (T_4) and to water depth during metamorphosis in Atlantic halibut, *Hippoglossus hippoglossus*, was investigated. Exogenous T_4 was given in daily doses of 0.1, 0.05 ppm or a control treatment to halibut larvae at 550 daydegrees (posthatch, premetamorphic) for 14 days. Water depths of 40 cm, 10 cm or 1.5 cm were used to rear halibut larvae from 590 daydegrees for 21 days. Halibut larvae given exogenous T_4 at 0.1 ppm had accelerated eye migration relative to MH in fish given 0.05 ppm and in control fish. Pigmentation was correlated with dosage after 14 days. The volume of thyroid tissue was expressed in a dose-dependent manner and exhibited a size-dependency within each treatment. However, the follicles were atypical with reduced colloid, increased lumen and low epithelial cells even in the control group. The results indicate that T_4 is a mediator in halibut metamorphosis. In the water depth experiment, only cortisol levels of larvae reared in 1.5 cm water were significantly affected after 21 days, but this was not correlated with metamorphic rate. Hormone profiles, morphological changes and size suggest the existence of a 'window of opportunity' for metamorphosis in halibut extending from about 16 mm and tapering off about 21 mm SL. The pooled hormone profiles indicate the commencement of a hormonal cascade similar to that of other flatfishes during metamorphosis. The results indicate that growth, neural and skeletal transformation, and pigmentation are biochemically separate processes in the metamorphosis of Atlantic halibut.

Introduction

Development of new organs and structures along with the losses of others, make metamorphosis a critical ontogenic interval in the adaptation of flatfish species such as Atlantic halibut, *Hippoglossus hippoglossus* L. to intensive aquaculture. However, there is great plasticity in the viable postmetamorphic forms of halibut (Pittman et al. 1998). Four variations in development have been tentatively identified: ambicolouration or malpigmentation, incomplete eye migration, anterior dorsal fin abnormalities, and elongated body shape, the latter three affecting growth during the juvenile period. To ensure seasonally independent delivery of high quality juveniles, the mechanisms that mediate metamorphosis in halibut need to be investigated.

Metamorphosis in flatfishes marks the transition to the definitive phenotype of the digestive system, the respiratory system, the visual (eye migration) and other neural systems, muscles, skin, pigmentation and behaviour (Evans & Fernald 1990, Yamano et al. 1991, Miwa et al. 1992, Matsumoto & Sekai 1992,

Yamano et al. 1994). The pituitary–thyroid axis mediate flounder metamorphosis, just as it does in amphibians (Allen 1916, Tata 1996).

Early studies have indicated that the thyroid organ in plaice, *Pleuronectes platessa*, and starry flounder, *Platichthys stellatus*, is activated during metamorphosis (Sclower 1930, Hoar 1951) and that the subsequent activation of the pituitary–thyroid axis for thyroid hormones (TH) induces metamorphosis in Japanese flounder, *Paralichthys olivaceus* (Miwa & Inui 1987a). The dominant hormones throughout metamorphosis in Japanese flounder are thyroxine (T_4) and cortisol (F) (Miwa & Inui 1988, Tagawa et al. 1990, de Jesus et al. 1991, 1993), where elevated F levels appear prior to elevation of T_4 levels. Exogenous F and T_4 have a synergistic effect on metamorphosis of Japanese flounder in vitro and increase the survival in other non-flatfish species (de Jesus et al. 1990, 1991, Brown & Kim 1995). Although the modulating role of environmental factors on the initiation and duration of metamorphosis in flounder has concentrated on temperature and/or T_4 tissue level (Chambers & Leggett 1987, Seikai et al. 1986, Tanangonan et al. 1989), environmental factors affecting the tissue level of F and its synergistic effect with T_4 have not been studied. The present study was designed to investigate the effects of exogenous T_4 and low water depth as a sublethal stressor on the metamorphosis in halibut, as measured by eye migration, growth, endocrinology and thyroid tissue variables.

Material and methods

The eggs were obtained from broodstock at the IMR, Austevoll Aquaculture Research Station (AARS), and reared through hatching and yolksac period according to Jelmert & Rabben (1987) and Harboe et al. (1994). In addition to using DHA Selco-enriched *Artemia* (INVE Aquaculture, Belgium) as first feed, a combination of natural zooplankton and *Artemia* were provided during a period of 10 days before initiation of metamorphosis to ensure normal development of larvae during metamorphosis (Næss & Lie 1998).

In experiment 1 on exogenous thyroxine, 200 larvae at an age of about 550 daydegrees (d°C) after hatching (66 days after hatching) from one egg group, were transferred to each of 9100 l circular black fiberglass tanks. The depth of the tanks were about 60 cm. The larvae were kept under continuous light (24L: 0D) according to IMR, AARS's own data. The water temperature was 12.5°C, with a maximum of 13.5°C and a minimum of 12.2°C, salinity was 32.3 ppt and the oxygen concentration was kept above 6.5 mg l^{-1} during the experiment.

Immediately after the transfer of larvae, two experimental concentrations of T_4, 0.1 and 0.05 ppm l-thyroxine (3-(4-hydroxy-3,5-diiodophenoxy)-3,5-diiodophenyl)-L-Alanine (Sigma Chemical Co, U.S.A.) were added to six tanks, while three tanks acted as a control group. The hormone powder was weighed into a syringe filled with approximately 5 ml of 9 mg ml^{-1} NaCl before the suspension was injected into 100 ml plasma bags (Pharmacia & Upjohn, Sweden). The water was stagnant for 22 h followed by 2 h continuous circulation of fresh seawater at rate of 1.3 l min^{-1}. The hormone doses were replenished after the daily water exchange.

Dead larvae, faeces and detritus were removed daily by siphon. Algae (*Tetraselmis* sp. and *Isochrysis galbana*) were added according to Naas et al. (1992) before supplying 1800–2000 *Artemia* nauplii $larvae^{-1}$ day^{-1}, and before replenishing the hormone.

Larvae were sampled on day 0 (n = 30), day 7 (n = 97–105) and day 14 (n = 176–185) from each treatment. The larvae were anesthesized by immersion in 8 ppt metomidate hydrochloride (Wild-life Pharmaceuticals, CO, U.S.A.) before measuring standard length (SL) from snout to end of notochord, myotome height (MH) (posterior to the urinary bladder) and eye migration (cf. Næss & Lie 1998) under a dissecting microscope (Leica Wild M3B, Switzerland). Eye migration was measured after an index of 6 positions; 0 = eyes symmetrical, 0.5 = the half of the migrating eye is visible over the right, 1.5 = the migrating eye is over the right, cornea visible, lens not visible, 2 = migrating eye on the dorsal midline, 2.5 = migrating eye between final position and dorsal midline, 3 = complete eye migration. Eight larvae from each treatment were fixed in Karnovsky's fluid (for a 100 ml solution: 80 ml buffer component pH 7.4, 10 ml glutaraldehyde (25%), 10 ml paraformaldehyde (10%) and 4 g sucrose for histological examination.

In experiment 2 on water depth, the effect of this environmental stressor was carried out using 400 larvae in each of six 1 × 1 m square, green fiberglass tanks and three water depths (1.5, 10 and 40 cm). The larvae were reared as described above but were slightly older (595 d°C, or 72 days after hatching) and originated from two egg groups. Larvae were exposed to continuous illumination and fed *Artemia* twice a day.

The water flow was adjusted to 0.02, 0.1 and 0.4 l min^{-1}, and mean temperature was 12.1, 11.9 and 11.7°C in the 1.5 cm, 10 cm and 40 cm tanks, respectively. The oxygen concentration was kept above 5 mg l^{-1} in all tanks. Husbandry was as in experiment 1. Larvae were sampled on day 0 (n = 25), day 14 (n = 60), and day 21 (n = 78–120) from each treatment. The larval SL, MH and eye migration were measured as in experiment 1, before freezing in liquid nitrogen and storage at −80°C until analysis of cortisol and thyroxine.

In the hormone assay, T_4 was extracted according to Tagawa & Hirano (1987), whereas extraction of F was done as described by Stensland (1995). Before extraction, a number of larvae (n = 10–20) from each tank were homogenised individually in ice-cold methanol for the T_4 extraction, and in PBS (10 mM phosphate buffer, pH 7.4, containing 1.05 M NaCl and 0.1% gelatine) for the F extraction by an ultrasound sonicator (Kontes). To estimate the extraction efficiency, about 4000 cpm of ^{125}I-labelled T_4 and about 2000 cpm ^{3}H-labelled hydrocortisone were added to a number of larva homogenates before T_4 extraction (n = 6) and F extraction (n = 6), respectively. Both extraction procedures were followed, before the recovery of the labelled hormones were estimated to 55.0 ± 2.4% for T_4 and 77.6 ± 2.7% for F. The whole body T_4 and F data reported in this study have not been corrected for the recoveries.

To measure the tissue level of T_4, an ELISA described by Cerdà-Reverter et al. (1996) adapted to blood plasma T_4 in sea bass, *Dicentrarchus labrax*, and sea bream, *Sparus aurata*, was used. Tissue levels of F were measured by RIA adapted to eggs, embryos and larvae of halibut according to Stensland (1995).

For the histological examination, larvae were fixed in Karnovsky's fluid, and rinsed in buffer, dehydrated in ethanol and embedded in Historesin™ (Reichert-Jung). Serial transverse sections, in which thickness varied between 3–60 μm within each larvae, were cut on a Reichert-Jung 2050 Supercut microtome. The variable thickness within a larva has no impact on the stereological measures of volumes and number. The sections were stained with toluidine blue. Estimates of the total volumes of epithelial cells, colloid and lumen for each fish were made by the Cavalieri method on all slices containing thyroid tissue (Cruz-Orive & Weibel 1990). The total number of follicles was counted using the optical disector. No correction for shrinkage was made.

In experiment 1, the number of larvae settled on the bottom was counted daily for each tank prior to feeding, and converted to percentages. During experiment 2, settlement was counted twice daily both in the morning prior to feeding and in the middle of the afternoon after feeding.

All statistical analyses were performed by Statistica™ 4.1. The SL and MH data were tested for normality using a Kolmogorov-Smirnov test (Zar 1996), and homogenity of variances (see Brown & Forsythe 1974). A two-factor nested ANOVA model I, where the replicates were nested within thyroxine (T_4) concentration groups, was applied to calculate the effect of T_4 concentrations on the myotome height at respective time periods. Before using a single factor ANOVA model I for testing differences among treatments, differences among replicates were tested for by either Student t-test (n = 2) or single factor ANOVA model I (n = 3). Non-significant replicates were pooled. Both ANOVAs were followed by a Tukey HSD multiple comparison test. Differences in number of settled larvae between the treatments were tested for with a Kolmogorov-Smirnov two sample test (Zar 1996).

Results

Survival was approximately 90% in experiment 1, but varied between 25–90% in experiment 2. Mortality was not related to treatment. MH in larvae given 0.05 ppm T_4 was significantly higher than control fish after 7 ($p < 0.05$) and 14 days ($p < 0.01$) and also significantly higher than larvae given 0.1 ppm after 14 days ($p < 0.01$). The MH of larvae reared under different water depths (experiment 2) was not significantly affected by the treatment (Table 1). The increase in MH over 14 days was greatest for the larvae in the hormone experiment (3 mm).

Mean SL increased from 16.1 mm to about 17.5 mm in experiment 1 whereas mean SL was stable at about 17.5 mm for the first two weeks of experiment 2 (Table 1). There were, however, significant differences between replicates in experiment 1, and the data were therefore not further analysed. The SL of larvae in 10 cm water depth was nonetheless significantly higher than that of groups of larvae at depths of 1.5 and 40 cm.

Within the interval of 12–24 mm SL, there was a linear relationship between SL and MH within all treatments (Figure 1a,b). MH was better related to SL in the 0.05 and 0.1 ppm groups ($r^2 = 0.80$, n = 180, $p < 0.001$ and $r^2 = 0.74$, n = 184, $p < 0.001$) than in the control group ($r^2 = 0.48$, n = 176, $p < 0.001$).

Table 1. Mean myotome height (MH) and standard length (SL) with SE over time in live metamorphosing halibut given exogenous thyroxine (experiment 1) or different water depths (experiment 2). Different letters indicate significant difference within an experiment at a given sampling day ($p < 0.05$ after 7 days, $p < 0.01$ after 14 days and $p < 0.005$ after 21 days). Replicates of SL in experiment 1 are pooled despite significant differences.

Experiment no.	Treatment	Variable (mm)	Days in experiment			
			0	7	14	21
1	0 ppm	SL	16.1 ± 0.2	—	17.2 ± 0.1	
		MH	2.85 ± 0.08	3.33 ± 0.05 ***b***	3.64 ± 0.04 ***c***	
	0.05 ppm	SL	16.1 ± 0.2	—	17.8 ± 0.1	
		MH	2.85 ± 0.08	3.56 ± 0.06 ***a***	4.06 ± 0.05 ***a***	
	0.01 ppm	SL	16.1 ± 0.2	—	17.5 ± 0.1	
		MH	2.85 ± 0.08	3.45 ± 0.06 ***ab***	3.85 ± 0.05 ***a***	
2	40 cm	SL	17.0 ± 0.2	—	17.9 ± 0.2	19.2 ± 0.2
		MH	3.4 ± 0.1	—	4.24 ± 0.1	4.69 ± 0.1 ***b***
	10 cm	SL	17.0 ± 0.2	—	17.9 ± 0.2	20.0 ± 0.2
		MH	3.4 ± 0.1	—	4.18 ± 0.1	4.89 ± 0.07 ***a***
	1.5 cm	SL	17.0 ± 0.2	—	17.4 ± 0.2	18.9 ± 0.2
		MH	3.4 ± 0.1	—	4.04 ± 0.09	4.76 ± 0.06 ***b***

The relationships between MH and SL in the water depth experiment were not affected by treatment ($r^2 = 0.87$, $n = 60$, $p < 0.001$ for 1.5 cm, $r^2 = 0.90$, $n = 60$, $p < 0.001$ for 10 cm and $r^2 = 0.85$, $n = 60$, $p < 0.001$ for 40 cm). A dose-dependent increase in ocular side pigmentation between the groups given exogenous hormone was seen after 14 days (Figure 2). There was no difference in pigmentation in groups given decreasing water depth.

In eye migration, a trend toward a dose response was found in experiment 1 despite differences between replicates (Figure 3). When stimulated by 0.1 ppm exogenous T_4, the eye migrates to the dorsal midline (i.e. index = 2) when the fish have a MH of 3.5–4.0 mm (Figure 4a). The number of larvae in which the migrating eye has reached the dorsal midline is highest in the 0.1 and 0.05 ppm groups. However, the migrating eye generally passes the dorsal midline (index > 2) when the MH is equal to or exceeds 4.5 mm (Figure 4b). There was no difference between the control and the 0.05 ppm T_4 treatment. No effect was found with varying water depth.

The thyroid follicles had a narrow perimeter of epithelial cells surrounding the colloid and lumen (Figure 5). The generally homogenous colloid had some vacuolisation, but the lumen of the follicle (surrounded by the epithelial cells but not containing colloid) could often be a significant portion of the entire follicle. None of the fish in any treatment had columnar epithelial cells in the thyroid follicles. Thyroid follicle number was positively related to fish size ($r^2 = 0.66$, $n = 12$, $p < 0.001$), but not with treatment ($r^2 = 0.04$, $n = 12$, $p > 0.05$), and ranged from 16 to 40 follicles per fish (Figure 6a).

The volume of thyroid tissue in the larvae expressed in a dose dependent manner, exhibited a size dependency within each treatment (Figure 6b). The volume was also positively related to size ($r^2 = 0.56$, $n = 12$, $p < 0.001$) (Figure 6a). The activity levels of this tissue, expressed by the ratio of the lumen to the total follicle, indicated that the thyroid of the 0.1 ppm group was most active, even when correcting for size differences (Figure 7a,b). There was in general a decrease in volume of colloid and an increase in volume of lumen with increasing dosage (Figure 7a). The ratio of epithelial cells to the follicle volume was not affected by treatment.

In the pooled water depths, larval tissue levels of T_4 were detectable (<10 ng larva^{-1}) at 72 days after hatching but increased with SL after 21 days (93 days after hatching). Tissue F was also detectable (<5 ng larva^{-1}) up to 14 days after transfer (86 days after hatching). At 21 days after transfer, the tissue F levels varied widely in fish up to 20 mm SL, whereas, in larger fish, F levels decreased and stabilized at <10 ng larva^{-1} (Figure 8a–d). The size range within the F peak after 14 days corresponds to a MH of about 2.5–5.5 mm, where the left eye migrates from initial position to the dorsal midline, whereas the slight increase in T_4 level may correspond to the further migration. However, larvae held

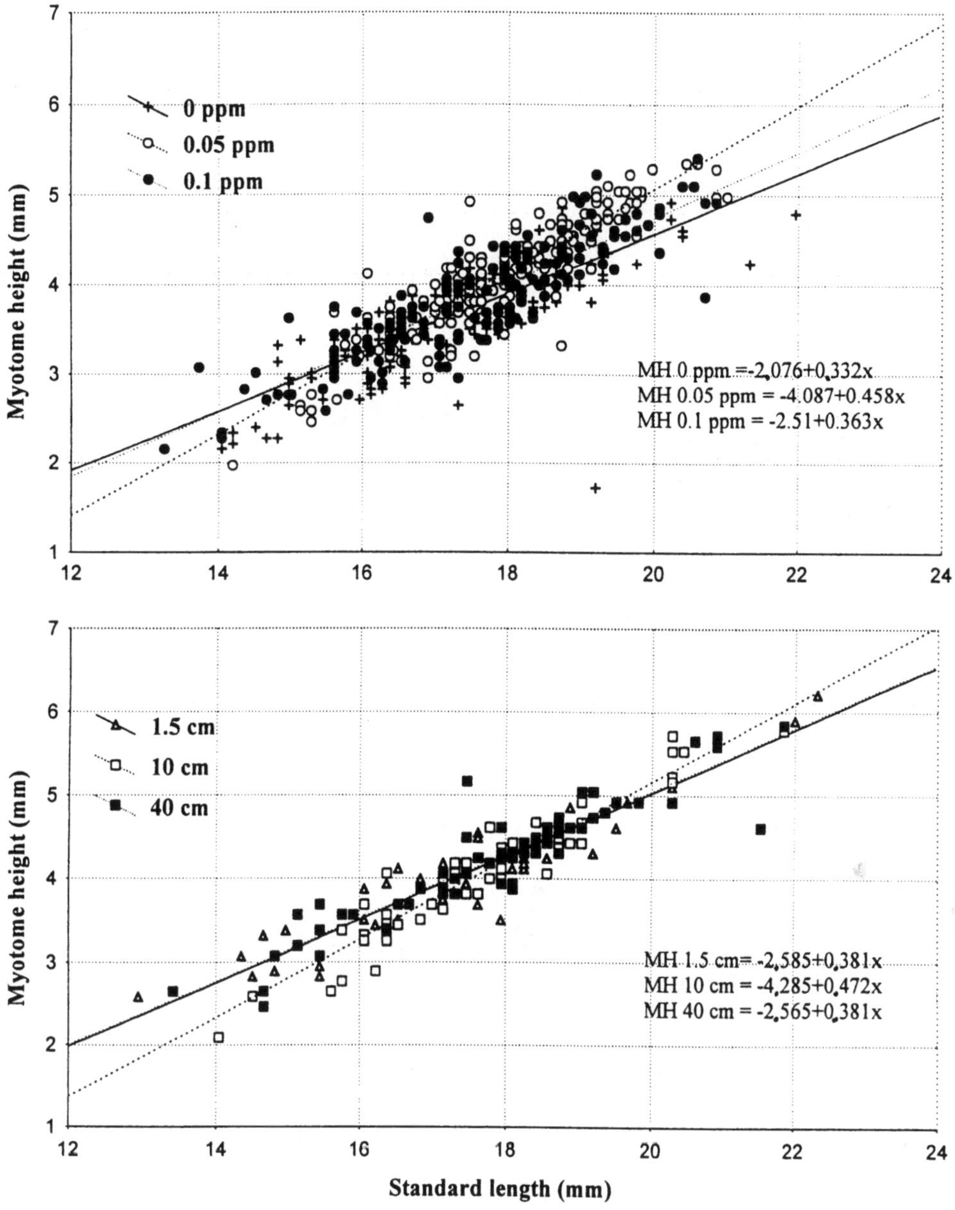

Figure 1. Myotome height (mm) as function of standard length (mm) in live metamorphosing halibut after 14 days given a – exogenous thyroxine (0.1 ppm, 0.05 ppm and control) or b – different water depths (1.5 cm, 10 cm and 40 cm).

in 1.5 cm water had significantly higher F levels than in the two other water depths ($p < 0.005$) (Figure 8c). Maximum F levels were 60 ng larva^{-1}, whereas maximum T_4 levels were 23 ng larva^{-1}.

After 9 days, the number of settled larvae in the 0.1 ppm exogenous thyroxine increased relative to the two other treatments and this trend lasted the duration of the experiment, although there was no increase in settlement over time in any tank. The percentage of settled larvae was not significantly different from beginning to end of the water depth experiment, nor was it different between morning and afternoon, or between tanks (Kolmogorov-Smirnov, $p > 0.05$). The mean percentage of settled larvae was 15% in the hormone experiment and 25% in the water depth experiment.

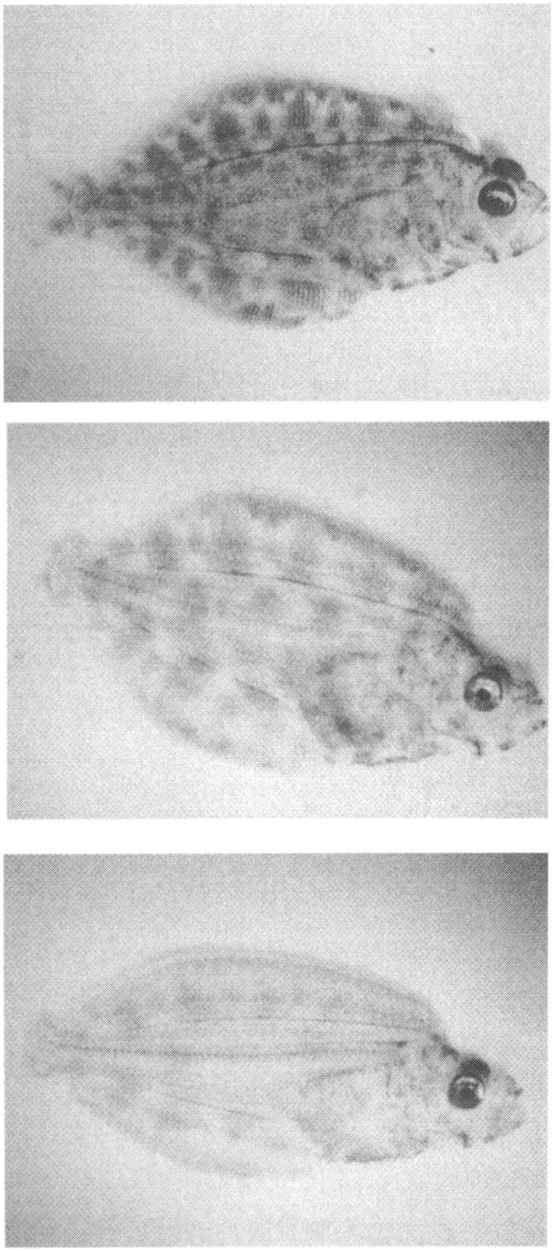

Figure 2. Photographs of representative larvae from 0.1 ppm (top), 0.05 ppm (middle) and control (bottom) after 14 days showing pigmentation effects of increasing dose of thyroxine.

Discussion

The partially positive effect of exogenous T_4 on the development of the thyroid follicle observed in the present study contrasts with other studies (Miwa & Inui 1987a,b, Leatherland 1994). Combined with the

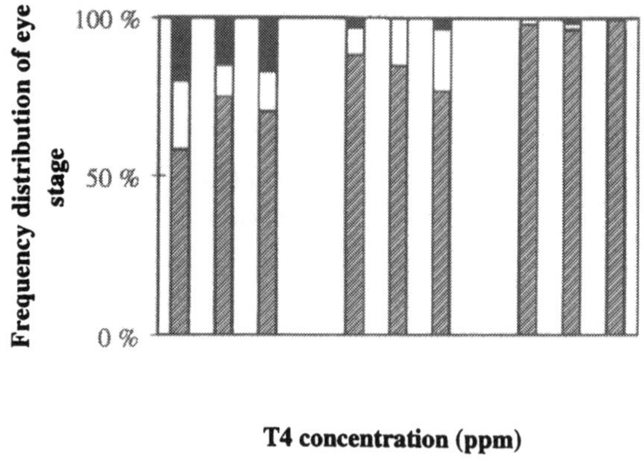

Figure 3. Eye migration as function of hormone dose (ppm). The sinistral eye above the dorsal midline = black bars, eye migrated to the dorsal midline = white bars; and sinistral eye on prospective abocular side of dorsal midline = striped bars.

inverse dose response of exogenous hormone on myotome height, the results suggest utilization of exogenous T_4 as an iodide source for further synthesis. In agreement with earlier work on other flatfishes (Inui & Miwa 1985, Miwa & Inui 1987a,b, Tanaka et al. 1995, 1996, Inui et al. 1995), our results suggest that activation of the pituitary–thyroid axis by exogenous T_4 induced metamorphosis in the halibut. However, the stimulated proliferation of the thyroid follicles and the apparent relative depletion of the colloid was contraindicated by the flattened follicular epithelial cells in all fish in all treatments.

Exogenous T_4 stimulated both muscle and skeletal growth in halibut, as in other teleost species (Donaldson et al. 1979, Lam 1980, Lam et al. 1985, Nacario 1983, Leatherland 1994, Takagi et al. 1994). Pigmentation and eye migration was also stimulated by exogenous T_4. However, when fed a diet of enriched *Artemia*, halibut larvae grow and attain full (ambi)pigmentation even though they display a lack of eye migration and free anterior pterygiophores (Pittman et al. 1998). Since T_4 also stimulates differentiation of gastric glands during metamorphosis in Japanese flounder (Miwa et al. 1992), an enhanced absorption of amino acids or peptides may take place, and result in higher growth. This would suggest that, although growth may be stimulated by TH, it is a separate process from that modulating neural changes and pigmentation during metamorphosis in halibut.

Eye migration in halibut begins normally at about 16 mm SL when MH increases from 3.5–4.5 mm, and preceding dorsal fin proximal pterygiophore migration (DFPM). The migration of the eye and DFPM influence

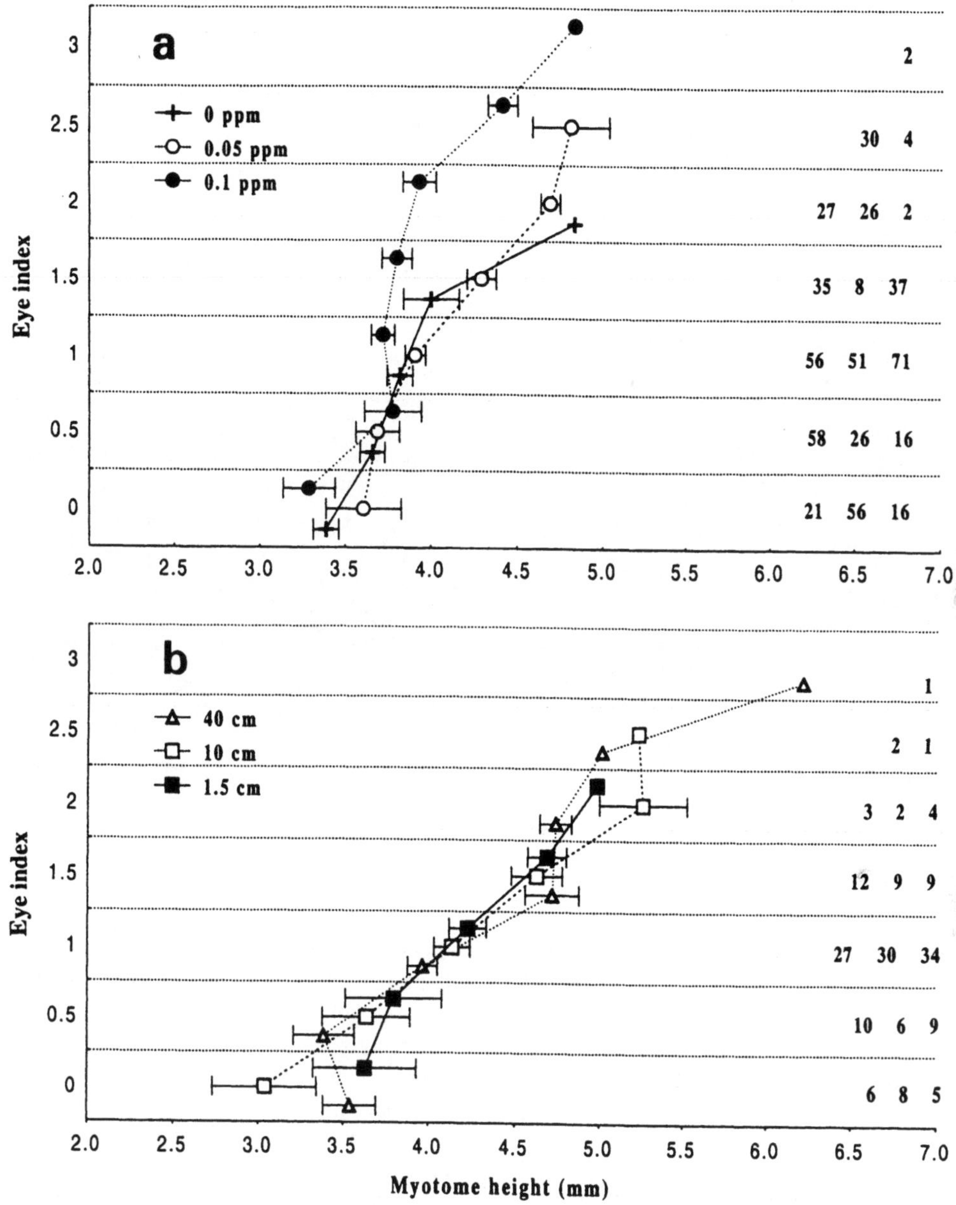

Figure 4. Eye index as function of myotome height (mm) (with SE bars) in live metamorphosing halibut given a – exogenous thyroxine (0.1 ppm, 0.05 ppm and control) or b – different water depths (1.5 cm, 10 cm and 40 cm). The number of larvae at each point in a given eye index is shown.

each other (Brewster 1987), such that there can be a physical blockage of eye extension if frontal plate ossification and pterygiophore movement occur first (Pittman et al. 1998). This may be partially regulated by various subtypes of TH receptors, as gene expression of thyroid hormone receptor subtype TRβ is observed in the cartilage cells and on the surface of the pterygiophores (Yamano & Miwa 1998). It is therefore likely that T_4 may be part of the underlying controlling mechanism of eye movement as well as that of DFPM.

Addition of T_4 affected pigmentation density in the larvae, but other factors are involved. Pigmentation in flatfish involves creation and dispersal of chromatophores, which develop in the layer below the dermis. In winter flounder, *Pleuronectes americanus*, there is adrenergic neuronal control of

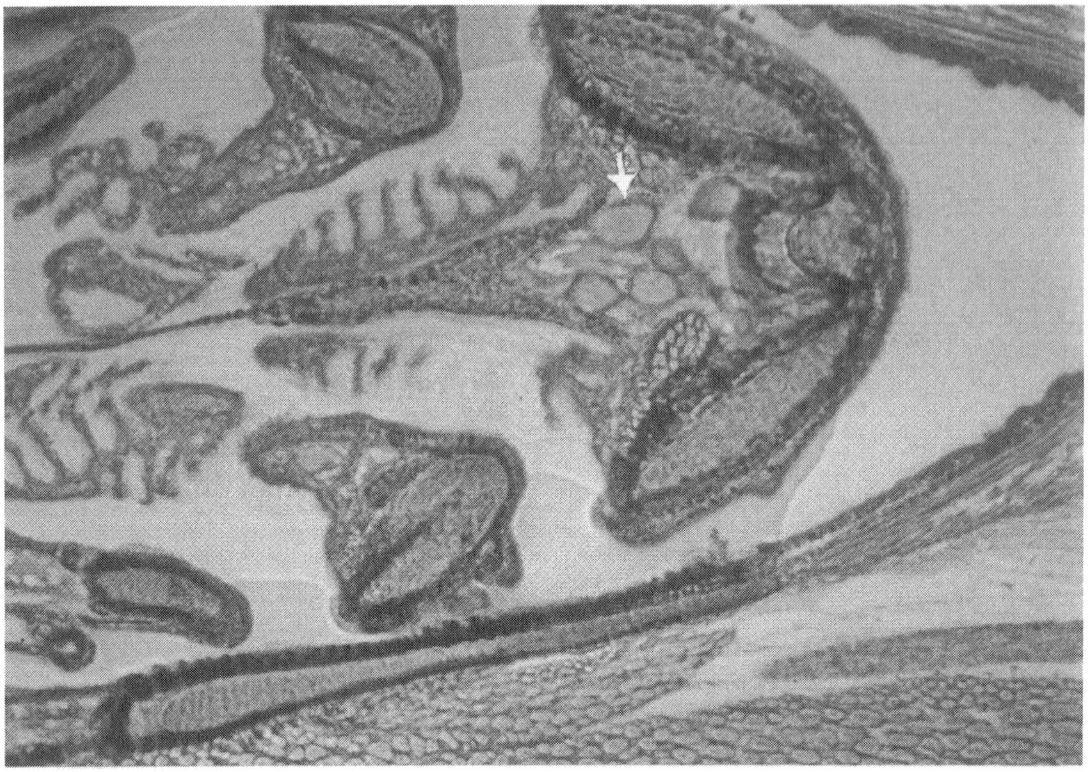

Figure 5. Thyroid tissue in a control larva after 14 days. The follicular epithelial cells are flattened (indicated by an arrow on the picture).

melanosome aggregation through α-adrenoreceptors on release of noradrenaline and of dispersion through β-adrenoreceptors (Mayo & Burton 1998). Mayo & Burton (op. cit.) propose a mechanism by which both dispersal and aggregation of melanosomes can be modulated by a single adrenergic innervation.

The thyroid follicles of the control group displayed flattened epithelial cells typical of inactive thyroids (Leatherland 1994), although larvae receiving exogenous hormone displayed indications of active glands. The dose response in increasing thyroid volume may be an effect of the stimulation of growth in general by T_4. However, it is the potential indication of activity in the thyroglobulin, despite low epithelial cell activity and the expectation of a negative feedback from high levels of exogenous hormone, that is most interesting.

T_4 production is dependent on iodide, which in fish is generally taken up from the ambient seawater and feed through gills and gut. Significantly, gills in halibut do not function until after metamorphosis (Pittman et al. 1990), although thyroid follicles have been histologically identified in halibut larvae with a mean SL of 13.9 mm (de Roode[1]). Intestinal function may therefore be particularly important in premetamorphic halibut. Iodide may have been a limiting factor in thyroid activity in the present study. The low dose of exogenous T_4 used may, as a consequence of this, have been a source of iodide for thyroglobulin as well as stimulating the pituitary–thyroid axis. However, the higher level of exogenous hormone may have been close to inducing negative feedback, agreeing with Miwa & Inui's (1987a) histological examination of the pituitary–thyroid axis of exogenous T_4 treated Japanese flounder, which resulted in flattened epithelial cells in the fish immersed with the highest dose.

The highly varying F level from 16–20 mm SL length, may be coincident to the period between initiation of eye migration and until the migrating eye has reached the dorsal midline. This, in addition to the

[1] de Roode, D. 1997. Metamorphosis in Atlantic halibut, *Hippoglossus hippoglossus*: effects of light regime, thyroid hormones and disease. Second Thesis Report for Environmental Biology at Utrecht University. 54 pp (mimeo).

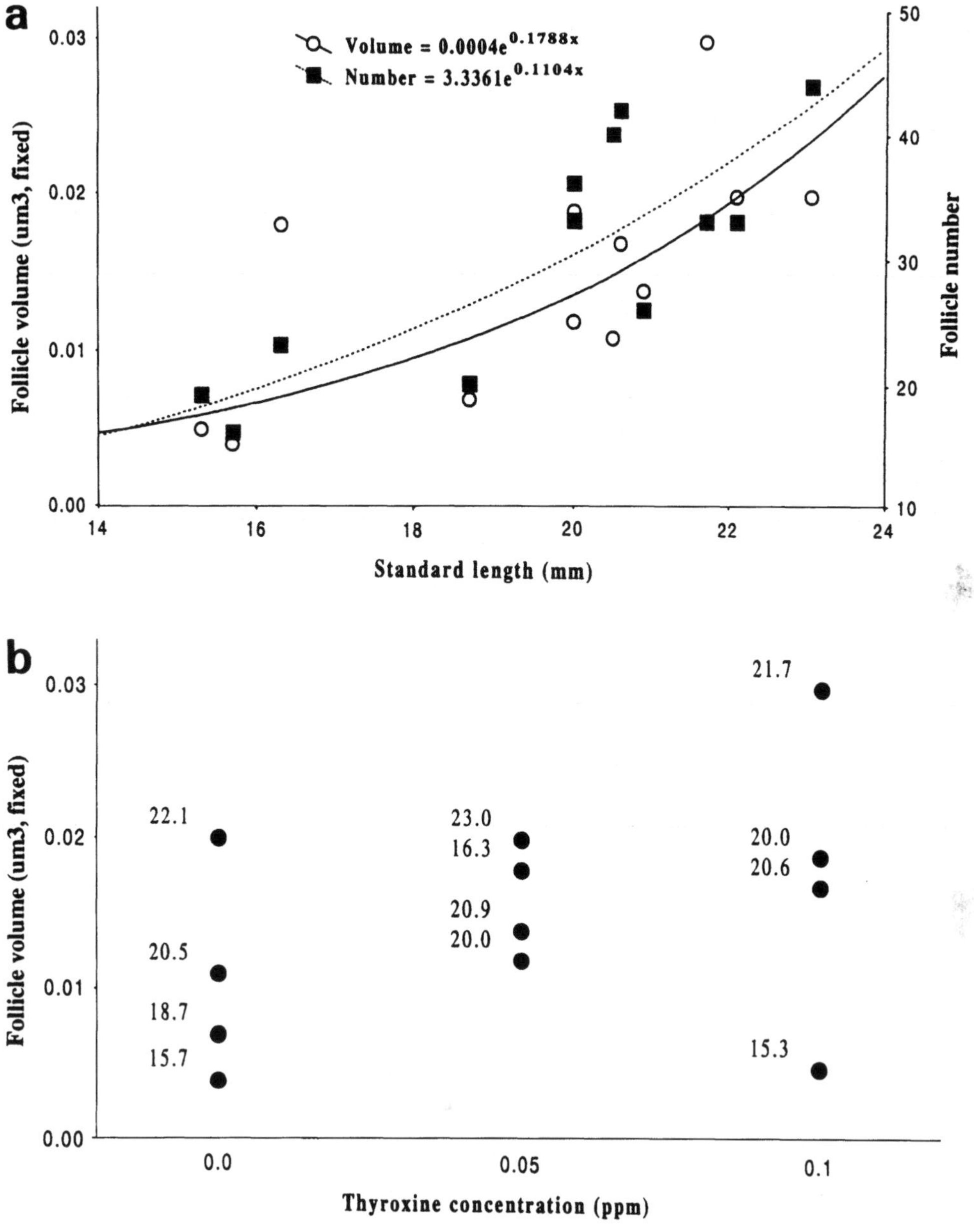

Figure 6. a – Number and volume of thyroid follicle (μm^3, fixed) as function of standard length (mm, live) in metamorphosing halibut. Regression equations is given for each line. b – Volume of follicle as function of dose (ppm) after 14 days treatment. Standard length for each larva is given beside the symbol.

further slight increase in T_4, may indicate the onset of a part of the same endocrinological cascade as in other flatfishes, where a cortisol peak precedes a T_4 peak during metamorphosis (de Jesus et al. 1991, 1993, Tanaka et al. 1995). The cascade or a part of it, may again respond to environmental and/or nutritional factors, of which water level is not a critical variable.

The results seem to support the hypothesis of a threshold size at which metamorphosis is possible, and suggest a 'window of opportunity' beginning at about 16 mm and tapering off between 20–22 mm SL for halibut. During this 'window of opportunity', the final phenotype characteristics of both pigmentation and morphology are determined. One of the

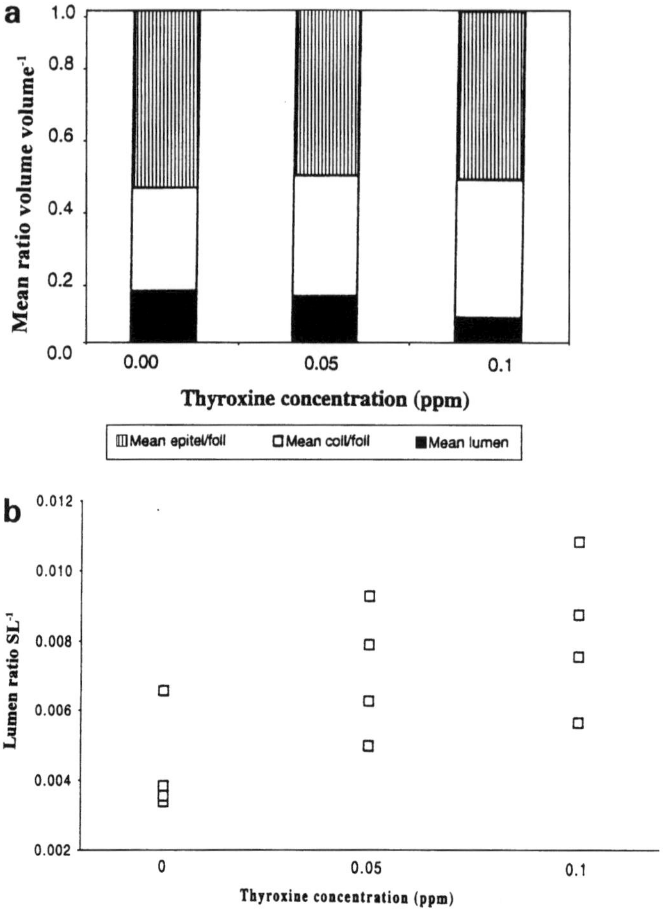

Figure 7. a – Mean ratio of the volume of epithelial cells, colloid or lumen in the total volume of the thyroid follicle (μm³) as function of exogenous l-thyroxine dose after 14 days. b – Lumen ratio over standard length as function of exogenous l-thyroxine dose after 14 days.

thyroxine precursors, the amino acid tyrosine, can also be converted to melanin as well as to adrenaline and noradrenaline. Thus tyrosine is important not only in the production of TH, but also for the creation, dispersal and aggregation of dark pigmentation. The conversion to melanine or T_4 may be a regulated process depending upon energy status and the amount of precursors physiologically available during 'window of opportunity'. However, since thyroid hormone induces both larval cell death and adult cell proliferation and differentiation via spatiotemporally differentiated receptors, the 'window of opportunity' may itself be in part genetically regulated (Ishizuya-Oka et al. 1997, Yamano & Miwa 1998).

The significant increase in tissue level of F in larvae after 21 days in 1.5 cm water did not affect the rate of metamorphosis of the larvae. Changes in tissue F concentrations in Japanese flounder appear to vary depending on the rearing conditions, but the effect on metamorphosis is still not clear (Tanaka et al. 1995). By adding exogenous F alone to premetamorphic larvae of Japanese flounder, neither the rate of the second dorsal fin resorption nor eye migration was affected. However, the addition of both F and T_4 did not affect metamorphosis in Japanese flounder in vivo (de Jesus et al. 1991). Since F has a role in mobilisation of energy resources, it could perhaps release energy needed for the transformation processes during climax metamorphosis. Any form of stressors to which premetamorphic larvae are exposed should therefore be minimized, to permit allocation of the stored energy to the transformation, regression and development of organs and tissues from embryonic anlagen or primordia.

In halibut, the lowest water level of 1.5 cm gave significantly higher F levels, but only after 21 days. At this point, MH was such that the halibut would be frequently

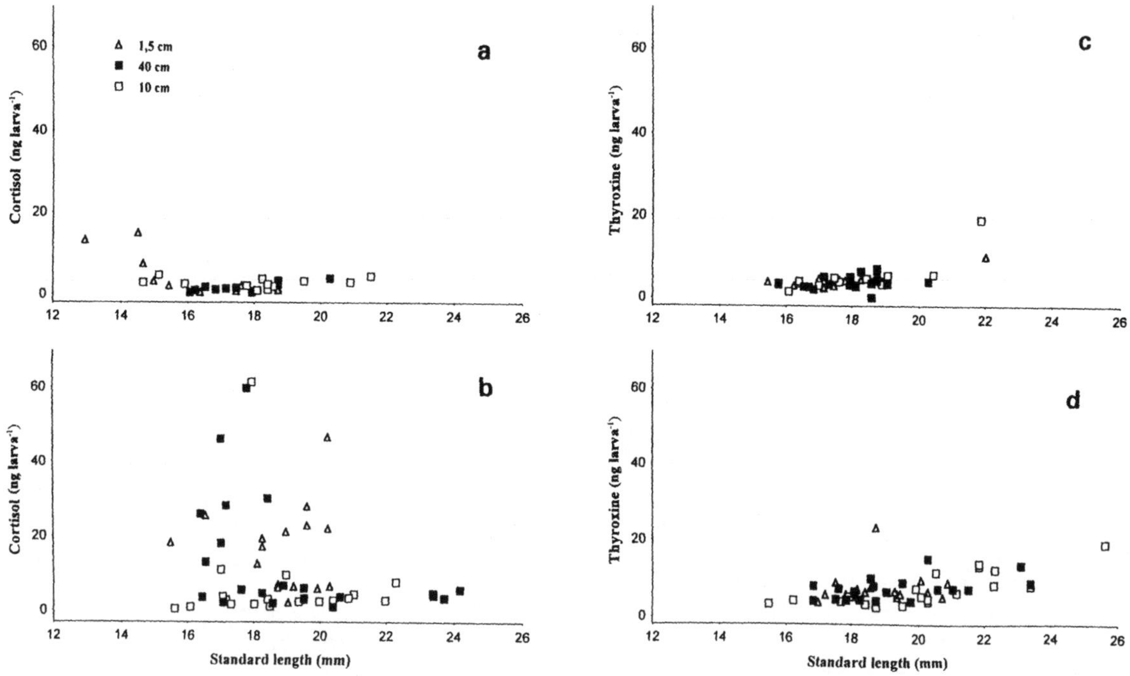

Figure 8. Tissue level of cortisol and thyroxine (mg larva^{-1}) as function of standard length (mm) after 14 (a, c) and 21 days (b, d).

in contact with either the surface or the substrate and normal feeding behaviour would be impaired. Higher mean temperature in the tanks than in those with higher water levels was expected to accelerate development, but this was not detected in the meristic data. Other workers have postulated that induced settlement may function in halibut (Kloksleth 1996), perhaps through contact with the substrate (Gibson & Batty 1990). The synergy of F and T_4 on metamorphosis has been implied by many workers, since F peaks precede those of T_4 in premetamorphic flatfish (de Jesus et al. 1990, Tanaka et al. 1995). Tissue F levels vary videly with rearing conditions (Tanaka et al. op. cit.). Since the difference in F levels occurred at the termination of the experiment (pro- and climax metamorphosis), the results suggest that cortisol can indeed be stimulated by water level; however, the timing of the effect occurs beyond a 'window' in which it could act synergistically with T_4 to influence metamorphosis.

Conclusions

(1) Thyroxine is a mediator of metamorphosis of the Atlantic halibut, affecting growth, eye migration and skeletal formation, pigmentation, and number and size of thyroid follicles. (2) The process of growth, skeletal and neural change and pigmentation appear to be biochemically and physiologically separate pathways in metamorphosing halibut. The conversion of tyrosine to either melanine or T_4 in fish may be a regulated process. (3) The results suggest a 'window of opportunity' for inducing or influencing metamorphosis in halibut, commencing around 16 mm SL and tapering off between 20–22 mm SL. (4) Intestinal function in the 'window of opportunity' may be important for the absorbtion of ions and/or essential aminoacids and thus the ability to complete metamorphosis in halibut. (5) Eye migration may be used as a measure of metamorphic interval in halibut, but settlement is not a useful indicator with these pelagic predators. (6) Low water depth does not affect developmental rate in early metamorphosis. (7) The pooled hormone levels indicate that the endogenous tissue levels of thyroxine and cortisol in the metamorphosing halibut larvae coincide with the endocrinological cascade of other flatfishes during metamorphosis.

Acknowledgements

The authors would like to thank Gert Flick, Catholic University of Nijmegen, for a constructive and timely comment; Anders Mangor-Jensen, Kjell Naas and Anders Jelmert for guidance in first feeding and experimental setup, Martin Lignell for establishment of the thyroid hormone ELISA at IMR, Austevoll Aquaculture Research Station (AARS), Linda Johansen for invaluable help in the laboratory and the technical staff at AARS for production of *Artemia* and algae, and finally Roger Bjugn, Pre-Clinical Institute, University of Bergen for guidance in stereology.

References cited

Allen, B.M. 1916. The results of extirpation of the anterior lobe of the hypophysis and of the thyroid of *Rana pipiens* larvae. Science (Washington, D.C.) 44: 755–758.

Brewster, B. 1987. Eye migration and cranial development during flatfish metamorphosis: a reappraisal (Teleostei: Pleuronectiformes). J. Fish Biol. 31: 805–833.

Brown, C.L. & B.G. Kim. 1995. Combined application of cortisol and triiodothyronine in the culture of larval marine finfish. Aquaculture 135: 79-86.

Brown, M.B. & A.B. Forsythe. 1974. Robust tests for the equality of variances. J. Amer. Stat. Assoc. 69: 364–367.

Cerdà-Reverter, J.M., S. Zanuy, M. Carrillo & O. Kah. 1996. Development of enzyme immunoassays for 3,5,3′-triiodo-L-tyronine and L-thyroxine: time course studies on the effect of food deprivation on plasma thyroid hormones in two marine teleosts, sea bass (*Dicentrarchus labrax* L.) and sea bream (*Sparus aurata* L.). Gen. Comp. Endocrinol. 103: 290–300.

Chambers, R.C. & W.C. Leggett. 1987. Size and age at metamorphosis in marine fish: an analysis of laboratory-reared winter flounder (*Pseudopleuronectes americanus*) with a review of variation in other species. Can. J. Fish. Aquat. Sci. 44: 1936–1947.

Cruz Orive, L.M. & E.R. Weibel. 1990. Recent stereological methods for cell biology: a brief survey. Am. J. Physiol. 258 (Lung Cell. Mol. Physiol. 2): L148–L156.

de Jesus, E.G., Y. Inui & T. Hirano, 1990. Cortisol enhances the stimulating action of thyroid hormones on fin-ray resorption of flounder larvae in vitro. Gen. Comp. Endocrinol. 79: 167–173.

de Jesus, E.G., T. Hirano & Y. Inui. 1991. Changes in cortisol and thyroid hormone concentrations during early development and metamorphosis in the Japanese flounder, *Paralichthys olivaceus*. Gen. Comp. Endocrinol. 82: 369–376.

de Jesus, E.G., T. Hirano & Y. Inui. 1993. Flounder metamorphosis: its regulation by various hormones. Fish. Physiol. Biochem. 11: 323–328.

Donaldson, E.M., U.H.M. Fagerlund, D.A. Higgs & J.R. McBride. 1979. Hormonal enhancement of growth. pp. 532–570. *In*: W.S. Hoar, D.J. Randall & J.R. Brett (ed.) Fish Physiology, Volume 3, Academic Press, New York.

Evans, B.I. & R.D. Fernald. 1990. Metamorphosis and fish vision. J. Neurobiology 21: 1037–1052.

Gibson, R.N. & R.S. Batty. 1990. Lack of substratum effect on the growth and metamorphosis of larval plaice (*Pleuronectes platessa*). Mar. Ecol. Prog. Ser. 66: 219–223.

Kloksleth, V.H. 1996. Adferd, vekst og overlevelse hos premetamorfosert kveite og piggvar i grunne lengdestrømsrenner. (Behaviour, growth and survival of premetamorphic halibut and turbot in shallow raceways.) MSc. Thesis in Aquaculture, Norwegian College of Fisheries, Tromsø. 74 pp.

Harboe, T., S. Tuene, A. Mangor-Jensen, H. Rabben & I. Huse. 1994. Design and operation of an incubator for yolk-sac larvae of Atlantic halibut. The Prog. Fish-Cult. 56: 188–193.

Hoar, W.S. 1951. Hormones in fish. Univ. Toronto Stud. Biol. Ser. 59, Publ. Ontario Fish. Res. Lab. 71: 1–111.

Inui, Y. & S. Miwa. 1985. Thyroid hormone induces metamorphosis of flounder larvae. Gen. Comp. Endocrinol. 60: 450–454.

Inui, Y., K. Yamano & S. Miwa. 1995. The role of thyroid hormone in tissue development in metamorphosing flounder. Aquaculture 135: 87–98.

Ishizuya-Oka, A., U. Shuichi & S. Yun-Bo. 1997. Temporal and spatial regulation of a putative transcriptional repressor implicates it as playing a role in thyroid hormone dependent organ transformation. Developm. Genet. 20: 329–337.

Jelmert, A. & H. Rabben 1987. Upwelling incubators for eggs of the Atlantic halibut (*Hippoglossus hippoglossus* L.). ICES C.M. 1987/F:20.

Lam, T.J. 1980. Thyroxine enhances larval development and survival in *Sarotherodon* (*Tilapia*) *mossambicus* Ruppel. Aquaculture 21: 287–291.

Lam, T., J.V. Juario & J. Banno. 1985. Effect of thyroxine on growth and development in post-yolk-sac larvae of milkfish, *Chanos chanos*. Aquaculture 46: 179–184.

Leatherland, J. F. 1994. Reflections on the thyroidology of fishes: from molecules to humankind. Guelph Ichthyol. Rev. 2: 1–64.

Matsumoto, J. & T. Sekai. 1992. Assymetric pigmentation and pigment disorders in Pleuronectiformes (flounders). Pigment Cell Res. Suppl. 2: 275–282.

Mayo, D.J. & D. Burton. 1998. Beta$_2$-adrenoceptors mediate melanosome dispersion in winter flounder (*Pleuronectes americanus*). Can. J. Zool. 76: 175–180.

Miwa, S. & Y. Inui. 1987a. Histological changes in the pituitary-thyroid axis during spontaneous and artificially induced metamorphosis of larvae of the flounder *Paralichthys olivaceus*. Cell Tissue Res. 249: 117–123.

Miwa, S. & Y. Inui. 1987b. Effects of various doses of thyroxine and triiodothyronine on the metamorphosis of flounder (*Paralichthys olivaceus*). Gen. Comp. Endocrinol. 67: 356–363.

Miwa, S. & Y. Inui. 1988. Thyroxine surge in metamorphosing flounder, *Paralichthys olivaceus*. Gen. Comp. Endocrinol. 70: 158–163.

Miwa, S., K. Yamano & Y. Inui. 1992. Thyroid hormone stimulates gastric development in flounder larvae during metamorphosis. J. Exp. Zool. 261: 424–430.

Nacario, J.F. 1983. The effect of thyroxine on the larvae and fry of *Sarotherodon niloticus* L. (*Tilapia nilotica*). Aquaculture 34: 73–83.

Næss, T & Ø. Lie. 1998. A sensitive period for the determination of pigmentation pattern in halibut juveniles: the role of diet. Aquacult. Res. (in press).

Naas, K.E., T. Næss & T. Harboe. 1992. Enhanced first feeding of halibut larvae (*Hippoglossus hippoglossus* L.) in green water. Aquaculture 105: 143–156.

Pittman, K., A.B. Skiftesvik & L. Berg. 1990. Morphological and behavioural development of halibut, *Hippoglossus hippoglossus* (L.) larvae. J. Fish Biol. 37: 455–472.

Pittman, K., A. Jelmert, T. Næss, T. Harboe & K. Watanabe. 1998. Plasticity of viable postmetamorphic forms of farmed Atlantic halibut (*Hippoglossus hippoglossus* L.). Aquacult. Res. 29: 949–954.

Sclower, A. 1930. Die Bedeutung der Schilddrüse für die Metamorphose des Aales und der Plattfishe (The involvement of the thyroid in the metamorphosis of eels and flatfishes). Forsch. Fortschr. Deutsch. Wiss. 6: 435–436.

Seikai, T., J.B. Tanangonan & M. Tanaka. 1986. Temperature influence on growth and metamorphosis in the Japanese flounder (*Paralichthys olivaceus*) in the laboratory. Bull. Japan. Soc. Sci. Fish. 52: 977–982.

Stensland, K. 1995. Cortisol og testosteron i egg og larver frå kveite (Cortisol and testosterone in eggs and larvae of Atlantic halibut). MSc. Thesis in General Aquaculture, Institute of Fisheries and Marine Biology, University of Bergen. 48 pp.

Tagawa, M. & T. Hirano 1987. Presence of thyroxine in eggs and changes in its content during early development of chum salmon, *Oncorhynchus keta*. Gen. Comp. Endocrinol. 68: 129–135.

Tagawa, M., S. Miwa, Y. Inui, E.G. de Jesus & T. Hirano. 1990. Changes in thyroid hormone concentrations during early development and metamorphosis in the flounder, *Paralichthys olivaceus*. Zool. Sci. 7: 93–96.

Takagi, Y., J. Hirano, H. Tanabe & J. Ymanda. 1994. Stimulation of skeletal growth by thyroid hormone administration in the rainbow trout, *Oncorhynchus mykiss*. J. Exp. Zool. 268: 229–238.

Tanaka, M., S. Kawai, T. Sekai & J.S. Burke. 1996. Development of the digestive organ system in Japanese flounder in relation to metamorphosis and settlement. Mar. Fresh. Behav. Physiol. 28: 19–31.

Tanaka, M., J.B. Tanangonan, M. Tagawa, E.G. de Jesus, H. Nishida, M. Isaka, R. Kimura & T. Hirano. 1995. Development of the pituitary, thyriod and interrenal glands and applications of endocrinology to the improved rearing of marine fish larvae. Aquaculture 135: 111–126.

Tanangonan, J.B., M. Tagawa, M. Tanaka & T. Hirano. 1989. Changes in tissue thyroxine level of metamorphosing Japanese flounder (*Paralichthys olivaceus*) reared at different temperatures. Nippon Suissan Gakkaishi 55: 485–490.

Tata, J.R. 1996. Metamorphosis: an exquisite model for hormonal regulation of post-embryonic development. Biochem. Soc. Symp. 62: 123–136.

Yamano, K., S. Miwa, T. Obinata & Y. Inui. 1991. Thyroid hormone regulates developmental changes in muscle during flounder metamorphosis. Gen. Comp. Endocrinol. 81: 464–472.

Yamano K., H. Takano-Ohmuro, T. Obinata & Y. Inui. 1994. Effect of thyroid hormone on developmental transition of myosin light chains during flounder metamorphosis. Gen. Comp. Endocrinol. 93: 321–326.

Yamano K. & S. Miwa. 1998. Differential gene expression of thyroid hormone receptor α and β in fish development. Gen. Comp. Endocrinol. 109: 75–85.

Zar, J.H. 1996. Biostatistical analysis, 3rd edition. Prentice Hall, Englewood Cliffs. 662 pp.

Environmental Biology of Fishes **56**: 67–77, 1999.

Early development of the sofie, *Chondrostoma toxostoma*

Rodolphe E. Gozlan[a,b], Gordon H. Copp[a] & Jean-Noel Tourenq[b]
[a]*Landscape & Ecology Research Group, Department of Environmental Sciences, University of Hertfordshire, College Lane, Hatfield, Herts AL10 9AB, U.K. (e-mail: g.h.copp@herts.ac.uk)*
[b]*Centre d'Écologie des Systèmes Aquatiques Continentaux, UMR C5576 – CNRS/UPS, Université Paul Sabatier, 118 route de Narbonne, Toulouse, 31062 France*

Received 3 February 1998   Accepted 2 December 1998

Key words: early intervals of life, saltatory ontogeny, southern European nase, eggs

Synopsis

The developmental biology of embryos, larvae and juveniles of sofie, *Chondrostoma toxostoma*, reared from artificial insemination, was examined under controlled laboratory conditions, using both in-vivo and preserved specimens. The complete remodelling process (metamorphosis) was described and interpreted according to the theory of saltatory ontogeny, providing a model within which the highly complex ecological niche and behaviour (habitat use, diet, etc.) of the free embryos and larvae can be evaluated. The sofie ova were relatively opaque and the free embryos presented a small yolksac. Consequently, the free embryo steps were short and the free embryos emerged rapidly from the substratum. The circulatory system appeared early and was simple, which suggests that sofie free embryos require well oxygenated water. The remodelling of the mouth from a superior to inferior position characterised changes in the sofie's diet during early life history, which is typical of the genus. Differences in development relative to the nase, *C. naus*, are discussed.

Introduction

The study of early ontogeny in fishes contributes to the understanding of a species' developmental biology (Prokeš & Peňáz 1978, Kováč 1992), growth bioenergetics (Holden & Bruton 1994), early life behaviour (Osse 1990, Kasumyan & Ponomarev 1990), as well as to species identification (Alderdice 1985) and to the development of recruitment models (Beyer 1989, Hoenig et al. 1990, Houde 1994, Bell et al. 1995). Knowledge of early ontogeny is of particular importance in identifying describers of river ecosystem function (Copp et al. 1991) and to the conservation of endangered fish species. One such example is the sofie, *Chondrostoma toxostoma* (Vallot, 1836), which has been replaced in many parts of its original range in south western Europe by the nase, *C. nasus* (L.), a species indigenous to the River Danube basin. In other parts of its range, the sofie is, like the entire genus, either endangered or locally extinct due to human alteration of river systems, e.g. pollution, water retention structures (Delfau 1979, Lusk 1995, Tourenq & Gozlan 1997).

The biology of the sofie has received only fragmentary study (Vallot 1836, Mathias 1921, Spillman 1961, Chappaz 1986, Maier et al. 1995) and its ontogeny has been completely ignored. The aim of the present study was to provide the first description of sofie early development, with the specific objective to interpret sofie development within the context of the theory of saltatory ontogeny (Balon 1985, 1990), including comparisons with the nase, *C. nasus*, and to determine the influence of environmental conditions on development (addressed in Gozlan et al. 1999 this volume). In order to understand better the relationship between form and function, early development, and in particular, the relative growth of different parts of the body are examined within the overall environmental context of the species' niche (Osse & Drost 1989).

Material and methods

In the laboratory on 20 May 1996, eggs were acquired from four females (mean TL = 19.15 mm, SE = 1.18) by gently stroking their abdomen. The female gametes were mixed directly in a bucket with the sperm of eight males (mean TL = 17.12 mm, SE = 0.53). The inseminated eggs were spread out in a tank on plastic artificial grass (50 cm × 30 cm × 30 cm), with continuous water flow (ground water source) and air bubblers ensuring good oxygenation of water surrounding the eggs. Each day, two thirds of the tank water was changed and dead eggs were removed using a micro-pump. Water temperatures in the field and in the tank were recorded every hour, from activation to the end of the study, using a 'Tiny talk' recorder. Five eggs were collected and preserved in 4% formalin every 6 h after activation, and then every 10 h until hatching. The larvae were fed on specially formulated crushed goldfish flakes with added vitamins and calcium.

Measurements were made to the nearest 0.1 mm using a binocular microscope fitted with an ocular micrometer. Lengths of specimens are given as notochord length (NL) for preflexion of the urostyle as well as in standard (SL) and total length (TL) as appropriate. Analysis of the different developmental phases was made on preserved and on living specimens following the model proposed by Balon (1985). The average heart beat rates were calculated on different specimens using a Nikon camera fitted to a microscope.

Results

Water temperature in the aquarium increased slowly from activation to the onset of exogenous feeding (Figure 1). Diel water temperature variations were more obvious in the field than in the laboratory, particularly after hatching. Some embryos presented abnormal spinal curvature, but mortality during the different developmental phases was relatively low, around 20%. When most of the sofie were in their 5th larva step, some specimens were still not eating, even though the mouth was well formed. Consequently, their overall development slowed, and a few days later they died; no cannibalism or aggressive behaviour was observed.

Embryo period

Cleavage phase

1st embryo step (beginning of step: activation). A few seconds after activation, the ova rapidly took on water, increasing in volume, creating the perivitelline space and at the same time becoming adhesive. Ova shape was spherical, with a yellow-olive colour; the envelope was hard and opaque. Egg diameter varied between 1.9 and 2.5 mm (mean = 2.1, SE = 0.19,

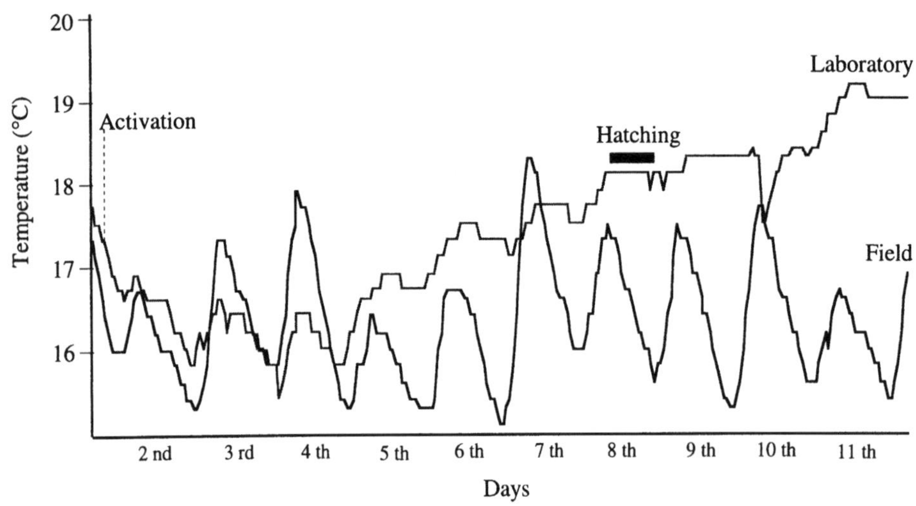

Figure 1. Incubation temperatures during sofie, *Chondrostoma toxostoma*, egg development in the laboratory (from activation to hatching) as well as the temperatures recorded in the field (River Lens) during the spawning time.

n = 38). Yolk diameter varied between 1.3 and 1.9 mm (mean = 1.54, SE = 0.22, n = 38), achieving 68.4% to 76% of the egg diameter.

2nd embryo step (beginning of step: start of cleavage). Observations up to the embryonic phase were described from preserved material extracted from the envelopes due to its opacity. At 15 h after activation, morula were discernible. The blastoderm became concave like a cap over the yolksac. Some morula presented an irregular shape and some were more rounded.

3rd embryo step (beginning of step: the epiboly). At 24 h after activation, the blastoderm covered a quarter to a half of the surface of the yolksac. The margin of the blastodisc showed an irregular shape.

4th embryo step (beginning of step: the equatorial epiboly). At 1 day and 15 h after activation, the cellular material covered more than half of the yolksac. At the end of this step, formation of the germ ring and the embryonic shield were visible.

Embryo phase

5th embryo step (beginning of step: germ ring closure). The ova envelope became less opaque, the germ ring was closed, and the embryonic shield became elongated, forming the anlagen of the future embryo. Cephalization became evident with anlagen of brain and optic vesicles visible. The embryo's length reached approximately half of the yolksac circumference.

At 2 days and 13 h after activation, the number of somites was 29. The heart, eye lens and brain were clearly visible; pigmentation was lacking, and a rudimentary notochord was visible (Figure 2a). The yolksac's form suggested the onset of its future division into two parts. The length of the embryo was approximately two-thirds of the yolksac circumference.

6th embryo step (beginning of step: onset of muscular contractions). At 3 days and 14 h after activation, the first myotome contraction was noticed. The auditory vesicles, olfactory pits and the anlagen of the gills were visible, as well as a caudal bud on the posterior edge of the caudal section, which increased in size during this step. Pigmentation was still lacking, and the blood was colourless. The head was bent down and still attached to the yolksac, which was clearly divided into two compartments: one round (anterior part), and the other cigar shaped (posterior part). Mesenchyme

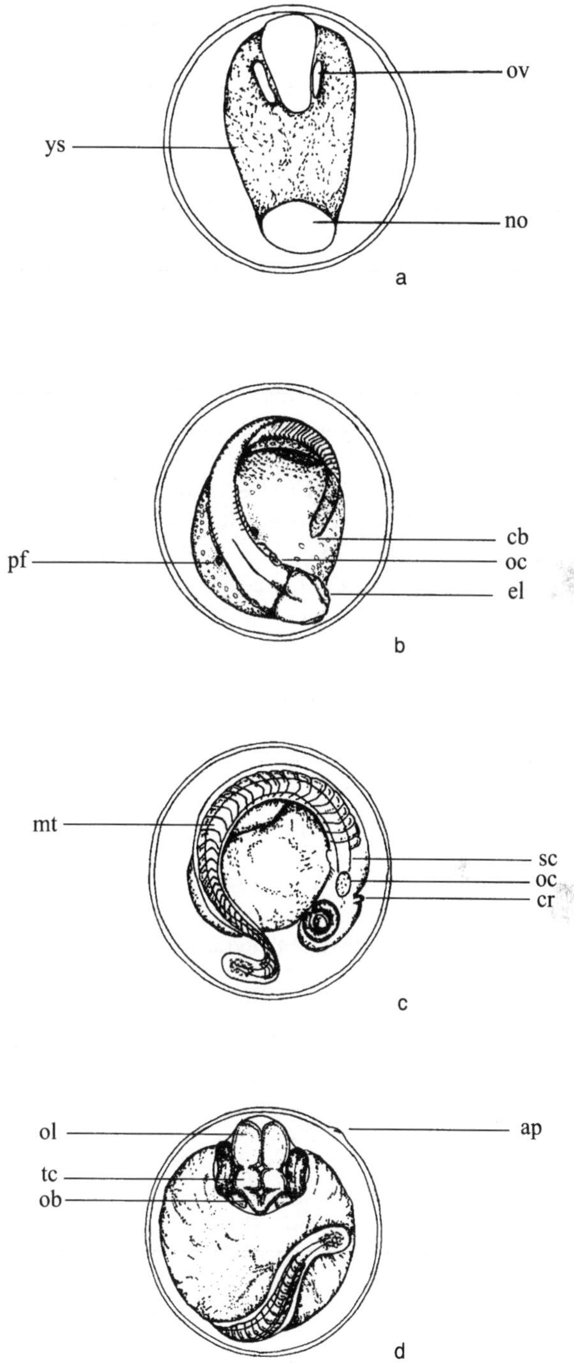

Figure 2. Morphological development of sofie, *Chondrostoma toxostoma*, eggs (a = 61:30 h after activation, b = 86 h after activation, c/d = 134 h after activation (ap = animal pole, cb = caudal bud, cr = cerebellum, el = eye lens, mt = myotomes, no = nothochord, ob = olfactory bulb, oc = otic capsula, ol = optic lobes, ov = optical vesicules, pf = pectoral fins, sc = spinal cord, tc = telencephalon, yc = yolksac).

concentrations consisted of two patches on the yolk-sac close to the embryonic shield, corresponding to the anlagen of the pectoral fins. The number of somites had increased to 41 (Figure 2b).

7th embryo step (beginning of step: pigmentation in the eyes). At 5 days and 14 h after activation, mesenchyme concentrations appeared in the caudal area; it was segmented and the tip was free. The embryo fully surrounded the yolksac (Figure 2c), the brain was complete (olfactory bulb, telencephalon, optic lobes, cerebellum, facial lobe, and spinal cord); the otoliths were clearly visible (mean diameter = 0.35 mm, SE = 0.1, n = 6); and the pectoral fins were differentiated and attached horizontally to the body. During this step, pigmentation increased in the eyes but was lacking on the body, and the embryos were still translucent. Average egg diameter was 2.3 mm (SE = 0.55, n = 9) and the periviteline space allowed the embryos to move freely from time to time by contracting their entire bodies. The number of somites increased to 42 (Figures 2c,d).

Free embryo phase
The free embryo phase began during the 8th embryonic step, with hatching of the first embryo, and ended with the onset of exogeneous feeding.

8th embryo step (beginning of step: onset of blood circulation, final somite number, straightening). At 6 days and 6 h after activation, blood circulation was evident but difficult to describe due to the envelopes and colourless blood. The head, still bent down, slowly separated from the yolksac during this step. On 27 May, the first embryos hatched 6 days and 20 h after activation (Figure 3a), and the last one 4 h later. The free embryos laid more or less on the bottom of the tank, but from time to time darted horizontally. Then, they fell back to the bottom and rested passively for some time. Locomotion resulted from whirling tail movements, and the embryos performed such horizontal darting movements when disturbed by direct touch. The free-embryos developed early photophobia and tended to congregate.

Blood circulated from the dorsal aorta to the caudal artery, which supplied the caudal vein. The caudal vein consisted of a very thin network of anastomosing (twisted) vessels (Figure 4), which near the prospective anal region turned to form a single subintestinal vein. Paired anterior cardinal veins, as well as the subintestinal vein, supplied the anterior vitelline veins on

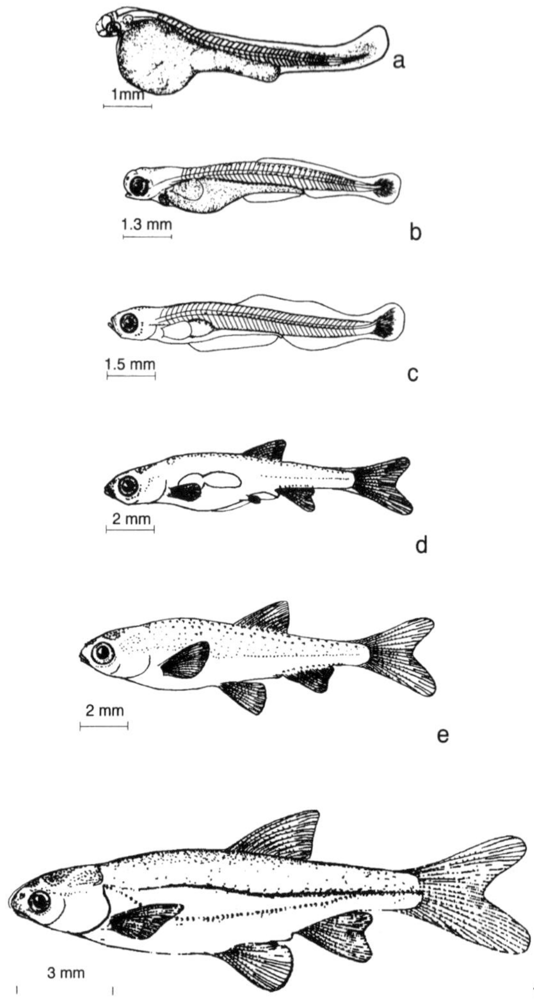

Figure 3. Morphological development of the sofie, *Chondrostoma toxostoma* (a – step ES8, age 6 days 20 h, TL = 6.1 mm; b – step ES9, age 8 days 20 h, TL = 7.5 mm; c – step LS2, age 16 days 20 h, TL = 9.3 mm; d – step LS5, age 50 days, TL = 13.2 mm; e – step LS6, age 59 days, TL = 15 mm; f = step LS7, age 63 days, TL = 16.5 mm).

both sides, descending on the yolksac. The anterior vitelline veins, formed by a single vessel on both hemispheres, joined at the anteroventral part of the yolksac to enter the heart. Mean heart rate was 89 times per minute (SE = 2.94, n = 9). The head was bent forward over the yolksac, but separated and slightly extended upward. The urostyle was straight and surrounded with a concentration of mesenchyme. No pigmentation was present on the body, except for the eyes. Blood elements

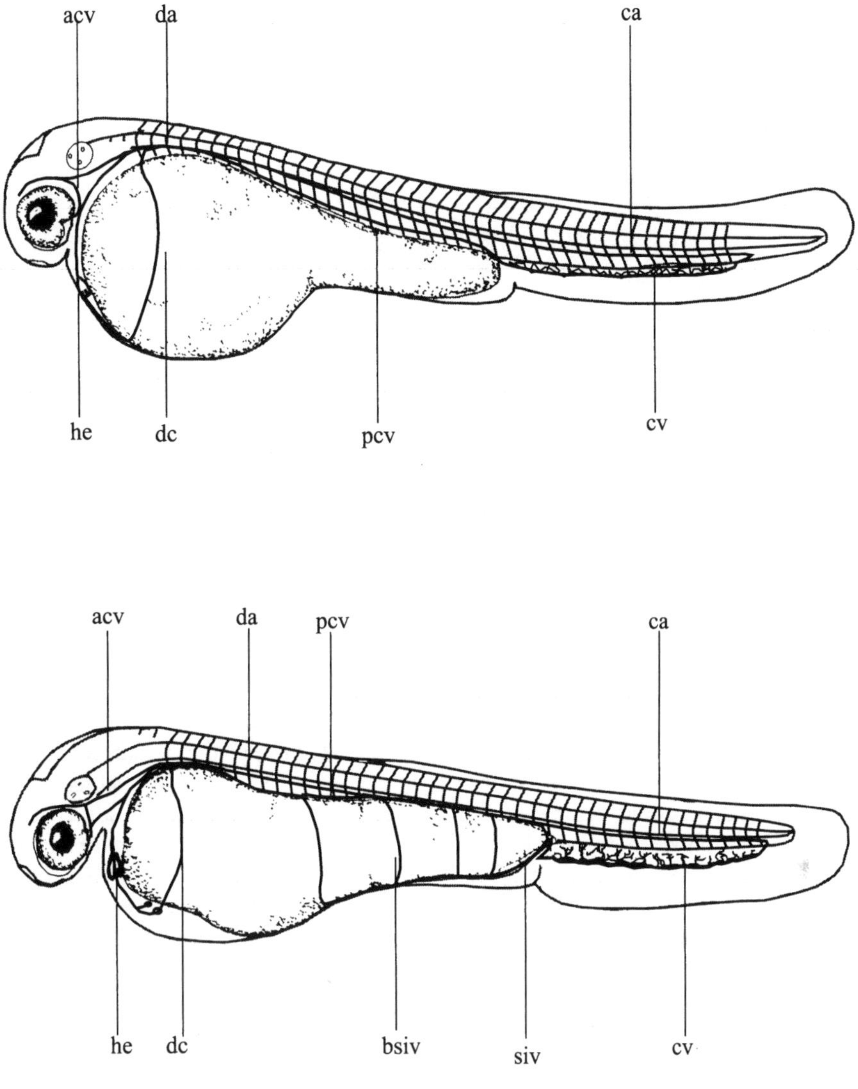

Figure 4. Vascular system of sofie, *Chondrostoma toxostoma*, free embryo, left lateral views. a – freshly hatched, at the start of 8th embryo step (SL = 5.3 mm, ca = caudal artery, da = dorsal aorta, acv = anterior cardinal vein, he = heart, dc = ductus cuvieri, pcv = posterior cardinal vein, cv = caudal vein). b – at the end of the 8th embryo step (SL = 5.8 mm, ca = caudal artery, pcv = posterior cardinal vein, da = dorsal aorta, acv = anterior cardinal vein, he = heart, dc = ductus cuvieri, bsiv = branch of subintestinal vein, siv = subintestinal vein, cv = caudal vein).

were still colourless and venation as well as pigmentation were absent on the yolksac. The embryos hid under the artificial substratum. The pectoral fins appeared as semicircular discs with a horizontal base. The embryos contained the final number of somites: 28 preanal and 14 postanal (Figure 3a). At the end of this step, branches of the subintestinal vein appeared on the oblong part of the yolksac (Figure 4b). The free embryos varied almost twofold in NL and TL (Table 1).

9th embryo step (beginning of step: onset of pectoral fins movement, filling of the gas bladder). On 29 May, 8 days and 20 h after activation, the embryos reached the surface and gulped in air in order to fill their swimbladders. They did not respond to light or to tapping on the side of the tank. Brown melanophores appeared at the dorsal edge of the first 10 myotomes, forming two parallel stripes, and on the ventral edge from the anus to the tail. Some melanophores were also visible at the

Table 1. Body length measurements of sofie, *Chondrostoma toxostoma*, at different steps of early ontogeny.

Developmental steps	Total length (mm)					Standard length (mm)				Notochord length (mm)			
	n	min/max	mean	SE	s^2	min/max	mean	SE	s^2	min/max	mean	SE	s^2
8th embryo step	18	4.1–8.1	6.32	0.21	0.79					4.0–7.8	6.11	0.20	0.75
9th embryo step	5	7.4–7.8	7.62	0.10	0.04					7.2–7.7	7.45	0.12	0.60
1st larva step	9	8.6–10.0	9.48	0.19	0.32					8.2–9.5	8.98	0.18	0.28
2nd larva step	5	9.1–10.1	9.60	0.18	0.16					8.7-9.5	9.10	0.15	0.11
3rd larva step	45	8.7–13.5	10.40	0.14	0.87	8.4–11.1	9.63	0.10	0.04				
4th larva step	5	10.9–13.5	12.32	0.44	0.95	10.0–11.1	10.72	0.22	0.25				
5th larva step	5	11.7–14.0	13.30	0.41	0.83	10.1–11.8	11.28	0.30	0.46				
6th larva step	5	12.2–15.6	14.34	0.57	1.62	10.6–12.6	11.84	0.33	0.55				
7th larva step	6	14.5–17.0	15.85	0.47	1.30	12.0–13.9	13.12	0.81	0.33				
8th larva step	8	16.1–21.8	19.16	0.77	4.78	13.0–17.6	15.18	0.57	2.58				

medulla oblongata, the optic lobes, as smaller dots at the cerebrum (Figure 5a) and as an intermittent line on the middle posterior part of the body along the future lateral line. The entire body cavity began to be covered with diffused black pigmentation from the pectoral fins to the future anus. In the lower part of the caudal fin, mesenchyme concentration increased. The pectoral fins increased in size and were still firm and moved around the base. The mouth, with the mandible immobile, was in a terminal position, although it was situated at the lower part of the head. The head was straight (Figure 3b). Intestine growth was considerable during this step and proportional to the decrease of the yolksac size. Heart beat rate increased to an average of 102 times per minute (SE = 3.53, n = 5). The NL and TL varied little (Table 1).

Larva period

1st larva step (beginning of step: onset of exogenous feeding). As soon as the first chamber of the swimbladder was functional (at 10 days and 20 h after activation) the fish were able to begin swimming actively and began to feed exogeneously. The mouth was mobile and still in a terminal position (mean gape = 0.3 mm, SE = 0.03, n = 10). Some melanophores appeared on the gill covers and in the heart region. The pigments at the medulla oblongata were more numerous. The five gill arches were well visible and the pectorals fins were vertical and still difficult to bend. Some lepidotrichia started to appear in the caudal mesenchyme. The gill covers were present on all larvae and were fully extended on the gills. Variation of NL and TL was almost twice that of the preceding step (Table 1).

At the end of this step (16 days after activation), pigmentation was more abundant at the dorsal edge of the first 17 myotomes and formed two parallel stripes, as well as an intermittent line on the middle posterior part of the body and on all of the dorsal part of the head (Figure 5b). Some very tiny brown dots appeared on the lips and the non-segmented part of the caudal fin where 7 to 11 downward pointing lepidotrichia were visible. Finfold differentiation continued with the appearance of unpaired fin areas. The yolksac, which had reduced considerably since the onset of exogenous feeding, disappeared by the end of this step. Simultaneously, the posterior chamber of the swimbladder became wider than the intestine, which increased considerably in depth and was more or less filled with food; the mouth was mobile. The oval-shaped pectoral fins increased slightly in length (mean = 0.9 mm, SE = 0.2, n = 8). The larvae had developed an escape behaviour for all type of disturbances (loud sounds, brutal changes in the light intensity and water surface disturbances).

2nd larva step (beginning of step: exclusively exogenous feeding). At 16 days and 20 h after activation, the overall shape of the fish was slightly concave, finfold decreased and some mesenchyme concentration was present in the future place of the dorsal and anal fin skeleton. The caudal end of the notochord was starting to point slightly upward, and from 13 to 15 lepidotrichia, obliquely pointed downward, were present. The lower jaw of the mouth was longer than the upper one (superior position, Figure 5c). Pigmentation increased on the tip of the snout, along the upper jaw and on the gill covers as well as on the dorsal part of the fish (head and dorsal edge of the somites), where the colour of the melanophore was darker (Figure 3c). The body ventral cavity was widely covered with black melanophores, prolonged in a stripe along the dorsal part of the intestine, followed by brown dots on

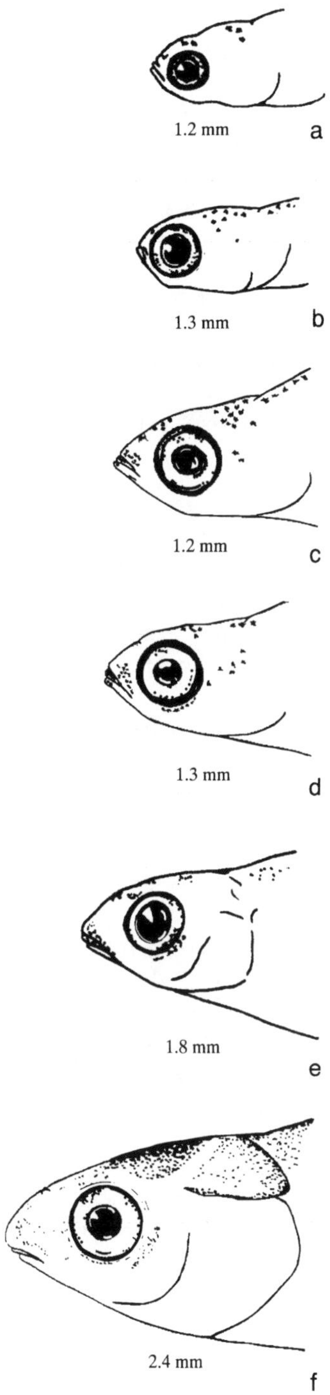

Figure 5. Morphological development of the sofie, *Chondrostoma toxostoma*, mouth (a – step ES9, age 8 days 20 h; b – step LS1; c – step LS2, age 16 days 20 h ; d – step LS3, age 28 days 20 h; e – step LS6, age 59 days; f – step LS8, age 68 days).

the ventral edge of the somites after the anus. The swimbladder and intestine continued growing. At the end of the step (28 days after activation), the mouth started to shift slightly from a superior position to a terminal position. The heart region was strongly pigmented, and the pectoral fins were of uniform length (mean = 1 mm, SE = 0.29, n = 4). The body was straight and the intestinal tract could still be followed, although pigmentation on the body's ventral cavity was more intense. Variation in NL and TL was similar to the preceding step (Table 1).

3rd larva step (beginning of step: formation of lepidotrichia in the dorsal and anal fins). At 28 days and 20 h after activation, the caudal end of the notochord was bent slightly, then strongly turned upwards at 45 degrees. The shape of the caudal fin presented a depletion, initialising the two final lobes of the caudal fin, which contained from 17 to 18 unbranched rays (approximately the final number of lepidotrichia). The first rays appeared in the dorsal and anal fins at the same time. Melanophores were progressively more numerous and dark, especially on the gills cover, the lips (Figure 5d), and the rays of the caudal fin. Variation in body length was slightly less than the preceding two steps (Table 1). The pectoral fins increased in length (mean = 1.2 mm, SE = 0.01, n = 24). From this step, the larvae formed a shoal in the middle of the water column with the biggest individuals in the lead; they seemed less frightened by different types of disturbances.

4th larva step (beginning of step: filling of the anterior chamber of the swimbladder). At 31 days and 20 h after activation, the anterior chamber of the swimbladder started to fill and was smaller than the posterior chamber. The caudal fin was clearly divided in two lobes, but it still remained a part of the finfold at the caudal base. Some unbranched rays were still present in the caudal and pectoral fin. In the dorsal fin, from 7 to 8 branched, unsegmented rays were present, with 10 to 11 in the anal fin. The final shape of these two fins was not properly defined due to the finfold connection with the caudal fin. The upper part of the mouth increased in size, emphasising its terminal position. The ventral fins appeared as a pair of buttons in the middle of the ventral finfold. Variation in body length increased considerably in this step (Table 1) as well as the pectoral fin length (mean = 1.44, SE = 0.07, n = 5) and mouth gape (mean = 0.5 mm, SE = 0.6, n = 5).

5th larva step (beginning of step: separation of the dorsal and anal fins). Separation of the dorsal and anal fins was completed and only small remains of the finfold were still present on the caudal part, 50 days after activation. The lower caudal lobe started to be longer than the upper. Ventral fin length increased, but did not exceed the finfold's outer edge. During this step, the volume of the anterior part of the swimbladder reached half that of the posterior. The latter had also increased in length and extended to the anterior edge of the dorsal fin. The mouth was still terminal and the pigmentation all over the body was more pronounced (Figure 3d). Variation in body length remained high (Table 1) as did that of the pectoral fins (mean = 1.8 mm, SE = 0.57, n = 7).

6th larva step (beginning of step: disappearance of remains of the finfold). At 59 days after activation, the anterior part of the swimbladder had reached two-thirds of the posterior chamber and was rounder in shape. The dorsal and anal fins were complete, with final ray counts of 8 and 11, respectively. The remains of connective tissue on the caudal peduncle had completely disappeared. The caudal fin had its lepidotrichia segmented and branched like the pectoral fins, which until this step had an oval fin shape and began to be more elongated. The ventral fins were not completely formed but extended beyond the finfold's outer edge. At the end of this step, the mouth started to be slightly inferior due to an increase in upper jaw length (Figure 5e). Body shape remained different from the juveniles and adults mainly due to the head, which had not completed its metamorphosis. Scales had still not appeared on the larvae, and despite an increased number of melanophores, overall pigmentation remained light. Variation in TL continued to increase, but that in pectoral fin length and SL remained almost constant (Table 1).

7th larva step (beginning of step: completion of the finfold differentiation). At 63 days after activation, the ventral fins were completely formed, and the process of finfold depletion had ended. The number of rays became definite: D III (7) 8, A III (9)10(11), V II (7)8(9), P I 14–15(16), C (17)18(19). Body shape was rounder than that of adults, and the mouth still did not resemble the inferior mouth of adults. Pigmentation was still lighter than in juveniles, but in some specimens a large dark stripe was present on the side of the body from the head to the tail and the fins were light yellow (Figure 3f). Variation in TL decreased slightly, whereas that in SL increased strongly (Table 1).

8th larva step (beginning of step: development of scales). At 68 days after activation, the first scales appeared on the body, situated ventrally under the lateral line behind the operculum. Soon thereafter, the entire body was covered with a layer of scales and the mouth was adult-like (Figure 5f). Complete scale cover occurred at about 15 mm SL (Figure 3f), but did not correspond to a stabilisation in relative growth, which was at a notably shorter mean SL than observed in nature (see Gozlan et al. 1999 this volume). Variation in TL increased significantly, whereas that of SL was much less (Table 1). The stabilization of relative growth, the end of the remodelling process (metamorphosis), and the start of the juvenile period of development is addressed by Gozlan et al. (1999 this volume).

Discussion

Some differences in early development can be identified between the sofie and the nase, using Peňáz's (1974b) description of the latter. The free embryo stages of both species have different circulatory systems, with the branches of the subintestinal vein on the yolksac of the sofie absent in the nase. During the 4th larva step, no typical dorsal tubercle was observed opposite the tip of a curved urostyle in the sofie. As soon as the ventral fin appears, a more anterior position can be noticed in the sofie. Pigmentation in the nase also seems to be much more pronounced than in the sofie (Figure 3d,e,f).

Sofie adults are nonguarding rock and gravel spawners with benthic larvae – A.1.3 (Balon 1990); they have the earliest spawning date of any cyprinid in the River Garonne basin, save that of chub, *Leuciscus cephalus* (L.). Adult sofie select spawning sites in small tributary streams to provide the optimal environmental requirements (temperature, dissolved oxygen, etc.) for embryonic development (Kryzhanovsky 1949, Balon 1975). Oxygen concentrations at the spawning sites are generally low, due to a very low water velocity and a high level of suspended organic matter. To assure good oxygen supply to the eggs, they are deposited on boulders in deep pools just downstream of riffles. This differs from the nase, which prefers shallow areas (20–30 cm) with a fairly strong water velocity (Peňáz 1974b, Keckeis et al. 1996).

Sofie larvae are thus exposed to critical discharges due to their position in the water column. This is one of the reasons why we assume that the adult stock migrates to spawn in small tributaries, as the discharge is 20 to 40 times lower (about 5 cm s^{-1}) than in the main channel. Moreover, water temperature in tributary pools is constantly fairly high during the spawning period (Figure 1) and facilitates rapid embryonic development. Providing that temperature variations are not too dramatic, the larvae are fairly tolerant to temperature fluctuations, particularly in the final stages of their development. In nase eggs, however, Peňáz (1974a) found that prolonged temperatures above 10°C can be 100% lethal.

As with dace, *Leuciscus leuciscus* (L.) (Kennedy 1969), chub (Economou et al. 1991) and nase (Peňáz 1974b), the sofie's ova are relatively opaque. Several morphological characters such as absence of melanophores on the yolksac, rapid growth of the operculum, early pigmentation of the eyes (around 134 h after activation), early development of photophobia and the position of the mouth seem to be some traits shared with the nase.

Sofie free embryos have a small yolksac (Figure 3a), which means that the free embryo steps are short, indicating a need to find food rapidly and thus to enter areas rich in plankton. In the wild, food availability does not appear to play a major role in the irregularity of recruitment; firstly, in vitro the larvae are able to withstand a few days without food, and secondly planktonic food is not a limiting factor in these areas of the Garonne basin, especially during the springtime (Eulin & Le Cohu 1998). As a consequence of this low amount of energy reserves, free embryos adapt by hiding in the gravel (photophobia) for a very short time only and become mobile (jerky movements) soon after emergence (presence of mobile pectoral since the 9th embryonic step). Early appearance of certain organs and structures (pectoral fins, eyes, sensory systems, skeletal) presumably enables the free embryos to avoid predators.

If sudden environmental changes appear during the remodelling process, then the recruitment of the new generation of sofie may be affected. Halačka & Lusk (1995) demonstrated that nase in their initial steps of development have a particularly low tolerance to high fluctuations of abiotic factors. The more advanced the development of the eggs is, the more the appearance of new organs enables the embryos to adapt to environmental conditions.

One example of this adaptation is the early appearance and rapid development of the circulatory system (Figure 4). During the 8th embryonic step, sofie blood is still colourless, but a well defined circulatory system is present. The appearance of subintestinal circulation branches on the oblong part of the yolksac occurs at the end of the 8th embryonic step, a development not recorded by Peňáz (1974b) for the nase; this may be used to differentiate the two species as free embryos. This event takes place at the same time as an increase in heart rate, probably in order to answer the needs of a new feeding mode, which starts with the 1st larva step. The rather rigid pectoral fins are often used for propulsion (Videler 1994) but by their vibrations (9th embryonic step) may provide a respiratory function (Kováč 1995).

Another example is flexion of the urostyle. Development of the caudal-body connection and the appearance of the first rays in the dorsal and anal fins (3rd larva step) increases the manoeuvrability of the larvae and corresponds to a change in their habitat use (water column) and social behaviour (creation of a small shoal). Nevertheless, the energetic cost of maintaining position in the water column may slow down a larva's growth (Table 1); this may have more serious consequences if a decrease in food supply occurs at the same time due to changes in environmental factors, such as has been reported for nase (Kamler et al. 1996). In the 3rd larva step, melanophores are more numerous and dark on the sofie's dorsal, whereas the ventral is still translucent white. This is characteristic camouflage pigmentation of pelagic fish. The upward-turning of the urostyle coincides with an increase in larva swimming capacity (Gozlan 1998), permitting the investigation of areas other than the spawning pool. Swimming ability in the sofie increases during the 4th larva step due to filling of the anterior part of the swimbladder. At this time, a large number of larvae were found downstream, near to the tributary connection with the river's main channel (Gozlan et al. 1999 this volume). The larvae's move into the main channel takes some months as they actively swim and do not drift (Gozlan 1998).

In the sofie, changes in mouth shape (form) reflect the phylogenetic position of the species and may be correlated with its changing diet (function) during early life history, as has been demonstrated for nase (Schneider 1992), whose prey spectrum quickly turns from rotifers to micro-crustaceans and chironomids. During the larva period (steps 1 to 8), mouth position in sofie is superior, which suggests a link with planktonic food consumption. During this period, small shoals of young sofie larvae can be found at the surface of pools in small tributaries (Gozlan et al. 1999 this volume). The end of the larva period is characterised by a change from terminal to the typical *Chondrostoma* inferior mouth; at this interval, the fish graze on the surface of rocks, highlighting the co-evolution between form and function of the mouth.

In conclusion, early development in the sofie demonstrates some deviations from that of the nase, in particular with respect to the circulatory system, the position of the ventral fin buds, pigmentation and spawning/incubation habitat. The appearance of adult structures in sofie did not coincide with the stabilisation of relative growth, which occurred much later, and the discussion of the transition from the larva to juvenile periods is addressed in detail by Gozlan et al. 1999 (this volume).

References cited

Abrahams, M.V. 1994. Risk of predation and its influence on the relative competitive abilities of two species of freshwater fishes. Can. J. Fish. Aquat. Sci. 51: 1629–1633.

Alderdice, D.F. 1985. A pragmatic view of early life history studies of fishes. Trans. Amer. Fish. Soc. 114: 445–451.

Balon, E.K. 1975. Terminology of intervals in fish development. J. Fish. Res. Board Can. 32: 1663–1670.

Balon, E.K. 1984. Reflections on some decisive events in the early life of fishes. Trans. Amer. Fish. Soc. 113: 178–185.

Balon, E.K. (ed.) 1985. Early life histories of fishes: new developmental, ecological, and evolutionary perspectives. Dr W. Junk Publishers, Dorchecht. 280 pp.

Balon, E.K. 1990. Epigenesis of an epigeneticist: the development of some alternative concepts of the early ontogeny and evolution of fishes. Guelph Icthyol. Rev. 1: 1–42.

Bell, K.N.L., P. Pepin & J.A. Brown. 1995. Seasonal, inverse cycling of length and age-at-recruitment in the diadromous gobies *Sicydium punctatum* and *Sicydium antillarum* in Dominica, West Indies. Can. J. Fish. Aquat. Sci. 52: 1535–1545.

Beyer, J.E. 1989. Recruitment stability and survival – simple size-specific theory with examples from the early life dynamics of marine fish. Dana 7: 45–147.

Chappaz, R. 1986. Étude piscicole de la retenue de Sainte Croix-Fontaine l'Evêque, croissance-alimentation-reproduction. Thèse de 3ème cycle, Université de Provence, St-Charles, Marseille. 193 pp.

Copp, G.H., J.M. Oliver, M. Peňáz & A.L. Roux. 1991. Juvenile fishes as functional describers of fluvial ecosystem dynamics: applications on the River Rhône, France. Regul. Rivers 6: 135–145.

Crozier, W.W. & G.J.A. Kennedy. 1995. The relationship between a summer fry (0+) abundance index, derived from semi-quantitative electrofishing, and egg deposition of Atlantic salmon, in the River Buch, Northern Ireland. J. Fish Biol. 47: 1055–1062.

Delfau, F. 1979. Étude écologique de la faune piscicole de la moyenne Garonne (entre Agen et Moissac). Diplome d'Étude Approfondie, Université Paul Sabatier, Toulouse. 140 pp.

Economou, A.N., C. Daoulas & T. Psarras. 1991. Growth and morphological development of chub, *Leusciscus cephalus* (L.), during the first year of life. J. Fish Biol. 39: 393–408.

Eulin, A. & R. Le Cohu. 1998. Epilithic diatom communities during the colonization of artificial substrates in the River Garonne (France). Comparaison with natural communities. Archiv. Hydrobiol. 143: 79–106.

Gozlan, R.E. 1998. Environmental biology of the sofie *Chondrostoma toxostoma* (Cypinidae), with emphasis on early development. PhD thesis, University of Hertfordshire, Hatfield. 210 pp.

Gozlan, R.E., G.H. Copp & J.-N. Tourenq. 1999. Comparison of growth plasticity in the laboratory and field, and implications for the onset of juvenile development in sofie, *Chondrostoma toxostoma*. Env. Biol. Fish. 56: 153–165 (this volume).

Halačka, K. & S. Lusk. 1995. Mortality in eggs of nase, *Chondrostoma nasus*, during incubation. Folia Zool. 44: 51–56.

Hoenig, J.M., P. Pepin & W.D. Lawing. 1990. Estimating relative survival rate for two groups of larval fishes from field data: do older larvae survive better than young? U.S. Fish. Bull. 8: 485–491.

Holden, K.K. & M.N. Bruton. 1994. The early ontogeny of the southern mouthbrooder, *Pseudocrenilabrus philander* (Pisces, Cichlidae). Env. Biol. Fish. 41: 311–329.

Houde, E.D. 1994. Differences between marine and freshwater fish larvae: implications for recruitment. ICES J. Mar. Sci. 51: 91–97.

Kamler, E., H. Keckeis & E. Bauer-Nemeschkal. 1996. Egg energy content and partitioning in a rheophilic cyprinid, Chondrostoma nasus (L.). Pol. Arch. Hydrobiol. 43: 273–281.

Kasumyan, A.O. & V.Y. Ponomarev. 1990. The ontogeny of feeding behavior in relation to natural chemical signals in cyprinid fishes. Voprosy Ikhtiologii 30: 447–456 (in Russian).

Keckeis, H., P. Frankiewicz & F. Sciemer. 1996. The importance of inshore areas for spawning nase *Chondrostoma nasus* (Cyprinidae) in free-flowing section of a large river (Danube, Austria). Arch. Hydrobiol. Suppl. 113: 51–64.

Kennedy, M. 1969. Spawning and early development of the dace *Leuciscus leuciscus* (L.). J. Fish Biol. 1: 249–259.

Kováč, V. 1992. Early development of the yellow pope, *Gymnocephalus schraetser*. Folia Zool. 41: 365–377.

Kováč, V. 1995. Reproductive behaviour and early development of the European mudminnow, *Umbra krameri*. Folia Zool. 44: 57–80.

Kryzhanovsky, S.C. 1949. Eco-morphological principles in the development of carps, loaches, and catfishes (Cyprinoidei, Siluroidei). Trudy Inst. Morph. Zhiv. Severtsova 1: 5–332 (in Russian).

Luczkovich, J., S.F. Norton & R.J.G. Gilmore. 1995. The influence of oral anatomy on prey selection during the ontogeny of two percoid fishes, *Lagodon rhomboides* and *Centropomus undecimalis*. Env. Biol. Fish. 44: 79–95.

Lusk, S. 1995. Influence of valley dams on the changes in fish communities inhabiting streams in the Dyje River drainage area. Folia Zool. 44: 45–56.

Maier, K.J., M. Zeh, J. Ortlepp & S. Zbinden. 1995. Distribution et reproduction des espèces du genre *Chondrostoma* en Suisse: le nase (*C. nasus*), la sofie (*C. toxostoma*), la savetta (*C. soetta*). Edited by the Office Fédéral de l'Environnement des Fôrets et du Paysage (O.F.E.F.P.) 53, Bern 53. 62 pp.

Mathias, P. 1921. Étude du genre *Chondrostoma* dans l'Europe occidentale et la région circum-méditerranéenne. Mémoires de la société zoologique de France 28: 1–52.

Osse, J.W.M. 1990. Form changes in fish larvae in relation to changing demands of function. Neth. J. Zool. 40: 362–385.

Osse, J.W.M. & M.R. Drost. 1989. Hydrodynamics and mechanics of fish larvae. Pol. Arch. Hydrobiol. 36: 455–465.

Paris, P. 1932. Faune de la Saône moyenne (2ème note). Cyclostomes et Poissons. Bulletin Scientifique de Bourgogne 2: 47–54.

Peňáz, M. 1974a. Influence of water tempareture on incubation and hatching in *Chondrostoma nasus* (Linneaus, 1758). Zoologické listy 23: 53–59.

Peňáz, M. 1974b. Early development of the nase carp, *Chondrostoma nasus* (Linnaeus, 1758). Zoologické listy 23: 275–288.

Prokeš, M. & M. Peňáz. 1978. The course of spawning, early development and longitudinal growth of the nase carp, *Chondrostoma nasus*, in the Rokytná and Jihlava rivers. Folia Zool. 27: 269–278.

Schiemer, F. & T. Spindler. 1989. Endangered fish species of the River Danube in Austria. Regul. Rivers 4: 397–407.

Schneider, A. 1992. Food availability and diet composition of 0+ *Chondrostoma nasus* in temporally variable environment. Program & Abstract, Ecotones UNESCO Map, Lunz, Vienna.

Spillman, C.J. 1961. Faune de France 65. Poissons d'Eau Douce. Éditions Paul Lechevalier. Paris. 303 pp.

Tourenq, J.N. & R.E. Gozlan. 1997. *Chondrostoma toxostoma*, une espèce en danger. Adour Garonne. Revue de L'Agence de l'Eau 71: 7–10.

Vallot, J.N. 1836. Histoire naturelle des poissons du département de la Côte d'Or. Mémoires de l'Académie des Sciences Arts et belles lettres de Dijon. 236 pp.

Videler, J.J. 1994. Fish swimming. Fish & Fisheries, Chapman & Hall, London. 241 pp.

Environmental Biology of Fishes **56:** 79–92, 1999.

The reproductive biology and early ontogeny of the mouthbrooding Banggai cardinalfish, *Pterapogon kauderni* (Perciformes, Apogonidae)

Alejandro Vagelli
New Jersey State Aquarium, 1 Riverside Drive, Camden, N.J. 08103, U.S.A. (e-mail: avagelli@njaquarium.org)

Received 31 October 1998 Accepted 1 March 1999

Key words: egg transfer, juvenile, precocial, oral incubation, osteology, direct development

Synopsis

Pterapogon kauderni differs from most apogonids in several aspects of its reproductive biology. A breeding pair often allows a secondary male to intervene during mating. Each clutch consists of about 40 eggs 3 mm in diameter, held together by chorionic filaments and is incubated in the buccal pouch of the male for approximately 19 days. *P. kauderni* is the first apogonid in which direct development is described. Embryos hatch at post-flexion stages measuring about 6 mm SL and remain in the male's oral cavity for another ten days. When released as juveniles they measure 8 mm SL, their large yolksac is almost entirely consumed, and they do not pass through a planktonic interval. After release, juveniles do not return to the male's mouth for refuge, and any association between the male and newly released juveniles ceases. Juveniles reach 30 mm TL after four months and mature in another five to seven months. The embryo and juvenile development is presented.

Introduction

The Banggai cardinalfish, *Pterapogon kauderni* Koumans, 1933, has become popular in the pet trade since its 'rediscovery' a few years ago. However, almost nothing is known about the population status of this species, which appears to be restricted to shallows off the Banggai Island, across from Central Sulawesi, in Indonesia. The only published works on *P. kauderni* since its original description (Koumans 1933) are the osteological-systematic work of Fraser (1972) based on one syntype, and the paper by Allen & Steene (1995) that provides some data on the ecology and reproduction of this species.

Soon after *P. kauderni* appeared in the aquarium trade, a breeding program was initiated at the New Jersey State Aquarium, with the goal of rearing this species in captivity and studying its reproductive biology and ontogeny. Owing to its restricted habitat and low fecundity, *P. kauderni* populations were suspected to be at risk if collecting for the aquarium trade persisted. This paper describes the reproductive biology of *P. kauderni* under laboratory conditions, focusing on mating behavior, embryo and juvenile development, and the behavioral association between juveniles and their male parent in this mouthbrooding species.

Material and methods

Holding system and general conditions

Fifteen specimens were commercially obtained and maintained in a 1170 l recirculating system consisting of two 400 l fiberglass tanks (119 × 58 × 58 cm deep) with a glass front, a 40 l conical tank, a 330 l sump (66 × 66 × 76 cm deep) and a 1/3 HP Doer TE-5.5MD-SC pump. The bottoms of the two 400 l tanks were covered with a 1 cm thick layer of crushed coral and sand. Several rocks, with plastic eel grass attached, and several live sea urchins *Diadema* sp. were added to serve as hiding places. Filtration consisted of a biofilter made with a 20 l bag filled with plastic media (bio-rings) submerged into the sump, an Aquanetics PQ15IL 15 w ultraviolet filter and a 10 l bag filled with activated carbon. Each 400 l tank was illuminated by two 40 w fluorescent tubes placed 20 cm above water level. A photoperiod of 12 h : 12 h (6:00–18:00 h light) was maintained. A 500 w Computherm heater kept the temperature at $25 \pm 0.5°C$. Moderate aeration was provided in both tanks. Artificial sea water, made at the aquarium, was used and a salinity range between 34–36‰ was maintained throughout the study. The pH varied between 7.8 and 8, while nitrites and ammonia remained at $0 \, mg \, l^{-1}$. Daily feeding (ad lib.) consisted of chopped mussels *Perna canaliculus* and krill *Euphasia pacifica* and *E. superba*. Once a breeding pair was established, the pair was isolated in one of the 400 l tanks to facilitate the collection of eggs used in the study. The rest of the fish were kept together in the second 400 l tank for observations on reproductive and social behavior.

Gametes and gonadal data

To determine the size and general morphology of male gametes, sperm were extracted by slightly pressuring its abdominal area while a 5 cm diameter petri dish containing sea water was placed in contact with the urogenital papilla. Spermatozoa were measured with an ocular micrometer mounted on an Olympus BH2 microscope with a 100× oil immersion lens. Photographs were taken with a 35 mm camera mounted on it. Ovaries from five females (2.5–5.0 cm SL) were obtained to calculate fecundity, gonadosomatic indices (GSIs) and to compare oocyte development among individuals of different sizes. All mature (>2.5 mm) and maturing oocytes (0.6–0.8 mm) were counted. Ten mature oocytes were measured to the nearest 0.1 mm utilizing polyester micrometer cover slips (18 × 18 mm) with subdivisions of 0.05 mm and weighed to the nearest 1 mg using a Metler AE 2000 analytical scale with a readability of 0.1 mg. The GSIs were calculated from the equation $W_1 W_2^{-1} \times 100$, where W_1 = weight of the gonad and W_2 = total body weight (Crim & Glebe 1990). All specimens were euthanized with MS222 (tricaine methane sulfonate).

Mating behavior and collection of eggs

Observations on courtship displays, egg transfers, and brooding of egg masses by the male were done over a twelve-month period during which more than twenty matings occurred. As soon as a female showed signs of maturity (enlarged pelvic region) and started separating a male from the other individuals, the pair's behavior was observed daily for several hours, generally from 10:00 to 15:00 h, until mating occurred. A compact VHS camcorder was used to record mating behavior and the transfer of eggs.

To obtain eggs for embryological studies, the brooding males were gently introduced into a large vinyl net immediately after the egg transfer. This sometimes induced the expulsion of the eggs, but on other occasions it was necessary to apply a little pressure on the opercular area to induce the release of the egg cluster.

To observe embryonic development, several eggs were separated from the cluster by cutting the chorionic filaments with tweezers and dissecting scissors. The egg clusters and lose eggs were placed into a 4 l beaker, which was submerged almost completely in the 40 l conical incubation tank, connected to the main system. The flow inside the beaker was provided by an 8 mm diameter plastic hose, and was strong enough to move the eggs gently around the bottom of the beaker. The incubation tank was covered to avoid direct light on the eggs and aeration was not provided. Eggs were checked daily and dead embryos were removed.

Description of early ontogeny

Following the life-history model of Balon (1975, 1990), early ontogeny was divided into two periods, i.e. *embryo*, which includes a series of three developmental phases: cleavage-egg, embryo and eleutheroembryo, and *juvenile*. The *cleavage-egg* phase begins with the activation of the egg and ends with the appearance of the embryonic shield. The *embryo* phase was divided into 14 chronological states (Crawford & Balon 1994) corresponding to the days of development 2, 3, 4, 5, 6, 7, 8, 9, 11, 12, 13, 15, 17, 19. This scheme was based on the appearance of the more conspicuous morphological structures. For embryological description, three to four eggs were sampled every 24 h during the 19 days of incubation. This sequence was repeated four times with eggs from four different females. The *eleutheroembryo* phase includes the time from hatching to release from male's mouth. The *juvenile* period starts at release (beginning of exogenous feeding) and ends at sexual maturity.

Egg and embryo development

Embryonic development was studied with an Olympus SZH (7.5–64×) dissection microscope and documented photographically using an Olympus PC 35 mm camera. Diameters of 15 eggs from 5 different females and embryos were measured to the nearest 0.1 mm using cover slips and a Vernier caliper with the same precision. Measurements of perivitelline space, and diameter of oil globules were made with a micrometer mounted on the dissecting microscope. Eggs were individually weighed (wet) to the nearest 1 mg. Embryos and juveniles were cleared and stained following the technique of Potthoff (1984). Description of early ontogeny was made on the basis of live specimens and microphotographs.

Free embryos and juveniles

After hatching, embryos were kept in a beaker until they were able to swim and to feed on exogenous food. They were then transferred to a separate holding tank. Juveniles were fed daily (ad lib.) with *Artemia salina* and were switched onto frozen food (krill) at about two months of age.

Results

Reproduction

Fecundity

Ovaries from the three larger females (3.5, 4.8 and 5.0 cm SL), showed three size classes of oocytes, (1) *previtellogenic oocytes* of about 0.1–0.3 mm in diameter, (2) *maturing oocytes* of 0.6–0.8 mm and (3) *mature ova* between 2.2–2.7 mm. Oocytes less than 2.5 mm are polyhedral (Figure 1) whereas oocytes larger than 2.5 mm are almost spheric and have smooth surfaces. The mean diameter and weight of 10 mature ooctyes from the 4.8 cm SL female were 2.7 mm (SD = 0.1) and 137.8 mg (SD = 11.7), respectively. There does not appear to be a correlation between SL and fecundity (in both size classes of vitellogenic oocytes). The ovaries from the juvenile females (2.4 and 2.6 cm SL) contained mostly previtellogenic oocytes of 0.1 mm in diameter or less, and few maturing oocytes of 0.35–0.5 mm (Table 1).

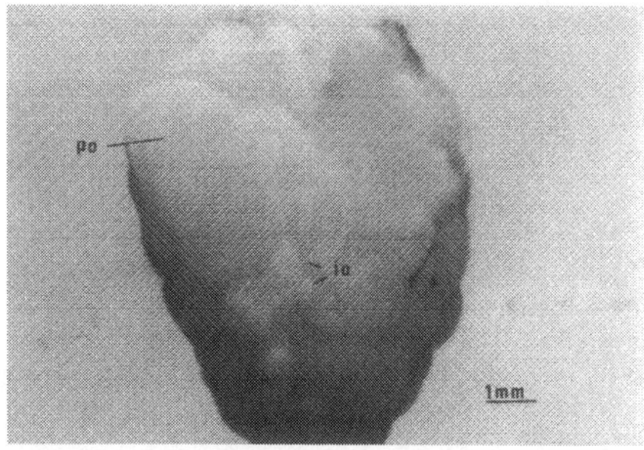

Figure 1. Ovary from a female 3.5 cm SL [po = polyhedral oocytes of about 2.2 mm. Oocytes of this size adhered very tightly to each other giving the ovary a very compacted structure. Immature ova (io) of about 0.8 mm are surrounded by the larger oocytes].

Table 1. Total number and size of vitellogenic ova found in the ovaries of five females with standard lengths (SL) 2.4–5.0 cm.

SL (cm)	2.4	2.6	3.5	4.8	5.0
Maturing oocytes (0.6–08 mm)	4	6	62	65	59
Mature preovulatory ova (2.2–2.7 mm)	0	0	60	57	62
G.I.	1.5	4.37		10.1	
Weight (g)	0.6	0.8		5.05	

The data from the female 4.8 cm and gonadosomatic index (G.I.) of 10.1 was obtained after she was displaying prespawning behavior for several hours.

Prespawning behavior

P. kauderni is a mouthbrooder in which transient pairs are formed and only the males incubate the eggs. Like other mouthbrooding apogonids, its reproduction involves an elaborate courtship that lasts normally several hours, but sometimes two to three days. Prespawning behavior comprised several courtship displays, some of which were similar to those described in several species of *Apogon* (Kuwamura 1983, 1985) and *Cheilodipterus* (Kuwamura 1987), i.e. female trembling, warping, side to side swimming, nuzzling and mouth-opening by the male.

Prespawning behavior was initiated also in this species by the female, who, after choosing a potential mate, isolated him from other individuals by creating a spherical spawning site of about 50–60 cm in diameter. This virtual site was generally maintained in one of the tank's halves, although it was 'carried out' by

Figure 2. 'Side by side trembling'. Female (left) and male anal and caudal fins in contact.

Figure 3. Male (in front) displaying 'mouth opening' behavior.

the pair wherever they were located. The most characteristic prespawning display was the 'side by side trembling' by the female; while the male remained almost motionless, the female, trembling vigorously, approached the male from behind, placed herself alongside him and started inclining outwards from her body's vertical plane until reaching an angle of about 30°. At this position the male and female's caudal and anal fins came into contact (Figure 2). Concomitantly, the female pushed the male sideways while he remained non-responsive to the female's display. This behavior lasted for about two or three seconds after which the female stopped trembling and moved away from the male 2–3 cm but immediately came back, trembling and pushing again. After repeating this behavior two or three times against one side of the male, the female swam backwards until she placed herself behind the male and started the same display against his other side.

'Side by side trembling' normally lasted several hours, but it was frequently interrupted by other individuals who approached the pair. This caused an immediate response, almost always by the female, consisting of a rapid and aggressive chase of the intruders out of the home site. 'Side by side trembling' occurred at different depths in tanks but it was generally observed near the bottom. The only sign of a male's receptiveness was a conspicuous darkness along the lower jaw and sporadic 'mouth opening' (Figure 3).

Spawning and eggs transfer

Pair formation and courtship usually began around 9:00–10:00 h and all observed matings occurred between 13:00 and 15:30 h. *P. kauderni* reproduced all year long under laboratory conditions. Signs of imminent spawning were an increase in the frequency of female approaches during 'side by side trembling', occasional trembling and opening of the male's mouth, and the pair's proximity to the bottom. Egg release occurred when both individuals were situated side by side, separated about 1 to 2 cm and a few centimeters from the bottom. Occasionally, at the moment of the egg release, the male was located a few centimeters ahead of the female. When about 3/4 of the egg mass protruded from the female urogenital papilla (which took between one and one and one half seconds) the male immediately turned around and gulped the clutch by pulling the eggs from the female (Figures 4, 5). The extraction of the eggs by the male required some amount of pulling and always some eggs were cut off from the clutch and remained attached to the female for a few seconds until they fell off. Before these loose eggs reached the bottom of the tank, they were eaten by other individuals. The male never attempted to recover them. Egg transfer was completed in no more than two seconds.

After the female chose and separated a receptive male from the rest of the school, the pair established a home site that was aggressively defended. However, often a secondary male (SM) was allowed to intervene in the courtship. Unlike other approaching individuals, which were immediately chased, the SM was allowed to approach the pair at a distance of about 5–6 cm. The SM followed the pair with a submissive posture, moved away from the pair and maintained its distance when the pair swam in his direction. The SM soon started to participate in defense of the site, alone or with one member of the pair. The SM displayed trembling behavior (rarely observed in the primary male (PM) and mouth opening to the female. When the PM responded to the

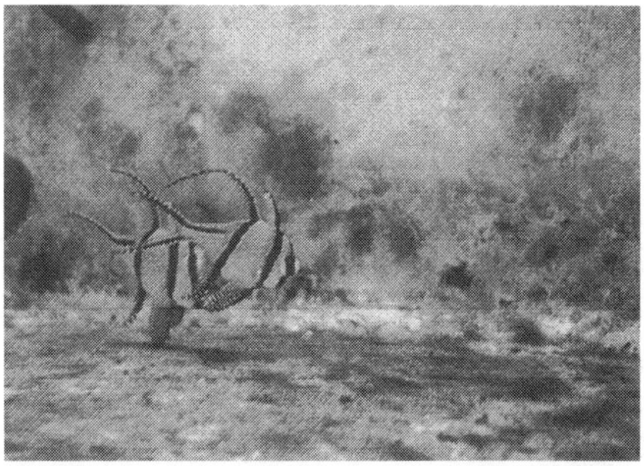

Figure 4. Male (closer to the bottom) at the moment of gulping the egg clutch.

Figure 5. Male (closer to the bottom) showing the enlarged buccal pouch immediately after the egg transfer.

female's displays by remaining close to her and swimming to the bottom, the PM mated and took the eggs while the SM did not intervene. However, when the PM did not show interest in the female's displays (by swimming away and remaining at the middle-top section of the tank for extended periods), the female started periodically leaving the PM and displaying mating behavior to the SM. The SM always responded to this by swimming to the bottom and frequently trembling. In six of the nine egg-transfer events observed, a SM was allowed to intervene, and on three occasions it was the SM that actually mated and took the eggs.[1]

Figure 6. Brooding male. The egg cluster occupies the entire buccal pouch.

Postspawning behavior

The female displayed three postspawning behaviors: (1) aggressive home site defense that included the immediate chase of any individual approaching the brooding male, (2) a more relaxed 'side by side trembling' (less frequent and less sustained), and (3) confining the male to a small space, generally against a rock or a side of the tank and always a few centimeters off the bottom. Such behaviors did not last more than twenty to thirty minutes.

During the first five to ten minutes after the egg transfer, the male frequently opened his mouth and rotated the egg mass, which was partially exposed, causing other individuals to approach and harass the male with clear intentions of eating the eggs. The brooding male, stopped eating and refused to take any food until the juveniles were released, approximately 30 days later (Figure 6).

Fertility

Four factors appeared to affect fertility: (1) In all six observed matings, between 10 and 20 eggs were lost during the egg transfers. (2) All five egg clusters extracted from the male's mouth within an hour after egg transfer, contained between three and six eggs that had not started cleavage (unfertilized eggs) or did not develop beyond the blastodisc interval. (3) Observations on eight males isolated in all-glass 50 l tanks after mating showed that between five to ten eggs (dead embryos) were expelled during the incubation period, particularly during the first ten days. (4) A high

[1] Kuwamura (1985) observed in the wild the presence of a SM of *Apogon niger* in a pair's home site (on two different occasions with two different pairs), however, it occurred 1 and 3 days after the female mated with the PM. In addition, a female 'was observed courting (but not mating) two different males successively in the same afternoon'.

percentage (60%) of the clutches produced in the community tank were spit out or swallowed by the brooding male. The probable reason for this is the aggressive harassment of the brooding male by other individuals, particularly during the first week of incubation.

Under laboratory conditions a fertility rate of about 40 to 60% is expected for an isolated pair and a rate of about 15 to 25% is expected in a community tank. The highest number of released juveniles by one male was 62 and the highest frequency of egg deposition by one female was one deposition every 25 to 30 days. Thus, a maximum fertility of about 750 juveniles would be expected per female per year assuming the female would spawn every month, would find a mate every time she is ready to spawn, and all its male partners incubate all the eggs and all the embryos successfully.

Embryo and juvenile development

Gametes

Ovulated oocytes measured between 2.7 and 3.0 mm in diameter. An average clutch consisted of 40 eggs that formed a round mass 1.5 cm in diameter (Figure 7). The eggs were held together by strong filaments that originated from a circular area of 0.5 mm in diameter on the chorion, and extended outwards entangling with filaments from other eggs. The eggs (Figures 8, 9) were generally spherical and, at release, were bright yellow-orange with a yolk full of small lipid globules. The yolk had a pyriform shape with a maximum vertical length of about 2.2 mm and a width of 2.8 mm. Two size classes of oil globules were present, one between 80 and 112 μm in diameter formed a cup-like aggregate in the upper part of the egg of approximately 2.6 × 0.8 mm height. The second size class consisted of a few scattered and very small oil globules of about 32 μm located below the large aggregate at about the egg's equator. The perivitelline space was widest at the animal pole, where it measured between 0.6–0.8 mm (20 min after release). At the bottom of the egg, the perivitelline space was about 0.1 mm. The spermatozoa were highly motile with an elongated head 4 μm long and a tail of 16 μm long.

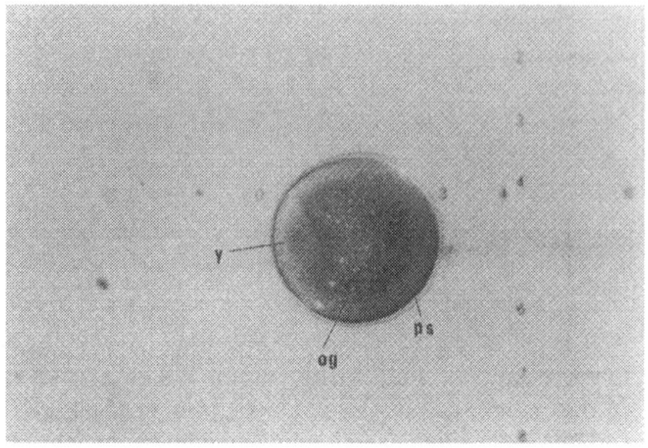

Figure 8. Egg just after activation (dorsal perspective). The perivitelline space (ps) is larger at the dorsal part of the egg. The viscous yolk (y) contains many oil globules (og). The scales are in mm.

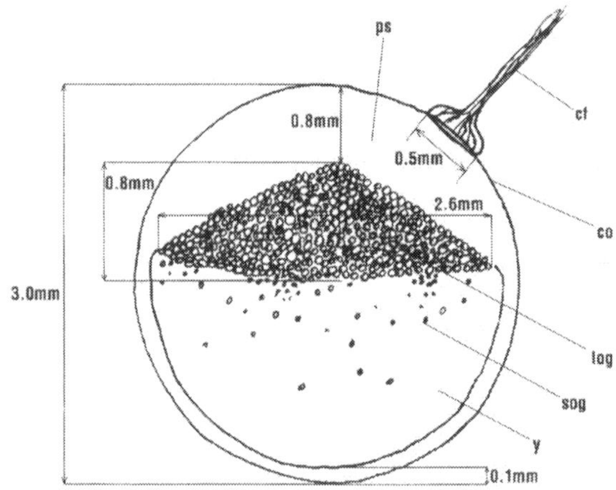

Figure 9. Egg before cleavage (lateral perspective) [log = large oil globules (80–112 μm), sog = small oil globules (32 μm), y = yolk, ps = perivitelline space, co = envelope, cf = chorionic filaments].

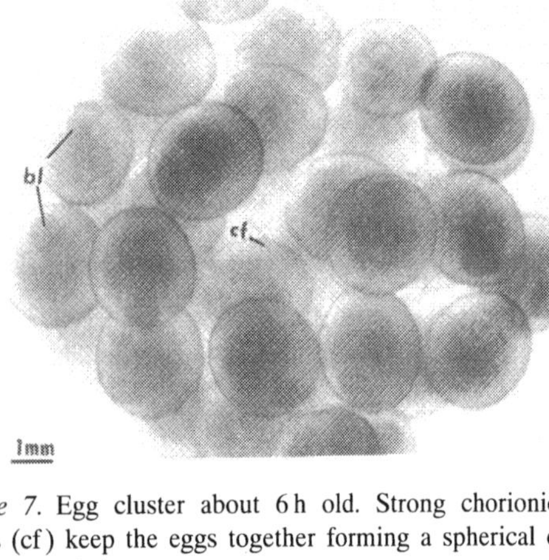

Figure 7. Egg cluster about 6 h old. Strong chorionic filaments (cf) keep the eggs together forming a spherical cluster (bl = blastodisc).

Summary of early ontogeny
Embryo period.
Phase: cleavage egg. – Cleavage was meroblastic and discoidal, and started at 3 to 4 h after activation and formation of the perivitelline space. About 6 h after activation the blastodisc consisted of 8 cells. At 24 h, the germ ring was visible, although at the animal pole the blastodisc remained high and was very convex.
Phase: embryo. – A series of chronological states was defined as a number of days (d) after activation.

2d: The germ ring covered up to 60% of the yolk. It was already possible to distinguish the beginning of lateral outgrowths on the head that represent the future optic vesicles. The embryo spanned the entire perivitelline space.

3d: Twelve to 15 somites were present; the notochord was evident. One third of the yolk was full of lipid droplets; a prominent head fold was present and optic vesicles were well differentiated. Primordia of the otic vesicles were visible.

4d: Thirty somites were visible. The most caudal 10 were not attached to the yolk forming a free tail-bud. Two otoliths were apparent in each otic vesicle and the eye lenses were visible. On the ventral side of the head the heart appeared as an unchambered conical tube beneath the lenses that extended rostrally to a distance equivalent to the optic vesicle diameter (Figure 10). The blood cells were not pigmented, and although no circulation was discernible at this state, the heart was beating at 110 times per minute.

5d: A common finfold was visible extending without indentations from the back of the head to the anus (Figure 11). The atrium and ventricle of the heart were differentiated, and the heart beat 140 times per minute. Blood circulation in the yolk was evident. Numerous exposed neuromasts covered the head and trunk without an apparent pattern of distribution. The olfactory vesicles appeared, and parachordals and trabeculae were the first elements of the chondrocranium to develop.

6d: The length of the embryo (2.5 mm) was equal to the yolk diameter, which was absorbed very little. Iridocytes appeared on the dorsal surface of the eyes.

7d: Blood was pigmented and the anterior vitelline veins and the sinus venosus were formed. The pectoral fin buds were present and melanophores appeared on the orbits.

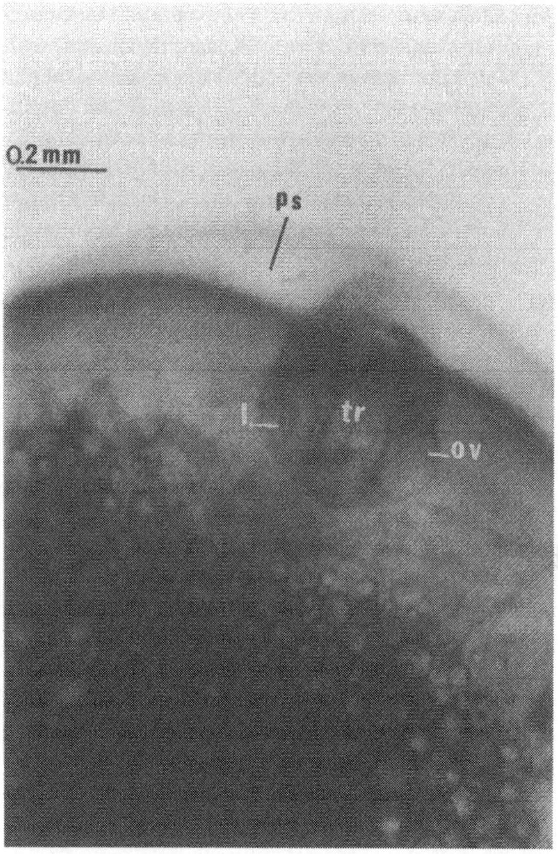

Figure 10. Embryo four days old (l = lens, ov = optic vesicle, ps = perivitelline space, tr = trabecula).

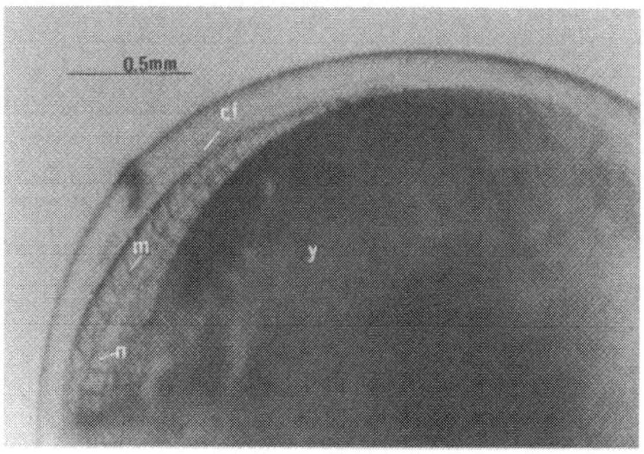

Figure 11. Embryo five days old (cf = common fin fold, m = myosepta, n = notochord, y = yolk).

8d: Blood circulation increased notably, it was visible in two branchial arches and in the eyes. The course of the main vitelline circulation on the yolk developed a distinct inverted-heart shape.

9d: The embryo measured 4 mm SL and yolk diameter was 2 mm. The first punctiform and scattered chromatophores appeared on the dorsal and ventral sides of the body. The circulation continued to increase with a large network of capillaries extending from the caudal vein particularly at the future location of the liver. The left anterior cardinal vein split in two before joining the right and posterior vitelline veins. Four to five fin rays started to develop in the ventral portion of the caudal fin fold. Several elements of the splachnocranium, including Meckel's cartilage, the ceratohyals and three ceratobranchials were present (Figures 12, 13).

11d: There was an increase in the pigmentation, chromatophores and iridophores, of the eyes. The urinary and swim bladders were visible. Many capillaries appeared on the future liver location and circulation expanded on the head and eyes. The heart was beating at 190 beats m^{-1}.

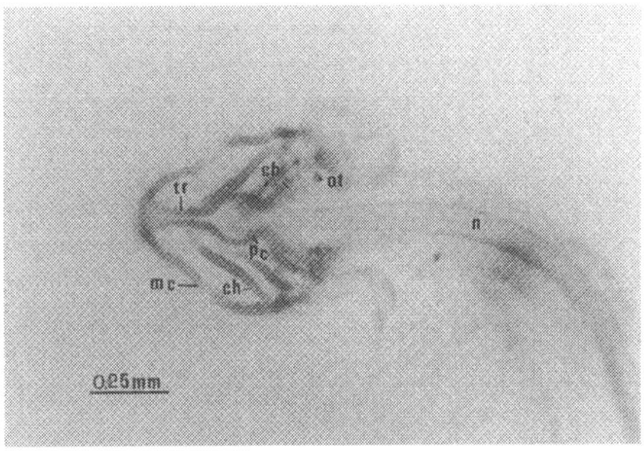

Figure 12. Embryo nine days old cleared and stained with alcian blue to show early chondrification (mc = Meckel's cartilage, n = notochord, tr = trabecula, pc = parachordal, ch = ceratohyal, cb = ceratobranchial, ot = otolith).

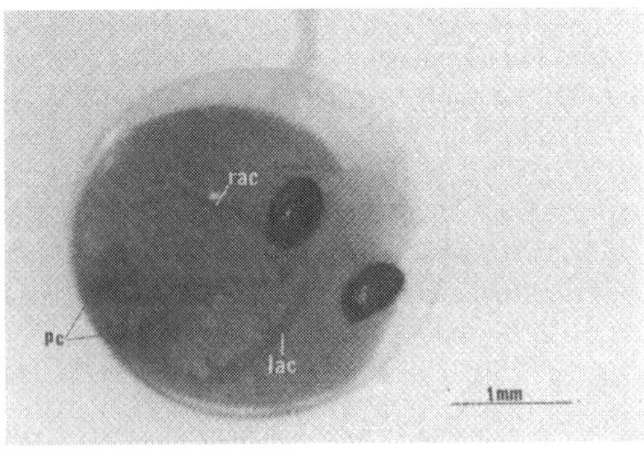

Figure 13. Embryo nine days old with characteristic inverted-heart shape vitelline circulation (lac = left anterior cardinal vein, rac = right anterior vitellin vein, pc = posterior cardinal veins).

12d: The flexion interval was reached; four hypurals (1–4) appeared. The liver and gall bladder were visible and the pigmentation on the body was represented by a few chromatophores (mostly erythrophores and some melanophores), some of which were clustered behind the head, and on the external dorsal surface of the swimbladder.

13d: At 4.7 mm SL, asymmetry developed on the anterior vitelline circulation. The left anterior cardinal vein increased notably in diameter while the right vein appeared very reduced. This asymmetry remained until the yolk was absorbed. The inverted-heart shape of the vitelline circulation disappeared. Chondrification of the neurocranium included the ethmoid plate, trabeculae, taeniae marginales, epiphysial bar and the otic capsulae. In the splanchnocranium, the cartilaginous palatoquadrate and hyosymplectic were well developed, but the maxillae and premaxillae stained slightly with alcian blue and looked almost transparent. All hyoid and branchial elements were chondrified, but no branchiostegal rays were developed. A fused coraco-scapular cartilage was connected to a long and thin cleithrum in the pectoral girdle. Only two proximal radials (still posteriorly connected) were formed. Two very small cartilage bars (separated by the vitelline sac) represented the future basipterygia of the pelvic girdle. In the axial skeleton, vertebrae were not yet chondrified, although superficial constrictions and blue stained segmentary areas were present along the notochord. Also, neural and haemal arches were present. Caudally, hypurals 1 and 2 started to fuse, 3 and 4 remained separated, and 5 was not yet visible. The first caudal fin rays were present, and the dorsal and anal fins started to develop proximal elements of pterygiophores. The embryo was capable of moving its pectoral fins (Figures 14, 15).

15d: A pigmented line was present on the dorsal internal surface of the body cavity, extending from the first to the sixth vertebra. In addition, several dark spots appeared on the dorsal surface of the head. There was a large increase in the network of capillaries in the yolk and in the hepatic portal system. A segmented series of vessels was visible between the developing vertebrae. The proximal

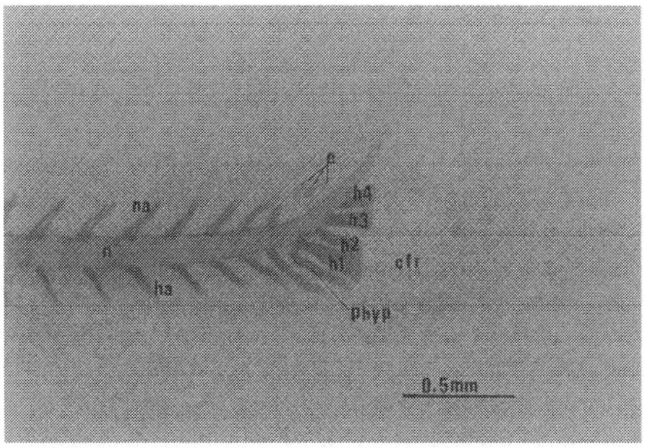

Figure 14. Caudal skeleton of an embryo 13 days old (n = notochord, na = neural arch, ha = haemal arch, h 1,2,3,4 = hypurals, e = epurals, phyp = parhypural, cfr = caudal fin rays).

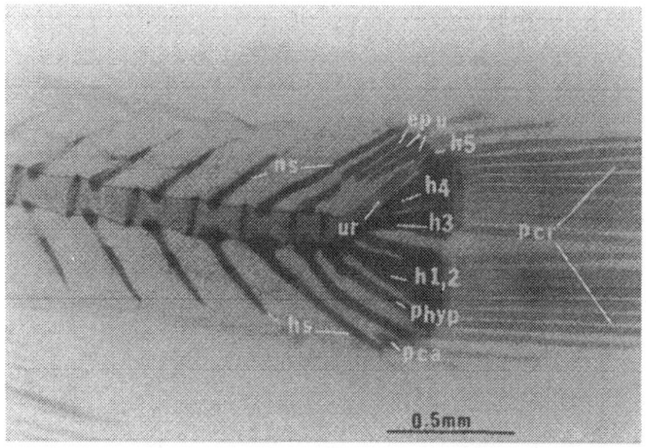

Figure 16. Caudal skeleton of an embryo 15 days old (epu = epural, hyp = hypural, hs = haemal spine, ns = neural spine, pca = procurrent cartilage, pcr = principal caudal rays, phyp = parhypural, ur = urostyle).

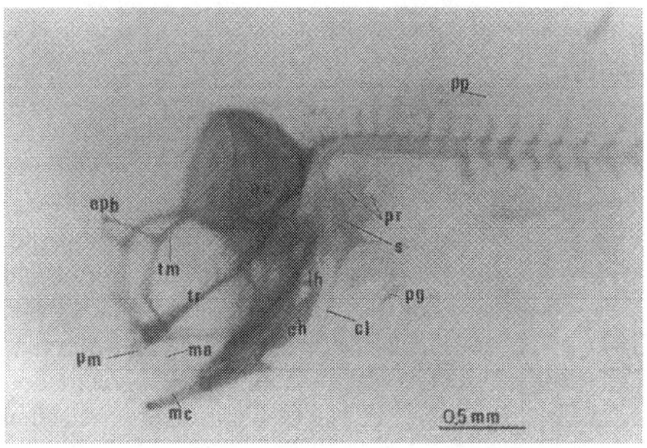

Figure 15. Embryo 13 days old (ac = auditory capsule, ch = ceratohyal, cl = cleithrum, epb = epiphyseal bar, ih = interhyal, ma = maxilla, pg = pelvic girdle, pm = premaxilla, pp = proximal elements of second dorsal fin pterygiophores, pr = proximal radial, s = coraco-scapular cartilage, tm = taenia marginalis, tr = trabecula).

and distal elements of the pterygiophores of the second dorsal and anal fins were formed. The pectoral girdle had four proximal radials, and about ten distal radials started developing. Vertebral elements started chondrifying, and blue stained rings appeared at the ends of each vertebra. A parhypural and three short epurals attached to a long ural centrum were visible on the hypural complex. Hypural 5 appeared (Figure16). Branchiostegal rays were attached to the ceratohyal and epihyal and were very lightly stained with alcian blue. The first rudiments of gillrakers appeared on all ceratobranchials. All fins presented unsegmented and unbranched rays, except the first dorsal fin, which developed only the first proximal elements. A few teeth were visible on the premaxilla and dentary.

17d: Melanophores were concentrated on both sides of the posterior upper part of the head forming two elongate blotches, and red pigment appeared in segmentary disposition among the myosepta. The yolk occupied about half the egg's volume. The circulation increased, particularly in the liver, where a dark-green gallbladder was very conspicuous.

19d: More advanced pigmentation was visible. On the head, many stellate melanophores formed two large and dark symmetrical areas, extending from almost the midline of the head to the eyes. A large concentration of pigment was present just behind the head, forming a ring that extended ventrally to the pectoral fin base. Pigmentation was also present on the body's midline, along the ventral side of the vertebral column, primarily in the caudal region and appeared segmented. Pigment was also visible on the base and rays of the pelvic fins. In the branchial region, several teeth started developing on ceratobranchials 5 and small tooth plates appeared on pharyngobranchials 2, 3 and 4.

Phase: eleutheroembryo. – *P. kauderni* hatched after 19–20 days of incubation at a postflexion state, and at 5.0 to 6.0 mm SL. Newly hatched embryos had a large bilobulated yolk and remained in the male's oral

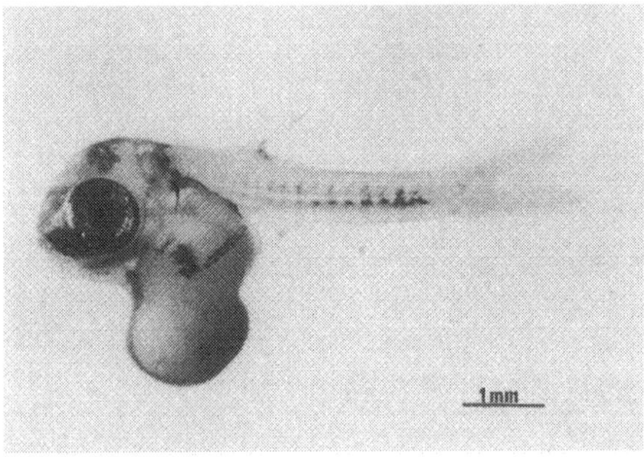

Figure 17. Newly hatched embryo (eleutheroembryo) showing a large bilobulated yolk and the beginning of juvenile pigmentation.

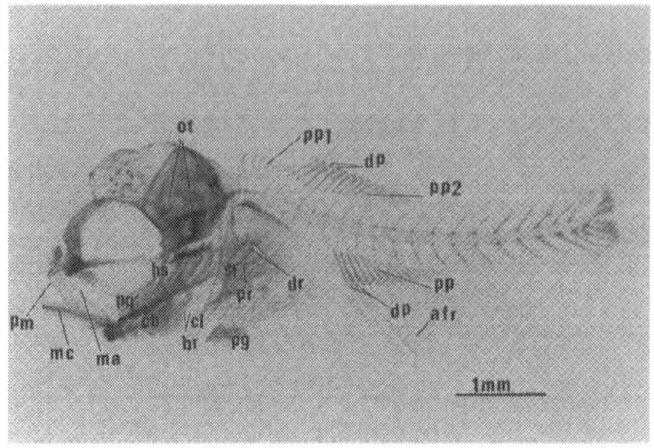

Figure 18. Embryo 19 days old (newly hatched) (afr = anal fin rays, br = branchiostegal rays, ch = ceratoyal, cl = cleithrum, dp = distal elements of second dorsal and anal fin pterygiophores, dr = distal radials, ma = maxilla, mc = Mekel's cartilage, hs = hyosymplectic, pg = pelvic girdle, pp = proximal elements of the anal fin pterygiophores, pp1, pp2 = proximal elements of the first and second dorsal fin pterygiophores, pq = palato-quadrate, sf = scapular foramen, ot = otoliths).

cavity feeding endogenously until released (Figure 17). In addition to the pigment distribution present at 19d, melanophores were also visible on several rays of the second dorsal fin. Iridophores covered some areas of the opercle, preopercle, and the upper section of the yolk, dorsal to the pelvic fins. At hatching, all principal caudal rays were present, although the hypurals remained separated. All basal elements and rays of the anal, pelvic, pectoral and second dorsal fins were present, and all proximal and few distal pterygiophores were formed on the first dorsal, although no fin rays were developed yet. The vertebrae were still in the process of chondrification (Figure 18). During the next several days, pigmentation increased very rapidly and two black vertical bands were completed. The first band ran from the dorsal surface of the head to the base of the eyes and covered the entire diameter of the lenses. The second band, encircled the body at the location of the pelvic fins. Pigmentation was also present along the anal and caudal fin rays.

Juvenile period.
The brooding male incubated the eleutheroembryos for a maximum of ten days. The release of most juveniles occurred between the 6th and 10th day after hatching, however, sometimes eleutheroembryos were expelled during the first few days. These 'premature juveniles' were not capable of swimming and generally showed signs of slow development. At normal release, juveniles were of 8 mm SL and their yolk was almost absorbed. By this age (approximately 30 days after activation) ceratobranchials 5 beared two rows of teeth, and tooth plates were well developed on pharyngobranchials 2, 3 and 4. With the reduction of the yolksac, the pelvic girdle finally joined the cleithrum. All proximal radials of the pectoral fins were separated and short rays appeared on the first dorsal fin. On the axial skeleton, the vertebrae were still not totally differentiated. Hypurals 3 and 4 were still separated, and 1 and 2 had not completed the fusion process. No ossification was apparent. However, several elements started developing, e.g., the vomer (still toothless), parasphenoid and ectopterygoids were present. Other structures had been present for several days (the opercular series, the otic, mandibular and maxillar elements), but did not retain alizarin red.

Immediately after release, juveniles swam together forming a small school and were capable of eating items as large as *Artemia* nauplii. At 1.8 cm SL the caudal skeleton was completed (Figure 19) and most of the neurocranium and splanchnocranium elements were calcified. Reproductive maturity was reached at approximately nine months at a SL of 3.5 cm.

Discussion

Some comparative aspects between P. kauderni *and others apogonids*

Apogonidae is one of several families of fishes in which oral incubation occurs (Breder & Rosen 1966, Oppenheimer 1970). Among the 22 valid genera of apogonids (Nelson 1994), 13 (about 50 species) are

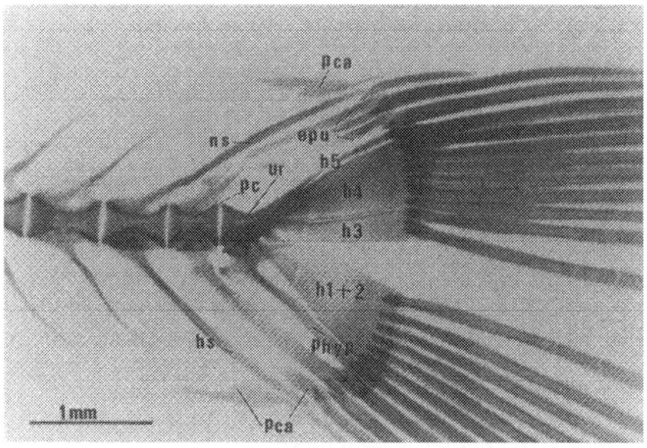

Figure 19. Caudal skeleton of a juvenile 18 mm SL (epu = epurals, the hypurals one and two (h1 + 2) are already fused, hypurals three and four (h3, h4) are still separated, h5 = hypural five, hs = haemal spine, ns=neural spine, pca = procurrent cartilage, pc = pleural centrum, phyp = parhypural, ur = urostyle).

Table 2. Genera of Apogonidae recorded as mouthbrooders.

Genus	Reference
Apogon	Chrystal et al. 1985, Ebina 1932, Garnaud 1950, Kuwamura 1985, Neira 1991, Smith et al. 1971
Apogonichthys	Petit 1931
Archamia	Lachner 1951
Astrapogon	Fraser 1972
Cheilodipterus	Fishelson 1970, Kuwamura 1987
Foa	Petit 1931
Glosamia	Whitley 1959
Phaeoptyx	Thresher 1984
Rhabdamia	Fowler & Bean 1930
Pterapogon	Allen 1993
Siphamia	Tominaga 1964
Spharaemia	Allen 1975
Vicentia	Hale 1947

described as mouthbrooders (Table 2). The Banggai cardinalfish, *P. kauderni*, differs from most apogonids in several aspects of its reproduction: it has the lowest recorded fecundity and its eggs, about 3 mm in diameter, are much larger than the eggs of most (usually less than 1mm) other apogonid species (Barlow 1981 after Chave 1971, Charney 1976, Allen 1993). Also, the clutch size of about 40 eggs differs from most apogonids which consist of hundreds or even thousands of eggs (Garnaud 1962, Neira 1991, Allen 1993). While recorded sizes of newly hatched embryos range from 1 mm for *Apogon affinis* (Smith et al. 1971) to 3.3 mm for *Spharaemia orbicularis* (Allen 1975), newly hatched embryos of *P. kauderni* measure about 6 mm SL (Table 3). Another aspect of the reproduction of *P. kauderni* that seems unique is the participation of a secondary male during courtship and mating. Yet, the most distinctive features of *P. kauderni* 's ontogeny are the retention of free embryos in the male's oral cavity for several days and the complete lack of a planktonic phase.[2]

Thus, *P. kauderni* is the first example of an apogonid with direct development. However, a comparison between the reproductive biology of *P. kauderni* and other apogonid species, raises the question of whether other apogonids may be precocial, and/or, incubate the free embryos. For example, although the eggs of *Apogon rueppelii* are reported to measure between 2.2 and 2.5 mm in diameter (Neira 1991), there is a disagreement on the reported size at which embryos of this species hatch. Chrystal et al. (1985) stated that after 16 days, embryos hatch at 7.8 mm, while Neira (1991) reported the size at hatching to be between 5.5 and 6.4 mm.

It is not clear whether the 'newly hatched larvae' described by Neira (1991) were obtained only by removing them from late stage eggs or from surface plankton samples also. In Chrystal et al.'s (1985) description, 'newly hatched larvae' are referred to as larvae expelled by males kept in aquaria. With respect to the 16 day period of incubation, this refers to the number of days it took males to expel the 'larvae', although the males in the mentioned study were already brooding when collected. Neira (1991) reported that *A. rueppelii* hatches at a postflexion 'stage' with a functional mouth and the elements of the second dorsal, anal and caudal fins completely formed. It is during the 'short planktonic life' when *A. rueppellii* starts to form the pectoral fin rays, the pelvic fins buds (complete by 7.7 mm) and the first dorsal fin; the remains of the yolksac are reabsorbed by 6.8 mm, and settlement occurs at sizes over 16 mm (Neira 1991). It seems more likely that this species, which produces eggs smaller than *P. kauderni*, would hatch at the 5.5–6.4 mm SL rather than at 7.8 mm.

[2] Ebina (1932, p.21) stated that 'fry' of *Apogon semilineatus* 'are still found in the mouth of adult for some time after hatching' A drawing of a newly hatched 'fry' shows it with a typical morphology of a planktonic-type larva, i.e. no fins developed (common finfold), preflexion state, very large eyes and mouth and very small yolksac. His study was based on material that came from 'a haul of a fishing net in their habitat' and no further comments nor data are given to support his statement with respect to 'fry' being retained 'for some time' in the male's mouth.

Table 3. Reproductive biology. Comparison among species of mouthbrooding apogonids for which data are available.

Species	Size of brooder TL (mm)	Eggs per clutch	Egg diameter (mm)	Time to hatching (days)	Size at hatching TL (mm)	Source
Apogon affinis	54.7–87.5	21000	0.35–0.40	—	1.0	Smith et al. (1971)
Apogon erythrinus	43.0	2600	0.49	—	—	Barlow 1981 (after Chave 1971)
Apogon imberbis	—	22137	0.5	8	2.0	Garnaud (1962)
Apogon lineatus	53.2–83.9	3160–13315	—	—	—	Neira 1991 (after Omori & Takahashi 1980)
Apogon maculatus	56.5–60.6	75–100	0.16–0.34	—	—	Charney (1976), Tresher (1984)
Apogon maculiferus	97	17000	0.43	—	—	Barlow 1981 (after Chave 1971)
Apogon menesemus	129	19500	0.47	—	—	Barlow 1981 (after Chave 1971)
Apogon niger	—	—	—	5	—	Kuwamura (1983)
Apogon notatus	—	Several thousands	—	8	—	Kuwamura (1983)
Apogon rueppellii	45–94	51–457	2.4	16	7.5–7.8 5.5–6.4	Chrystal et al. (1985) Neira (1991)
Apogon semilineatus	70–100	—	0.58–0.6	—	2.3	Ebina (1932)
Apogonichthys waikiki	35–43	1500–5100	0.59–0.61	—	—	Barlow 1981 (after Chave 1971)
Cheilodipterus lineatus	—	—	2.0–2.5	—	—	Fishelson (1970)
Phaeoptyx conklini	34.8–42.4	—	0.19–0.31	—	—	Charney (1976)
Foa brachygramma	52	4800	0.44	—	—	Barlow 1981 (after Chave 1971)
Foa madagascariensis	41	—	0.5	—	—	Petit (1931)
Pterapogon kauderni	35–50 (SL)	40–50	2.8–3.0	19	6–6.5	This study
Spharaemia orbicularis	62–89	6100–1170	0.6–0.7	8	3.3	Allen (1975)
Siphamia corallicola	28	162	—	—	—	Allen (1975)
Vicentia conspersa	95	150	4.5	—	—	Hale (1947)

Since the newly 'hatched larvae' of 7.8 mm are 'larvae' expelled from males kept in aquaria, the difference between the two reported sizes could account for the post-hatching incubation in the male's mouth. Another likely species to have direct development is *Vicentia conspersa*, which has the largest eggs of any apogonid recorded to date (4.5 mm), and in which the clutch is incubated in the male's mouth (Hale 1947).

Allen & Steene (1995, p. 8) mentioned the possibility that juveniles of *P. kauderni* would seek refuge in the parent's mouth: 'Apparently the young fish are protected by the male parent for the first few weeks. Presumably they progressively spend more time outside the mouth with increasing growth'. They also stated that after being placed in a plastic bag an incubating male regurgitated its eggs which 'were very large compared to other apogonids and did not appear to be bound together in a single egg ball'. Allen & Steene's (1995) speculation about male-juvenile interaction was based on observations of two males, which after being collected, expelled a total of 26 juveniles of about 10–11 mm TL. In the present study, the newly released juveniles, of about 10 mm TL, were not observed to retreat into the parent's mouth after being expelled. Once free, juveniles swim together forming a tight school around different structures in the tank, such as rocks with attached plastic eel grasses, or among the spines of the sea urchins *Diadema* sp. Similarly, the male does not show interest in the newly released juveniles, and usually stays in a different section of the tank. Regarding the egg cluster's appearance of not being bound together, it is not clear why at least some eggs would have remained attached by the chorionic filaments. It is possible the clutch was at a very late developmental state when the eggs' filaments become weaker and the eggs were easier to separate.

Insemination is presumed to occur during the first second after egg release, before the male turns to grab the eggs in its mouth. The precise moment of sperm release could not be determined. Observations of mating and egg transfer in slow-motion videos, did not show evidence of sperm release (whitish cloud of spermatic fluid). No body contractions, spasms, or trembling by the male were observed during the short time (1–2 s) that the entire mating process lasted, from the eggs release to the rapid turn and egg gulping by the male. A similar situation was reported by Kuwamura (1983) during his field studies on the reproduction of *Apogon notatus.*

One interesting feature of embryo incubation is the male's ability to detect dead embryos inside the eggs and to expel them. Perhaps, dead embryos cause the dilution of their egg's envelopes and chorionic filaments. Thus, the male's indication of having dead embryos would be loose eggs, which could be easily separated from the egg-cluster and expelled. Also, males are capable of detecting and expelling malformed and slow-developing embryos. Here, the male could sense the smaller size of the embryos, their lower activity or a combination of both factors.

In accordance with the life-history model of Balon (1990), direct ontogeny is an advanced type of development. By eliminating the most vulnerable period (larva) and avoiding the 'costly metamorphosis', precocial species have a survival advantage. The combination of fewer eggs per clutch (however with a larger volume of yolk), prolonged embryonic development and the sessile state of free embryos, all predisposes for further parental protection. Also, the degree of parental care should increase with the energy invested into individual offspring (Crawford & Balon 1996). The correlation between these factors is evident in *P. kauderni* in which an elevated energy allocation per offspring (long-elaborated courtship displays, defending of spawning sites, production of large ova with highly dense yolk, and long period of fasting and egg-embryo care by the male) is associated with an advanced type of parental care. As a result, in comparison with most altricial marine species, a high percentage of *P. kauderni*'s eggs would be expected to reach the juvenile period. However, this reproductive style, which includes the retention of eggs in the parent's oral cavity, a small number of eggs and the elimination of the larva period, also means the sacrifice of all planktonic dispersal. The low fecundity of *P. kauderni* and its lack of planktonic dispersal could explain its very restricted geographical distribution, i.e. the south coast of Banggai Island (Allen & Steene 1995).

The combination of its low fecundity and complex reproductive behavior, together with its restricted distribution, makes *P. kauderni* potentially vulnerable to commercial collecting. It is possible to breed and raise this species in captivity, and efforts should be made to reduce the capture of wild specimens.

Acknowledgements

I would like to express my gratitude to Jackie Webb for her helpful criticism and her detailed suggestions that greatly improved this manuscript. I thank Judy Wellington and Robert Fournier for their careful revision of the draft. I also thank Jennifer Warholak for helping me with the scanning of the photographs and graphics.

References cited

Allen, G. 1975. The biology and taxonomy of the cardinalfish *Sphaeramia orbicularis* (Pisces; Apogonidae). J. Proc. Roy. Soc. West. Austral. 58(3): 86–92.

Allen, G. 1993. Cardinalfishes (Apogonidae) of Madang Province, Papua New Guinea, with descriptions of three new species. Revue fr. Aquariol. 20: 9–20.

Allen, G.R. & R.C. Steene. 1995. Notes on the ecology and behaviour of the Indonesian cardinalfish (Apogonidae) *Pterapogon kauderni* Koumans. Revue fr. Aquariol. 22(1-2): 7–9.

Balon, E.K. 1975. Terminology of intervals in fish development. J. Fish. Res. Board Can. 32: 1663–1670.

Balon, E.K. 1990. Epigenesis of an epigeneticist: the development of some alternative concepts on the early ontogeny and evolution of fishes. Guelph Ichthyol. Rev. 1: 1–42.

Barlow, G.W. 1981. Patterns of parental investment, dispersal and size among coral-reef fishes. Env. Biol. Fish. 6: 65–85.

Breder, C. & D. Rosen. 1966. Modes of reproduction in fishes. American Natural History Press, Garden City. 941 pp.

Charney, P. 1976. Oral brooding in the cardinalfishes *Phaeoptyx conklini* and *Apogon maculatus* from the Bahamas. Copeia 1976: 198–200.

Chrystal, P.J., I.C.Potter, N.R. Loneragan & C.P. Holt. 1985. Age structure, growth rates, movement patterns and feeding in an estuarine population of the cardinalfish *Apogon rueppellii*. Mar. Biol. 85: 185–197.

Crawford, S.S. & E.K. Balon. 1994. Alternative life histories of the genus *Lucania*: 1. Early ontogeny of *L. parva*, the rainwater killifish. Env. Biol. Fish. 40: 349–389.

Crawford, S.S. & E.K. Balon. 1996. Cause and effect of parental care in fishes. An epigenetic perspective. pp. 53–107. *In*: J.S. Rosenblatt & C.T. Snowdon (ed.) Parental Care: Evolution, Mechanisms, and Adaptative Significance, Advances in Study of Behavior 25, Academic Press, San Diego.

Crim, L. & B. Glebe. 1990. Reproduction. pp. 529–547. *In*: C.Schreck & P. Moyle (ed.) Methods for Fish Biology, Amer. Fish. Soc., Bethesda.

Ebina, K. 1932. Buccal incubation in the two sexes of a percoid fish, *Apogon semilineatus*.T. & S. J. Imp. Fish.Inst., Tokyo 27: 19–2.

Fishelson, L. 1970. Spawning behavior of the cardinalfish, *Cheilodipterus lineatus*, in Eilat (Gulf of Aqaba, Red Sea). Copeia 1970: 370–371.

Fraser, T. 1972. Comparative osteology of the shallow water cardinal fishes (Perciformes:Apogonidae) with reference to the systematics and evolution of the family. Ichthyol. Bull. J.L.B. Smith Inst. Ichthyol. 34: 1–105.

Fowler, H. & B. Bean.1930. The fishes of the families Amiidae, Chandidae, Duleidae, and Serranidae, obtained by the United States Bureau of Fisheries steamer 'Albatros' in 1907 to 1910, chiefly in the Phillipine islands and adjacent seas. Bull. U.S. Nat. Mus. 100(10): 30.

Garnaud, J. 1950. La reproduction et l'incubation branchiole chez *Apogon imberbis* G. et L. Bull. Inst. Océanogr. Monaco 977: 1–10.

Garnaud, J. 1962. Monographie de l'Apogon méditerranéen, *Apogon imberbis* (Linne) 1758. Bull. Inst. Océanogr. Monaco. 1248: 1–83.

Hale, H. 1947. Evidence of the habit of oral gestation in a south Australian marine fish *(Apogon conspersus* Klunzinger). The South Australian Naturalist 24(3): 1–3.

Koumans, F.P. 1933. On a new genus and species of Apogonidae. Zool. Med. Mus. Leiden 16: 78.

Kuwamura, T. 1983. Spawning behavior and timing of fertilization in the mouthbrooding cardinalfish *Apogon notatus*. Japan. J. Ichthyol. 30: 61–71.

Kuwamura, T. 1985. Social and reproductive behavior of three mouthbrooding cardinalfishes, *Apogon doederlini*, *A. niger* and *A. notatus*. Env. Biol. Fish. 13: 17–24.

Kuwamura, T. 1987. Night spawning and paternal mouthbrooding of the cardinalfish *Cheeilodipterus quinquelineatus*. Japan. J. Ichthyol. 33: 431–433.

Lachner, E. 1951. Studies of certain apogonid fishes from the Indo-Pacific, with descriptions of three new species. Proc. U.S. Nat. Mus. 101(3290): 581–610.

Neira, F. 1991. Larval development of the oral brooding cardinalfish *Apogon rueppelli* (Teleostei: Apogonidae) in western Australia. Rec. West. Aust. Mus. 15: 573–584.

Nelson, J. 1994 Fishes of the world, 3rd edition. J. Wiley & Sons, New York. 600 pp.

Oppenheimer, J. 1970. Mouthbrooding in fishes. Anim. Behav. 18: 493–503.

Petit, M. 1931. Une espèce nouvelle du genre *Foa* présentant un cas d'incubation bucco- branchiale. Bull. Mus. Hist. Nat. Paris 3(1): 91–95.

Potthoff, T. 1984. Clearing and staining techniques. pp. 35–37. *In*: H. Moser (ed.) Ontogeny and Systematics of Fishes, American Society of Ichthyologists and Herpetologists, Special Publication No. 1, Lawrence.

Smith, C.L., E.H. Atz & J.C. Tyler. 1971. Aspects of oral brooding in the cardinalfish *Cheilodipterus affinis* Poey (Apogonidae). American Museum Novitates 2456: 1–11.

Thresher, R. 1984. Reproduction in reef fishes. T.F.H. Publications, Neptune City. 399 pp.

Tominaga, Y. 1964. Notes on the fishes of the genus *Siphamia* (Apogonidae), with a record of *S. versicolor* from the Ryukyu Islands. Japan. J. Ichthyol. 12: 10–17.

Whitley, G. 1959. The freswater fishes of Australia. Biogeogr. Ecol. Aust. Monographiae Biol. 8: 136–149.

Environmental Biology of Fishes **56:** 93–102, 1999.

The onset of the juvenile period in carp, *Cyprinus carpio*: a literature survey

Lorenzo Vilizzi[a] & Keith F. Walker
River Murray Laboratory, Department of Zoology, University of Adelaide, SA 5005, Australia
[a]*Present address: Landscape & Ecology Research Group, Department of Environmental Sciences, University of Hertfordshire, Hatfield, Hertfordshire AL10 9AB, UK (e-mail: l.vilizzi@herts.ac.uk)*

Received 30 July 1997 Accepted 18 July 1998

Key words: Cyprinidae, early ontogeny, life-history model, metamorphosis, larva, morphology, differential growth, alprehost

Synopsis

Literature data on the relative growth of body parts, functional morphology, feeding behaviour and differential growth capacity in 0+ carp, *Cyprinus carpio*, were relatively consistent with respect to the onset of the juvenile period in this species, which occurs at 20–25 mm standard length (20–30 days after hatching), when the process of metamorphosis (larva–juvenile transition) is regarded as complete and the definitive phenotype is attained. However, changes in metabolic rate, development of social interactions and shifts in habitat use suggest that carp are not 'small adults' until they attain a greater size and age. Problems relative to the global heterogeneity of carp stocks, including wild and domesticated morphs, and the importance of an holistic approach to the study of metamorphosis in carp, and fish in general, are also discussed.

Introduction

With such an embarass de richesses, *this chapter can be little more than a potpourri of citations. My files are bulging...*
S. J. Gould (1977)
in 'Ontogeny and Phylogeny'

Copp & Kováč (1996) highlighted the fact that, within the context of the life-history model (Balon 1975, 1990), the start of the juvenile period in fish with indirect development remains ill defined. They argued that, in metamorphic species, the process of remodelling, the stabilisation of allometric growth and changes in physiology, behaviour and habitat use may be crucial in defining the larva–juvenile transition, and that this transition is inherently unclear because metamorphosis in fish generally is progressive and less dramatic than in, say, metamorphic invertebrates (Flegler-Balon 1989).

The larva–juvenile transition was equated by Copp & Kováč (1996) with the onset of the juvenile period, contrary to Balon (1984a) who identified it with metamorphosis, a remodelling process involving 'extensive changes, from an unfishlike appearance into an adultlike condition', where 'temporary organs are replaced by definitive organs' while other characters 'persist from the larval period and disappear only later' (Balon 1975, p. 1664). In fish with direct development (ametamorphic), on the other hand, the embryo develops directly into the definitive phenotype, a juvenile. And as the definitive organs are formed throughout embryogenesis, there is no need to remodel larval structures (e.g. Balon 1984a,b, 1985a, 1990, 1991, Flegler-Balon 1989). Therefore, according to Balon's (1984a) interpretation, the larva–juvenile transition in metamorphic fish would be well defined, since the process of metamorphosis has been described in many species (e.g. Bertin 1958, Just et al. 1981).

The relative duration of metamorphosis varies considerably among fish with indirect development, being a threshold, step or longer interval (Flegler-Balon 1989). Thus in the eel, *Anguilla anguilla*, metamorphosis lasts four months, whereas in the flounder, *Pseudopleuronectes herzensteini*, it is completed in one week (Balon 1985a). Fish with a protracted metamorphic interval have been sometimes referred to as 'prejuveniles' in the literature (e.g. Lewis et al. 1972, Brothers & McFarland 1981), but more simply called 'metamorphosing larvae' in Balon's life-history model (Balon 1984b, 1985a, 1986a, 1990). Flegler-Balon (1989) stated that the end of metamorphosis 'marks the beginning of the juvenile period' (p. 78), and argued that 'what one author considers a larva, is a juvenile in someone else's opinion' (p. 72). This represents a dilemma long recognised by many authors (Flegler-Balon 1989), one that lead Wald (1982) to state that 'with metamorphosis it is easy to know where to start, but hard to know where to stop'. Wald's (1982) statement, however, may not be entirely true, as even the onset of metamorphosis can be difficult to locate. Thus Bertin (1958, p. 1815) defined metamorphic fish as 'hémimétaboles', and observed that the three essential criteria for metamorphosis, namely (1) hystolysis of some larval characters, (2) new formation of some adult characters, and (3) persistence of common larval/adult characters (Geigy & Portmann 1941, Bertin 1958, see also Flegler-Balon 1989), 'empiétent en général tellement les uns sur les autres qu'il est difficile d'établir une limite exacte entre les larves et les postlarves [Balon's metamorphosing larvae]'. This contention was also supported by Flegler-Balon (1989, p. 78), who wrote that 'metamorphosis starts some time during the larval period' (see also her figure 4). In their comparative study on the early ontogeny of the rainwater killifish, *Lucania parva*, Crawford & Balon (1994, p. 387) noted that 'quantitative measures of relative body shape indicated that *L. parva* reached the juvenile or definitive phenotype at different ages, depending on the particular system of interest'; however 'by the completion of finfold resorption [...], the juvenile *L. parva* resembled the adults in most respects'. The completion (and the onset) of the metamorphic process, therefore, remain ill defined. However, this probably is a consequence of the different criteria used to determine when metamorphosis in fish (and, perhaps, other animals) can be regarded as complete. It need not be a flaw in Balon's life-history model.

In the following discussion, the expression 'onset (or start) of the juvenile period' is preferred to 'larva–juvenile transition'. The latter term, in concert with Balon (1984a), is taken as synonymous with metamorphosis. Copp & Kováč (1996) reconsidered metamorphosis in the roach, *Rutilus rutilus*, and identified an interval at the end of the larva period wherein fish of 15–40 mm standard length (SL) (conventionally 'juveniles' or 'small adults') showed protracted allometric growth in several characters (Kováč & Copp 1996), related to adaptations for locomotion of larvae, visual acuity and microhabitat use. The onset of the juvenile period was alternatively equated with the disappearance of larval characters, the appearance of adult structures and the stabilisation of allometric growth in most characters.

A survey of literature on the early life history of carp, *Cyprinus carpio* L., suggests that published data on morphometrics, metabolism, functional morphology, intra-specific competition and habitat use may shed light on the timing of the onset of the juvenile period in this species. These are examined below.

Key studies

Gradual versus saltatory ontogeny

Studies on the early development of fish have viewed ontogeny either as a continuous, incospicuous accumulation of small changes (gradual ontogeny), or as a series of steps and thresholds (saltatory ontogeny). Work on carp also has followed these models, with gradualistic (e.g. Verma 1970, 1971, Hoda & Tsukahara 1971, Ahmed et al. 1989) as well as saltatory studies (e.g. Smirnov 1955, Balon 1958a, Peňáz et al. 1976, Peňáz et al. 1983 and references therein). These latter works traditionally have followed the original formulation of the theory of stepwise development as a sequence of 'etaps' and 'leaps' (Kryzhanovsky et al. 1953, Vasnetsov 1953, see also Smirnov et al. 1995), not even remotely akin to the stabilised, self-organising energy states (steps) and integrative actions of developmental reorganisation (thresholds) of Balon's (1986b, 1990, Crawford & Balon 1996) theory of saltatory ontogeny. As a result, the Vasnetsov–Kryzhanovsky version of stepwise development has since been the only available paradigm in ecomorphological studies of carp, including investigations on diet (Adzhimuradov 1972, Kamler et al. 1990, Vilizzi 1998a) and growth (Szlamińska et al. 1989). Only recently did Van Snik et al. (1997) venture a functional interpretation of morphometric changes in carp larvae in light of Balon's saltatory model. However, in spite of evidence for homeorhetic development in the characters examined, the authors eventually resorted to a gradualistic explanation, apparently misled by a misinterpretation of the concept of homeorhesis (Balon 1990). A re-evaluation of carp early development according to the homeorhetic model, as advocated by Balon (1995a), therefore would contribute towards a better understanding of morphological development, ontogenetic shifts in resource use, and evolutionary trajectories (Kováč 1994), let alone shed light on the relationship between the two loosely-defined subspecies *C. c. carpio* and *C. c. haematopterus* (Paaver & Tammert 1993, Balon 1995b).

Criteria for the onset of the juvenile period in carp

Throughout the present paper, the type of length measured (namely, standard length: SL or total length: TL) is given whenever possible, as this was often not indicated in the references consulted. Smallwood & Smallwood (1931, p. 220) described two intervals for carp 'larvae' in North America, namely an inactive period 'when the yolk supply is still adequate' and an active period 'when important morphological changes take place'. With regard to the onset of the juvenile period they wrote (p. 222): 'it is difficult to state just when the larval stage ends, but in two weeks the body takes on the characteristic hump just back of the head [peculiar to domesticated carp (Balon 1974, 1995a,b)], the fins have become formed, and the pigmentation begins to have the golden tints that are so characteristic of the fry stages'. Sarig (1966, p. 3:10) reported that 'at the size of 2 cm the fingerlings already resemble adult carp in growth patterns, nutrition, local movements and schooling'. McCrimmon & Swee (1967) observed that scale formation was initiated in fish of 16–18 mm TL and complete by 25 mm, and the fins, swimbladder and mouth parts were fully developed in 'young' 21 mm fish. Rhouma (1975, p. 108) described the acquisition of juvenile characters thus: 'la deuxième semaine la longeur a été de 23 mm, les alevins ont eu la totalité de leurs nageoires et quelques écailles sont apparues à l'implantation de ses nageoires'.

Balon (1958a) found that in Danubian wild carp the onset of the juvenile period was marked by almost complete scale cover (see also Balon 1958b), formation of the fin rays and disappearance of the fin fold. In domesticated pond carp, Peňáz et al. (1983, p. 19) defined the beginning of the juvenile period (19.1 mm TL, age 21 days, at 25°C) as 'the end of metamorphosis, the appearance of scale-cover and the attainment of a body shape similar to that in adulthood'. The first juvenile step was marked by the appearance of scale cover (19.1–30.2 mm TL, age 21–29 days), and the second by the presence of complete scale cover and fusion of the nasal septum (30.2 mm TL, age 26 days). However, Copp & Kováč (1996) argued that attributes like the onset of scale cover, whilst a key to recognition of juveniles, do not necessarily indicate a threshold in the sense of the saltatory model, because no allowance is made for other potential changes in physiology, behaviour and niche breadth. Some of these changes are examined below.

Relative growth (Figure 1)

Hoda & Tsukahara (1971) investigated the relative growth of body parts and organs in carp 4–300 mm SL. For seven characters (snout length, head length, body depth, caudal peduncle height, length of snout to insertion of dorsal fin, pectoral fin length, pharyngeal arch length), a shift from positive allometry to isometry was apparent at 21.4 mm SL (range 17–24 mm). Allometric growth in eye diameter, mouth gape and opercular opening changed from positive to negative at similar lengths (20, 17, 20 mm SL, respectively). The growth of the intestine was triphasic (negative, positive, positive), with inflection points in the log–log relationship with SL at 9.5 and 22.5 mm. Hoda & Tsukahara (1971, their figure 13A) also illustrated a 'young' one-month-old individual of 25 mm TL.

Oikawa & Itazawa (1984a,b, 1985) reported that body height and width increased isometrically at SL > 18 mm, but were positive allometric at smaller lengths. Osse (1990, p. 367) observed that 'it is quite striking that at a [standard] length of some 21 mm, that is at an age of about one month post-hatch (at 23°C), the generally positive allometry of these dimensions changes into approximately isometric growth'. Osse (1990) also reported of a break in the log–log relationship with weight at 19 mm SL (see also Szlamińska et al. 1989). This value is comparable to the 16.8 mm SL estimated by Vilizzi (1998b) for feral carp in the River Murray, Australia, but considerably lower than the 26.9 mm SL, obtained by the same author, unconstrained by any *a priori* definition of developmental intervals.

Metabolic rate

Post & Lee (1996) reanalysed literature data on the metabolic ontogeny of some teleost fishes, including carp. Collected data of mass-specific respiration rate

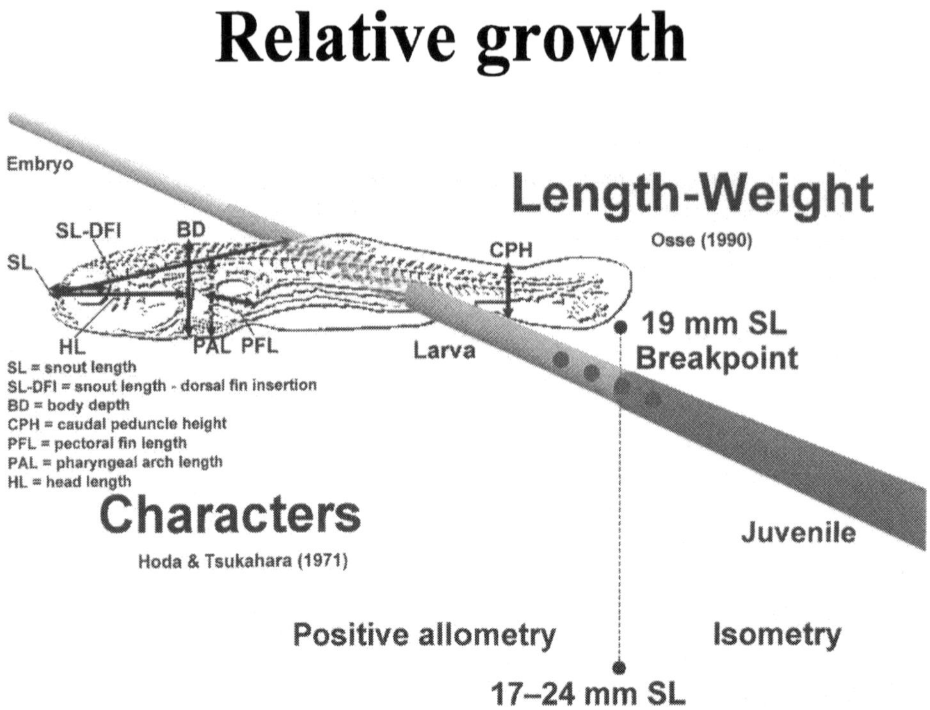

Figure 1. Illustration of patterns of relative growth in carp, based on literature data. Along the ontogenetic trajectory from embryo to larva to juvenile, changes in the growth of morphometric characters and in the length–weight relationship are indicated (dashed line).

vs. mass indicated an inflection point in the log–log relationship of the two variables at 0.290 g, 'which is substantially larger than size at metamorphosis for common carp' (Post & Lee 1996, p. 914). Although these authors did not provide an indication of size at metamorphosis, based on data by Peňáz et al. (1983), the weight of carp at metamorphosis would fall somewhere between that of step 6 larvae (33.1–110.0 mg) and that of step 1 juveniles (110.0–525.2 mg), thus in accordance with Post & Lee's (1996) findings.

Functional morphology (*Figure* 2)

Osse (1990) proposed a functional interpretation of changes in carp larvae and juveniles, based primarily on the morphometric data of Hoda & Tsukahara (1971), with reference to locomotion, feeding and respiration. The functional demands of feeding, in particular, warrant attention. Thus, changes in the position of the mouth, from sub-ventral in free embryos to terminal in (feeding) larvae, and from terminal to downwardly projected in 'juveniles' of 20 mm SL (Hoda & Tsukahara 1971, their figure 21), mirror the transitions from endogenous to exogenous nutrition, and from zooplanktivory to benthivory, respectively (e.g. Adzhimuradov 1972, Osse 1990, Vilizzi 1998a). Support comes from Hoda & Tsukahara (1971), who recorded an increase in mouth protrusibility at 20–25 mm SL. Osse (1990) also attributed the positive allometric growth of the pectoral fins and mouth gape (to 20 mm and 17 mm SL, respectively), followed by negative allometry, to the functional requirements of feeding. Rapid growth of the pectoral fins and the mouth gape would increase the chance of prey capture by reducing the head yaw caused by angular recoil, and by increasing the radius of the mouth relative to the target (Drost 1987). Finally, a corollary to the pattern of relative growth in body depth and caudal peduncle height, observed by Hoda & Tsukahara (1971), is that the body shape (body length/body depth) and peduncle depth factor (peduncle depth/body depth) (*sensu* Webb & Weihs 1986) attain adult values at SL about 5–6 times the average length at hatching (4.49 mm SL: Peňáz et al. 1983). There are clear indications, therefore, of changes in body form and function at 20–25 mm SL, and these appear to be correlated with changes in feeding and locomotion.

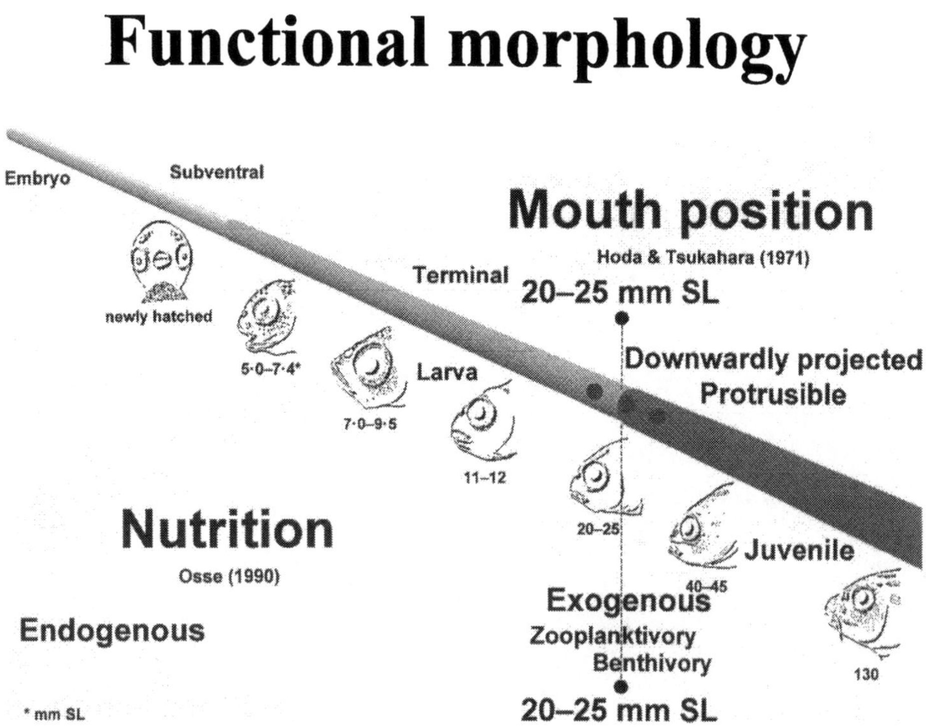

Figure 2. Illustration of patterns of functional morphology in carp, based on literature data. Along the ontogenetic trajectory from embryo to larva to juvenile, changes in feeding habits and in the position of the mouth are indicated (dashed line).

Social interactions and 'shooting' (Figure 3)

Aquarium experiments by Panyushkin (1989) showed that social interactions among carp are apparent even in the first month of life, at 15–30 mm TL. Persistent aggregations, however, are identifiable only after the fish attain 35–70 mm TL, and are associated mainly with foraging and reactions to external stimuli (e.g. fright responses).

A peculiarity of aquarium and pond carp is the appearance of so-called 'tobi koi' or 'shoot carp' under certain conditions of stocking density and food quality and availability. This phenomenon was investigated by Nakamura & Kasahara (1955, 1956, 1957, 1961), and its evolutionary significance and implications for aquaculture and breeding programs were reviewed by Wohlfarth (1977). Shooting is indicated by a strong positive skew in the frequency distribution of length. When food is scarce, the fast-growing fish dominate, slowing or even suppressing the growth of other fish in the stock. Shooters do not develop when individual fish are grown in isolation, or when larger fish or fast-growing varieties are added early during growth of the original stock, so that its frequency distribution remains symmetrical.

Occasional references to divergent growth patterns appear in other studies. Smallwood & Smallwood (1931, p. 219) observed that 'one lot of carp were hatched [. . .] and retained in the same aquarium. These were regularly fed on daphnia and in a month some of the fry were twice the size of the smaller ones'. Żarnecki et al. (1961) conducted a series of experiments on 'quick growing' carp, with the aim to improve methods of selection in aquaculture. In that study, fast growers were observed starting from the fourth week of life. McCrimmon (1968, p. 9) stated that 'differences in individual growth, which are influenced by water temperature, stocking density and availability of food, become prominent by the 12th week of life'. Ivanova (1978) monitored the growth of 'under-yearlings' fed either natural or supplemented natural diets. 'Quick developers' were not reported under the first feeding regime, although the length distribution of fish reared at a density 1.5 times normal does appear slightly right-skewed (Ivanova 1978, her figure 2). In the second regime, shooters did appear.

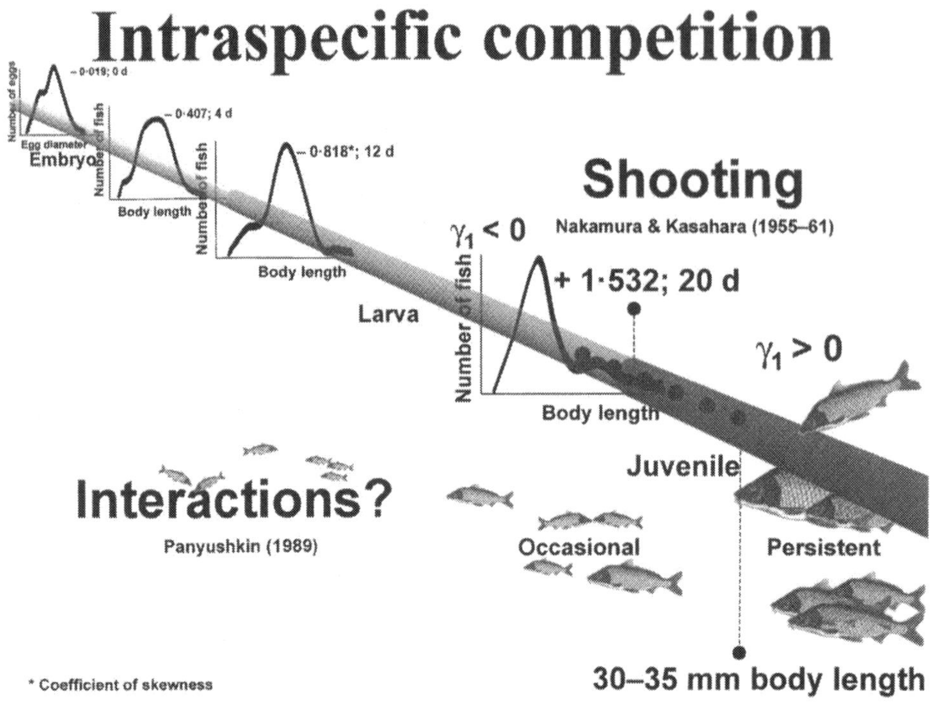

Figure 3. Illustration of processes of intra-specific competition in carp, based on literature data. Along the ontogenetic trajectory from embryo to larva to juvenile, the occurrence of intra-specific interactions and changes in length–frequency distribution are indicated (dashed line).

The time of appearance of the shooters is significant in the context of the onset of the juvenile period. Belyaev (1976) provided evidence of growth variability in carp 1–6 days old, but this is unsupported by other work. Nakamura & Kasahara (1955) showed that positively skewed body-length distributions appeared in a population fed on Cladocera for 20 days, although the distributions of the eggs and younger larvae had been symmetrical. If allowance is made for the free embryo phase (about 2 days: Peňáz et al. 1983, Balon 1995a), then the age of first appearance of fast-growing individuals, under pond conditions, would correspond to the accepted time for the onset of the juvenile period.

Habitat use (*Figure* 4)

Little is known of the habitat use by carp larvae and juveniles, unlike some other European cyprinids (e.g. Copp 1989, 1992, Copp et al. 1994, Garner 1996, 1997). Casual observations by Beckman & Elrod (1971) for Lake Oahe, and Sheaffer & Nickum (1986) for backwaters of the upper Mississippi River showed that 0+ carp were most abundant in shallow areas associated with flooded vegetation. A more intensive study in Illinois by Richardson (1913) concerned 10–20 mm carp from consecutive spawning seasons in shallow, vegetated waters. Richardson (1913) noted that the affinity of ‘hatchlings and fingerlings’ for these areas made them vulnerable to changes in water level. Sigler (1958, p. 9) considered freshly-flooded, vegetated areas as an ideal habitat where ‘carp hatch and live from 2 weeks to 2 months’, and noted that the young fish tend to abandon their nursery areas after attaining lengths of 75–100 mm (see also Johnson & Dropkin 1994). Circumstantial evidence for this behaviour is provided also by Reynolds (1983) in Australia: from observations in the Murray and Murrumbidgee rivers he speculated that ‘smaller fish’ (< 200 mm) move out of backwaters to colonise new areas.

Discussion

The foregoing comments indicate significant changes in carp of 20–25 mm SL (20–30 days after hatching), which corresponds to published reports on the onset of the juvenile period in this species. After the onset,

Habitat use

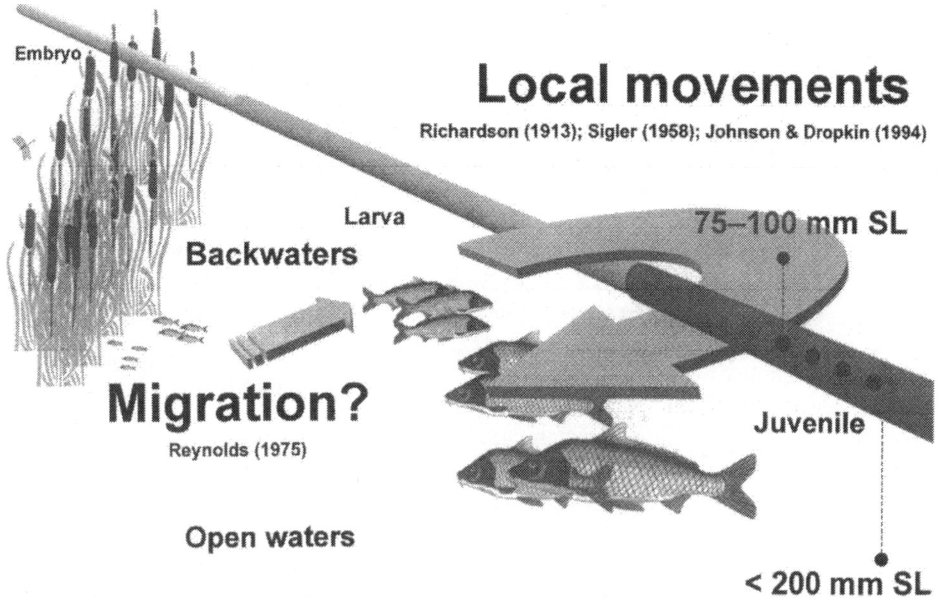

Figure 4. Illustration of patterns of habitat use in carp, based on literature data. Along the ontogenetic trajectory from embryo to larva to juvenile, the occurrence of migratory and local movements is indicated (dashed line).

the carp resemble adults in morphology, feeding habits, locomotion and, perhaps, differential growth capacity. Adult modes of life, in regard to metabolic rate, habitat use and social behaviour, would be achieved at a greater size and age.

The global genetic heterogeneity of carp stocks (Balon 1995b), including domesticated and feral forms (e.g. Olaniyan 1961, McCrimmon 1968, Toor & Chauan 1975, Welykochatko 1976, Shearer & Mulley 1978, Fitzmaurice 1983, Johal et al. 1984, Moyle 1984, Wohlfarth 1984, Prochelle & Campos 1985, Krupka et al. 1989, Brumley 1991, Costa-Pierce et al. 1993, Paaver & Tammert 1993, Coates & Ulaiwi 1995, Kålås & Johansen 1995, Moreau & Costa-Pierce 1997), may obscure the larva–juvenile transition, notwithstanding variation within local populations (Balon 1985a, 1993), which may include altricial and precocial forms (Balon 1995a,b). For instance, Balon (1995a, p. 8) postulated differences in the early ontogeny of wild carp from the Danube, Volga and Amur rivers, and ascribed 'the enormous variation' in forms to 'states of reversal from the domesticated pond form to the feral form'.

More attention should be devoted to the implications of shooting for the dynamics of carp stocks. Wohlfarth (1977) highlighted its potential value, whereby it would promote reproduction of the faster-growing individuals. Differential growth capacity would allow young 'shooters' to abandon their nursery grounds earlier, increasing their chances of survival (with consequences for recruitment), and enable stocks to disperse and colonise new areas more rapidly. Shooting may also play a key role in the over-winter survival of 0+ fish, representing an alternative or complementary means to overcome the constraints imposed by small body size (Miller et al. 1988, but see Litvak & Leggett 1992). In terms of 'alprehost' theory (Balon 1983, 1985b, 1990), the appearance of fast-growing individuals, under conditions favouring high intra-specific competition, would indicate the acquisition of a precocial life style. As explained by Wolfarth (1977), in subsistence farming, shooting may be of advantage, as it would ensure the availability of a constant supply of large fish. Conversely, in intensive aquaculture this phenomenon would have undesirable effects due to market demands for fish of uniform size, so that 'it is disproportionally important to supply adequate amounts of feed during nursing, when the aim is producing a uniform size of fish' (Wolfarth 1977, p. 38). Sibling cannibalism (van Damme et al. 1989) may also occur under such conditions. On the other hand, in a non-competitive environment, a heterochronous shift towards altriciality would be possible, thus leading to the appearance of fish undesirably (from an aquaculturist's point of view) 'old for their size' (Noakes & Balon 1982). Incidentally, this may explain why length–frequency distributions for feral carp larvae and juveniles of different age and developmental state in the River Murray, Australia, were always normally distributed—possibly a response to large availability of food under non-competitive conditions (Vilizzi 1997). In conclusion, published data and casual observations indicate that shooting is a juvenile phenomenon, although its morphological, functional and social advantages remain unclear.

The importance of an holistic approach to the study of metamorphosis, whereby physiological and biochemical processes are also taken into consideration, was shown by Forstner et al. (1983), and, more recently, has been emphasised by Copp & Kováč (1996). To this end, further investigations into the early ontogeny of carp should address more closely the 'metabiotic' aspects (Flegler-Balon 1989) of the larva–juvenile transition, so that anatomical and physiological data (e.g. Drost & van den Boogaart 1986, Alami-Durante et al. 1997) can be integrated into ecomorphological studies.

Acknowledgements

This paper is drawn from a Ph.D. project by L.V., supervised by K.W. We are grateful to our colleague Jim Puckridge for his constructive criticism on an early draft. This work was partly supported by research grants from the Department of Environment and Natural Resources, Adelaide, made available by Anne Jensen, and from the Mark Mitchell Foundation.

References cited

Adzhimuradov, K.A. 1972. The food of juvenile carp [*Cyprinus carpio* (L.)] in early development stanzas in bodies of water of the Arakum (Terek River Delta). J. Ichthyol. 12: 981–986.

Ahmed, A.T.A., G. Mustafa & H. Farida. 1989. Study on breeding and early stages in the development of the common carp, *Cyprinus carpio* (L.). Bangladesh Journal of Agriculture 14: 151–158.

Alami-Durante, H., B. Fauconneau, M. Rouel, A.M. Escaffre & P. Bergot. 1997. Growth and multiplication of white skeletal muscle fibres in carp larvae in relation to somatic growth rate. J. Fish Biol. 50: 1285–1302.

Balon, E.K. 1958a. Development of the Danubian carp during the prelarva and larva period. Biologické práce 4 (6): 5–54 (in Slovak).

Balon, E.K. 1958b. Die Entwicklung der Beschuppung des Donau-Wildkarpfen. Zool. Anz. 160: 68–73.

Balon, E.K. 1974. Domestication of the carp *Cyprinus carpio* L. Royal Ontario Mus. Life Sci. Misc. Publ., Toronto. 37 pp.

Balon, E.K. 1975. Terminology of intervals in fish development. J. Fish. Res. Board Can. 32: 1663–1670.

Balon, E.K. 1983. Epigenetic mechanisms: reflections on evolutionary processes. Can. J. Fish. Aquat. Sci. 40: 2045–2058.

Balon, E.K. 1984a. Patterns in the evolution of reproductive styles in fishes. pp. 35–53. *In*: C.W. Potts & R.J. Wootton (ed.) Fish Reproduction: Strategies and Tactics, Academic Press, London.

Balon, E.K. 1984b. Reflections on some decisive events in the early life of fishes. Trans. Amer. Fish. Soc. 113: 178–185.

Balon, E.K. 1985a. The theory of saltatory ontogeny and life history models revisited. pp. 13–30. *In*: E.K. Balon (ed.) Early Life History of Fishes: New Developmental, Ecological and Evolutionary Perspectives, Dr W. Junk Publishers, Dordrecht.

Balon, E.K. 1985b. Reflections on epigenetic mechanisms: hypotheses and case histories. pp. 239–270. *In*: E.K. Balon (ed.) Early Life History of Fishes: New Developmental, Ecological and Evolutionary Perspectives, Dr W. Junk Publishers, Dordrecht.

Balon, E.K. 1986a. Types of feeding in the ontogeny of fishes and the life-history model. Env. Biol. Fish. 16: 11–24.

Balon, E.K. 1986b. Saltatory ontogeny and evolution. Rivista di Biologia/Biology Forum 79: 151–190.

Balon, E.K. 1990. Epigenesis of an epigeneticist: the development of some alternative concepts on the early ontogeny and evolution of fishes. Guelph Ichthyol. Rev. 1: 1–42.

Balon, E.K. 1991. Probable evolution of the coelacanth's reproductive style: lecithotrophy and orally feeding embryos in cichlid fishes and in *Latimeria chalumnae*. Env. Biol. Fish. 32: 249–265.

Balon, E.K. 1993. Dynamics of biodiversity and mechanisms of change: a plea for balanced attention to form creation and extinction. Biol. Cons. 66: 5–16.

Balon, E.K. 1995a. Origin and domestication of the wild carp, *Cyprinus carpio*: from Roman gourmets to the swimming flowers. Aquaculture 129: 3–48.

Balon, E.K. 1995b. The common carp, *Cyprinus carpio*: its wild origin, domestication in aquaculture, and selection as colored nishikigoi. Guelph Ichthyol. Rev. 3: 1–55.

Beckman, L.G. & J.H. Elrod. 1971. Apparent abundance and distribution of young-of-the-year fishes in Lake Oahe, 1965–69. Reservoir Fisheries and Limnology, American Fisheries Society Special Publication 8: 333–348.

Belyaev, V.I. 1976. Relationship between length, weight, and adjusted body weight in carp larvae in a spawning pond. Sov. J. Ecol. 7: 565–567.

Bertin, L. 1958. Larves et métamorphoses. pp. 1813–1834. *In*: P.-P. Grassé (ed.) Traité de Zoologie, Vol. 13, Agnathes et Poissons, fasc. 3, Masson et Cie., Paris.

Brothers, E.B. & W.N. McFarland. 1981. Correlations between otolith microstructure, growth and life history transitions in newly recruited French grunt [*Haemulon glavolineatum*

(Desmarest), Haemulidae]. Rapp. P.-V. Reun. Cons. Int. Expl. Mer. 178: 369–374.

Brumley, A.R. 1991. Cyprinids of Australasia. pp. 264–283. *In*: I.J. Winfield & J.S. Nelson (ed.) Cyprinid Fishes: Systematics, Biology and Exploitation, Chapman and Hall, London.

Coates, D. & W.K. Ulaiwi. 1995. A simple model for predicting ecological impacts of introduced aquatic organisms: a case study of common carp, *Cyprinus carpio* L., in the Sepik-Ramu River Basin, Papua New Guinea. Fish. Manag. Ecol. 2: 227–242.

Copp, G.H. 1989. The habitat diversity and fish reproductive function of floodplain ecosystems. Env. Biol. Fish. 26: 1–27.

Copp, G.H. 1992. Comparative microhabitat use of cyprinid larvae and juveniles in a lotic floodplain channel. Env. Biol. Fish. 33: 181–193.

Copp, G.H. & V. Kováč. 1996. When do fish with indirect development become juveniles? Can. J. Fish. Aquat. Sci. 53: 746–752.

Copp, G.H., G. Guti, B. Rovný & J. Černý. 1994. Hierarchical analysis of habitat use by 0+ juvenile fish in Hungarian/Slovak flood plain of the Danube River. Env. Biol. Fish. 40: 329–348.

Costa-Pierce, B.A., J. Moreau & R.S.V. Pullin. 1993. New introductions of the common carp (*Cyprinus carpio* L.) and their impact on indigenous species in sub-Saharan Africa. Discovery and Innovation 5: 211–221.

Crawford, S.S. & E.K. Balon. 1994. Alternative life histories of the genus *Lucania*: 1. Early ontogeny of *L. parva*, the rainwater killifish. Env. Biol. Fish. 40: 349–389.

Crawford, S.S. & E.K. Balon. 1996. Cause and effect of parental care in fishes. Advances in the Study of Behavior 25: 53–107.

Drost, M.R. 1987. Relation between aiming and catch success in larval fishes. Can. J. Fish. Aquat. Sci. 44: 304–315.

Drost, M.R. & J.G.M. van den Boogaart. 1986. The energetics of feeding strikes in larval carp, *Cyprinus carpio*. J. Fish Biol. 29: 371–379.

Fitzmaurice, P. 1983. Carp (*Cyprinus carpio* L.) in Ireland. Irish Fisheries Investigations, Series A 23: 5–10.

Flegler-Balon, C. 1989. Direct and indirect development in fishes—examples of alternative life-history styles. pp. 71–100. *In*: M.N. Bruton (ed.) Alternative Life-History Styles of Animals, Kluwer Academic Publishers, Dordrecht.

Forstner, H., S. Hinterleitner, K. Mähr & W. Wieser. 1983. Towards a better definition of "metamorphosis" in *Coregonus* sp.: biochemical, histological, and physiological data. Can. J. Fish. Aquat. Sci. 40: 1224–1232.

Garner, P. 1996. Microhabitat use and diet of 0+ cyprinid fishes in lentic, regulated reach of the River Great Ouse. J. Fish Biol. 48: 367–382.

Garner, P. 1997. Habitat use by 0+ cyprinid fish in the River Great Ouse, East Anglia. Freshwater Forum 8: 2–27.

Geigy, R. & A. Portmann. 1941. Versuch einer morphologischen Ordnung der tierischen Entwicklungsgänge. Die Naturwissenschaften 29: 734–743.

Gould, S.J. 1977. Ontogeny and phylogeny. The Belknap Press of Harvard University Press, Cambridge. 501 pp.

Hoda, S.M.S. & H. Tsukahara. 1971. Studies on the development and relative growth in the carp, *Cyprinus carpio* (Linné). Journal of the Faculty of Agriculture, Kyushu University 16: 387–509.

Ivanova, Z.A. 1978. Growth variability in the carp, *Cyprinus carpio*, in Siberian waters. J. Ichthyol. 18: 45–55.

Johal, M.S., J. Novák & O. Oliva. 1984. Notes on the growth of the common carp (*Cyprinus carpio*) in northern India and in central Europe. Věst. Česk. Společ. Zool. 48: 24–38.

Johnson, J.H. & D.S. Dropkin. 1994. Biology of subyearling carp (*Cyprinus carpio*) in the Juniata River, Pennsylvania. J. Penns. Acad. Sci. 68: 34–36.

Just, J.J., J. Kraus-Just & D.A. Check. 1981. Survey of chordate metamorphosis. pp. 265–326. *In*: L.I. Gilbert & E. Frieden (ed.) Metamorphosis, a Problem in Developmental Biology, Plenum Press, New York.

Kålås, S. & R. Johansen. 1995. The common carp (*Cyprinus carpio* L.) in Norway. Fauna Norvegica, Serie A 16: 19–28.

Kamler, E., M. Szlamińska, A. Przybył, B. Barska, M. Jakubas, M. Kuczyński & K. Raciborski. 1990. Developmental response of carp, *Cyprinus carpio*, larvae fed different foods or starved. Env. Biol. Fish. 29: 303–313.

Kováč, V. 1994. Early ontogeny of three *Gymnocephalus* species (Pisces: Percidae): reflections on the evolution of the genus. Env. Biol. Fish. 40: 241–253.

Kováč, V. & G.H. Copp. 1996. Ontogenetic patterns of relative growth in young roach *Rutilus rutilus* (L.): within-river basin comparisons. Ecography 19: 153–161.

Krupka, I., J. Meszáros, P. Ráb & V. Šlechtová. 1989. Toward the preservation of endangered fish gene pools—example of wild Danube carp (*Cyprinus carpio carpio*). Práce VÚRH Vodňany 18: 27–33.

Kryzhanovsky, S.G., N.N. Disler & E.N. Smirnova. 1953. Ecomorphological principles of development in percids. Trudy Inst. Morph. Zhiv. Severtsova 10: 3–138 (in Russian).

Lewis, R.M., E.P.H. Wilkens & H.R. Gordy. 1972. A description of young Atlantic menhaden, *Brevoortia tyrannus*, in the White Oak River estuary, North Carolina. U.S. Fish. Bull. 70: 115–118.

Litvak, M.K. & W.C. Leggett. 1992. Age and size-selective predation on larval fishes: the bigger-is-better hypothesis revisited. Mar. Ecol. Prog. Ser. 81: 13–24.

McCrimmon, H.R. 1968. Carp in Canada. Fish. Res. Board Can. Bull. 165. 93 pp.

McCrimmon, H.R. & U.B. Swee. 1967. Scale formation as related to growth and development of young carp, *Cyprinus carpio* L. J. Fish. Res. Board Can. 24: 47–51.

Miller, T.J., L.B. Crowder, J.A. Rice & E.A. Marschall. 1988. Larval size and recruitment mechanisms in fishes: toward a conceptual framework. Can. J. Fish. Aquat. Sci. 45: 1657–1670.

Moreau, J. & B. Costa-Pierce. 1997. Introduction and present status of exotic carp in Africa. Aquacul. Res. 28: 717–732.

Moyle, P.B. 1984. America's carp. Nat. Hist. 93 (9): 43–51.

Nakamura, N. & S. Kasahara. 1955. A study on the phenomenon of the tobi koi or shoot carp I. On the earliest stage at which the shoot carp appears. Bull. Japan. Soc. Sci. Fish. 21: 73–76 (in Japanese; English translation in Bamidgeh 29: 41–44, 1977).

Nakamura, N. & S. Kasahara. 1956. A study on the phenomenon of the tobi koi or shoot carp II. On the effect of particle size and quantity of food. Bull. Japan. Soc. Sci. Fish. 21: 1022–1024 (in Japanese; English translation in Bamidgeh 29: 45–47, 1977).

Nakamura, N. & S. Kasahara. 1957. A study on the phenomenon of the tobi koi or shoot carp III. On the results of culturing the modal group and the growth of fry reared individually. Bull. Japan. Soc. Sci. Fish. 22: 674–678 (in Japanese; English translation in Bamidgeh 29: 48–52, 1977).

Nakamura, N. & S. Kasahara. 1961. A study on the phenomenon of the tobi koi or shoot carp IV. Effects of adding a small number of larger individuals to the experimental batches of carp fry and culture density upon the occurrence of shoot carp. Bull. Japan. Soc. Sci. Fish. 27: 73–76 (in Japanese; English translation in Bamidgeh 29: 53–56, 1977).

Noakes, D.L.G. & E.K. Balon. 1982. Life histories of tilapias: an evolutionary perspective. pp. 61–82. *In*: R.S.V. Pullin & R.H. Lowe-McConnell (ed.) The Biology and Culture of Tilapias, ICLARM Conf. Proc. 7, Manila.

Oikawa, S. & Y. Itazawa. 1984a. Allometric relationship between tissue respiration and body mass in the carp. Comp. Biochem. Physiol. 77A: 415–418.

Oikawa, S. & Y. Itazawa. 1984b. Relative growth of organs and parts of the carp, *Cyprinus carpio*, with special reference to the metabolism-size relationship. Copeia 1984: 800–803.

Oikawa, S. & Y. Itazawa. 1985. Gill and body surface areas of the carp in relation to body mass, with special reference to the metabolism-size relationship. J. Exp. Biol. 117: 1–14.

Olaniyan, C.I.O. 1961. On the introduction of the common carp, *Cyprinus carpio* L., into Nigerian waters and its possible effect on the hydrography of the region. Journal of the West African Science Association 7: 74–92.

Osse, J.W.M. 1990. Form changes in fish larvae in relation to changing demands of function. Neth. J. Zool. 40: 362–385.

Paaver, T. & M. Tammert. 1993. Genetic variability of bream and wild carp of Lake Yaskhan. Limnologica 23: 279–283.

Panyushkin, S.N. 1989. Social relations in juvenile carp, *Cyrinus carpio* [sic], maintained in a confined space. J. Ichthyol. 29: 161–166.

Peňáz, M., S. Lusk & M. Prokeš. 1976. Changes in wet weight, dry matter content and energetic value of eggs, embryos and larvae of the carp, *Cyprinus carpio*. Folia Zool. 25: 81–90.

Peňáz, M., M. Prokeš, J. Kouřil & J. Hamáčková. 1983. Early development of the carp, *Cyprinus carpio*. Acta Sci. Nat. Brno. 17: 1–39.

Post, J.R. & J.A. Lee. 1996. Metabolic ontogeny of teleost fishes. Can. J. Fish. Aquat. Sci. 53: 910–923.

Prochelle, O. & H. Campos. 1985. The biology of the introduced carp *Cyprinus carpio* L., in the River Cayumapu, Valdivia, Chile. Studies on Neotropical Fauna and Environment 20: 65–82.

Reynolds, L.F. 1983. Migration patterns of five fish species in the Murray-Darling River system. Aust. J. Mar. Freshwater Res. 34: 857–871.

Rhouma, A. 1975. L'élevage de la carpe en Tunisie. Quelques aspects de sa biologie. Bulletin Institute National Scientifique et Technique d'Océanographie et de Pêche Salambô 4: 97–113.

Richardson, R.E. 1913. Observations on the breeding habits of the European carp in the vicinity of Havana, Illinois. Bull. Ill. St. Lab. Nat. Hist. 9: 387–405.

Sarig, S. 1966. Synopsis of biological data on common carp (*Cyprinus carpio* L.), 1758 (Near-East and Europe). FAO Fisheries Synopsis 31.2. 35 pp.

Sheaffer, W.A. & J.G. Nickum 1986. Backwater areas as nursery habitats for fishes in Pool 13 of the Upper Mississippi River. Hydrobiologia 136: 131–140.

Shearer, K.D. & J.C. Mulley 1978. The introduction and distribution of the carp, *Cyprinus carpio* Linnaeus, in Australia Aust. J. Mar. Freshwater Res. 29: 551–563.

Sigler, W.F. 1958. The ecology and use of carp in Utah. Utah State University, Logan Agricultural Experiment Station Bulletin 405. 63 pp.

Smallwood, W.M. & M.L. Smallwood 1931. The development of the carp, *Cyprinus carpio*. J. Morph. Physiol. 52: 217–231.

Smirnov, A.I. 1955. Some pecularities of carp ontogeny. Vopr. Ikthiol. 3: 69–76 (in Russian).

Smirnov, S.A., A.P. Makeyeva & A.I. Smirnov. 1995. Development of ecomorphology of fishes in Russia. Env. Biol. Fish. 44: 23–33.

Szlamińska, M., E. Kamler, P. Bergot, B. Barska, M. Jakubas, M. Kuczyński & K. Raciborski. 1989. The course of the relationship between body weight and length during development of carp (*Cyprinus carpio* L.) larvae. Pol. Arch. Hydrobiol. 36: 149–160.

Toor, H.S. & K.S. Chauhan 1975. Studies on the biology of the exotic fish *Cyprinus carpio* Linnaeus from Punjab waters. 1. Stock composition. Journal of Research Punjab Agricultural University 12: 276–285.

van Damme, P., S. Appelbaum & T. Hecht. 1989. Sibling cannibalism in koi carp, *Cyprinus carpio* L., larvae and juveniles under controlled conditions. J. Fish Biol. 34: 855–863.

van Snik, G.M.J., J.G.M. van den Boogaart & J.W.M. Osse. 1997. Larval growth patterns in *Cyprinus carpio* and *Clarias gariepinus* with attention to the finfold. J. Fish Biol. 50: 1339–1352.

Vasnetsov, V.V. 1953. Etaps in the development of bony fishes. pp. 207–217. *In*: E.N. Pavlovskiy (ed.) Otcherky po Obshtch. Vopr. Ikthiol., AN SSSR Press, Moscow–Leningrad (in Russian).

Verma, P. 1970. Normal stages in the development of *Cyprinus carpio* var. *communis* L. Acta Biol. Acad. Sci. Hung. 21: 207–218.

Verma, P. 1971. The early development of *Cyprinus carpio* var. *communis* (Linn.). Acta Anat. 80: 388–417.

Vilizzi, L. 1997. Age, growth and early life history of carp, *Cyprinus carpio* L., in the River Murray, Australia. Ph.D. Thesis, University of Adelaide, Adelaide. 231 pp.

Vilizzi, L. 1998a. Observations on ontogenetic shifts in the diet of 0+ carp, *Cyprinus carpio*, from the River Murray, Australia. Folia Zool. 47: 225–229.

Vilizzi, L. 1998b. Age, growth and cohort composition of 0+ carp in the River Murray, Australia. J. Fish Biol. 52: 997–1013.

Wald, G. 1982. Metamorphosis: an overview. pp. 1–39. *In*: L.I. Gilbert & E. Frieden (ed.) Metamorphosis, a Problem in Developmental Biology, Plenum Press, New York.

Webb, P.W. & D. Weihs. 1986. Functional locomotor morphology of early life history stages of fishes. Trans. Amer. Fish. Soc. 115: 115–127.

Welykochatko, T. 1976. Biology of the carp in Brazil. The Annals of Zoology, Agra 12: 53–65.

Wohlfarth, G.W. 1977. Shoot carp. Bamidgeh 29: 35–40.

Wohlfarth, G.W. 1984. Common carp. pp. 375–80. *In*: I.L. Mason (ed.) Evolution of Domesticated Animals, Longman, London.

Żarnecki, S., J. Rychlicka, Z. Rychlicki & W. Karbowski. 1961. The increase of growth of 'quick growing' carps in groups mixed with average fry. Rocz. Nauk Roln. 78-B-2: 343–355 (in Polish).

Part 2. Organism-environment relationships

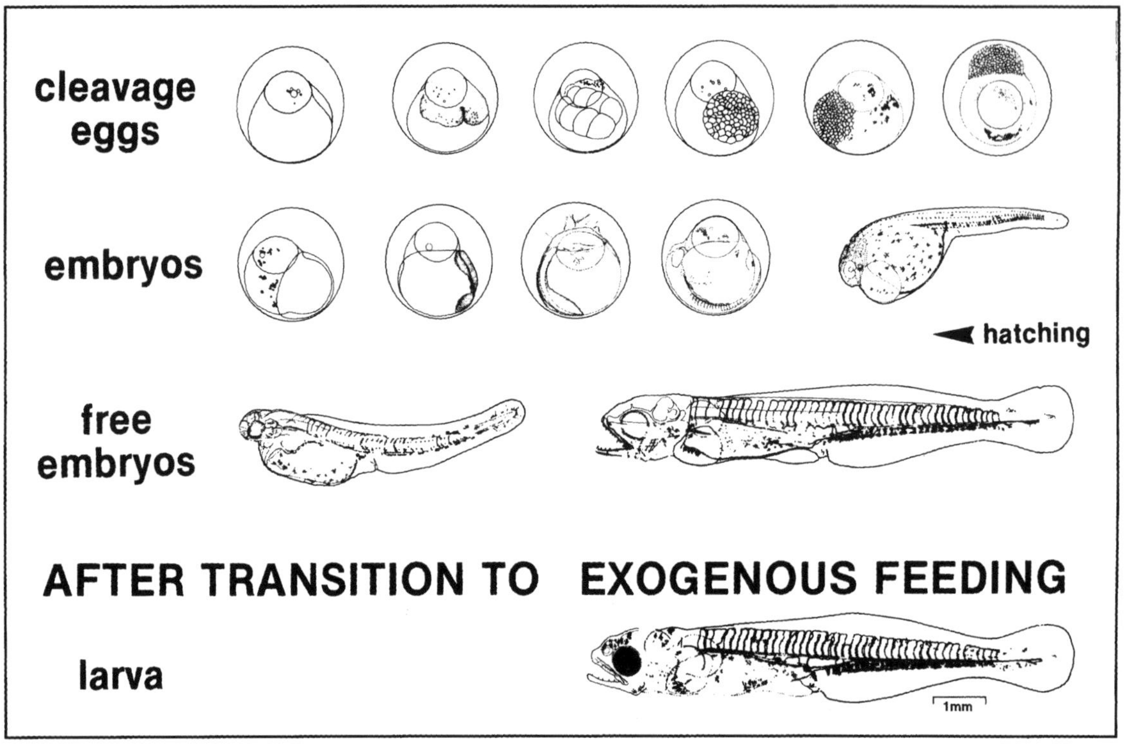

Stizostedium vitreum

Environmental Biology of Fishes **56:** 105–115, 1999.

Morphometry of the stone loach, *Barbatula barbatula*: do mensural characters reflect the species' life history thresholds?

Vladimír Kováč[a], Gordon H. Copp[b] & Malcolm P. Francis[c]
[a] *Comenius University, Faculty of Natural Sciences, Institute of Ecology, Mlynská dolina B2, SK-842 15 Bratislava, Slovakia (e-mail: kovac@nic.fns.uniba.sk)*
[b] *Landscape & Ecology Research Group, Department of Environmental Sciences, University of Hertfordshire, College Lane, Hatfield, Herts AL10 9AB, UK*
[c] *National Institute of Water and Atmospheric Research, P. O. Box 14-901, Kilbirnie, Wellington, New Zealand*

Received 8 July 1997 Accepted 3 September 1998

Key words: ecomorphology, saltatory ontogeny, larva-to-juvenile transition, growth variability, breakpoints in development of mensural characters

Synopsis

Growth variability in 23 mensural characters was examined in 387 specimens of stone loach, *Barbatula barbatula*, from England. The standard length (SL) of the specimens ranged from 15.3 to 115.4 mm. We tested the hypothesis that body proportions change abruptly, rather than gradually, at certain intervals of ontogeny by fitting linear, quadratic and split linear curves to plots of each variable against SL. Based on patterns of allometric growth, two groups and two subgroups of mensural characters have been found. Three characters were best explained by a linear regression, indicating isometric growth. Eight characters were best explained by a quadratic curve, indicating gradual allometry. The remaining 12 characters were best explained by a split regression, indicating mainly isometric growth with abrupt allometry occurring at a specific SL (breakpoint). The first shift in morphometric values (a transformation of the head; breakpoints in three characters) occurred at 26–35 mm SL, the second (a change in fin shape and size as well as body form; breakpoints in six characters) at 36–47 mm SL. The coincidence of shifts in body morphology with those in microhabitat use (between the respective size classes) suggests that thresholds (though not as sudden as those between embryo and early larva steps) do occur during this interval of stone loach life history. We suggest that the larva period ends with the completion of the first shift in relative growth (i.e. not later than at 35 mm SL, depending on individual variability), and that the second shift in morphometric values reflects a threshold between the first and the second step of juvenile period. The importance of changes in external morphology decreased as the fish grew bigger and older.

Introduction

The morphometry of fishes is amongst the most easily perceivable means of assessing the evolutionary adaptation of a species to its environment. For example, differences in the orientation and structure of the head have been used to indicate the feeding style and instream position of riverine fishes (Gatz 1979, Watson & Balon 1984). Similarly, variations in body form can be attributed to swimming abilities associated with endurance, power and speed (Webb & Weihs 1986). Though less frequently examined in studies of early development in fishes, morphometric characters have been suggested as a potential means for identifying thresholds in the life history of fishes (e.g. Crawford & Balon 1994, Copp & Kováč 1996).

Although relative growth has been investigated in many fish species, most attempts to identify these patterns (isometry, allometry, inflexion points) have been based on empirical (visual) perception rather than appropriate statistical analysis (e.g. Hoda & Tsukahara 1971, Holčík & Skořepa 1971, Oikawa & Itazava 1984, Copp & Kováč 1996, Kováč & Copp 1996). Recently, new approaches to resolve this problem have been presented (Sagnes et al. 1997, van Snik 1997), with the latter demonstrating more objectivity in the determination of developmental shifts than the former, but still without testing whether alternative equations (quadratic, simple linear, etc.) provide a more significant model. A variety of criteria have been used to determine developmental intervals, such as the presence or absence of certain structures (Hoda & Tsukahara 1971, Oikawa & Itazava 1984), with morphometric analyses focused on the differences between these intervals following only thereafter.

To avoid this bias, which might result from empirical, visual estimates of patterns in relative growth and/or from pre-determination of developmental intervals by other than morphometric criteria, we propose as an alternative to use statistical analyses to identify the saltatory pattern in relative growth (isometric and/or allometric pattern, too). Thus, the inflexion points (hereafter called breakpoints) in morphometric characters demonstrating a saltatory pattern of development are identified independent of any other criteria. This allows us to examine the changes in external shape of fish with regard to its growth objectively. Furthermore, we examine the development of morphometric characters not only during a selected interval of life, but throughout almost the entire life history of the species.

In a recent study of relative growth in the roach, *Rutilus rutilus* (Kováč & Copp 1996), we found that stabilisation (i.e. reduction of growth variability) of morphometric characters related to swimming ability and visual perception did not coincide with the achievement of all criteria given for the start of the juvenile period (Balon 1956, 1981, 1990). Rather, stabilisation of relative growth in nine characters occurred later, coinciding with a shift in microhabitat, the achievement of adult body shape and a temporary stabilisation in visual acuity (Copp & Kováč 1996). We proposed that the end of metamorphosis in roach might occur later than reported by Balon (1956), coinciding with a functional change in life history. We note now, however, that the size discrepancy at the onset of juvenile development between Balon's (1956) and our study (Kováč & Copp 1996) is probably due to the laboratory and field origins, respectively, of the specimens (see Gozlan et al. 1999 this volume).

The aim of the present study was to determine (1) whether changes in the relative growth of stone loach, *Barbatula barbatula* (see Kottelat 1997 for current nomenclature) appear abruptly, and (2) whether these morphological changes coincide with changes in microhabitat, so that they can be used to define ontogenetic thresholds of life history. Thus, an underlying objective of the present investigation is to determine (3) whether the start of the juvenile period occurs later than previously believed in stone loach.

Material and methods

Specimens of stone loach were collected by electrofishing and originate from three locations in England. The majority, consisting of 276 specimens (mostly juveniles), were taken in August and September 1990 within the catchment of the River Great Ouse (Copp 1992), with 53 specimens subsequently collected in September 1995 from the River Rib near Barwick (National Grid Reference: TL 386 188), 15 specimens in October 1995 from the River Lee at Woolmers Park near Essendon (NGR: TL 288 100), and 43 specimens in the same period in the River Hiz at Ickleford near Hitchin, Herts (NGR: TL 186 317) to provide additional specimens throughout the range of size classes. The fish were killed with an overdose of benzocaine and then preserved in 4% formaldehyde. The life-history terminology used in the text (embryo, larva, juvenile, adult, senescent) follows Balon (1975, 1990).

In the laboratory, 25 mensural characters (Table 1) including total length (TL) and standard length (SL) were measured using vernier callipers to the nearest 0.05 mm on 387 specimens of stone loach as described by Holčík (1989). The dorsal (D), anal (A), pectoral (P), ventral (V) and caudal (C) fins are denoted by their respective abbreviations.

To examine the patterns of relative growth, 23 mensural characters were plotted against SL. However, in contrast to many previous studies (e.g. Řepa 1969, Holčík & Skořepa 1971, Copp & Kováč 1996), raw data were used instead of relative values expressed in % SL because plots of ratios of isometric variables against SL can produce spurious curvilinear trends in the absence of allometry (Marr 1955, Atchley et al. 1976, Jackson et al. 1990). A previous study (Kováč 1987) showed that, although significant differences occurred between males and females in several characters, there was considerable overlap between the male and female character ranges. Therefore, sexual dimorphism was not expected to affect the results and so sex was not determined. To allow future inter-population

Table 1. Mean, standard deviation (SD), range, coefficient of variation (CV) and number of specimens (n) for mensural characters of stone loach from England. Standard error of the 23 characters expressed in % SL ranged from 0.002 to 0.005.

No	Character	Mean	SD	Min	Max	CV	n
	Total length (mm)	55.15	27.46	17.65	136.90	49.78	384
	Standard length (mm)	45.87	23.04	15.30	115.40	50.23	387
	in % of standard length						
1	Head length	26.27	1.70	22.08	31.43	6.46	386
2	Preorbital distance	11.32	0.72	8.26	13.43	6.37	386
3	Eye diameter	5.98	1.07	3.65	8.41	17.94	385
4	Postorbital distance	11.80	1.11	9.28	15.16	9.38	385
5	Interorbital distance	8.06	0.98	5.44	10.54	12.15	386
6	Head depth	14.31	1.09	11.35	17.76	7.59	385
7	Predorsal distance	51.98	1.44	47.58	56.48	2.77	386
8	Preventral distance	54.05	1.58	49.20	58.43	2.92	386
9	Preanal distance	74.68	1.86	69.08	85.48	2.48	385
10	Postdorsal distance	35.73	2.10	29.83	40.77	5.88	384
11	P–V distance	30.30	1.78	23.78	34.75	5.88	386
12	V–A distance	22.05	1.30	18.40	25.53	5.89	381
13	Body depth	16.15	1.36	11.76	20.21	8.41	387
14	Body width	12.60	1.37	8.14	16.44	10.89	386
15	Caudal peduncle length	17.28	1.67	11.31	23.39	9.69	387
16	Minimum body depth	10.68	0.93	6.86	13.24	8.66	385
17	Dorsal fin length	12.95	1.14	9.85	17.38	8.82	385
18	Anal fin length	9.59	1.26	6.69	13.73	13.12	383
19	Pectoral fin length	20.44	1.81	15.22	25.80	8.86	384
20	Ventral fin length	15.19	1.32	9.17	19.35	8.70	383
21	Caudal fin length	20.47	1.78	15.23	25.48	8.72	381
22	Dorsal fin depth	21.64	1.69	16.98	26.79	7.80	379
23	Anal fin depth	17.00	1.31	13.84	21.04	7.72	379

comparisons, 23 mensural characters were expressed in % SL and evaluated subsequently using variation analysis (Snedecor 1946).

The null hypothesis we wished to test was that body proportions do not change abruptly rather than gradually at life-history thresholds and/or transitions. If this hypothesis is rejected, then changes in morphometric characters could be used to determine the length at which fish undergo life-history thresholds and/or transitions.

Isometric growth produces a straight line with positive slope when a mensural character is plotted against some measure of overall body size, such as SL. Allometric growth occurs when the mensural character and SL vary in proportion over the whole range of SL, thus producing a non-linear relationship between the character and SL. Usually, allometric variation occurs gradually, resulting in a smooth curve – concave upwards if the character grows more rapidly than SL, and concave downwards if it grows more slowly. Such gradual allometric variation can be represented by a quadratic curve.

An abrupt transition in the relationship between a character and SL at a life history transition implies the existence of two intervals of isometric growth, each of which has a different proportional relationship between the character and SL. An abrupt transition can be represented by a 'split' linear regression (e.g. Nickerson et al. 1989) composed of two linear limbs (one representing the first life history interval and the other representing the second life history interval) that meet at some value of SL called the breakpoint. The breakpoint is the estimated length at which the transition occurs.

We tested each of the 23 mensural characters (hereafter called the dependent variables) for gradual allometry and abrupt transition. Our null hypothesis was that growth was isometric, and therefore best described by a simple linear regression. The first alternative hypothesis was that growth was gradual allometric, and best described by a quadratic equation. The second alternative hypothesis was that growth occurred in two different isometric stages and was best described by a split linear regression. We fitted simple linear, quadratic and split linear regression models to plots of the dependent variables against SL, and tested the quadratic and split linear models for significant improvements in fit over the simple linear model. The second alternative hypothesis, that body proportions change abruptly, was only accepted if the split linear fit was significantly better than both the simple linear and quadratic fits.

A simple linear regression has the form:

$$Y = aX + b. \tag{1}$$

Parameters a and b were estimated by least squares linear regression.

A quadratic equation has the form:

$$Y = cX^2 + dX + e. \tag{2}$$

A split regression can be imagined as two simple linear regressions, fitted to different non-overlapping data ranges, that meet at a breakpoint:

$$Y = \begin{cases} f(X - p) + h & \text{for } X < p \\ g(X - p) + h & \text{for } X \geq p, \end{cases} \tag{3}$$

where f and g are slope parameters for the two limbs of the regression, and h and p are the Y-axis and X-axis co-ordinates of the breakpoint, respectively. The parameters for equations (2) and (3) were estimated by least squares using the curve fitting routine in the Sigmaplot statistical and graphing package (Kuo 1994).

Testing whether a quadratic or split regression represents a significant improvement in fit over a simple linear regression is analogous to testing for an improvement in fit that results from the addition of extra parameters in a multiple regression model. We therefore used F tests as described by Sokal & Rohlf (1981):

quadratic versus linear:

$$F_{1,n-4} = \frac{(R_Q^2 - R_L^2)/1}{(1 - R_Q^2)/(n-4)},$$

split linear versus linear:

$$F_{2,n-5} = \frac{(R_S^2 - R_L^2)/2}{(1 - R_S^2)/(n-5)},$$

and split linear versus quadratic:

$$F_{1,n-5} = \frac{(R_S^2 - R_Q^2)/1}{(1 - R_S^2)/(n-5)},$$

where R_L^2, R_Q^2 and R_S^2 are the coefficients of determination for the linear, quadratic and split regressions respectively, and n is the sample size. Our total sample size was 387 fish, but occasional missing data meant that effective sample sizes varied from 379 to 387.

To assess the integrated transformations in body morphology, we transformed our data into their natural logarithms to account for size-related effects, retained the data of specimens for which all 23 characters were available (364 specimens × 23 characters), and treated the resulting data set with double-centred principal components analysis (PCA) using the ADE software package (Chessel & Dolédec 1993) as per Sagnes et al. (1997).

Environmental data collected during sampling of the River Great Ouse (Copp 1992) were re-analysed to test whether significant shifts in microhabitat use by stone loach coincided with observed shifts in relative growth. At each sampling point, fifteen environmental variables were measured qualitatively or by semi-quantitative classes. Of the fifteen, nine variables varied little with respect to stone loach microhabitat and thus only six variables were considered: *depth* (0–10, 10–20, 20–50, 50–100, 100–150, >150 cm), *substrate* (leaf bed or organic mud, <0.06 cm; mineral mud or sand and mud, <0.06 cm; sandy silt, sand or silty sand, 0.06–0.2 cm; gravel and sand or gravel and mud, 0.2–2.0 cm; pebbles, pebbles and sand or pebbles and gravel, 2.0–6.0 cm; rocks, rocks and mud, rocks and sand or blocks, >6.0 cm), *Myriophyllum* spp./*Elodea* sp. (absent, some, dense), *filamentous algae* (absent, some, dense), *ligneous debris/roots* (absent, some, dense), and *water velocity* (null, weak, medium, strong). Copp (1992) provided a full description of the methods used for assessing microhabitat character. We tested for differences in microhabitat use between four size classes of fish, as derived from the results of regression analysis, using the Mann-Whitney U-test, with the frequency distributions for specimens from a given size class compared with those of the other size class.

Results

The SL of loach specimens ranged from 15.3 to 115.4 mm, and total length (TL) ranged from 17.65

Table 2. Linear, quadratic and split linear regression statistics for mensural characters in stone loach from England. R^2 = coefficient of determination, Q = quadratic regression, L = linear regression, S = split linear regression, SE = standard error of the breakpoint, NS = not significant.

Character	R^2 linear	R^2 quadratic	R^2 split linear	F-test Q/L	p	F-test S/Q	p	F-test S/L	p	Best model	Breakpoint	SE	n
1 Head length	0.9910	0.9914	0.9913	17.77	<0.01	−3.64	NS	6.96	<0.01	Q	—	—	386
2 Preorbital distance	0.9843	0.9850	0.9850	17.83	<0.01	−0.68	NS	8.54	<0.01	Q	—	—	386
3 Eye diameter	0.9170	0.9281	0.9314	58.82	<0.01	18.25	<0.01	39.87	<0.01	S	32.3	1.257	385
4 Postorbital distance	0.9760	0.9761	0.9771	1.59	NS	16.32	<0.01	8.99	<0.01	S	32.8	2.642	385
5 Interorbital distance	0.9476	0.9477	0.9499	0.73	NS	16.93	<0.01	8.85	<0.01	S	28.5	2.335	386
6 Head depth	0.9823	0.9831	0.9827	18.04	<0.01	−8.45	NS	4.57	<0.05	Q	—	—	385
7 Predorsal distance	0.9965	0.9967	0.9968	23.15	<0.01	12.52	<0.01	18.19	<0.01	S	73.6	3.512	386
8 Preventral distance	0.9965	0.9965	0.9965	0.32	NS	8.03	<0.01	4.09	<0.05	S	29.0	3.409	386
9 Preanal distance	0.9982	0.9982	0.9982	0.00	NS	−1.10	NS	−0.55	NS	L	—	—	385
10 Postdorsal distance	0.9920	0.9925	0.9925	25.33	<0.01	0.89	NS	13.11	<0.01	Q	—	—	384
11 P–V distance	0.9924	0.9925	0.9927	5.08	<0.05	7.96	<0.01	6.57	<0.05	S	82.6	5.170	386
12 V–A distance	0.9875	0.9882	0.9882	22.36	<0.01	−0.02	NS	11.14	<0.01	Q	—	—	381
13 Body depth	0.9812	0.9812	0.9812	0.00	NS	−0.39	NS	−0.19	NS	L	—	—	387
14 Body width	0.9694	0.9698	0.9695	5.06	<0.05	−3.35	NS	0.83	NS	Q	—	—	386
15 C peduncle length	0.9679	0.9682	0.9683	3.61	NS	1.23	NS	2.42	NS	L	—	—	387
16 Min. body depth	0.9858	0.9863	0.9877	13.91	<0.01	43.02	<0.01	29.17	<0.01	S	88.7	2.106	385
17 D fin length	0.9692	0.9702	0.9713	12.79	<0.01	13.90	<0.01	13.56	<0.01	S	32.5	2.168	385
18 A fin length	0.9332	0.9422	0.9444	59.01	<0.01	15.00	<0.01	38.10	<0.01	S	37.2	1.548	383
19 P fin length	0.9573	0.9585	0.9586	10.99	<0.01	0.97	NS	5.98	<0.05	Q	—	—	384
20 V fin length	0.9773	0.9794	0.9797	38.53	<0.01	5.24	<0.05	22.10	<0.01	S	43.8	2.974	383
21 C fin length	0.9742	0.9750	0.9749	12.06	<0.01	−1.39	NS	5.30	<0.05	Q	—	—	381
22 D fin depth	0.9734	0.9751	0.9770	25.60	<0.01	31.51	<0.01	29.60	<0.01	S	37.7	1.860	379
23 A fin depth	0.9744	0.9766	0.9772	35.26	<0.01	10.58	<0.01	23.37	<0.01	S	40.6	2.477	379

to 136.9 mm. Most of the 23 characters expressed in % SL showed limited variation (Table 1). Twelve of the characters were best described by split linear regressions; eight characters were best described by quadratic regressions, and the remaining three characters, i.e. preanal distance, body depth and caudal peduncle length, were best described by simple linear regressions (F-tests, Table 2, see also Figure 1). Thus, only three mensural characters developed isometrically, i.e. in proportion to SL. Twelve mensural characters were characterised as isometric with an abrupt change at certain SL, and eight mensural characters as allometric. This means that in the twelve characters the growth rate was initially the same as that of SL, however it changed suddenly at a particular range of SL (given by the breakpoint and the range of SE), to continue again with the same rate as SL, whereas in the eight characters the growth rate was always different from that of SL.

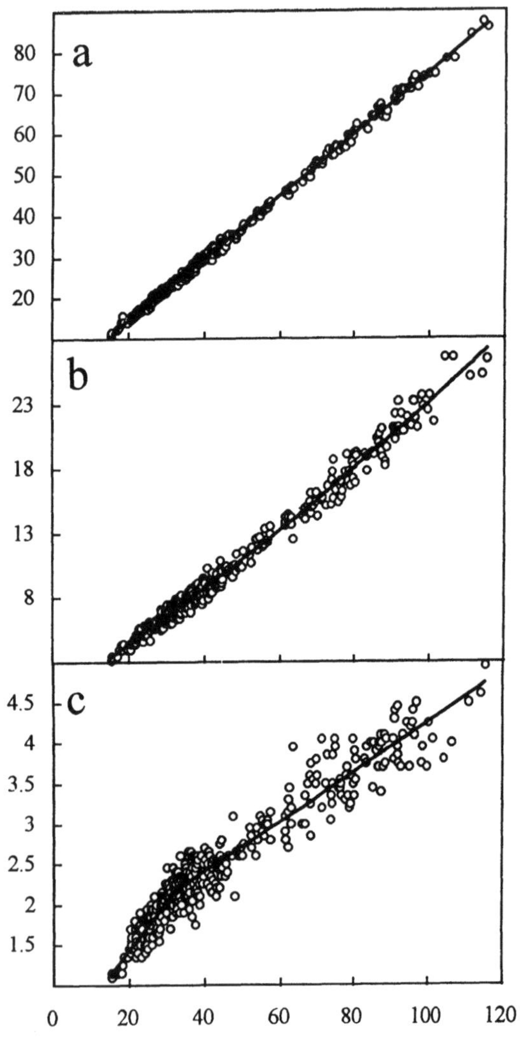

Figure 1. Examples of plots for mensural characters of stone loach from England against SL (in mm): a – preanal distance (linear regression, isometric growth), b – V–A distance (quadratic regression, allometric growth), c – eye diameter (split linear regression, isometric growth with abrupt change).

The twelve mensural characters with abrupt isometric growth can be divided into two groups, based on non-overlap of the standard error bars (Figure 2). Group 1 consisted of nine characters (eye diameter, postorbital distance, interorbital distance, preventral distance, D fin length, A fin length, V fin length, D fin depth and A fin depth). In this group, all the breakpoints fell within the range between 28 and 44 mm of SL (Table 2). Group 2 was composed of three characters (predorsal distance, P-V distance and minimum body depth) with breakpoints in the range from 73 to 89 mm of SL.

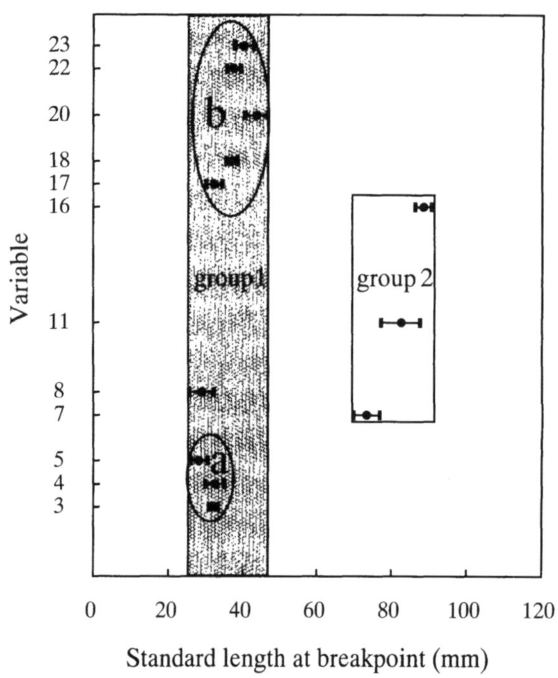

Figure 2. Estimates of the breakpoints and their standard errors (see Table 2) for the morphometric characters demonstrating significant abrupt changes in slope when plotted against SL. Two groups of characters are identified based on non-overlapping standard error intervals. Ovals indicate subgroups of characters within group 1 (a – head characters, b – fin characters). The numerical sequence of variables is the same as in Table 1.

Double-centred PCA of the reduced, and natural log transformed, data set revealed a progressive morphological transformation from the <26 mm through 26–35 mm SL (sub-group 1a) to the 36–47 mm SL (sub-group 1b) size classes, with a more distinct composite morphology in specimens >47 mm SL (Figure 3a). The two smaller size classes (<26 mm and 26–35 mm SL) demonstrated little or no overlap (95% ellipses) with the >47 mm SL group, with sub-group 1b (36–47 mm SL; Figure 2) appearing to be a transitional size class. The morphological characters contributing most to the observed changes were two head characters (eye diameter, interorbital distance), body width and depths, anal-fin length, and caudal-peduncle

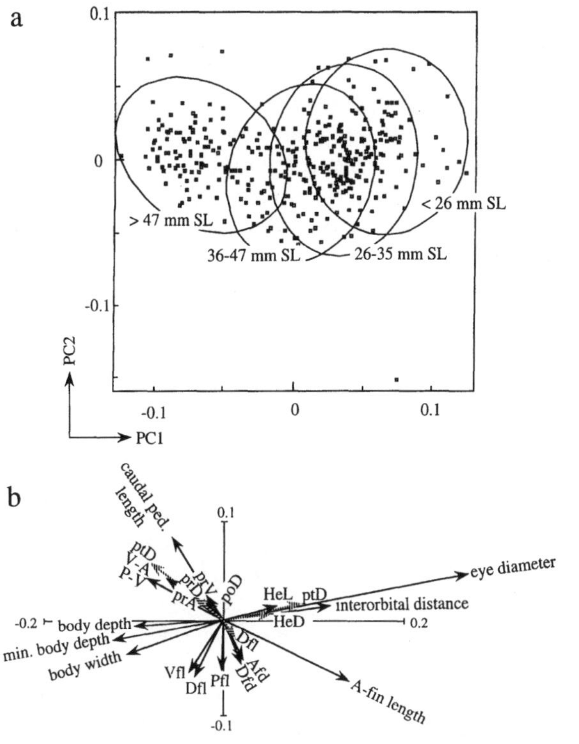

Figure 3. Double centred principal components analysis biplot of 364 stone loach based on 23 natural log transformed morphological measurements, with 95% ellipses, a – for specimens of the four size classes delimited using the results of regression analyses (Table 2), and b – the correlation vectors for the 23 morphological characters (the longer the length of vector, the more important the contribution to the trace). Abbreviations: HeL = head length, poD = preorbital distance, ptD = postorbital distance, HeD = head depth, prD = predorsal distance, prV = preventral distance, prA = preanal distance, ptD = postdorsal distance, P–V = P-V fin distance, V–A = V-A fin distance, Dfl = D fin length, Pfl = P fin length, Vfl = V fin length, Dfl = C fin length, Dfd = D fin depth, Afd = A fin depth.

length (Figure 3b) of which four characters demonstrated significant breakpoints (Table 2).

Corresponding with the abrupt changes in growth of group 1 were shifts in microhabitat use. Stone loach of <26 mm SL mainly occupied areas of silty sandy bottom at 10–20 cm depth, usually with no flow (or weak) and little vegetation but some filamentous algae and ligneous debris. Stone loach of subgroup 1a (26–35 mm SL, Figure 2) occupied similar areas to <26 mm SL specimens but with significantly (Mann-Whitney U-test, $p < 0.05$) shallower water depths (more frequently at 0–10 cm depth) and less ($p < 0.05$) ligneous debris (usually absent); subgroup 1a also frequented weakly flowing areas with larger substrata (i.e. sand with gravel), but neither difference was significant.

More dramatic was the microhabitat shift between stone loach of subgroups 1a and 1b; fish of subgroup 1b (36–47 mm SL) continued the shift to significantly ($p < 0.05$) shallower waters (mainly at 0–10 cm) but over larger ($p < 0.05$) substrata (i.e. gravel with pebbles) with less ($p < 0.05$) filamentous algae and higher ($p < 0.05$) water velocities (weak-to-medium). The microhabitat use of larger stone loach (>47 mm SL) resembled that of the 36–47 mm SL specimens with respect to substratum (gravel and pebbles) and ligneous debris (usually absent, but occasionally some), but they shifted back to significantly ($p < 0.05$) deeper (10–20 cm), faster (moderate) flowing waters with significantly ($p < 0.05$) less ligneous debris (usually absent) but greater amounts of *Myriophyllum* spp./*Elodea* spp.

Discussion

Central to the theory of saltatory ontogeny is that ontogeny is a sequence of longer stabilised states punctuated by rapid changes in 'integrative actions', or thresholds (Balon 1975, 1981, 1985, 1990). The life history model for fishes recognises a hierarchy of three different types of developmental intervals separated by thresholds: periods, phases and steps. Different morphological and/or physiological criteria have been used to identify the thresholds, but morphometric patterns in fish development have been ignored until recently (Crawford & Balon 1994, Copp & Kováč 1996, Kováč & Copp 1996). Therefore, a question emerges: can mensural characters be used to identify some thresholds in the life history of fish?

For stone loach ranging from 15.3 to 115.4 mm SL, twelve mensural characters demonstrated significant breaks in relative growth (Figure 2). In terms of the theory of saltatory ontogeny (Balon 1990), these 'breakpoints' could potentially indicate two kinds of thresholds: (1) the threshold between periods, e.g. between the larva and the juvenile period; or (2) the thresholds between steps within a period. Nevertheless, the breakpoints can be considered as developmental thresholds only if they coincide with each other and/or with other morphological, physiological, ecological or behavioural changes in the ontogeny of the species. In other words, if shifts in relative values of mensural characters do indicate any thresholds, then mensural characters with breakpoints occurring at a similar size must be associated with some common ecological, physiological and/or behavioural function.

Not all of the breakpoints in the twelve mensural characters appeared at the same SL. The range of SL at which the slopes changed significantly was quite wide (28–89 mm). However, two groups of breakpoints were distinguished among the twelve characters (Figure 2). With respect to whether mensural characters reflect life-history thresholds, it is very important that a large group of nine characters had breakpoints within a relatively narrow range of SL (28–44 mm; Figure 2). The large number of characters in this group suggests that this range of SL represents a life history interval, in which major remodelling of external body shape of the loach occurs. Moreover, two distinct and very consistent subgroups can be distinguished within the group 1 of characters (Figure 2). Subgroup 1a consisted of three head characters, with the breakpoints falling in a very narrow range (28.5–32.8 mm SL). Similarly, subgroup 1b was composed of five fin characters, and again with the breakpoints falling in a relatively narrow range (32.5–43.8 mm SL). In other words, major remodelling of the head and the fin apparatus in stone loach occurs rapidly, within a relatively short range of body size.

Most of the characters of group 1 are related to locomotion. The dorsal, anal and ventral fins attain their greatest relative size (Table 3) before their respective breakpoints (Table 2). Subsequently, the relative size of the fins slowly decreases, perhaps due to mechanical abrasion, as stone loach are bottom dwellers and have frequent contact with abrasive elements of the river bottom (sand, fine gravel and small cobbles). An interesting relation arises in relative positions of dorsal fin and ventral fins with respect to the longitudinal axis of the body. Both the predorsal distance and the preventral distance show abrupt isometric growth, however,

Table 3. Mean relative size (expressed as % of SL) for abruptly developing characters before and after the breakpoint.

Character	Relative size (% SL)	
	Before breakpoint	After breakpoint
3 Eye diameter	6.07	3.08
4 Postorbital distance	11.66	9.42
5 Interorbital distance	9.39	6.18
7 Predorsal distance	52.49	47.30
8 Preventral distance	57.56	53.06
11 P–V distance	32.58	36.32
16 Min. body depth	12.45	7.50
17 D fin length	15.39	11.55
18 A fin length	10.54	6.28
20 V fin length	17.51	13.97
22 D fin depth	24.85	18.31
23 A fin depth	18.35	14.22

the breakpoint occurs at about 74 mm SL in the former, and at about 29 mm SL in the latter. Therefore, based on relative size (Table 3) it appears that stone loach smaller than about 29 mm SL have the dorsal fin in a slightly more anterior position compared to ventral fins, specimens from 29 to 74 mm SL have all of these fins approximately in the same position, and the specimens bigger than about 74 mm SL have the dorsal fin in a slightly more posterior position with respect to ventral fins.

The coincidence of shifts in body morphology (Table 2) with those in microhabitat use (between the respective size classes) suggests that thresholds (though not as sudden as those between embryo and early larva steps) do occur during this interval of stone loach life history. But the question remains whether these thresholds separate two developmental steps within the juvenile period, or two periods (larva and juvenile). The first shift in morphometric values (occurring at 26–35 mm SL) represents a transformation of the head (characters 3–5), but microhabitat use before (<26 mm SL) and after (26–35 mm SL) the onset of this shift in relative growth differed little (water depth, amount of ligneous debris). The second shift in relative growth (at 36–47 mm SL) represents a change in fin shape and size as well as body form (characters 8, 17, 18, 20, 22 and 23), which probably improves swimming capabilities (thrust, manoeuvrability); microhabitat use before (26–35 mm SL) and after (36–47 mm SL) this second shift in relative growth differed more comprehensively, involving water depth, type of bottom substratum, water velocity and amount of filamentous

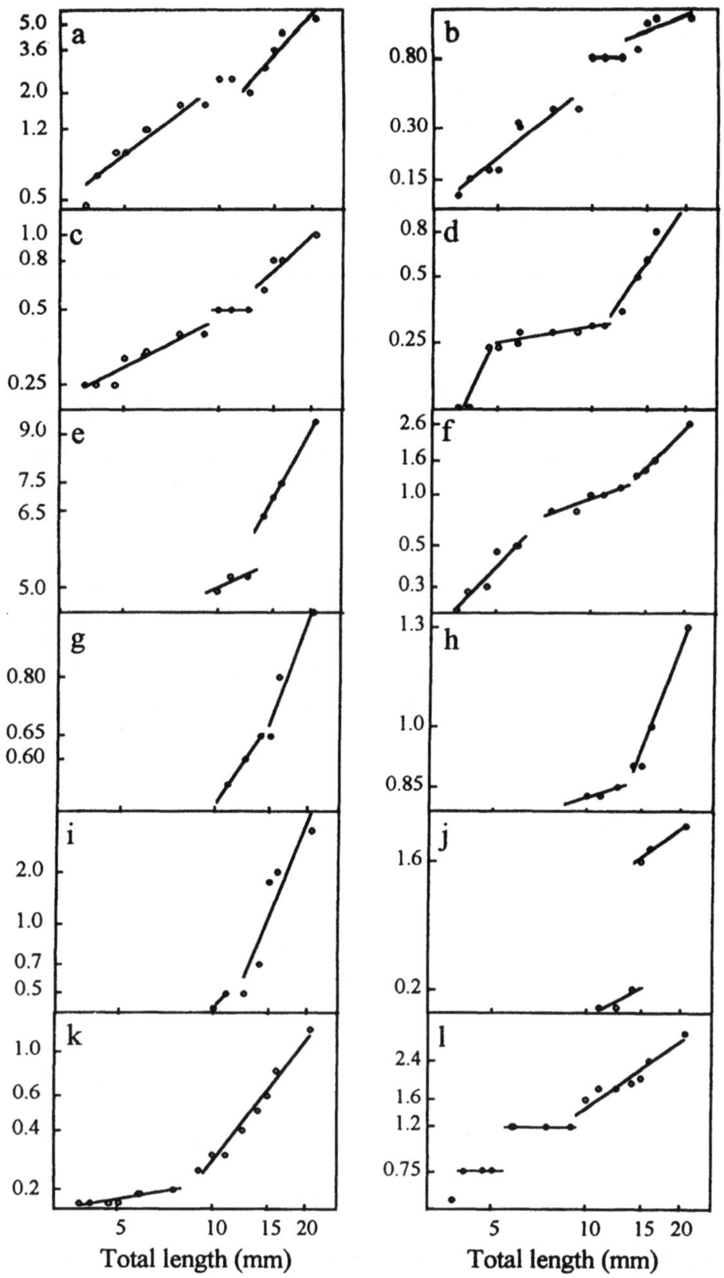

Figure 4. The development of mensural characters of loach from the River Raba (Poland) demonstrating significant breaks during embryo and larva periods (data from Starmach 1966): a – head length, b – preorbital distance, c – eye diameter, d – interorbital distance, e – postdorsal distance, f – body depth, g – A fin length, h – D fin length, i – D fin depth, j–A fin depth, k-minimum body depth, l–P fin length. Data are plotted on a double logarithmic grid.

algae. And although a significant change in microhabitat use (water depth and velocity, amount of vegetation and filamentous algae) was observed after (>47 mm SL) this second shift in relative growth (subgroup 1b), the underlying microhabitat character of specimens established with specimens 36–47 mm SL remained essentially shallow, lotic areas over gravel with some association with submerged vegetation or algae. This microhabitat profile appears to be characteristic of larger 'adult-like' (150 mm SL) stone loach elsewhere in the UK (Copp et al. 1994, Watkins et al. 1997) and in France (Mastrorillo et al. 1996).

The size at which the second transformation ends (47 mm SL) comes too late to suggest that this represents the transition to juvenile development (specimens may already be sexually mature at 52 mm SL; Mills et al. 1983). Similarly, the comprehensive morphology of the stone loach revealed by PCA (Figure 3) suggests that the morphology of specimens 26–35 mm SL resembles little that of >47 mm SL stone loach (limited overlap of ellipses); but the 36–47 mm SL size class appears as a transitional interval, probably the first juvenile step, after which only minor and unassociated changes in relative growth occur. This assumption is supported by the transitional microhabitat use of 36–47 mm SL stone loach, which is essentially similar to that of those > 47 mm SL, but still intermediate between the previous and subsequent size classes. Thus, we suggest that the larva period ends with the completion of the first shift in relative growth, i.e. not later than at 35 mm SL, depending on individual variability.

We suggest further that the subsequent stabilisation of the three mensural characters of group 2 (Figure 2) are minor adjustments during the juvenile period. The range of SL at which the breakpoints in these characters occurred is rather wide and inconsistent (73.6–88.7 mm). Moreover, in this case the changes in the proportional development do not seem to be associated with any important morphological, physiological, ecological or behavioural changes in the ontogeny of the species. Therefore, we do not consider these breakpoints to reflect any kind of threshold.

A brief analysis of available data on the development of mensural characters during embryo and larva periods in stone loach (Starmach 1966) revealed obvious and sudden changes in twelve of the fifteen mensural characters studied (Figure 4). Unfortunately, Starmach (1966) did not provide any comments on the developmental steps and thresholds, and therefore it is not possible to identify the exact thresholds with which these morphometric changes correspond. However, nine of them, connected with body shape and head and fin proportions, occurred at the same size approximately (12.5 mm TL, Figure 4a–i) and five of them also at 9 mm TL (Figure 4a–c, k and l). Such sudden morphological changes must be associated with new organ-to-organ and/or organism-to-environment interactions, i.e. they reflect thresholds between two different stabilised states (life-history intervals; Balon 1990).

Stone loach achieve a more-less definitive form at about 86 to 91 mm SL (when the final breakpoint in a mensural character occurred), though such a definitive appearance theoretically does not exist in this species, as some mensural characters show positive or negative allometry throughout the ontogeny. However, the maximum SL of a fish is limited by death, which actually means that the definitive appearance does exist in practice. Specimens of 86–91 mm SL may be: (1) immature, i.e. juveniles (Mills & Eloranta 1985); (2) at the start of the adult period of life (Skryabin 1993); or (3) already sexually mature (i.e. in adult period of life, Mills et al. 1983). However, the last threshold within the juvenile period in stone loach probably occurs not later than about 47 mm SL, when a notable stabilisation in the relative growth coincides with significant changes in microhabitat. Thus, the development of external morphology follows a similar pattern as that observed in other morphological, physiological, behavioural and another developmental events that occur within the periods in the life of fish: it decreases as the fish grows. In other words, the most significant changes, or thresholds (reflecting the highest number of coincident completion of structures and changing functions, Balon 1990) occur in stone loach ontogeny mainly during the embryo and larva periods. During the juvenile period, only one threshold probably takes place, and in the later development, the changes in mensural characters are not associated with other developmental events, and so they are not strong enough to be considered as thresholds.

Acknowledgements

The research was undertaken during a Royal Society-funded research visit by VK to the University of Hertfordshire on material collected during the Fishmongers' Company Postdoctoral Fellowship to GHC. We thank C. Flegler-Balon for her valuable comments on earlier versions of this manuscript, as well as M. Watkins, T. Bennetts, V. Jubb and J. Černý for assistance in the field and J. Lathen for assistance in laboratory and R.I.C.C. Francis for statistical advice. This paper was partly supported by Slovak SGA, Project No 1/6172/99.

References cited

Atchley, W.R., C.T. Gaskins & D. Anderson 1976. Statistical properties of ratios. I. Empirical results. Systematic Zoology 25: 137–148.

Balon, E.K. 1956. Neres a postembryonálny vývoj plotice (*Rutilus rutilus* ssp.) (Spawning and postembryonic development of roach) Biologické práce 2(13): 7–60 (in Slovak).
Balon, E.K. 1975. Terminology of intervals in fish development. J. Fish. Res. Board Can. 32: 1663–1670.
Balon, E.K. 1981. Saltatory processes and altricial to precocial forms in the ontogeny of fishes. Amer. Zool. 21: 573–596.
Balon, E.K. 1985. The theory of saltatory ontogeny and life history models revisited. pp. 13–30. *In*: E. K. Balon (ed.) Early Life Histories of Fishes, Dr W. Junk Publishers, Dordrecht.
Balon, E.K. 1990. Epigenesis of an epigeneticist: the development of some alternative concepts on the early ontogeny and evolution of fishes. Guelph Ichthyol. Rev. 1: 1–48.
Chessel, D. & S. Dolédec. 1993. ADE Version 3.6: HyperCard© Stacks and MicroSoft QuickBasic© Programme library for the analysis of environmental data. Université Lyon I, Villeurbanne.
Copp, G.H. 1992. An empirical model for predicting the microhabitat of 0+ juveniles in lowland streams. Oecologia 91: 338–345.
Copp, G.H. & V. Kováč 1996. When do fish with indirect development become juveniles? Can. J. Fish. Aquat. Sci. 53: 746–752.
Copp, G.H., S. Warrington & Q. de Bruine. 1994. Comparison of diet in bullhead, *Cottus gobio*, and stone loach, *Barbatula barbatula*, in a small English lowland river. Folia Zoologica 43: 171–176.
Crawford, S. & E.K. Balon 1994. Alternative life histories of the genus *Lucania*: 2. Early ontogeny of *L. goodei*, the bluefin killifish. Env. Biol. Fish. 41: 331–367.
Gatz, A.J. 1979. Community organization in fishes as indicated by morphological features. Ecology 60: 711–718.
Gozlan, R.E., G.H. Copp & J.-N. Tourenq. 1999. Comparison of growth plasticity in the laboratory and field, and implications for the onset of juvenile development in sofie, *Chondrostoma toxostoma*. Env. Biol. Fish. 56: 153–165 (this volume).
Jackson, D.A., H.H. Harvey & K.M. Somers 1990. Ratios in aquatic sciences: statistical shortcomings with mean depth and the morphoedaphic index. Can. J. Fish. Aquat. Sci. 47: 1788–1795.
Hoda, S.M.S. & H. Tsukahara. 1971. Studies on the development and relative growth in the carp, *Cyprinus carpio* (Linne). Journal of the Faculty of Agriculture, Kyushu University 16: 387–509.
Holčík, J. (ed.) 1989. The freshwater fishes of Europe, Vol 1/II. AULA, Wiesbaden. 469 pp.
Holčík, J. & V. Skořepa 1971. Revision of the roach, *Rutilus rutilus* (Linnaeus, 1758), with regards to its subspecies. Annotationes Zoologicae et Botanicae 64: 1–60.
Koblitskaya, A.F. 1981. Identification key for young stages of freshwater fish. Leg. I. Pishtsch. Promyshlennost, Moscow. 208 pp. (in Russian).
Kottelat, M. 1997. European freshwater fishes. An heuristic checklist of the freshwater fishes of Europe (exclusive of former USSR), with an introduction for non-systematists and comments on nomenclature and conservation. Biologia (Bratislava) 52 (supplement 5): 1–271.
Kováč, V. 1987. Morphology of Slovak and Mongolian populations of the stone loach, *Noemacheilus barbatulus* (Linnaeus, 1758) with notes on its systematics. Acta Facultatis Rerum Naturalium Universitatis Comenianae – Zoologia 29: 79–129.
Kováč, V. & G.H. Copp 1996. Ontogenetic patterns of relative growth in young roach *Rutilus rutilus* (L.): within-river basin comparisons. Ecography 19: 153–161.
Kuo, J. 1994. Sigmaplot scientific graphing software. Transforms and curve fitting. Jandel Scientific, San Rafael.
Marr, J.C. 1955. The use of morphometric data in systemetic, racial and relative growth studies in fishes. Copeia 1955 23–31.
Mastrorillo, S., F. Dauba & A. Belaud. 1996. Utilisation des microhabitats par le vairon, le goujon et la loche franche dans trois rivières du sud-ouest de la France. Ann. Limnol. 32: 158–195.
Mills, C.A. & A. Eloranta. 1985. Reproductive strategies in the stone loach *Noemacheilus barbatulus*. Oikos 44: 341–349.
Mills, C.A., J.S. Welton & E.L. Rendle. 1983. The age, growth and reproduction of the stone loach *Noemacheilus barbatulus* (L.) in a Dorset chalk stream. Freshwater Biology 13: 283–292.
Nickerson, D. M., D. E. Facey & G. D. Grossman 1989. Estimating physiological thresholds with continuous two-phase regression. Physiological Zoology 62: 866–887.
Oikawa, S. & Y. Itazawa. 1984. Allometric relationship between tissue respiration and body mass in the carp. Comp. Biochem. Physiol. 77A: 415–418.
Řepa, P. 1969. Über Morfologie, proportionales Wachstum und Organogenese der Frühstadien des Flussbarsches (*Perca fluviatilis* L. 1758). Acta Univ. Carolinae-Biologica 1: 61–92.
Sagnes, P., P. Gaudin & B. Statzner. 1997. Shifts in morphometrics and their relation to hydrodynamic potential and habitat use during grayling ontogenies. J. Fish Biol. 50: 846–858.
Skryabin, A.G., 1993. The biology of stone loach *Barbatula barbatulus* in the rivers Goloustnaya and Olkha, East Siberia. J. Fish Biol. 42: 361–374.
Snedecor, G.W. 1946. Statistical methods. Iowa State College Press, Ames. 485 pp.
Sokal, R.R. & F.J. Rohlf 1981. Biometry. The principles and practice of statistics in biological research, 2nd edition. W. H. Freeman, San Francisco. 859 pp.
Starmach, J. 1966. Rozród oraz rozwój embryonalny i larwalny u śliza (*Nemachilus barbatulus* L.) (Reproduction and embryonic and larval development of the stone loach). Acta Hydrobiol. 8: 111–122.
van Snik, G.M.J., J.G. van den Boogaart & J.W.M. Osse. 1997. Larval growth patterns in *Cyprinus carpio* and *Clarias gariepinus* with the attention to the finfold. J. Fish Biol. 50: 1339–1352.
Watkins, M.S., S. Doherty & G.H. Copp. 1997. Microhabitat use by 0+ and older fishes in a small English chalk stream. J. Fish Biol. 50: 1010–1024.
Watson, D.J. & E.K. Balon 1984. Ecomorphological analysis of fish taxocenes in rainforest streams of northern Borneo. J. Fish Biol. 25 : 371–384.
Webb, P.W. & D. Weihs. 1986. Functional locomotor morphology of early life history stages of fishes. Trans. Amer. Fish. Soc. 115: 115–127.

Environmental Biology of Fishes **56:** 117–128, 1999.

Correspondence between ontogenetic shifts in morphology and habitat use in minnow *Phoxinus phoxinus*

Predrag D. Simonović[a], Paul Garner[b], Edward A. Eastwood[c], Vladimír Kováč[d] & Gordon H. Copp[c]
[a]*University of Belgrade, Faculty of Biology, Institute of Zoology, Studentski trg 16, 11001 Belgrade, Yugoslavia (e-mail: pedja@bf.bio.bg.ac.yu)*
[b]*Freshwater Biological Association, Ferry House, Far Sawrey, Ambleside, Cumbria, UK. Current address: Relum Ltd, Carlton Park Industrial Estate, Saxmundham, Suffolk, IP17 2NL, UK*
[c]*Landscape & Ecology Research Group, Department of Environmental Sciences, University of Hertfordshire, College Lane, Hatfield, Herts. AL10 9AB, UK*
[d]*Comenius University, Faculty of Natural Sciences, Institute of Ecology, Mlynská dolina B2, 842 15 Bratislava, Slovakia*

Received 3 June 1998 Accepted 1 December 1998

Key words: developmental shifts, habitat use, mensural characters

Synopsis

The morphology of minnows *Phoxinus phoxinus* from two rivers at the south-east of England was analysed on mensural characters and qualitative traits (occurrence of appendages, pattern of pigmentation and scalation). Eight developmental intervals were identified from the qualitative traits, and bivariate and multivariate analysis revealed that allometric growth occurs occasionally during ontogeny, mainly in earlier developmental periods. Body shape is under the influence of rapid increase during development, mainly due to the quick growth in early developmental intervals. The exclusion of 'general size', remarkable in early developmental intervals, revealed several 'shape' characters that mainly determine minnow morphology by changing the mode of variability during development (e.g. caudal characters, maximum body height, belly length and ventral head length). The tail certainly undergoes the most consistent and most durable change, its characters being the most variable and the most discriminative for developmental periods from larvae to adults. The most prominent alteration in the overall shape development occurs at the transition from larva to juvenile, both regarding the number of characters involved into the change and their variability. This transition takes place at about 28 mm SL, which corresponds to a significant change in microhabitat use in both the rivers Lee and Frome, characterized by shifts by 0+ juveniles to significantly deeper waters than those occupied by larvae, with increased amounts of either submerged filamentous algae (Lee) or vegetation (Frome).

Introduction

The study of morphology throughout an organism's ontogeny, either on the longitudinal or the cross-sectional data set (Cock 1966), can include different levels of organisation (e.g. organism, organ systems, organs, tissues and cells), which play particularly prominent roles in certain developmental intervals. However, to identify shifts in resource use, a good starting point is at higher levels of organisation (organism, organ systems), which are linked to functional changes that underly these shifts.

Morphological development plays an important role in the early ontogeny of stream-dwelling fishes such as the grayling, *Thymallus thymallus* (L.) (Sagnes et al. 1997, Sempeski & Gaudin 1995) and the roach, *Rutilus rutilus* (L.) (Copp & Kováč 1996). Of particular importance to cyprinid fishes (Copp & Kováč 1996, Kováč &

Copp 1996; also G.H. Copp unpublished, Gozlan et al. 1999 this volume) and some other families (Kováč et al. 1999 this volume) is the question of when metamorphosis ends and juvenile development begins in fishes that undergo indirect development (Balon 1999 this volume). This is an important consideration when using ontogenetic scales for interspecific comparisons of morphological development as well as in the identification of ontogenetic shifts in resource use. We suspect this to be true of the European minnow, *Phoxinus phoxinus* (L.), a common cyprinid species in lakes and rivers of the palearctic region (Howes 1985) that distinguishes itself from other cyprinids (Mills 1987) by having a life history more akin to that of other small, benthic species such as the bullhead *Cottus gobio* (L.) and the stone loach *Barbatula barbatula* (L.). The minnow demonstrates great plasticity in habitat use and in its biology, being able to reproduce both in early summer and late fall (Mills 1987) as well as to subsist under both natural (Garner 1997b) and human-induced environmental perturbations (Mastrorillo et al. 1996). The short life span and great environmental plasticity of the minnow suggests that it may undergo a relatively short metamorphosis compared to other cyprinids, perhaps possessing unique or unusual patterns of relative growth to facilitate its pioneer-style life history (Mills 1987, Mastrorillo et al. 1996).

The aim of the present study was to examine the relationship between morphological development and resource use in the minnow, with the specific objectives of: (1) identifying a morphometric–habitat–behavioural basis for distinguishing the larva period of development in the minnow, and (2) to determine at what point metamorphosis ends and juvenile development begins. Mensural characters were examined to determine their overall variation and their discriminant power for distinguishing particular intervals of development. We have not directly measured the hydrodynamic and biomechanical roles of changes in these characters, rather the patterns were interpreted based on general knowledge of their function.

Material and methods

Fish and habitat use (environmental) data were collected from two river systems (rivers Lee and Frome, UK) using point abundance sampling by electrofishing (see Copp & Garner 1995, Garner 1997a). A total of 710 samples were collected from the River Lee at Woolmers Park (NGR: TL 288 100) in Hertfordshire (England) between May and October 1995 by Watkins et al. (1997). For a more detailed description of the site, see also Copp & Bennetts (1996). Fourteen environmental variables were measured at each sampling point: distance from bank (in m); depth (cm); channel width (m); bed slope (depth/distance from bank); clay, silt, sand, gravel and pebbles (as a % of point sample area); submerged macrophytes (%); filamentous algae (%); number of submerged branches; tree roots (%); number of overhanging trees & bushes; water temperature °C; dissolved oxygen (mg l^{-1}); and water velocity (cm s^{-1}). Seasonal variation in the abundance of filamentous algae and in water temperature were eliminated by using the deviates from the weekly mean rather than the raw measurements.

A total of 225 point samples were collected from the River Frome (NGR SY869 869) in Dorset between 17–19 July 1997, using a portable, battery-powered (pulsed DC at 200 volts, 200 pulses per minute and 3/4 pulse width) electrofishing apparatus (Electracatch, Wolverhampton, UK). For a full description of the study sites refer to Ibbotson et al. (1994). Upon capture, fish were overdosed with carbon dioxide and frozen. At each sampling point from the River Frome, the following habitat variables were recorded: depth of the water column (m) and mean column velocity (cm s^{-1}) were measured quantitatively and converted to categories; percentage of *Ranunculus* sp., reeds, and ligneous material, and substratum composition (modified Wentworth scale, 1 = leaf litter, 2 = silt, 3 = mud, 4 = sand, 5 = gravel, 6 = pebbles, 7 = boulders, 8 = bedrock) were visually estimated.

Morphological analysis

The specimens (River Lee, $n = 158$; River Frome, $n = 44$) were first examined separately by origin for three groups of qualitative traits derived from Koblitskaya (1981) as well as descriptions of other cyprinids (Prokeš & Peňáz 1978, 1979, Krupka 1988, Economou et al. 1991): (1) the absence or presence of appendages (median finfold, dorsal, anal, pectoral, ventral and caudal fins), (2) body pigmentation patterns (no pigmentation, single melanophores, rows of distinct melanophores, dark narrow band along the flank and band of large ocellate spots along the flank), and (3) scale coverage (no scales, less than 50% of body scaled, between 50% and 75% of body scaled, and fully scaled). These data were subjected to chi-square (χ^2) analysis (Sokal & Rohlf 1981) to identify significant associations between qualitative characters and fish size, which was determined by centroid cluster analysis (Rohlf 1988) of χ^2 distances between them; this revealed eight and six groupings of specimens from the rivers Lee and Frome, respectively, that we refer to henceforth as developmental intervals (sensu Balon 1990).

Correspondence between these eight intervals of development and ontogenetic intervals described elsewhere (e.g. Prokeš & Peňáz 1978, Copp 1992) was then established: (1) free-embryo (FE, sensu Balon 1990) referred to specimens with predominant yolk-sack not showing food items in the gut; (2) 'young larvae' referred to the conventional larva steps (L1–L3) described elsewhere, from the onset of exogenous feeding to just prior to differentiation of the swim bladder; (3) as 'middle larvae' referred to larva intervals L4–L6, from differentiation of the swim bladder to reduction of the finfold, muscular differentiation and pelvic fins surpass the preanal finfold; (4) 'older larvae' referred to specimens at the end of metamorphosis, from complete differentiation of the fins and the nasal septa to the onset of bifurcation of the fin rays and initial appearence of scales; (5) 'young juveniles' referred to specimens possessing complete scale cover, nasal septa, fully defined and bifurcated fins/rays; (6) 'older juveniles' referred to specimens more developed and approaching maturation; and specimens thought to be sexually mature (Mills 1987) were categorised as (7) 'young adults' and (8) 'older adults'.

Ten continuous mensural characters (linear distances between eight homologous points) common to all developmental intervals in the minnow were measured (Figure 1) using an Image Analysis system (Acorn Risc PC 900 in conjunction with a Watford Electronic's Archimedes Video Digitiser, fed from a Sony video camera Model XC-75CE, equipped with a SIGMA 50 mm macro lens), supported with a software package (FishMetric) specifically designed by one of us. The standard length (SL) of minnows in each developmental interval from the two rivers was compared using the Student's t-test. Subsequently, the Lee and Frome data were analysed separately.

Ontogenetic trajectories on log-transformed data from the Lee and Frome were calculated for each interval on the scores derived from principal component analysis (PCA), obtained from the correlation matrix of log-transformed data (Joliceur 1963), by bivariate regression of PC2 to PC1 and tested for subsequent stages by Tukey's q-test (Zar 1984). The analysis of shape changes through particular intervals of development was conducted on the data set by factor analysis (FA), applying the varimax rotation (Sneath & Sokal 1973), with size-effects in all developmental intervals removed by changing the initial raw data with residuals from the bivariate linear regression of each character versus SL (Sneath & Sokal 1973); multiple analysis of variance (MANOVA) was used to test for differences between the developmental intervals identified in the initial χ^2 analysis of qualitative characters.

Changes in the proportional growth of the minnows were tested for using split linear regression analysis of the factor scores (axes 1 and 2) plotted against SL to determine whether a split regression represented a significant improvement in fit over the simple linear regression (Kováč et al. 1999 this volume). The split regression can be imagined as two simple linear regressions, fitted to different non-overlapping data

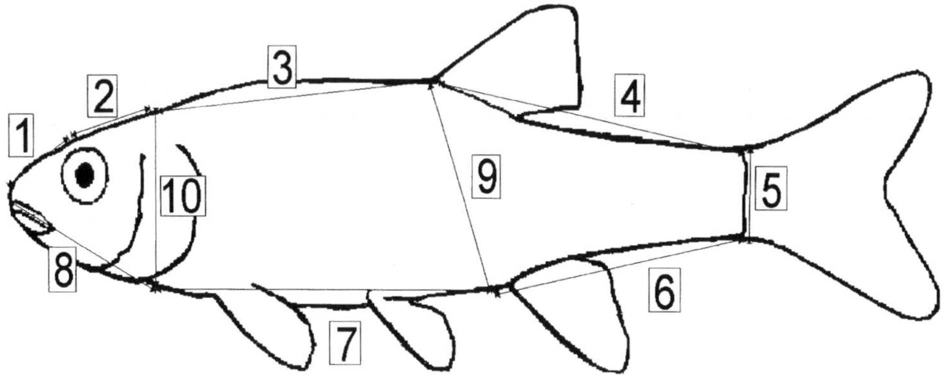

Figure 1. Linear distances of mensural characters between homologous points (1, snout length; 2, nape length; 3, predorsal length; 4, dorsal tail length; 5, tail height; 6, ventral tail length; 7, belly length; 8, ventral head length; 9, mid-body height; 10, nape-level height).

ranges, that meet at a point called the breakpoint. A split regression has the form:

$$Y = \begin{cases} c(X - p) + e & \text{for } X < p \\ d(X - p) + e & \text{for } X \geq p, \end{cases}$$

where c and d are slope parameters for the two limbs of the regression, and e and p are the Y-axis and X-axis coordinates of the breakpoint, respectively. The parameters c, d, e and p were estimated by least squares using the curve fitting routine in the Sigmaplot statistical and graphing package (Kuo 1994). The F-test described in Sokal & Rohlf (1981) was used to test whether a split regression represents a significant improvement in fit over a simple linear.

Microhabitat analysis

The microhabitat data from the River Lee (Watkins et al. 1997) and the River Frome were re-analysed to compare the mean values of each habitat variable for specimens categorised by the regression and factor analyses as free embryos + larvae (<27mm SL), juveniles (27–57 mm SL) and adults (>57 mm SL) using analysis of variance (ANOVA) and the Kruskal–Wallis test. Suitability scores were calculated for the River Frome data as the ratio of the frequency of habitat used to that available, and normalised to 1 for comparison with previous work (Garner & Clough 1996).

Results

Free-embryos from the River Lee possessed no pigmentation (Table 1). At the onset of exogenous feeding, young larvae (7.8–12.6 mm SL) began to possess single melanophores. Middle and older larvae (12.6–27.2 mm SL) had rows of distinct melanophores. Young and older juvenile minnows (27.2–57.3 mm SL) possessed a narrow, dark longitudinal band along the flank and around the snout. Specimens > 57.3 mm SL, presumed to be adults (Mills 1987, gave 50 mm fork length as the size-at-maturity for minnow), had a broad band of large ocellate spots along the flank.

Minnows from the River Frome followed the same pattern of pigmentation and the appearance of appendages, however, following the larva period significant differences for mean SL were found in young and older juveniles, as well as in the adult developmental intervals (Table 2). Therefore, the two minnow samples were subsequently analysed separately.

Clustering of the χ^2 distances of qualitative characters (Table 3) for River Lee minnows revealed four groupings of developmental intervals (Figure 2): (1) free embryos; (2) young and middle larvae; (3) older larvae, young and older juveniles; and (4) the adult intervals. The χ^2 distances (Table 4) and clustering (Figure 3) of the Frome minnows were different, with the following groupings: (1) middle larvae; (2) older larvae and young juveniles; (3) older juveniles, and (4) adults (younger + older combined). There were no free embryos and young larvae in the River Frome samples, therefore graphic illustrations of the factor analysis results are given for Lee specimens only.

Table 1. Grouping of minnow specimens from the River Lee into developmental intervals according to the absence and presence of qualitative traits: F – finfold; D – dorsal; C – caudal; A – anal; P – pectoral; V – ventral fins; nsc – no scales; hsc – half of body scaled; tsc – 3/4 of body scaled; wsc – whole body scaled; nco – no coloration; lco – larva coloration; jco – juvenile coloration; aco – adult coloration.

	Interval	SL range	Appendages						Scales				Coloration			
			F	D	C	A	P	V	nsc	hsc	tsc	wsc	nco	lco	jco	aco
1	Free embryos	7.4–7.8	4	0	0	0	4	0	4	0	0	0	4	0	0	0
2	Young larvae	9.9–11.6	8	8	7	7	8	0	3	5	0	0	0	8	0	0
3	Middle larvae	12.0–12.5	4	4	4	4	4	0	0	0	2	2	0	4	0	0
4	Older larvae	13.6–27.2	0	30	30	30	30	30	0	0	0	30	0	0	33	0
5	Young juveniles	27.2–35.0	0	38	38	38	38	38	0	0	9	29	0	0	38	0
6	Older juveniles	35.0–44.9	0	38	38	38	38	38	0	0	6	32	0	0	38	0
7	Young adults	44.9–57.3	0	27	27	27	27	27	0	0	0	27	0	0	11	16
8	Older adults	57.7–69.2	0	9	9	9	9	9	0	0	0	9	0	0	0	9

Table 2. Mean (M), standard deviation (s), number of specimens per interval (n), Student's t-value (t, unequal variances option), degrees of freedom (df) and probability (p) for comparisons of standard lengths (in mm) of minnows from the rivers Lee and Frome at particular developmental intervals (see Table 1).

Developmental interval	River Lee $M \pm s$ (mm)	n_L	River Frome $M \pm s$ (mm)	n_F	t	df	p<
Middle larvae	12.3 ± 0.23	4	12.5 ± 0.00	1	—	—	
Older larvae	21.9 ± 3.32	30	21.0 ± 4.19	9	0.619	11	n.s.
Young juveniles	31.7 ± 2.20	38	34.0 ± 2.24	10	2.823	14	0.02
Older juveniles	39.0 ± 2.77	38	42.8 ± 2.68	10	3.964	14	0.01
Young & older adults	53.0 ± 6.63	36	58.8 ± 4.35	14	3.365	36	0.01

Table 3. The number of specimens (n), and the χ^2 distances between developmental intervals of minnows from the River Lee on their discrete characters with the means (M) and standard errors of means (SE) for standard length (SL) in mm for each interval and significant differences between them.

	$M \pm$ SE	Free embryo $n = 4$	Young larvae 8	Middle larvae 4	Older larvae 30	Young juvenile 38	Older juvenile 38	Young adults 27	Older adults 9
Free embryos	7.6 ± 0.09	0							
Young larvae	10.8 ± 0.16	5.480	0						
Middle larvae	12.3 ± 0.12	5.169	3.453	0					
Older larvae *	21.9 ± 0.61	13.128	10.798	9.162	0				
Young juveniles ***	31.7 ± 0.36	14.642	11.895	9.236	2.862	0			
Older juveniles ***	39.0 ± 0.45	14.642	11.895	9.314	2.793	0.865	0		
Young adults	49.8 ± 0.65	12.513	10.357	6.742	4.971	6.107	5.879	0	
Older adults	62.8 ± 1.15	7.864	7.157	5.596	6.245	7.045	6.973	2.298	0

* = $p < 0.05$; *** = $p < 0.001$.

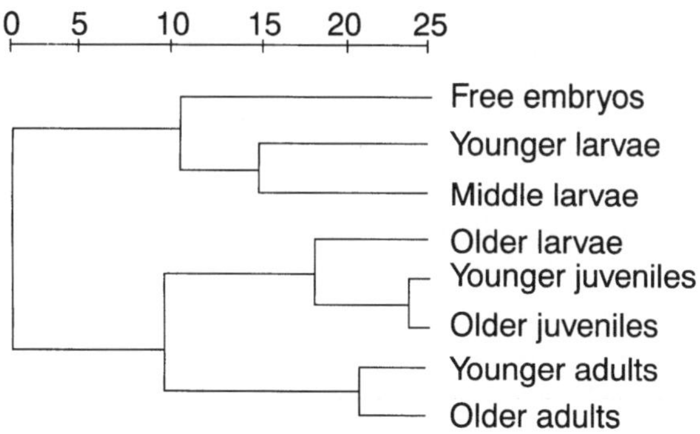

Figure 2. Dendogram of χ^2 distances between developmental intervals of River Lee minnows based on qualitative traits.

Corroboration of morphological distinction between particular developmental intervals in the River Lee minnows was apparent in the significant differences in the ontogenetic trajectories of older larvae and young juveniles ($b_{ol} = -0.62 \pm 0.562$, $b_{yj} = 1.93 \pm 0.731$; $q = 3.156$; $k = 7$; df $= 138$; $p < 0.05$) and those of older juveniles and young adults ($b_{oj} = 2.76 \pm 1.374$, $b_{ya} = -2.09 \pm 0.909$; $q = 4.470$; $k = 7$; df $= 138$; $p < 0.02$). With the size-effect retained in the character set, the small amount of variability explained by 'shape-axes' of PC2 ($\lambda_2 = 0.252$, 2.3%) and PC3 $\lambda_3 = 0.136$, 1.2%), in relation to PC1 ($\lambda_1 = 10.382$, 94.4%), illustrates its huge influence. In Frome specimens, middle larvae were represented by only one fish, which was, thus, omitted from comparison of ontogenetic trajectories. However, only older larvae differed from young juveniles ($b_{ol} = -0.088 \pm 0.165$, $b_{yj} = 2.635 \pm 1.525$, $q = 2.273$, $k = 4$, df $= 35$, $p < 0.05$). The shape was again enormously impacted with the general size ($\lambda_1 = 10.435$, 94.9%; $\lambda_2 = 0.254$, 2.3%).

Elimination of size effects in the Lee specimens, and ordination of 'size-independent' characters, revealed different mensural characters as being the most variable (Figure 4). Thus, the first Factor $\lambda_1 = 1.731$, 62.49%) was weighted as highly positive for nape level

Table 4. The number of specimens (n), and the χ^2 distances between developmental intervals of minnows from the River Frome on their discrete characters with the means (M) and standard error of means (SE) for standard length (SL) in mm.

	$M \pm$ SE	Middle larvae $n = 1$	Older larvae 9	Young juveniles 10	Older juveniles 10	Young & older adults 14
Middle larvae	12.5 ± 0.00	0				
Older larvae	21.0 ± 1.40	5.477	0			
Young juveniles	34.0 ± 0.71	5.745	0.000	0		
Older juveniles	42.8 ± 0.85	5.745	2.809	2.928	0	
Young & older adults	58.8 ± 1.16	6.708	4.796	4.899	2.592	0

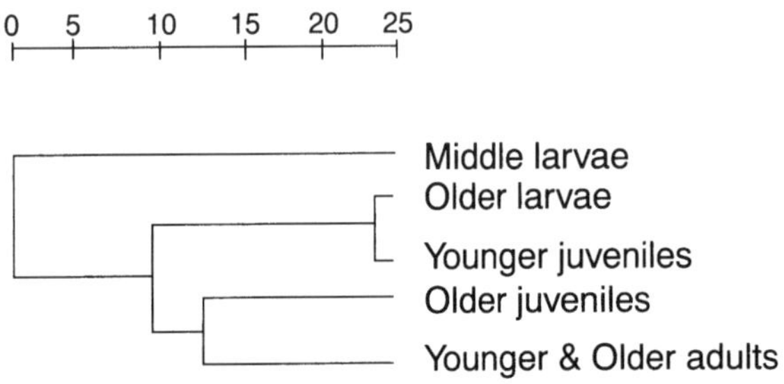

Figure 3. Dendogram of χ^2 distances between developmental intervals of River Frome minnows based on qualitative traits.

height (Figure 5), along with belly length, whilst highly negative for ventral tail length; this is most prominent for older larvae and least so far for free embryos. The second Factor $\lambda_2 = 0.885$, 31.92%) was weighted as highly positive for ventral head length along with mid-body height and caudal peduncle, whereas belly length was highly negative; this was prominent in many developmental intervals, but least so in middle and older larvae, prior to the transition to juvenile development (Figures 4, 5). Significant differences in variability between intervals was observed (Hotteling's $T^2 = 2.771$, df $= 35 - 745$, $p < 0.001$). The composite patterns of ontogenetic changes in relative growth of Lee specimens, expressed by the factorial analysis scores for axes 1 and 2 displayed significant changes in slope, with breakpoints occurring at about 28 and 24 mm SL, respectively (Figure 6).

In the ordination of 'size-independent' characters in Frome specimens (figure not shown), the first Factor ($\lambda_1 = 4.985$, 49.9%) was explained with snout length, caudal peduncle, mid-body height, and nape level height (illustration not presented here). The second 'size-independent' Factor ($\lambda_2 = 1.214$, 12.1%) was explained with predorsal length and ventral head length. Similarly, the third Factor ($\lambda_3 = 0.989$, 7.7%) was heavily weighted with dorsal tail length and ventral tail length. All three factors were positively weighted. The most variable were older larvae, mainly along Factor 1, but also along Factor 2. However, no differences were found between developmental intervals (Hotteling's $T^2 = 0.673$, df = 20–134). It was, however, possible to distinguish intervals of older larvae and juveniles by Factor 2 only ($p < 0.05$).

In the River Lee (Table 5), the microhabitat use of larvae (free embryos to older larvae combined) differed greatly from that of juveniles (younger and older combined), with juveniles occupying significantly deeper, wider and cooler areas with less silt and higher amount of sand and filamentous algae than larvae. Microhabitat use by adults (young and older combined) differed from that of juveniles, with adults occupying significantly faster flowing areas, further from the bank and over coarser bottom substrata. Habitat use by larvae (<27 mm SL), juvenile (>27 <51 mm SL) and adult (>51 mm SL) minnows from the River Frome differed significantly (Kruskal–Wallis test, $p < 0.05$) in a number of ways (Table 6). The minnows moved into deeper water with variable velocities and the presence of *Ranunculus fluitans* as juveniles, before returning to shallow, fast flowing, open water habitats as adults (Figure 7). Mean velocity increased significantly as the fish progressed from larva, through adult intervals. This is also reflected in the shift from predominantly mud and silt substrata to gravel and pebble, which co-varies with velocity.

Discussion

The observed qualitative and mensural characters generally corresponded with the description in Koblitskaya (1981) for free embryos (i.e. her 'pre-larvae') and

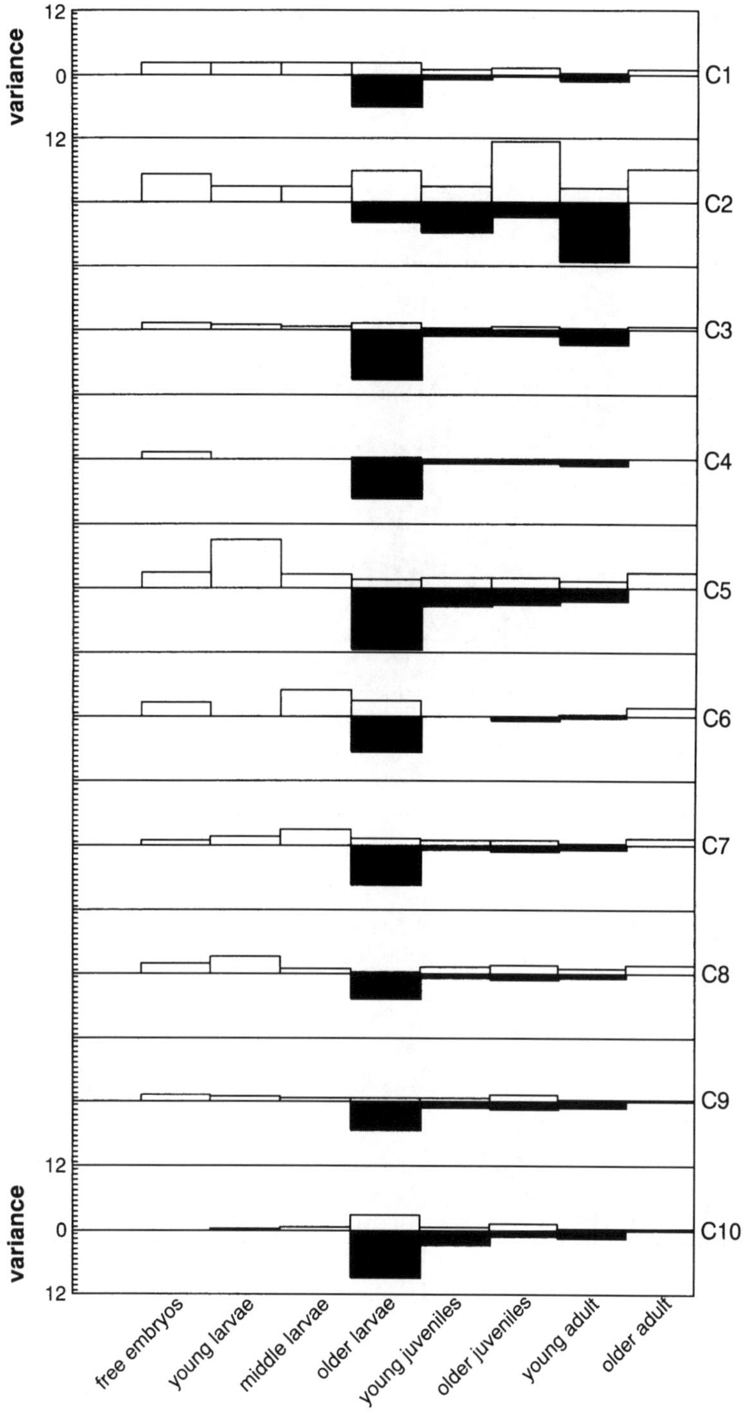

Figure 4. Variation in the ten mensural characters in the eight developmental intervals of minnows from the River Lee (open columns) and River Frome (black columns).

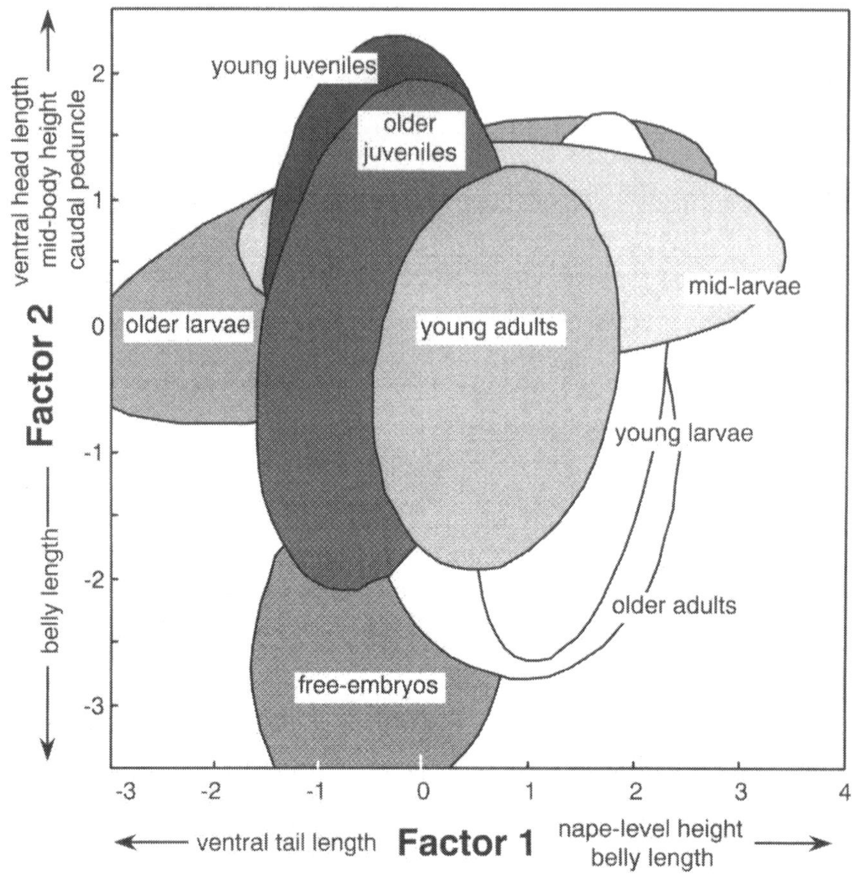

Figure 5. Scatterplot of axes 1 and 2 factor scores from factor analysis of 10 mensural characters for minnows from the River Lee.

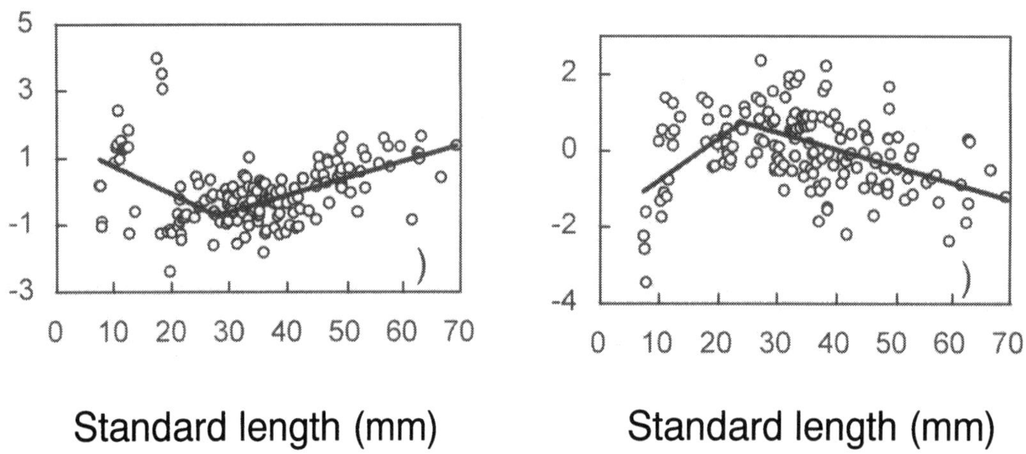

Figure 6. The relationship between factorial scores and standard length in Lee minnows, demonstrating significant changes in the slope using split-linear regression, with factor 1 accounting for nape level height, belly length, and ventral tail length, and factor two for ventral head length, mid-body height, caudal peduncle and belly length.

Table 5. Mean and standard error (SE) for minnow larvae (young, middle and older combined; $n = 39$), juveniles (younger and older combined; $n = 81$) and adults (young and older combined; $n = 156$) from River Lee, with respect to the fourteen environmental variables measured by Watkins et al. (1997), with probabilities (p) for statistical comparisons with analysis of variance (F) and Kruskal–Wallis (K–W) tests.

Variable	Larvae		Juveniles		Adults		Test	Statistic	p<
	mean	(SE)	mean	(SE)	mean	(SE)			
Distance from bank (m)	0.34	(0.020)	0.39	(0.023)†	0.45	(0.013)	*F*	7.276	0.001
Water depth (cm)	26.87	(1.627)*	34.70	(1.510)	36.61	(0.803)	*F*	11.784	0.001
Channel width (m)	5.99	(0.108)*	6.49	(0.132)	6.76	(0.104)	*F*	8.782	0.001
Bed slope	0.26	(0.018)	0.34	(0.037)†	0.24	(0.011)	K–W	6.007	0.05
% silt/mud	59.49	(5.999)*	35.12	(3.629)†	25.22	(2.140)	K–W	32.523	0.001
% sand	19.74	(3.769)*	34.32	(2.586)†	24.07	(1.377)	K–W	18.031	0.001
% gravel	18.20	(3.559)	25.68	(2.668)†	46.12	(2.232)	K–W	45.164	0.001
% pebbles	2.56	(1.409)	3.52	(1.219)	4.07	(0.774)	K–W	3.424	n.s.
% algae	0.51	(0.513)*	6.42	(1.289)	11.57	(1.823)	K–W	10.754	0.05
% submerged veg.	1.28	(1.282)	5.56	(2.034)	5.64	(1.350)	K–W	3.404	n.s.
Number overhang branches	0.38	(0.079)	0.51	(0.061)	0.55	(0.044)	K–W	2.806	n.s.
Dissolved oxygen (mg l^{-1})	10.21	(0.239)	10.59	(0.313)	10.52	(0.180)	*F*	0.360	n.s.
Water velocity (cm s^{-1})	3.46	(0.653)	3.85	(0.484)†	6.22	(0.402)	K–W	20.269	0.001
Deviation from weekly mean °C	0.37	(0.060)*	−0.06	(0.033)	−0.05	(0.032)	*F*	21.999	0.001

* for significant ($p < 0.05$; Fisher PLSD or Mann–Whitney) differences between juveniles and larvae, and † for significant differences between juveniles and adults.

Table 6. Mean and standard error (SE) for minnow larvae (<27 mm SL, $n = 108$), juveniles (>27 mm SL, $n = 125$) and adults (>51 mm SL, $n = 20$) with respect to the seven environmental variables measured by Garner & Clough (unpublished) in the River Frome during July 1996, with probabilities (p) for statistical comparisons with Kruskal–Wallis (K–W) tests.

Variable	Larvae		Juveniles		Adults		K–W	p<
	mean	(SE)	mean	(SE)	mean	(SE)	statistic	
Water depth (cm)	40.5	(2.04)*	53.4	(2.31)	45.0	(4.29)	16.8	0.001
Water velocity (cm s^{-1})	1.9	(0.45)*	12.8	(1.14)†	35.9	(4.82)	107.0	0.001
Substratum composition	3.7	(0.07)*	4.1	(0.11)†	4.8	(0.11)	36.9	0.001
% *Ranunculus fluitans*	8.2	(2.03)*	24.6	(2.89)†	4.5	(2.73)	27.8	0.001
% reeds	2.5	(0.71)	1.1	(0.67)	0.0	(0.00)	10.4	n.s.
% ligneous material	2.6	(1.10)	0.3	(0.24)	2.0	(1.37)	5.0	n.s.

* for significant (Mann–Whitney) differences between juveniles and larvae, and † for significant differences between juveniles and adults.

young larvae (her 'early larvae'). However, deviation from Koblitskaya (1981) became more apparent in the latter part of the larva period and in particular with respect to the start of the juvenile period. Similarly, the developmental intervals defined using the analytical approach in the present study only roughly approximate Balon's (1990) intervals. Cross-sectional data sets (Cock 1966), such as we used, are dependent on catch, i.e., a sufficient number of fish for statistical analysis as well as a sufficient size range for interval delimitation. Whereas, studies of reared specimens permit continuous survey and measurement of individuals from the same brood stock.

The grouping of minnows into developmental intervals based on qualitative characters did not conform completely with that by mensural characters. From qualitative characters, young and middle larvae in the River Lee (Figure 2) could not be distinguished, even

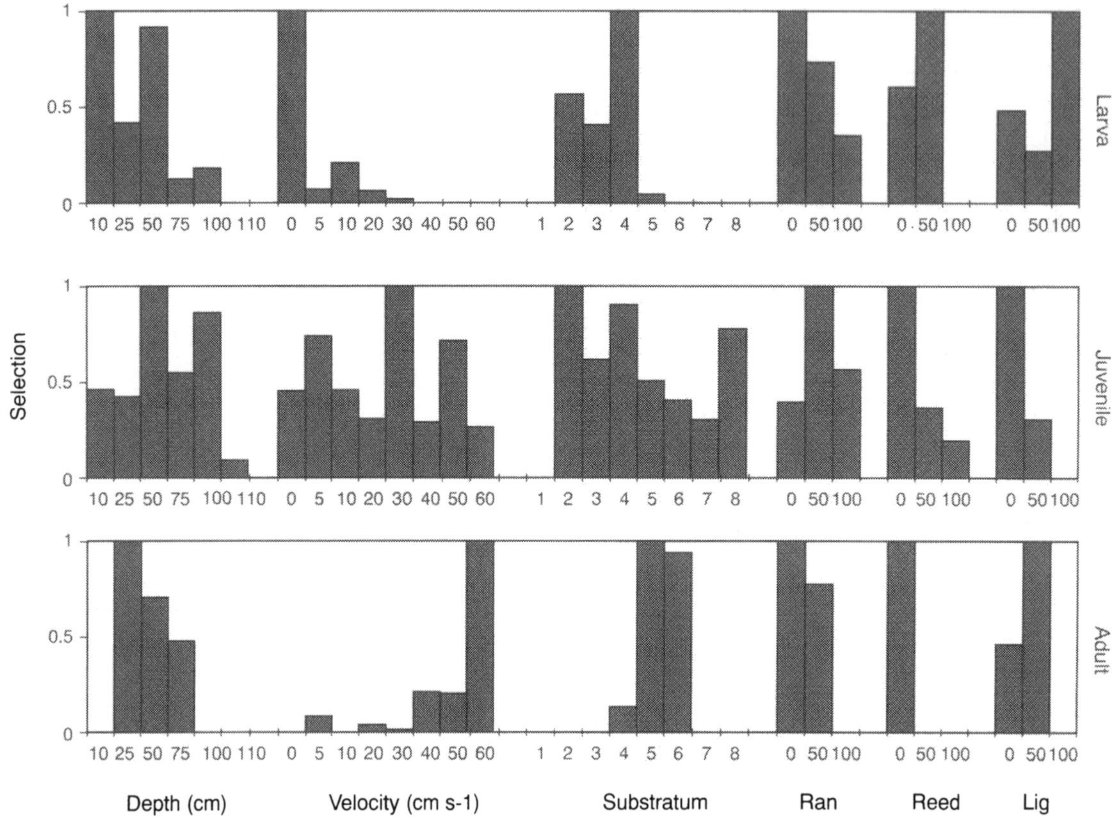

Figure 7. Normalized habitat suitability scores for minnow larvae, juveniles and adults in the River Frome during July 1996. For descriptions of life-history intervals see text. 1 = most suitable habitat, 0 = never used. Categories were: water depth (cm), water velocity (cm s^{-1}), substratum composition (Wentworth scale), percentage of *Ranunculus fluitans*, percentage of reeds, percentage of ligneous material.

though other developments separating these intervals (e.g. differentiation of the swim bladder) are important events that result in shifts in resource use (e.g. Copp 1992). Middle and older larvae were distinguishable in both rivers (Figures 2, 3), but older larvae in both the rivers Lee and Frome could not be distinguished from young juveniles using qualitative characters. With respect to relative growth, changes observed in certain mensural characters suggest that the remodelling process continues in older larvae in both rivers, and thus these specimens are probably correctly categorised as larvae rather than juveniles. This discrepancy suggests a lack of synchrony in development of different kinds of morphological traits (i.e., qualitative and mensural).

Belly length was the most prominent mensural character to discriminate free embryos and young larvae from all older intervals. This probably reflects an accelerated development of digestive organs, particularly the intestine and hence intensified feeding and growth. The adaptive role of dorsal and ventral tail length, ventral head length and nape-level body height in determining body shape (hydrodynamics) during ontogeny was apparent, in particular in the shift from juvenile to adult development (Figures 4, 5). The difference between free embryos and young larvae with respect to ventral tail length could be interpreted as an upward flexing of the spine and the formation of the caudal fin skeleton (Koblitskaya 1981). Whereas, the difference detected later, between older larvae with younger and older juveniles and younger and older adults, could be considered as further strengthening of the tail and an enhancement of swimming ability during adult life (Peňáz 1975, Webb & Weihs 1986). Indeed, the highly positive and synchronous allometric growth of dorsal and ventral tail lengths in middle larvae of the River Frome illustrates the relative strengthening of their tail and improvement of their swimming ability.

Differences between the ontogenetic trajectories of older larvae and young juveniles, as well as the great variability in older larvae, could also be considered as a consequence of the increased developmental instability in the oldest larvae prior to their shift to the next (juvenile) developmental period, with its drop in morphometric change. Although the transition of morphological characters expressed dynamic changes between developmental intervals, it did not occur for all characters involved at the same time (Figure 6). The covariance matrix from Factor analysis for particular groups revealed an association between different character sets (Figure 5). Thus, in free embryos, the association occurred between most of characters, and particularly for ventral head and tail lengths and mid-body height; in young larvae, the most prominently associated were caudal peduncle and ventral head and belly lengths; in middle and older larvae, that prominence was mainly in the belly and ventral tail lengths. Beyond these intervals, no association was detectable, until the oldest adults. The prominent covariation between characters in early developmental periods up to and including the older larvae, versus no prominent covarying characters in juveniles and adults, emphasizes the boundary between the larva and juvenile developmental periods, which we suggest occurs later than previously reported for the species (Koblitskaya 1981). This also reveals the similarity of pattern in juvenile and adult morphology, and accounts for the fact that these intervals are often explained in terms of reproductive capability only.

Despite some contradictory evidence in qualitative characters, the transition from larva to juvenile development in the minnow is prominent and relatively comprehensive. It consists mainly of the immense alteration of the morphology of the tail as the main propulsive organ. The tail also appears to undergo the longest period of morphology change: from the embryos, when its skeleton support forms, through the larvae and juveniles, when the tail strengthens, to the adult period. Prominent changes to other body regions are both less pronounced and of shorter duration than in the tail, occurring only in particular developmental intervals. Thus, the prominent elongation of the belly occurs only at the onset of exogenous feeding in early larvae. The remarkable increase in body height in adults probably facilitates the increased 'demand for space' of the ripening gonads. The change in ventral head length in juveniles and adults, relative to early intervals, may be associated with as yet unknown feeding and respiration functions.

Central to the theory of saltatory ontogeny is that ontogeny is a sequence of longer stabilized states punctuated by rapid changes in 'integrative actions', or thresholds (Balon 1990). For minnows with SL ranging from 7.4 to 69.2 mm, significant shifts in relative growth of six mensural characters were apparent in the factorial composites thereof at about 24 and 28 mm SL (Figure 6). Nevertheless, the breakpoints could be considered as developmental thresholds only if they coincide with each other and/or with other morphological, physiological, ecological or behavioural changes in the ontogeny of the species. In other words, if shifts in the growth of mensural characters would indicate any thresholds, then mensural characters with breakpoints occurring at a similar size should have been associated with some common ecological, physiological or behavioural function (Kováč et al. 1999 this volume).

The six mensural characters fit this condition. During the SL interval between 24 and 28 mm snout length, ventral tail length and ventral head length stopped growing faster than SL, whereas, in the contrary, belly length started to grow faster. Thus, the remodelling of overall external shape of body in minnows (Figure 6) coincided with significant shifts in habitat from slower, shallower waters with little aquatic vegetation as larvae to faster flowing, deeper waters with greater amounts of submerged vegetation or algae (Tables 6, 7, Figure 7). This shift could be attributable to changes in feeding and locomotion abilities that the fish have undergone concurrent with this life-history threshold between larva and juvenile development. Further studies on the early development of minnow, in particular comparing the ontogeny of natural and laboratory-reared specimens, is needed to identify why this species is a particularly adaptable and pioneering species amongst European cyprinids.

Acknowledgements

This study was supported by the Royal Society Ex-agreement exchange programme (PS) and the Freshwater Biological Association (PG). We thank these organizations for their support, as well as S. Newby and K. Grey for assistance in the field, and K. Haynes for laboratory assistance.

References cited

Balon, E.K. 1990. Epigenesis of an epigeneticist: the development of some alternative concepts on the early ontogeny and evolution of fishes. Guelph Ichthyol. Rev. 1: 1–48.

Balon, E.K. 1999. Alternative ways to become a juvenile or a definitive phenotype (and on some persisting linguistic offenses). Env. Biol. Fish. 56: 17–38 (this volume).

Cock, A.G. 1966. Genetical aspects of metrical growth and form in animals. Quat. Rev. Biol. 41: 131–190.

Copp, G.H. 1992. Comparative microhabitat use of cyprinid larvae and juveniles in a lotic floodplain channel. Env. Biol. Fish. 33: 181–193.

Copp, G.H. & T.A. Bennetts. 1996. Short-term effects of removing riparian and instream cover on barbel (*Barbus barbus*) and other fish populations in a stretch of English chalk stream. Folia Zool. 45: 283–288.

Copp, G.H. & P. Garner. 1995. Evaluating the microhabitat use of freshwater fish larvae and juveniles with point abundance sampling by electrofishing. Folia Zool. 44: 145–158.

Copp, G.H. & V. Kováč. 1996. When do fish with indirect development become juveniles? Can. J. Fish. Aquat. Sci. 53: 746–752.

Economou, A.N., C. Daoulas & T. Psarras. 1991. Growth and morphological development of chub, *Leuciscus cephalus* (L.), during the first year of life. J. Fish Biol. 39: 393–408.

Garner, P. 1997a. Determination of length bias and calculation of minimum sample sizes required for length and density estimation when using point sampling by electrofishing. J. Fish Biol. 50: 95–106.

Garner, P. 1997b. Effects of variable discharge on the velocity use and shoaling behaviour of *Phoxinus phoxinus*. J. Fish Biol. 50: 1214–1220.

Garner, P. & S. Clough. 1996. Habitat use by adult dace (*Leuciscus leuciscus* (L.)) in a side channel of the River Frome, England. Fish. Manag. Ecol. 3: 349–352.

Gozlan, R.E., G.H. Copp & J.-N. Tourenq. 1999. Comparison of growth plasticity in the laboratory and field, and implications for the onset of juvenile development in sofie, *Chondrostoma toxostoma*. Env. Biol. Fish. 56: 153–165 (this volume).

Howes, G.J. 1985. A revised synonymy of the minnow genus *Phoxinus* Rafinesque, 1820 (Teleostei: Cyprinidae) with comments on its relationships and distribution. Bull. Br. Mus. Nat. Hist. (Zool.) 48: 57–74.

Ibbotson, A., P. Armitage, W. Beaumont, M. Ladle & S. Welton. 1994. Spatial and temporal distribution of fish in a small lowland stream. Fish. Manag. Ecol. 1: 145–156.

Joliceur, P. 1963. The multivariate generalisation of the allometry equation. Biometrics 19: 497–499.

Koblitskaya, A.F. 1981. Key for identifying young freshwater fishes. Light and Food Industrial Publ. House, Moscow (in Russian).

Kováč, V. & G.H. Copp. 1996. Ontogenetic patterns of relative growth in young roach *Rutilus rutilus*: within-river basin comparisons. Ecography 19: 153–161.

Kováč, V., G.H. Copp & M.P. Francis. 1999. Morphometry of the stone loach, *Barbatula barbatula*: do mensural characters reflect the species' life history thresholds? Env. Biol. Fish. 56: 105–115 (this volume).

Kuo, J. 1994. Sigmaplot scientific graphing software. Transforms and curve fitting. Jandel Scientific, San Rafael.

Krupka, I. 1988. Early development of barbel [*Barbus barbus* (Linnaeus, 1758)]. Práce ústavu Rybárstva a Hydrobiológie 6: 115–138.

Mastrorillo, S., F. Dauba & A. Beland. 1996. Microhabitat use by minnow, gudgeon and stone loach, in three rivers in Southwestern France. Ann. Limnol. 32: 185–195.

Mills, C.A. 1987. The life history of the minnow *Phoxinus phoxinus* (L.) in a productive stream. Freshwat. Biol. 17: 53–67.

Peňáz, M. 1975. Diel lokomotorische Aktivitat larvaler und juveniler Elritzen (*Phoxinus phoxinus*). Zoologické Listy 24: 263–276.

Prokeš, M. & M. Peňáz. 1978. The course of spawning, early development and longitudinal growth of the nase carp, *Chondrostoma nasus*, in the Rokytná and Jihlava Rivers. Folia Zool. 27: 269–278.

Prokeš, M. & M. Peňáz. 1979. Reproduction and early development of the gudgeon, *Gobio gobio*. II. Larval and juvenile period. Folia Zool. 28: 55–64.

Rohlf, F.J. 1988. NTSYS-pc: numerical taxonomy and multivariate analysis system. Version 1.50. Exeter Publishing Ltd. & Applied Biostatistics Inc., New York.

Sagnes, P., P. Gaudin & B. Statzner. 1997. Shifts in morphometrics and their relation to hydrodynamic potential and habitat use during grayling ontogenesis. J. Fish Biol. 50: 846–858.

Sempeski, P. & P. Gaudin. 1995. Habitat selection by grayling (*Thymallus thymallus*). II. Preliminary results on larval and juvenile daytime habitats. J. Fish Biol. 47: 345–349.

Sneath, P.H. & R.R. Sokal. 1973. Numerical taxonomy. W.H. Freeman & Co., San Francisco. 573 pp.

Sokal, R.R. & F.J. Rohlf. 1981. Biometry. W.H. Freeman & Co., San Francisco. 776 pp.

Zar, J.H. 1984. Biostatistical analysis. Prentice Hall Inc., Englewood Clifts. 375 pp.

Watkins, M.S., S. Doherty & G.H. Copp 1997. Microhabitat use by 0+ and older fishes in a small English chalk stream. J. Fish Biol. 50: 1010–1024.

Webb, P.W. & P. Weihs. 1986. Functional locomotor morphology of early life history of fish. Trans. Amer. Fish. Soc. 115: 115–127.

Environmental Biology of Fishes **56:** 129–151, 1999.

Fish, flows and flood plains: links between freshwater fishes and their environment in the Murray-Darling River system, Australia

Paul Humphries[a], Alison J. King[a] & John D. Koehn[b]
[a]*Cooperative Research Centre for Freshwater Ecology, Department of Biological Sciences, Monash University, c/- Murray-Darling Freshwater Research Centre, GPO Box 921, Albury, NSW 2640, Australia (e-mail: hump@mdfrc.canberra.edu.au)*
[b]*Freshwater Ecology Division, Marine and Freshwater Research Institute, Department of Natural Resources and Environment, 123 Brown Street, Heidelberg, Victoria 3084, Australia*

Received 11 August 1998 Accepted 18 December 1998

Key words: floods, low flow recruitment hypothesis, recruitment, life cycles, spawning, lowland rivers, fish larvae

Synopsis

Knowledge of the biology of native fishes of the Murray-Darling Basin is based largely on studies conducted under hatchery conditions and on a limited number of recreationally important species. From observations that increases in water level in aquaculture ponds initiate spawning in some species, and from limited studies of wild fishes and studies in overseas floodplain river systems, a perception has emerged of the importance of flooding and the flood plain in the life cycles of Murray-Darling fishes in general. However, there is little confirmatory evidence of the use of temporary floodplain habitats by larvae, juveniles or adults. The significance of in-channel habitats, especially for rearing, has received little attention. Murray-Darling fish species can be placed into three life history modes, based mainly on spawning style and time and developmental intervals of larvae at first feeding. Fish in each group may be able to take advantage of floods if the timing is right and prey are plentiful, however, the larvae of some species are able to recruit under non-flood conditions within the main river channel. This forms the basis of the 'low flow recruitment hypothesis', which attempts to explain why some species spawn during the warmest months and lowest flows and how they are able to recruit under these conditions. This hypothesis is then placed in the context of the current state of knowledge of the relationships between flow and the biology of Murray-Darling fishes, specifically cues for spawning, movement and recruitment. The lack of widespread evidence for floodplain use by any life history interval of fish may be due to a paucity of study, however, there are some fundamental factors, such as the predictability of timing and duration of high flow events as well as the lack of coincidence of high flows and high temperatures in some regions of the Basin, which may be important in determining the use of floodplain habitats by fish.

Introduction

The Murray-Darling Basin (Figure 1) drains an area of 1.073 million km^2, with the Murray and Darling rivers alone comprising approximately 5300 river km. Interannual flow variability, especially within the drier and temperate regions of the basin, is extremely high (Walker 1986, Walker et al. 1995, Puckridge et al. 1998). It is this highly variable flow, together with Australia's long isolation from the rest of the world, which is thought to have played a major role in maintaining a depauperate native fish fauna (Lake 1971, Harris 1984, Allen 1989). Indeed only 26 native species from 12 families spend their entire lives within the Murray-Darling Basin, and of these, only 10 are endemic (Cadwallader & Lawrence 1990). Despite the

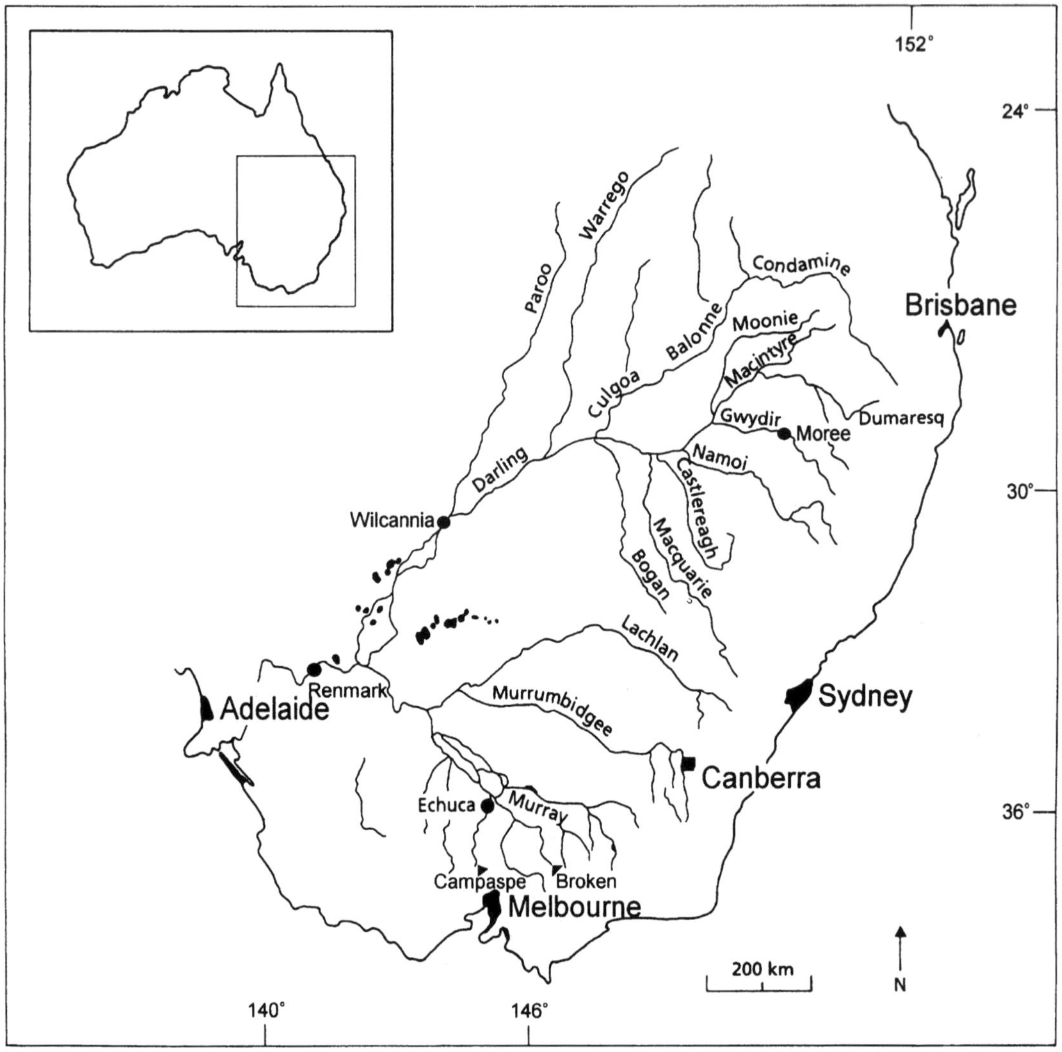

Figure 1. The Murray-Darling Basin.

paucity of species, the fish fauna exhibits a diverse range of sizes, forms and life cycle modes (Merrick & Schmida 1984). For example, some gudgeon species (Eleotridae) rarely exceed 50 mm in length and 2 g in weight and live for little more than two years, whereas Murray cod, *Maccullochella peelii peelii* (Mitchell), has been known to grow to 2 m in length, can weigh in excess of 100 kg and live for more than 30 years (McDowall & Fulton 1996).

Murray-Darling fish, such as Murray cod, have traditionally played a major role as a food source and as a cultural icon for indigenous Australians (Lawrence 1971). These traditions were carried over to early European settlers (see Rowland 1989). Early commercial operations exploited the bountiful fishery resource, especially Murray cod, golden perch, *Macquaria ambigua*, and Macquarie perch, *Macquaria australasica* (Cuvier)[1] (Cadwallader 1978, Rowland 1989) and exploitation continues mainly through an enormously popular recreational fishery. Despite sporadic observations and anecdotal evidence, it was not until the late 1940s that the first systematic survey was done of the Murray-Darling fish fauna by J.O. Langtry.[1] Unfortunately this was well after the introduction of a

[1] Cadwallader, P.L. 1977. J.O. Langtry's 1949–50 Murray River investigations. Fisheries and Wildlife paper 13, Victoria. 70 pp.

number of exotic fish, including Eurasian perch, *Perca fluviatilis*, tench, *Tinca tinca*, brown trout, *Salmo trutta*, and rainbow trout, *Oncorhynchus mykiss*. Langtry's study and past fisheries catch data suggested that intense commercial fishing, river regulation, barriers to fish movement and the introduction of exotics, had already contributed to a decline in the abundance and range of native fish[1] (Whitely 1937, Rowland 1989). Carp, *Cyprinus carpio*, have since been introduced and now make up a large proportion of fish populations.

Much initial work on the breeding biology of Murray-Darling native fish was performed in the 1960s and 1970s, in aquaculture rearing ponds (Lake 1967a,b, Llewellyn 1971, 1973, 1974). Conclusions from this work, combined with hypotheses from studies overseas (Welcomme 1985, Junk et al. 1989), have resulted in speculation that over-bank flows and inundation of the flood plain are important for Murray-Darling fish, both as a cue for final maturation and spawning of some species (Lake 1967a), and for the subsequent survival of larvae and successful recruitment into juvenile stocks[2] (Lake 1967a, Arumugam & Geddes 1987, Lloyd et al. 1989, Puckridge & Walker 1990, Rowland 1992). Indeed, Harris & Gehrke (1994) have proposed the 'flood recruitment model' to describe how some species, such as golden perch, respond to rises in flow and flooding and how flooding provides the high densities of food for recruitment of this species.

Recent work on wild populations, however, has begun to put into question some of the hatchery-derived assumptions of the nature of the relationship between the biology of Murray-Darling fishes and their environment. Studies of life history traits[3] (Humphries 1995), spawning cues (Mallen-Cooper et al. 1995), movement and migration[3,4] habitat use[3,4] (Gehrke 1991) and recruitment (Mallen-Cooper et al. 1995), have provided evidence that traits vary widely among species and may not even be consistent for one species across different regions. It is becoming increasingly apparent that reliance on hatchery-based studies as a major source of information is an unsound foundation upon which to base river management and that a better understanding of how fish interact with, and respond to, their environment is needed. Furthermore, whilst not a novel idea, the evidence is mounting that many species spawn independently of high flows and flooding and can successfully recruit during low flow periods (Humphries unpublished data).

Few studies of wild populations have found evidence that fish adults and larvae use non-permanent floodplain habitats.[5] The fact that the Murray-Darling Basin is not homogeneous climatically, but includes a range of zones, some where the highest flows occur in summer and some where the highest flows occur in winter and early spring, may mean that the flood plain is inundated at times not necessarily optimal for the breeding of fish or for recruitment of young. It is logical to infer, then, that a variety of life history and recruitment styles may have evolved to deal with a range of conditions encompassed in the Murray-Darling Basin and that some involve use of the flood plain and others do not. Spawning of adults or rearing of larvae on the flood plain may be advantageous if flooding coincides with warm temperatures, and flooding initiates blooms of microscopic animals, as some suggest (Arumugam & Geddes 1987). However, if flooding and warm temperatures do not coincide, as for the southern Mediterranean region of the Murray-Darling Basin, it may be more advantageous for a fish to spawn within the main channel of the river. Indeed, periods of low flow during summer, in rivers in these regions, are more predictable from year to year and generally of longer duration than floods at this time of the year.

The aim of this paper is to assess the adequacy of current ideas relating to the importance of flooding and the flood plain in the biology of Murray-Darling fishes and to propose the 'low flow recruitment hypothesis' as a way of explaining how some native species are able to spawn and recruit during periods when flows

---

[2] Geddes, M.C. & J.T. Puckridge. 1988. Survival and growth of larval and juvenile native fish – the importance of the floodplain. pp. 101–114. *In*: Proceedings of the Workshop on Native Fish Management, Murray-Darling Basin Commission, Canberra.

[3] Koehn, J. & S. Nicol. 1999. Habitat and movement requirements of fish. *In*: R.J. Banens & R. Lehane (ed.) Riverine Environment Research Forum, Proceedings of the Riverine Environment Research Forum of MDBC Natural Resource Management Strategy Funded Projects, Brisbane.

[4] Koehn, J.D. 1997. Habitats and movments of freshwater fish in the Murray-Darling Basin. pp.27–32. *In*: R.J. Banens & R. Lehane (ed.) Riverine Environment Research Forum, Proceedings of the Inaugural Riverine Environment Research Forum of MDBC Natural Resource Management Strategy Funded Projects, Attwood.

[5] McKinnon, L.J. 1997. The effects of flooding on fish in the Barmah forest. pp. 1–8. *In*: R.J. Banens & R. Lehane (ed.) Riverine Environment Research Forum, Proceedings of the Inaugural Riverine Environment Research Forum of MDBC Natural Resource Management Strategy Funded Projects, Attwood.

are low, but temperatures are high. We begin by categorising Murray-Darling fishes into one of four life history modes in an attempt to understand how different species achieve successful recruitment under what are often highly variable hydrological conditions. We then describe the 'low flow recruitment hypothesis' and provide some justification for it from our own observations and from the literature. We then compare the conditions experienced by Murray-Darling fishes with those of other floodplain systems throughout the world and question whether concepts such as the 'flood pulse concept' of Junk et al. (1989) are appropriate for all regions of a climatically heterogeneous river basin like the Murray-Darling Basin. Finally, we suggest subjects for future study and emphasise the need to be vigilant against complacency that our knowledge of the relationships between fishes and the Murray-Darling riverine environment is adequate to manage this large and complex system.

Life cycles of Murray-Darling Basin fishes and influences on recruitment

Murray-Darling fishes have been grouped previously into four categories related to their spawning style, association or lack of association with flooding and significance for recruitment.[6] The 'flood recruitment model' (Harris & Gehrke 1994) has attempted to deal with the group of Murray-Darling fishes for which flow appears to initiate spawning and for which the flood plain may be important as a habitat of larvae, but this model is inadequate for many species of Murray-Darling fishes for which flow may not play a major role. We propose four categories of Murray-Darling fishes based on life history traits and emphasising spawning style and the importance of the developmental interval of larvae at first feeding as well as the differential ability that this confers on larvae, within each group, to ingest prey of various sizes. In constructing these categories, we also consider that the timing and duration of the spawning period is crucial to understand various styles for successful recruitment. Another important trait is the occurrence of parental care of eggs and larvae. Finally, adopting Balon's (1984) methodology of describing the early life stages of fish, it is important to determine whether species go through a larva period or metamorphose directly from an embryo to a juvenile.

Mode 1. Includes the largest species in the Murray-Darling Basin (Table 1). These species: tend to spawn once only in late spring/early summer in a relatively short breeding period; have demersal eggs which number in the thousands to tens of thousands; have young that hatch well developed, with pigmented eyes: exhibit parental care after hatching; at the time they lose their yolksac and begin feeding, generally at about 20 days of age, have developed all their fins and can swim well; and have large gapes (>1 mm) at first feeding. Spawning may or may not coincide with high flows, but is not apparently initiated by it. According to Balon's (1984) definition, each species in Mode 1 develops straight from the embryo to the juvenile, bypassing the larva period.

We suggest that, because of the advanced developmental interval at which fish in Mode 1 first begin feeding, they are able to pursue and ingest relatively large prey items, including early instars of macroinvertebrates and large potamoplankton. These are in high abundance within the main channel of rivers, especially in backwaters and slow-flowing or still sections of large lowland rivers (Thorp et al. 1994, King unpublished data). Although the breeding season of these species encompasses at least part of the time when floods may occur in some parts of the Murray-Darling Basin, we suggest that floods are not, as a rule, important for maintaining populations. However, if floods coincide with spawning, it may be that food on the flood plain proper is washed into the river channel from backwaters, billabongs (oxbow lakes) and anabranches, might enhance conditions for species in this group and greater recruitment would result. It is also possible that large floods at the time of spawning may actually displace eggs, embryos and larvae from nests or nurseries and cause higher than normal mortalities.

Mode 2. Fishes also grow to a relatively large size as adults (Table 1). These species: spawn once only, between late spring and early autumn and can apparently delay spawning until the appropriate conditions occur; some species, such as golden perch, are known to resorb gonads if the right conditions do not occur;

[6] Lloyd, L., J. Puckridge & K. Walker. 1989. The significance of fish populations in the Murray-Darling system and their requirements for survival. pp. 86–99. *In*: T. Dendy & M. Coombe (ed.) Conservation and Management of the River Murray System – Making Conservation Count, Proceedings of the Third Fenner Conference on the Environment, Canberra.

Table 1. Life cycle styles for several Murray-Darling Basin fishes.

Variable	Mode 1	Mode 2	Mode 3a	Mode 3b
Duration of spawning	short	variable	long	short
Spawning style	Single spawning, approx. same time each year	Single spawning, timing, can delay	Protracted, serial or repeat	Single spawning
Spawning time	October–December	October–March	September–March	Late winter or summer
Cues for spawning	Circannual and min. temp.	Rising water level? and min. temp.	Uncertain	Uncertain
Number of eggs	1000s–10 000s	100 000s	100s–1000s	100s–1000s
Type of egg	Demersal	Semi-buoyant or planktonic	Planktonic or demersal	Planktonic or demersal
Parental care of embryo/larva	Yes	No	No	No
Incubation period	10+ days	Hours	<10 days	<10 days
Size of embryo at hatching	6–9 mm	3–6 mm	3–4 mm	2–7 mm
Time to first feeding	ca. 20 days	ca. 5 days	ca. 3 days	ca. 3 days
Development of embryo/larva at first feeding	Advanced, large gape, well-formed fins, highly mobile	Undeveloped, small gape, limited mobility	Undeveloped small gape, limited mobility	Undeveloped, small gape, limited mobility
Examples of species	Murray cod[(a,e,l,m,u,v)] Trout cod[(i,j)] Freshwater catfish[(f,l,m)] River blackfish[(k)]	Golden perch[(c,g,l,m)] Silver perch[(l,m)]	Australian smelt[(a,n,q,r)] Flat-headed gudgeon[(a,n)]	Carp gudgeons[(a,b,l,m,n)] *Galaxias olidus*[(a,s)] *G. rostratus*[(n)] Crimson-spotted rainbowfish[(a,d,p,t)] Southern pygmy perch[(h,n,o)]

[a]Humphries unpublished data, [b]Anderson et al. 1991, [c]Arumugam & Geddes 1987, [d]Backhouse & Frusher 1980, [e]Cadwallader et al. 1979, [f]Davis 1977, [g,2], [h]Humphries 1995, [i]Ingram & Rimmer 1992, [j]Ingram & Douglas 1995, [k]Jackson 1978, [l]Lake 1967a, [m]Lake 1967b, [n]Llewellyn 1971, [o]Llewellyn 1974, [p]Milton & Arthington 1984, [q]Milton & Arthington 1985, [r]Milward 1969. [s]O'Connor & Koehn 1991, [t]Reid & Holdway 1995, [u]Rowland 1983, [v]Rowland 1992.

eggs number in the hundreds of thousands and are either semi-buoyant or planktonic; incubation is very short, a matter of hours in some cases; embryos hatch at a relatively small size, are not well developed and do not have pigmented eyes; there is no parental care of either eggs, embryos or larvae; larvae first start feeding after about five days, and the gapes of first feeding larvae are moderate in size ($\sim 500\,\mu$m). Spawning has been thought to be linked to rises in flow and later flooding. Species in this group have a definite larva period.

Mode 2 fish produce large numbers of eggs and must experience very high mortality of eggs, embryos and larvae. This is consistent with general patterns observed in marine and freshwater invertebrates and vertebrates that have planktonic larvae. It is also consistent with life in an unpredictable environment, where the densities of food required for larvae may be extremely patchy in space and time. Species in this group attempt to cope with climatic and hydrological unpredictability by apparently tracking key environmental variables, such as temperature and flow rises. These fish are able to delay spawning and may not spawn at all if the conditions are not right. Gape size and undeveloped fins at first feeding would probably restrict the size of prey caught and ingested to rotifers and small planktonic crustacea. If zooplankton blooms do occur during flooding, as under aquaculture conditions,[2] than this might ensure high enough densities for larvae of this group to utilise. Otherwise, other habitats that might provide high densities of prey would be backwaters and potentially impoundments, where water residence time is relatively long.

Mode 3a. Fishes are mostly small, have either protracted, serial or repeat spawning and the breeding season extends from early spring to early autumn (Table 1). Spawning of these species seems unrelated to flow, although breeding may occur during high flows; there may be a threshold temperature above which fish can begin spawning; eggs number in the hundreds to thousands and are mostly demersal and adhesive; incubation is less than 10 days; embryos hatch at a small size, are not well developed but do have pigmented eyes; there may be parental care of eggs (flathead gudgeons, *Philypnodon grandiceps*), but not of embryos and larvae; larvae first start feeding within 2–3 days after hatching, and the gapes of first feeding larvae are small ($\sim$200 μm). Species in this group have a definite larva period.

The key feature of Mode 3a is that fishes are able to spawn over an extended period. To date, there is not sufficient knowledge of their reproductive styles to determine whether each species is: a protracted spawner, a serial spawner, a repeat spawner or simply that females come into maturity and spawn haphazardly throughout the breeding season. Despite this uncertainty, species such as flathead gudgeons and Australian smelt, *Retropinna semoni*, spawn, and recruitment occurs, for approximately six months of the year, although the most successful recruitment times may be limited (Milton & Arthington 1985), and it is likely that in most years larvae have to contend with both high and low flows. The developmental stages and extremely small gape size of fish in this group at first feeding, means that they are restricted to ingesting very small prey and possibly even phytoplankton. Thus, they are reliant on high densities of minute prey to get them through the critical stages of first feeding.

Mode 3b. Fishes are similar to those in Mode 3a in regard to numbers and type of eggs, incubation period, size at hatching and developmental stages at first feeding (Table 1). However, they differ markedly in that they are single spawners and their breeding season is short, typically two months or less. Species either spawn in late winter/early spring or in summer, and gapes of first feeding larvae in this group are small (approximately 200 μm). Spawning is apparently unrelated to high flows and indeed, for some species, takes place during periods of low flow. Species in this group also have a definite larva period.

Since Mode 3b fish spawn for only a short period and have small gapes, they are very much dependent on high densities of microinvertebrates or algae at first feeding. Many galaxiids (*Galaxias*) spawn in later autumn/early winter and larvae spend some months at sea during the winter (McDowall & Fulton 1996). This is thought to have been a style to ensure adequate food for the early intervals of life. The spawning time of freshwater species of galaxiids and of normally diadromous species have shifted to early spring, presumably for similar reasons (Humphries 1989). Some Mode 3b species only spawn in summer and during low flows. In the 1995/1996 and 1996/1997 breeding seasons, crimson-spotted rainbowfish, *Melanotaenia fluviatilis*, larvae have been collected from the Broken River, north-eastern Victoria only in December, January and February and carp gudgeons, *Hypseleotris* spp., have been collected from the Campaspe, north-eastern Victoria and Broken Rivers only in the summer months (Humphries unpublished data).

'Low flow recruitment hypothesis'

Spawning of adults and recruitment into juvenile stocks of flathead gudgeon, Australian smelt, crimson-spotted rainbowfish, and three species of carp gudgeon can occur in mid summer, when the prospect of flooding is remote, but the predictability of high temperatures and low flows are high (Milton & Arthington 1984, 1985, Humphries unpublished data). These species have small gapes at first feeding and fall into our Mode 3a and 3b, initially requiring extremely small prey. Thus, they must encounter high densities of small prey during periods of low flow in the main stem of rivers.

The 'low flow recruitment hypothesis' postulates that some species of fishes take advantage of the extended low flow period of the rivers in certain regions of the Murray-Darling Basin to spawn, because of the concentrations of appropriately-sized prey. This prey is of sufficient size and density to allow larvae of these species to make the transition from endogenous to exogenous feeding without recourse to the flood plain.

As flows decrease over late spring and into summer and temperatures warm, a smaller volume of water would concentrate the prey to such an extent that densities may increase to a level which would be sufficient for feeding of larvae. This would most likely occur in backwaters, pools and other still habitats which are common during low flow periods in lowland rivers in many regions of the Murray-Darling Basin. Justification for this hypothesis is based on previous studies in rivers and observations detailed below.

Zooplankton biomass has been shown to be positively correlated with water residence time (Basu & Pick 1996), negatively correlated with flow (Pace et al. 1992, Thorp et al. 1994) and positively correlated with temperature (Thorp et al. 1994) in the main stem of large rivers. Similar results have been found for phytoplankton (Wehr & Thorp 1997), since there is a direct relationship between biomass of zooplankton and chlorophyll a (Basu & Pick 1997). Ferrari et al. (1989) showed that under summer low flow conditions, the zooplankton community of the Po River tended to become highly stable and very productive with flooding creating a destabilising effect, leading to increased diversity but reduced density of rotifers. Thus, the greatest biomass of zooplankton in the main stem of rivers will tend to occur during the lowest flows of each year (Ferrari et al. 1989, Pace et al. 1992, Thorp et al. 1994, Basu & Pick 1996). This is also presumably the explanation for the occurrence of maximum densities of zooplankton given by Welcomme (1985) in some tropical systems at low flow.

Although the above studies have shown that the biomass of zooplankton, by definition occurring in the water column, is greatest at times of low flow, in some cases densities are relatively low (Thorp et al. 1994). We suggest, however, that there is an alternative food source, potentially more abundant and of an ingestable size for fish larvae, which also occurs in slow-flowing or still habitats in the main stems of lowland rivers. This food resource consists predominantly of rotifers and benthic microcrustaceans, including cladocerans, copepods and their nauplii and ostracods. The benthic microcrustacean fauna of lotic systems has received little attention relative to its lentic counterpart (e.g. Williams 1982). However, recent evidence suggests that it may play important roles in riverine food webs (Perlmutter & Meyer 1991). Indeed, benthic microcrustaceans are a significant component of both invertebrate (Hildrew et al. 1985, Lancaster & Robertson 1995) and fish diets (Rundle & Hildrew 1992). Robertson (1990) found peaks in abundance of chydorids in summer in the benthos of the River Thames at levels of 50 000 individuals per m^2, which coincided with maximum daily production and biomass (Robertson 1995). The maximum densities of three chydorid species each reached between 2000 and 8000 individuals m^{-2} at this time (Robertson 1995).

Collections of benthic microfauna from the Campaspe and Broken rivers, northern Victoria, suggests that there is an extremely rich and abundant fauna occurring on or near the substratum (King & Humphries unpublished data). This fauna is not sampled adequately using conventional zooplankton sampling techniques and therefore is easily missed. Many of these taxa are entirely benthic and typically associated with lentic water bodies such as billabongs (R.J. Shiel personal communication) and many fall into the size range able to be ingested by fish larvae with gapes of 200 μm. There are no data on the densities of benthic microcrustaceans and rotifers from backwaters of the main stems of Murray-Darling rivers and how these are affected by declining discharge and increasing temperatures. However, studies overseas have suggested that such littoral habitats support greater densities of potamoplankton than main channel habitats and may support even greater densities of benthic organisms (Thorp et al. 1994).

The use of in-channel habitats by fish embryos and larvae has been documented in other large rivers throughout the world. In general, fish larvae have been shown to use still or slow flowing littoral areas, backwaters and shallow embayments as nursery habitats (Sager 1987, Schiemer & Spindler 1989, Schiemer et al. 1991, Sempeski & Gaudin 1995, Watkins et al. 1997, Sempeski et al. 1998). Their need for low velocity habitats is thought to be related to their poor swimming ability (Lightfoot & Jones 1996, Mann & Bass 1997), with some researchers noting an ontogenetic habitat shift for faster velocity water as the larvae develop (Schiemer & Spindler 1989, Schiemer et al. 1991, Sempeski & Gaudin 1995, Mann & Bass 1997, Watkins et al. 1997).

There are clearly also potential disadvantages with spawning and recruiting during low flow periods. As discharge decreases and temperature warms, there is potential for pools to stratify and dissolved oxygen levels to fall to lethal levels for fishes. With evaporation and smaller volumes of water, concentrations of solutes, such as salt, may also increase to harmful, if not lethal levels. Extended periods of dry, warm weather would increase the risk of desiccation, although in most large lowland rivers, large pools would remain as refuges for fishes. However, the decreasing volume of water and concentration of food would also mean a concentration of aquatic predators and potentially easier access to small fish by aerial and terrestrial predators (Copp 1992). Competition for resources may also be exacerbated, since a decrease in the volume of water may reduce overall area of available habitat and therefore increase the amount of overlap in microhabitat use by larvae and juveniles (Copp 1992).

Murray-Darling fishes and the riverine environment

There is a current perception that for some species of Murray-Darling fishes, increases in flow may provide a cue to initiate maturation and spawning, and flooding may open up important spawning and rearing habitat[2,6] (Lake 1967a, Harris & Gehrke 1994). For others, despite the fact that they spawn independent of high flows, flooding may indirectly benefit them through the input of nutrients or food. However, it is also apparent that there are many species for which there has been no relationship established between high flows and maturation, spawning and rearing of young. Despite this diversity of styles, the management literature often lumps Murray-Darling fishes together when making recommendations for managing rivers within this system[6]. This is highly unsatisfactory, especially, as often happens, the significance of in-channel processes are downplayed or even ignored. If the 'low flow recruitment hypothesis' has some validity, than the lack of recognition of the importance of in-channel habitats may be detrimental to recruitment for several native species. If we are ever to manage Murray-Darling rivers in an ecologically sustainable way, than it is imperative to understand the role that flow, both within and outside the channel, plays in the biology of native fishes. In this section, we examine the evidence for the influence of flow and other environmental variables on maturation, spawning and movement of Murray-Darling fishes and what role the flood plain plays in the biology of fishes in general. We begin by examining whether flow has a role in the cuing of fish to mature and spawn, and what implications this has for the conditions likely to be encountered by larvae and juveniles in the wild. We then look at the question of the movement of adult and fish larvae in relation to flow and postulate on the adaptive significance of this. We then proceed with an examination of the nature of Murray-Darling Basin flood plain and the evidence for its direct use by fishes, what role the flood plain plays here and overseas for fishes and ask whether riverine ecosystem theories, such as the 'flood pulse concept', are appropriate for all regions of the Murray-Darling Basin.

Cues for spawning

The role of the environment in influencing gametogenesis, maturation, ovulation and spawning of fishes has received much attention (see Jobling 1995). Considerable progress in the understanding of the effects of photoperiod, temperature, food availability and water level changes has been made by experimental investigations (see reviews: de Vlaming 1972, Bye 1984). Such controlled studies have so far been lacking for Murray-Darling fishes, with our knowledge of spawning cues relying heavily on data gained from hatchery studies in aquaria and ponds (Table 2), concentrating mostly on a few commercial or recreationally important species such as Murray cod and golden perch (Lake 1967a,b, Llewellyn 1971). Hatchery studies, however, cannot hope to mimic fully conditions in the wild, particularly the types and quantities of available food, diel variation in temperature, a lentic rather than lotic environment and the fact that a water level change in ponds is unlikely to be an adequate substitute for river flow variations. Despite these concerns, information gained from aquaculture studies can indicate which variables may play a role in initiating spawning.

Only temperature, flow changes or fluctuations in water level and photoperiod have been considered as potential cues associated with spawning in Murray-Darling fishes, although Lake (1967a) mentioned the presence of food for young fishes as being coincident with the onset of spawning in several species in ponds (Table 2). Variables such as moon phase and chemical cues have not been studied.

A change in water level has been suggested as a cue for spawning in five of the 13 species studied (Table 2). This has been taken as circumstantial evidence for the significance of a flow rise, which would presumably then lead to flooding in a main stem river. This apparently confirmed Langtry's observations that golden perch in the Murray River spawned in spring if the weather was warm, when the first water level rise occurred[3]. Langtry notes, however, that golden perch can spawn in response to both a rise and a fall in river height, and observed that fish spawned all together, for a short time, in the river proper. Mackay (1973) documented spawning of golden perch in January and February in a tributary of the Lachian River during extensive flooding and found that in the absence of flooding, few fish spawned. His results suggested that spawning is tied to flooding and that the lack of flooding in any year would coincide with poor recruitment.

Table 2. Evidence from scientific literature on the importance of temperature, flow or water level change and photoperiod as spawning cues for Murray-Darling native fish. Note that gaps in the table may indicate the documented lack of significance of a specific variable as a cue or simply a lack of sufficient data to assess the significance of that variable. ⇑ = increase; ⇓ = decrease; min = threshold minimum; min, ⇑ = threshold minimum followed by an increase; Vic = Victoria, Qld = Queensland, SA = South Australia, NSW = New South Wales, Tas = Tasmania.

Species	Temp	Flow/ water level	Photo period	Study environment	Reference
Murray cod	min	⇑, slight		Pond	Lake 1967a
(*Maccullochella peeli peeli*)	min	⇑⇓		Murray R	[1]
	min			Pond	Cadwallader et al. 1979
	min, ⇑	No	Likely	Pond	Rowland 1983
		No	Likely	Broken R, Vic	Humphries unpub. data
	⇑			Pond	[16]
Trout cod	min		Likely	Pond	Ingram & Rimmer 1992
(*Maccullochella macquariensis*)					
Golden perch	min	⇑		Pond	Lake 1967a
(*Macquaria ambigua*)		⇑		Trib. of Lachlan R	Mackay 1973
	min	⇑ or ⇓		Murray R	[1]
Macquarie perch	min			Lake Eildon	Wharton 1968
(*Macquaria australasica*)	min			Lake Eildon	Cadwallader & Rogan 1977
Silver perch (*Bidyanus bidyanus*)	min	⇑		Pond	Lake 1967a
Spangled perch	min			Pond	Llewellyn 1973
(*Leiopotherapon unicolor*)		⇑		Black R, NE Qld	Beumer 1979
Bony herring (*Nematalosa erebi*)		No		Cooper Ck, SE Qld	[2]
	min	No		Lower River Murray, SA	Puckridge & Walker 1990
Australian smelt	min			Pond	Llewellyn 1971
(*Retropinna semoni*)	min			Goulburn R, Vic	Hume et al. 1983
	min			Moggill Ck, SE Qld	Milton & Arthington 1985
Freshwater catfish	min			Pond	Lake 1967a
(*Tandanus tandanus*)	min	No		Gwydir R, NSW	Davis 1977
Crimson-spotted rainbowfish	min			Aquaria	Backhouse & Frusher 1980
(*Melanotaenia fluviatilis*)	min, ⇑	No		Enoggera Ck, SE Qld	Milton & Arthington 1984
Southern pygmy perch	min			Pond	Llewellyn 1971
(*Nannoperca australis*)	⇑	No		Pond	Llewellyn 1974
	min			Macquarie R, Tas.	Humphries 1995
Southern purple-spotted gudgeon (*Mogurnda adspersa*)	min			Pond/aquaria	Llewellyn 1971
Flathead gudgeon (*Philypnodon grandiceps*)	min			Pond/aquaria	Llewellyn 1971
Western carp gudgeon (*Hypseleotris klunzingeri*)	min			Pond	Lake 1967a
Murray jolytail (*Galaxias rostratus*)	min			Pond	Llewellyn 1971
Murray hardyhead (*Craterocephalus fluviatilis*)	min			Pond	Llewellyn 1971
Olive perchlet (*Ambassis agassizii*	min	⇑		Pond	Cadwallader & Backhouse 1983
	min, ⇑	No		Brisbane R, SE Qld	Milton & Arthington 1985

Cadwallader, P.L. & G.J. Gooley. 1985. Propogation and rearing of Murray cod *Maccullochella peeli* at the warmwater fisheries station pilot project Lake Charleygrark. Fisheries and Wildlife Service, Department of Conservation, Forests and Lands, Melbourne. 189 pp.

The 'flood recruitment model' (Gehrke 1994, Harris & Gehrke 1994) emphasised the flood stimulus as the key to spawning and the resorption of gonad if this did not occur. By contrast, recent work has indicated that strong year classes of golden and silver perch in one section of the Murray River were associated with spring flows that were contained within the river channel and poor year classes were associated with high spring flows that inundated flood plains[7].

Lake (1967a) suggested that Murray cod may require a rise in water level as a spawning cue, and Langtry[1] observed that they may spawn on either a rise or a fall in water level[3]. However, later pond experiments indicated that spawning was induced by a rise in water temperature during spring, and that an associated rise in water level was not required (Rowland 1983, Cadwallader & Lawrence 1990). Although 'minimum' temperatures of around 20°C are often mentioned for breeding, spawning has been observed to occur at lower temperatures[4].

Beumer (1979) found that spangled perch, *Leiopotherapon unicolor*, require increasing water temperature and day length to initiate gonad development, with spawning probably being induced by the onset of the 'wet' season. It has been suggested that bony herring, *Nematalosa erebi*, spawn in backwaters during floods[8], however, more recent evidence from south-east Queensland and the lower River Murray suggests that this species spawns well before a rise in flow and that it is the juveniles only which are able to utilise the newly inundated flood plain[2] (Puckridge & Walker 1990).

In general, workers on Murray-Darling fishes have tended to assign a minimum or threshold temperature as a cue for the spawning, whereas the important component of flow as a cue tends to be a rise or rarely a fall. This may reflect the nature of the experiments, since studies in ponds typically manipulated water levels, but were unable to control for temperature, which was apparently increasing during the trials (Lake 1967a). To our knowledge, no spawning cue experiments have manipulated one variable while controlling for others. Thus, it is difficult to determine exactly the relationship between gametogenesis, final stages of maturation, ovulation and release of eggs and environmental variables, such as temperature, water level rises, food densities and photoperiod.

Despite these uncertainties, it is likely that temperature does play a significant role in the onset of gonadal development and in final maturation and spawning of many Murray-Darling fishes. Much of the Basin experiences distinct seasonal patterns in temperature which would be predictable from year to year and it is these types of patterns of which fish tend to make use to initiate gametogenesis (Bye 1984).

Movement

Rises in flow have long been thought to initiate upstream pre-spawning movements in several of the larger Murray-Darling fish species (Cadwallader & Rogan 1977, Reynolds 1983, Llewellyn & MacDonald 1980). The hypothesis has been that fishes move upstream in response to increasing flows and then move laterally on to the flood plain to spawn or that they spawn mid-channel and free embryos are washed downstream to inundated nursery areas (Lake 1967b). Such upstream movement of adults would compensate for the downstream drift of embryos and larvae and prevent larvae from being washed down to unfavourable areas such as the lower reaches of the Murray River or even out to sea.

Although the migratory nature of some fish species in the Murray-Darling Basin has long been recognised by Aborigines and professional fishermen[7], our knowledge of movement is generally poor. Studies have tended to concentrate on large fishes and on large-scale migrations (Koehn & O'Connor 1990). Several studies have tagged large numbers of fishes and recorded subsequent recaptures[5] (Llewellyn 1968, Reynolds 1983), whereas more recently, monitoring of fishways[7] and radiotracking have provided insights into both small and larger-scale fish movements[3,4].

Adult golden perch have been shown to move large distances both upstream and downstream[5] (Llewellyn 1968, Reynolds 1983), with maximum reported distances of 1000 km upstream (Reynolds 1983) and 900 km downstream[5]. Large numbers of juvenile golden perch have been recorded moving upstream through the fishway at Torrumbarry Weir on the Murray River, possibly recolonising from rearing habitats[7]. Results from radiotracking indicated that this species is highly mobile in the Murray River for much of the

[7] Mallen-Cooper, M., I.G. Stuart, F. Hides-Pearson & J.H. Harris. 1995. Fish migration in the Murray River and assessment of the Torrumbarry fishway. Final report for the Natural Resources Management Strategy Project N002, NSW Fisheries, Sydney. 149 pp.

[8] Llewellyn, L.C. 1983. The distribution of fish in New South Wales. Australian Society for Limnology Special Publication 7. 23 pp.

year[3]. Llewellyn (1968) suggested that the upstream movement of golden perch was related to warm water temperatures and increases in river height, whereas Reynolds (1983) was uncertain of the cause of movement, although large-scale movements were coincident with large floods. Adult silver perch have been reported moving more than 200 km (Reynolds 1983) and juveniles of this species were also recorded passing upstream through the Torrumbarry Weir in large numbers[7].

Reynolds (1983) concluded from his tagging study, that freshwater catfish, *Tandanus tandanus*, and Murray cod were non-migratory and that these species were essentially sedentary. It has been found recently, however, that Murray cod can make relatively large (up to 120 km) upstream movements prior to spawning before making a return journey[4]. Movement of Murray cod was largely seasonal and in an overall upstream direction. Once preceded by an upstream movement, however, movement occurred both upstream and downstream. Distances moved were greater in an unregulated than a regulated river and were greater during high flow than low flow years[3]. Such movements may indicate behaviour to compensate for the drift of fish larvae, but there is no direct evidence for this. The distance that larvae drift is likely to be related to discharge conditions at the time and so, if there is some cue to which a fish may respond, fishes may vary the distance that they migrate prior to spawning to compensate.

Koehn (1986) suggested that river blackfish, *Gadopsis marmoratus*, undertook only limited movements and had a small home range in the order of 10–20 m. Trout cod, *Maccullochella macquariensis*, showed similar limited movements and were territorial around home areas of large woody debris. Movement of this species appeared to be related to season and flow rather than water temperature[3].

A recent study of a fishway on the Murray River found that increases in water levels, both large and small stimulated movement[7]. Migration was primarily seasonal, but river flow was found to be an important stimulus for movement for many species. In many instances large numbers of small fishes moved through the fishway following changes in river levels of less than 0.2 m per day.

Those species of Murray-Darling fishes for which there is evidence of large-scale upstream movements, are known to have drifting eggs, embryos or larvae[3] (Humphries unpublished data). The large distances sometimes travelled by adult golden and silver perch may relate to their relatively small and buoyant eggs, embryos and larvae which may be expected to drift greater distances than Murray cod larvae, which are relatively much larger. The absence of golden perch above some artificial barriers from rivers within their range suggests that movement both upstream by juveniles and adults and downstream by eggs, embryos and larvae may be a significant component of their overall life history style[9]. Aside from recent stockings, silver perch have been similarly absent in the Murray River above the barriers that form Lake Mulwala and Lake Hume (Koehn unpublished data). Although there have been few recorded captures of the free embryos or larvae of golden and silver perch, these buoyant embryos and larvae will drift and the eggs of these species have been collected recently in drift samples[3]. Similarly, until recently Murray cod larvae had not been collected in the wild[3] (Humphries unpublished data), and this reflects the paucity of studies that have focussed on fish embryos and larvae of the Murray-Darling Basin.

There is thus, to date, evidence of large-scale movements (10's of km) for only three of the 26 freshwater species of Murray-Darling fishes. The smaller species of fish have been largely ignored in movement studies and this is clearly a major gap in our knowledge. We can make guesses as to why some individuals of the three species migrate large distances, but we are a long way from being able to provide a definitive answer to such an important question. In tropical floodplain rivers elsewhere in the world, it has been postulated that adult fish of some species migrate upstream to spawn in temporary habitats that are potentially important to the early intervals of fishes (P.B. Bayley personal communication). Subsequent drift of juveniles takes them to more substantial floodplain habitats, where food is plentiful and predators are scarce. Whilst we have some knowledge of longitudinal movements of Murray-Darling fishes in response to changes in flow, lateral movement into the flood plain remains a vexed question.

[9] Brumley, A.R., A.K. Morison & J.R. Anderson 1987. Revision of the conservation status of several species of warmwater native fish after surveys of selected sites in northern Victoria (1982–1984). Arthur Rylah Institute for Environmental Research Technical Report Series No.33, Melbourne.

The floodplain environment and its use by fishes

Floodplain rivers have longitudinal, as well as lateral linkages. The periodic pulsing of flows into the flood plain is thought to underpin the food webs of many flood plain rivers (Junk et al. 1989). Fish are typically prominent higher-order consumers within such food webs. The flood plain can be defined in an ecological sense as: 'areas that are periodically inundated by the lateral overflow of rivers or lakes, and/or by direct precipitation or groundwater; the resulting physicochemical environment causes the biota to respond by morphological, anatomical, physiological, phenological and/or ethological adaptations and produce characteristic community structures' (Junk et al. 1989, p. 112). This definition stresses the importance of periods of wetting and drying as being of primary significance in defining the flood plain, in contrast to other definitions, which may include permanently inundated areas. Welcomme (1985) describes several different habitat types within the floodplain environment of rivers, which include flooded grassland, lagoons and depressions, lakes, flooded forest and flood areas outside the main flood area.

Fish may potentially use a range of floodplain habitats, such as ephemeral anabranches, permanent and ephemeral billabongs and the flood plain proper, but since some of the habitats are temporary and some permanent, each may be used by different species or life intervals for different purposes and varying durations. Thus, permanent billabongs may have permanent fish communities, and may or may not experience immigration and emigration during connection with the river channel. Temporary floodplain habitats, on the other hand, can only be used either while there is connection with the river channel or while water persists in that habitat. Since no Murray-Darling fishes have desiccation resistant life stages, it is likely that fish using temporary floodplain habitats move into that habitat as long as the connection between it and the river channel persists, but move out before the connection is severed.

There are few published studies that have documented the occurrence of native Murray-Darling fishes in floodplain habitats. Some data are available on the fish faunas of billabongs[10] and from the Barmah Forest in the south-eastern region of the Basin[4], but virtually nothing about the use of the flood plain proper. Several species have been recorded from billabongs, with some species abundant in permanent waters, including various species of carp gudgeon, crimson-spotted rainbowfish and Australian smelt[10] (Cadwallader & Lawrence 1990, Balcombe unpublished data). It is likely that most of these species can complete their entire life cycle within billabongs and do not require connection with the river proper. Indeed, larger native fish species, with the exception of bony herring, seem to prefer mainstream habitats rather than billabongs. Golden perch, flathead gudgeon and Murray cod also have been collected from billabongs, but generally prefer lotic environments (Cadwallader & Lawrence 1990). Bony herring have been observed spawning in shallow backwaters during floods[8] and, although flooding did not coincide with the occurrence of larvae, juveniles of this species colonised floodplain habitats in Cooper Creeks in large numbers.[2] Freshwater catfish have been recorded as spawning in flooded and shallow portions of main rivers or in quiet backwaters (Lake 1967a). Langtry[1] documented spawning of Murray cod in the main channels and major anabranches of the Murray River. He cites fishermen saying that they have never caught juvenile Murray cod in billabongs. No published work has recorded the presence of larvae of any species of Murray-Darling native fishes in temporary billabongs or the flood plain proper.

Gehrke (1990a,b, 1991, 1992, 1994, Gehrke et al. 1993) has attempted to elucidate the role of the floodplain in the ecology of some species of Murray-Darling fishes through a series of studies, which have included experiments in artificial flood plains. Whilst acknowledging the lack of empirical data on floodplain use by fishes, Gehrke has nevertheless attempted to determine the suitability of the floodplain environment for golden perch, silver perch, and gudgeon larvae (Gehrke 1990b, 1991, 1992). His results indicated that, although golden perch larvae are attracted to chemical leachates from river redgum wood, a common riparian and floodplain tree, these leachates can sometimes be lethal to fish larvae and that the poor water quality of the floodplain environment may have a greater effect on the distribution of golden perch larvae than the abundance of food (Gehrke 1990b, 1991). Low dissolved oxygen levels and high concentrations of tannins and lignins on the flood plain make it an unattractive environment for young stages of this and potentially other species.

[10] Hume, D.J., A.R. Fletcher & A.K. Morison. 1983. Final Report: Carp Program 10. Arthur Rylah Institute for Environmental Research, Melbourne. 213 pp.

McKinnon (1995) has similarly shown that low oxygen levels and adverse water quality conditions associated with inundation of a floodplain forest can be harmful for aquatic species, such as the Murray crayfish, *Euastacus armatus*. McKinnon & Shepheard (1995) also reported a large fish kill which resulted from the return of oxygen depleted water from pastured flood plain. Anecdotal evidence from several sources indicated that rises in water levels and floods earlier this century in the Lachlan River coincided with staining of the water, large fish kills and subsequent low abundance of fishes for many years (Roberts & Sainty 1996). These examples suggest that flooding and floodplain habitats may not always provide conditions suitable for fish and fish larvae.

Despite the paucity of data which links Murray-Darling fishes to the floodplain environment, apart from permanent billabongs, the flood plain has been proposed by many researchers in Australia as playing a significant role in the life cycle of several species of Murray-Darling fish[2] (Lake 1967a, Arumugam & Geddes 1987, Rowland 1992, Gehrke 1990a,b, 1991, 1992, 1994). This suggestion has come about for several reasons. It stems partly from patterns observed in floodplain systems throughout the world (Welcomme 1985), partly from the perceived applicability of the 'flood pulse concept' (Junk et al. 1989) to the Murray-Darling Basin and partly from the perception that for successful recruitment of fishes at the larva period, the only environment that would provide sufficient sizes and concentrations of food would be newly inundated flood plain[2]. We will examine each of these issues below, assess the quantity and quality of evidence in support of their applicability to the Murray-Darling Basin and, where they are deemed wanting, provide some alternative explanations.

The role of the flood plain in the biology of fishes

A variety of species of fishes in floodplain river systems, such as the Amazon and the Mississippi–Missouri rivers, make use of the flood plain at times of high flow, for feeding and/or spawning. According to Welcomme (1985), all insect and plant foods consumed by fishes originate from overhanging vegetation, and he cites examples from the Amazon, Zaire and Mekong River systems[11] (Roberts 1973, Goulding 1980, 1981). Indeed, several families of fishes in the Amazon rely heavily on fruits and seeds at times of high flows (Goulding 1980). Fruits, seeds and flowers are important to species of fishes in other floodplain systems as well (see Welcomme 1985). Detritovores and predators are also well represented in floodplain systems (see Lowe-McConnell 1975, Welcomme 1985) and take advantage of flooding to gain access to accumulations of detritus on the floodplain, for the former group[12] (Almeida 1980, Bowen 1984), and to make use of the large amount of terrestrial insects falling from forests for the latter group (Goulding 1980).

The majority of Murray-Darling Basin fishes are opportunistic carnivores, with few known to consume plant material or detritus (Merrick & Schmida 1984). Furthermore, the floodplain environments encountered by Murray-Darling fishes do not support the types of plants that have large quantities of edible fruits or seeds. Unlike rivers in the world whose riparian trees and shrubs are dominated by deciduous species, annual litter fall in much of the Murray-Darling peaks in summer (Lake 1995). No fish species are known to feed on fallen leaves or the seeds and flowers of riparian plants. Nevertheless, there would be a large quantity of terrestrial insects potentially available to fishes should they use the flood plain. While the importance of this type of food for Murray-Darling fishes is largely undocumented, some species of freshwater fish will take advantage of this resource when it is available (e.g. Cadwallader et al. 1980). It would be likely that, within a forested floodplain environment there would be more terrestrially-derived food available to fishes than there would be in the main stem river, because of the amount of vegetation per unit area of water. It is also likely that flood waters would wash floodplain food into the main river channel and other habitats[5] and so make it potentially available to fishes that do not directly use the flood plain themselves. There is no doubt that the

[11] Geisler, R., H.A. Knoppel & H. Sioli. 1973. The ecology of freshwater fishes in Amazonia; present status and future tasks for research. pp. 144–162. *In*: Applied Sciences and Development, Vol. 2, Institute for Scientific Cooperation, Tübingen (cited *In*: R.L. Welcomme. 1985. River fisheries. Food and Agriculture Organization of the United Nations, FAO Fisheries Technical Paper 262).

[12] Santos, G.M. 1981. Estudos de alimentacao e habitos alimentares de *Schizodon fasciatus* Agassiz, 1829, *Rhytiodus microlepis* Kner, 1859 e *Rhytiodus argenteofuscus* Kner, 1859, do lago Janauaca (Osteichthyes, Characoidei, Anostomidae). Acta Amazonica 11: 267–283 (cited *In*: W.J. Junk, P.B. Bayley & R.E. Sparks. 1989. The flood pulse concept in river-floodplain systems. Can. Sp. Publ. Fish. Aquat. Sci. 106: 110–127).

periodic connection between the main channel of rivers and their flood plains has an extremely important role in riverine trophic processes (see Junk et al. 1989), but the nature of the relationship between fish feeding and the flood plain in the Murray-Darling is unresolved.

Many species of fish in floodplain systems spawn during high flows, thus ensuring a variety of spawning sites and an abundance of food for their larvae (Bayley 1983, 1988a, Holland et al. 1983, Welcomme 1985). River fishes are typically seasonal in their breeding habits, with temperature and flow seeming to be the two major factors that dictate when fish spawn. In temperate systems, flow and temperature both have significant roles in the timing of spawning. In the tropics, however, where seasonally contrasting temperatures are not as marked, flow, and especially flooding, is the dominant factor. In most of the temperate systems that Welcomme (1985) uses as examples, temperature and flooding tend to coincide. This is also the case in the examples provided within the framework of the 'flood pulse concept' (Junk et al. 1989). The link between flooding and breeding of river fishes is well documented for tropical rivers in Asia, Africa, South America and Northern Australia (Welcomme 1985). The relationship is so tight, that in some cases where the flood wave takes time to move down a river, fish spawning follows the wave downstream (Welcomme 1985). It is important to note, however, that in cases where high temperatures and flows do not coincide, temperature will often be the dominant variable that influences the timing of spawning. An example of this is from the Okavango Delta, where for the most part the flood occurs at the coldest time of the year, and so the fish spawn during low water (Welcomme 1985). Although spawning modes of flood plain fishes are diverse, the use of the flood plain seems for the majority to be initiated by adults. Adults move on to the flood plain as the flood rises and spawn in a variety of habitats (Welcomme 1985, Lowe-McConnell 1975). The adults place their larvae in optimal habitats, rather than relying on the larvae to get there themselves.

The spawning styles of Murray-Darling fish species are varied, yet again our knowledge is mostly from aquaculture studies (Lake 1967a, Llewellyn 1973, Cadwallader et al. 1979, Koehn & O'Connor 1990, Ingram & Rimmer 1992) conducted on only a few species. What happens in the wild is much less certain. Langtry[1] describes his observations of spawning Murray cod and notes depressions made '... in pans and on mud banks situated on the downstream side of a bend, sheltered from the main current' ([1]p. 28). He also notes that golden perch spawn in the river or in lagoons when the water is rising, but that in the Murrumbidgee River, both Murray cod and golden perch are thought not usually to spawn in the anabranch-billabong systems adjacent to this river. Some suggestions by Australian workers have been that fishes spawn in the main stem of rivers and either the eggs, embryos or larvae are washed on to the flood plain as the flood waters rise. No Murray-Darling fishes have been observed spawning on the flood plain proper and we reiterate that no larvae have ever been collected from seasonally inundated floodplain habitats.

The 'flood pulse concept' (FPC) and environmental prerequisites for use of the flood plain

The FPC states that: '... in unaltered large river systems with floodplains in the temperate, subtropical, or tropical belt, the overwhelming bulk of the riverine animal biomass derives directly or indirectly from production within the floodplains and not from downstream transport of organic matter produced elsewhere in the basin' (Junk et al. 1989, p. 112). This is an alternative to the 'river continuum concept' (Vannote et al. 1980), where the emphasis is on longitudinal transport of organic material, not lateral. Many river fisheries are dominated by those species that seasonally colonise the flood plain (Bonnetto et al. 1969, Welcomme 1979, Bayley 1981, 1983, Goulding 1981). The FPC emphasises the role of the flood plain in the ecology of fishes in providing a benign spawning environment and bountiful food for juveniles, adults, and larvae. However, periods of low flow are also common in floodplain rivers in all climatic zones and fishes in these rivers exhibit a diversity of life history styles, some of which involve spawning within the main river channel (Junk et al. 1989, Junk et al. 1997).

Nevertheless, one of the main points of the FPC is that the best conditions for fish spawning is when high flows and high temperatures coincide (Junk et al. 1989). An example is of the relative dominance of spring vs. summer spawners being controlled by the timing and duration of flooding in Missouri floodplain forests (Finger & Stewart 1987). If floods are of only short duration during the warm period of the year, then recruitment can suffer (Junk et al. 1989). The interannual predictability of floods is critical. For fish to make use of flood plains for spawning purposes, floods need to be at the same time each year or fish have to have a flexible life cycle, such as proposed for golden perch in the Murray-Darling Basin. Puckridge et al. (1998) showed clearly that flow variability, at a number of temporal scales, elicits major biological responses by a variety of riverine fishes throughout the world. These responses can be proximate responses (poor recruitment, local extinctions, different levels of migration) or ultimate ones (long breeding seasons, flexible life histories, wide physiological tolerances), depending on the nature of the flow variability.

Fish tend to spawn during the warmest months of the year, partly because rates of egg, embryo and larva development are positively correlated with temperature, and partly because it is at this time of the year that in temperate systems food for larvae and juveniles is most abundant (Jobling 1995). Rates of development and growth are critical to the survival of larvae, since the larger a fish is, the greater the swimming speed and the greater the volume of water able to be searched for food (Bone et al. 1995). Furthermore, the longer an individual spends as a highly vulnerable larva, the greater the risk of predation by larger fish (Jobling 1995). A number of studies have indeed found that larger larvae are able to respond better to an attack from a predator and that the probability of capture decreases with increasing larva size (Butler & Pickett 1988, Fuiman 1989, Margulies 1990). It has even been suggested that survival in the early intervals of life is a direct function of growth rate (Houde 1987). On the other hand, others have suggested that larger larvae are more conspicuous to predators and provide more net energy gain than small larvae (Fuiman 1989, Litvak & Leggett 1992).

High flows in the Murray-Darling Basin do not necessarily coincide with high temperatures. In the south-eastern temperate regions of this river system, flows tend to peak in mid-winter to early spring, before the water has warmed significantly (Figure 2). Flooding may occur during this time, whereas temperatures usually peak well after flows have declined dramatically,

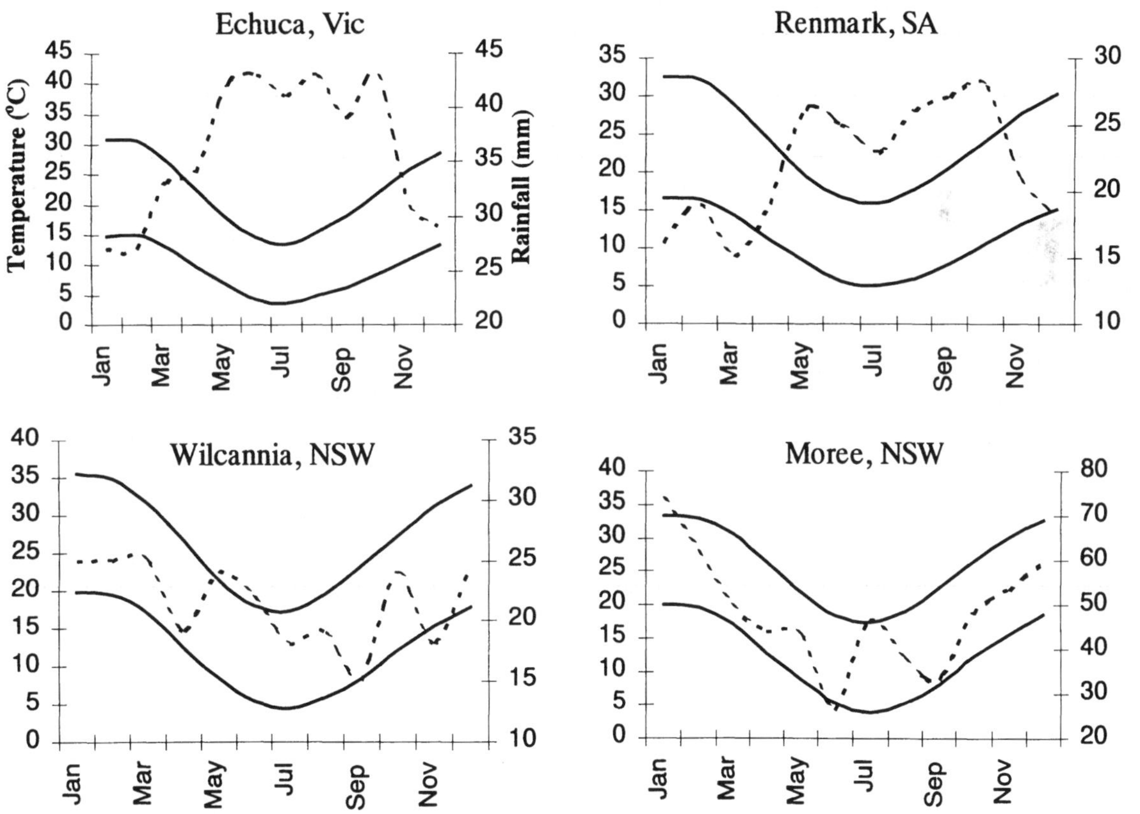

Figure 2. Average monthly rainfall and temperature for four localities in the basin. Solid line = maximum/minimum temperature, dashed line = rainfall. Echuca and Renmark = Murray River, southern Murray-Darling Basin, Wilcannia = Darling River, central-west Murray-Darling Basin, Moree = Gwydir River, north-eastern Murray-Darling Basin. See Figure 1.

certainly well below flood levels. It is clear, however, that as the flood wave passes down the Murray River, there is a shift in the timing of peak flows to later in spring. Therefore, in the lower Murray, floods may coincide with higher temperatures. This is also true of the sub-tropical regions of the Darling River and their major tributaries like the Gwydir, where floods tend to occur during summer when temperatures are high.

Thus, in regions of the Murray-Darling Basin where high flows and high temperatures coincide, fish have optimal conditions for breeding. However, in those regions of the basin where temperature and flow do not coincide, there may be a conflict. Clearly, evidence from studies to date have emphasised temperature more than flows in cues for spawning of native Murray-Darling fishes (Table 2). Furthermore, there is some evidence to suggest that temperature is the overriding parameter that determines timing of spawning elsewhere in the world when this non-coincidence occurs (Welcomme 1985). Since it is not possible to treat the Murray-Darling Basin as climatically homogeneous, it is possible that fish which occur in different regions of this system may have adapted to disparate climatic and hydrological influences and so may make use of different cues and environments for spawning.

The high interannual variability of discharge in rivers in the Murray-Darling Basin (Walker 1992, Walker et al. 1995, Puckridge et al. 1998) has been suggested as one of the contributing factors for the depauperate freshwater fish fauna (Harris 1984, Allen 1989). This extreme variability makes the Murray-Darling Basin a harsh environment for fishes and may limit the diversity of successful life history styles. If the appropriate environmental conditions, a flood, are unpredictable both interannually and seasonally, then it would be an advantage to have a life history style that would allow either a flexible spawning time, an extended spawning period so that a subset of the progeny are likely to encounter favourable conditions, or considerable longevity and so reduce the need to breed every year.

An alternative to relying on breeding during a highly unpredictable event is to breed under other conditions which are more predictable. This forms the basis of the 'low flow recruitment hypothesis' proposed earlier. While the timing of floods is often unpredictable, the timing of low flows can often be relatively predictable[13]. Spawning during stable low-flow times has also been suggested to be adaptive for crimson-spotted rainbowfish (Milton & Arthington 1984) and Australian smelt (Milton & Arthington 1985) in coastal streams in the Brisbane area.

The other aspect of unpredictability is that of flood duration. It is risky for a fish to move on to the floodplain to spawn if the high flows are potentially of only short duration. This is especially so if spawning involves nest building, adhesive eggs and/or protracted larva period or parental care. In floodplain rivers in other regions of the world, especially in the tropics, the flood plain may be reliably inundated for six months (Goulding 1980, Welcomme 1985) and the risk of desiccation is slight compared with the conditions encountered throughout much of the Murray River (Walker 1986).

Food for fish larvae

The significance of high concentrations of food for fish larvae during the critical transition from endogenous to exogenous feeding has received much attention (see Bone et al. 1995). It is at this time that mortality due to starvation is thought to be at its highest. Fish larvae are generally visual predators and when small, can only search a relatively small volume of water. Furthermore, their gapes are generally very small, often in the order of 200–500 μm (Table 3). Thus, to get through this critical transition interval, larvae must generally encounter dense concentrations of appropriately sized prey taxa. This subject has been the focus of many studies and much speculation in marine systems (e.g. Frank 1988, Miller et al. 1988, Pepin & Shears 1995, Gotceitas et al. 1996), with several hypotheses proposed that relate to the relationship between first-feeding larvae and their food, the most notable of which is the match/mismatch hypothesis (Cushing 1990). This hypothesis states that recruitment is governed by the relative timing of the occurrence of larvae relative to the occurrence of their prey. The key assumption is that fish spawn, and therefore larvae occur, at approximately the same time each year, but that the peak prey density is more influenced by relatively unpredictable environmental conditions. If it is possible to predict when and where high densities of food will occur, say on newly inundated flood plain, then some of the uncertainties in matching progeny with prey are alleviated.

The hypothesis proposed by several workers has been that the main stem of rivers in the Murray-Darling Basin generally do not support high densities

[13] Nathan, R.J. & P.E. Weinmann. 1992. Low flow atlas for Victorian streams. Long term planning guidelines background report, L3. Department of Conservation and Natural Resources, Melbourne. 42.

Table 3. Gape of larvae of some Murray-Darling Basin fish at first feeding (SL = standard length, TL = total length).

Species	Length at first feeding (mm)	Gape (mm)	Reference
Australian smelt	4.1–8.8 (SL)	0.16–0.52	Humphries & King unpublished data
Flathead gudgeon	3.7–5.3 (SL)	0.24–0.51	Humphries & King unpublished data
Murray cod	8.5–10.0 (SL)	1.0–1.2	Humphries & King unpublished data
Crimson-spotted rainbowfish	3.2–5.6 (SL)	0.2–0.3	Humphries & King unpublished data
Golden perch	4.7–4.9 (TL)	0.5	Arumugam & Geddes (1987)
Silver perch	4.6–5.4 (TL)	0.4	Arumugam & Geddes (1987)

of zooplankton and that the blooms of zooplankton that are thought to occur in newly inundated flood plain may provide the best environment for rearing of young fish[2,6] (Lake 1967a, Arumugam & Geddes 1987, Rowland 1992). To date, the diets of the larvae of only a few species have been studied (Lake 1967a, Arumugam & Geddes 1987, Gehrke 1992, Rowland 1992) with only one of these conducted in a natural system (Gehrke 1992). Concentrations of between 100 and 1000 individuals per litre have variously been proposed as being necessary for the survival of first-feeding larvae (Bone et al. 1995). Rowland (1992) found that hatchery-reared Murray cod survived well if exposed to 250 zooplankton per litre when they first started feeding, but that a delay of only a few days in encountering food increased the mortality rate of larvae exposed to 250 zooplanktors per litre relative to a group exposed to 3000 zooplanktors per litre.

The importance of the flood plain in providing high concentrations of prey for larvae rests on the assumption that inundated flood plain supports such densities. Welcomme (1985) provides numerous examples of the density of zooplankton in floodplain rivers. In some cases, densities were highest at high flows[14] (Welcomme 1985) and in others densities were highest at low flows[15] (Holden & Green 1960, Welcomme 1985). There is generally a direct relationship between water residence time and zooplankton biomass (Basu & Pick 1996) and therefore lentic water bodies will tend to be more productive than rivers (Pace et al. 1992, Thorp et al. 1994). Inundation of the flood plain can release nutrients and the resting stages of a variety of invertebrates (Crome 1986, Crome & Carpenter 1988, Boulton & Lloyd 1992, Tan & Shiel 1993), but responses to changing water levels are not simple. Crome (1986) reported that the concentration effects of declining water levels meant that many groups of microcrustacea peaked in a swamp immediately before drying, whilst Crome & Carpenter (1988) showed that chironomids bloomed following the subsequent inundation, presumably as a result of nutrient release. Tan & Shiel (1993) showed that inundation of a billabong resulted in increases in the densities of some rotifer taxa, whereas this was not the case for the zooplankton fauna in general. High densities of zooplankton in the newly flooded Barmah Forest were suggested to have originated upstream and were washed into the forest by the flood waters[5]. The inundation of aquaculture ponds to provide food for hatchery-bred native fish larvae has been used as a model for inundation of the

[14] CECOAL, 1977. Estudios ecologicos en el area de Yacyreta: informe de avance, 2. Argentina, Corrientes, Centro de Ecologia Aplicada del Litoral (cited *In*: R.L. Welcomme. 1955. River fisheries. Food and Agriculture Organization of the United Nations, FAO Fisheries Technical Paper 262).

[15] Arias, P.A. 1977. Evaluacion limnologica de las planicies inundables de la cuena norte del Rio Magdalena. Proyecto para el desarollo de la pesca continental en Colombia, Catagena, Inderena-FAO (cited in: R.L. Welcomme 1985. River fisheries. Food and Agriculture Organization of the United Nations, FAO Fisheries Technical Paper 262).

flood plain[2]. They found that there was a succession of phytoplankton and zooplankton taxa (from rotifers through to large zooplankters) through time. They suggest that a similar pattern of emergence and succession in flood plains would provide ideal nursery conditions for the larvae of native fishes. There is currently insufficient data to compare densities of microinvertebrates in the main stem of rivers with those of inundated flood plain and results of floodplain sampling do not provide enough evidence to indicate that blooms after flooding are consistent throughout the basin, nor that the densities of zooplankton are high enough to provide for the requirements of fish larvae. However, anecdotal evidence suggests that the microfauna of backwaters and similar habitats in lowland rivers are sufficiently dense and small to be useful for fish larvae and, therefore, the 'low flow recruitment hypothesis' should receive serious consideration.

The other assumption relating to the feeding of fish larvae is that they all require extremely small prey items to get them through the critical interval from endogenous to exogenous feeding. Whilst many species have small gapes, this is by no means universal (Table 3). It has already been suggested that the well developed larvae of Mode 1 fish, such as Murray cod, may be able to consume relatively large prey at first feeding. A lack of empirical data on the diets of fish larvae in the wild, however, again leads to a reliance on results from aquaculture studies.

Lake (1967a) described the food of several species of native fishes reared in ponds. Some species, such as silver perch and western carp-gudgeons begin feeding after only a few days on algae, before switching to microcrustacea (Table 4). Others, such as freshwater catfish and Murray cod, have large yolksacs and begin feeding much later at relatively large sizes on zooplankton and chironomids (Rowland 1992). The only study performed under natural conditions found that eleotrid larvae less than 5 mm in length consumed only rotifers, whereas larger larvae switched to larger prey, such as calanoid copepods and cladocerans (Gehrke 1992).

Gape size is a major limiting factor in the consumption of prey for a wide range of species of marine and freshwater fishes (Bone et al. 1995). This may mean that there is a reduction or entire elimination of the dependence of those species of Murray-Darling Basin fishes that, as first-feeding larvae, have large gapes, on the very small prey which have generally been assumed to be significant in getting larvae through the critical endogenous/exogenous feeding interval. This may explain why, for species like Murray cod and freshwater blackfish, spawning may be unrelated to flooding, since the larvae may be able to obtain sufficient food in the main stem of rivers via relatively large zooplankton, benthic microfauna, and smaller instars of benthic macroinvertebrates.

Concluding remarks

The fish faunas of most major river systems in the world have been affected adversely by river regulation, barriers to movement, riparian clearing, habitat alteration, the introduction of exotic species and declining water

Table 4. Food of the larvae of some Murray-Darling fish.

Species	Age at feeding (days)	Length at feeding (mm)	Prey consumed	Reference
Silver perch	6*		phytoplankton	Lake (1967a)
	7		copepods & cladocerans	
Catfish	19*		zooplankton	Lake (1967a)
Murray cod	27*	12–14	chironomids, *Daphnia* & copepods	Lake (1967a)
Western carp-gudgeon	5*	3.4	Algae	Lake (1967a)
Golden perch			*Artemia* nauplii	Arumugam & Geddes (1987)
Gudgeons	<5		rotifers	Gehrke (1992)

*First feeding.

quality. Attempts to redress many of these problems are underway, but to be successful, a good understanding of the relationship between fishes and their environment is vital. This paper has identified three life history modes of Murray-Darling native fish species, some of which may have strong links with the flow environment and others which may not. There is clearly a group of species whose spawning period is unrelated to high flows and flooding and seems to occur during the low flow period in summer. We have proposed the 'low flow recruitment hypothesis' as an explanation for why these species may spawn and how they can recruit during the warm, low flow periods of some Murray-Darling Basin lowland rivers. The widespread acceptance of the importance of flooding and the flood plain for successful recruitment of Murray-Darling fishes and virtual disregard of a rearing role for the in-channel environment has led us to review the status of our knowledge of the biology of Murray-Darling fishes, especially the presumed links with flow and the flood plain. Information on the use of floodplain habitats, and particularly the flood plain proper during inundation, by all stages of Murray-Darling Basin fishes, is either totally lacking, based on studies of limited taxonomic or geographic scope, or based solely on aquaculture studies. It should be a priority for future research to address these gaps in our knowledge. Some suggestions for research are: properly controlled and rigorously designed experiments investigating spawning cues, preferably in conjunction with long-term intensive studies of fish breeding in the wild; further studies into the movement of free embryos, larvae, juveniles and adults of small and large Murray-Darling fish species; sampling of floodplain habitats for emerging zooplankton, zoobenthos and fish larvae over large spatial scales; investigation of the size and composition of the diets of fish larvae in a range of habitats and under a variety of temperature and flow conditions; and studies of the population dynamics of small and large species of Murray-Darling fishes. In addition, our categories of life history modes of Murray-Darling fishes are based on insufficient data and require additional information, refinement and evaluation. The 'low flow recruitment hypothesis' is also as yet untested. Research is required into the relative importance of in-channel versus floodplain habitats as nursery areas, and also the influence of fluctuations in water level on the density and composition of zooplankton and benthic microfauna in these nursery habitats. One could conceive of a rigorous test of the 'low flow recruitment hypothesis' involving investigation of recruitment patterns of fishes in the different life history modes under a variety of experimental flow releases. It should be possible to hypothesize that those species that recruit during low flow times will recruit well if low flows are maintained for several months during the summer, whereas the opposite will occur if flows fluctuate or are stable and high during this same period.

We are a long way from being able to state definitively what the significance of flow and the flood plain are to Murray-Darling fishes, let alone what the major environmental variables that influence recruitment are. It is incumbent on fisheries biologists to be cautious when ascribing unfettered significance of flooding and the flood plain to the biology of Murray-Darling fishes, since this has likely encouraged complacency in managers that our knowledge of the ecology of fishes is adequate. It may have also diverted attention from other important habitats such as those within the main channel of lowland rivers, as well as relegated to low priority other natural components of the hydrological cycle such as low flows. Indeed, river regulation that involves using rivers as irrigation conduits during periods when flows would normally have been low, has undoubtedly affected adversely those species which use in-channel habitats for rearing. Current stream ecosystem discussions constantly reinforce the importance of longitudinal as well as lateral and vertical linkages (Bayley 1988b, Junk et al. 1989, Thorp & Delong 1994, Ward & Stanford 1995, Neiff 1996) and the need to understand the significance of many facets of the hydrology of rivers (Walker et al. 1995, Richter et al. 1996, Puckridge et al. 1998). Until we accept for ourselves, and convince resource managers, of the importance of the entire riverine ecosystem and its associated hydrology, there is a danger of establishing priorities based on inadequate knowledge and potentially condemning some of our native fish species to extinction in the process.

Acknowledgements

We would like to thank the many people who discussed various aspects of this paper during its preparation, in particular Luciano Serafini and Jane Growns and in general staff at the Murray-Darling Freshwater Research Centre. For comments on drafts of the paper, we are grateful to Jane Growns, Simon Nicol, Andrew Boulton and Sam Lake. This paper was written while

AK was funded by the Land and Water Resources Research and Development Corporation through Environment Australia.

References cited

Allen, G.R. 1989. Freshwater fishes of Australia. T.F.H. Publications, Neptune City. 240 pp.

Almeida, R.G. 1980. Aspectos taxonomicoa e habitos alimentares de tres especies de *Triportheus* (Pisces: Characoidei, Characidae), do lago do Castanho, Amazonas. M.S. Thesis, Instituto Nacional de Pesquisas da Amazonia, Manaus. 104 pp. (cited in: Junk, W.J., PB. Bayley & RE. Sparks. 1989. The flood pulse concept in river-floodplain systems. Can. Sp. Publ. Fish. Aquat. Sci. 106: 110–127).

Anderson, J.R., J.S. Lake & N.J. Mackay. 1991. Notes on reproductive behaviour and ontogeny in two species of *Hypseleotris* (=*Carassiops*) (Gobiidae: Teleostei). Aust. J. Mar. Freshw. Res. 22: 139–145.

Arumugam, P.T. & M.C. Geddes. 1987. Feeding and growth of golden perch larvae and fry (*Macquaria ambigua* Richardson). Trans. Roy. Soc. S. Aust. 111: 59–65.

Balon, E.K. 1984. Patterns in the evolution of reproductive styles in fishes. pp. 35–52. *In*: G.W. Potts & R.J. Wootton (ed.) Fish Reproduction: Strategies and Tactics, Academic Press, London.

Basu, B.K. & F.R. Pick. 1996. Factors regulating phytoplankton and zooplankton biomass in temperate rivers. Limnol. Oceanogr. 41: 1572–1577.

Basu, B.K. & F.R. Pick. 1997. Phytoplankton and zooplankton development in a lowland, temperate river. J. Plank. Res. 19: 237–253.

Bayley, P.B. 1981. Fish yield from the Amazon in Brazil: comparison with African river yields and management possibilities. Trans. Amer. Fish. Soc. 110: 351–359.

Bayley, P.B. 1983. Central Amazon fish populations. Biomass, production and some dynamic characteristics. Ph.D. Thesis, Dalhousie University, Halifax. 330 pp. (cited in: R.L. Welcomme 1985. River fisheries. Food and Agriculture Organization of the United Nations, FAO Fisheries Technical Paper 262).

Bayley, P.B. 1988a. Factors affecting growth rates of young tropical floodplain fishes: seasonality and density-dependence. Env. Biol. Fish. 21: 127–142.

Bayley, P.B. 1988b. Understanding large river-floodplain ecosystems. BioScience 45: 153–158.

Backhouse, G.N. & D.J. Frusher. 1980. The crimson-spotted rainbowfish, *Melanotaenia fluviatilis* (Castelnau 1878). Vict. Nat. 97: 144–148.

Beumer, J.P. 1979. Reproductive cycles of two Australian freshwater fishes: the spangled perch, *Therapon unicolor* Günther, 1859 and the East Queensland rainbowfish, *Nematocentris splendida* Peters, 1866. J. Fish Biol. 15: 111–134.

Bone, Q., N.B. Marshall & J.H.S. Blaxter. 1995. Biology of fishes, 2nd edition. Blackie Academic and Professional Publishers, Glasgow. 332 pp.

Bonetto, A.A., W. Dioni & C. Pignalberi. 1969. Limnological investigations on biotic communities in the middle Parana river valley. Verh. int. Ver. theor. angew. Limnol. 17: 1035–1050.

Boulton, A.J. & L.N. Lloyd. 1992. Flooding frequency and invertebrate emergence from dry floodplain sediments of the River Murray, Australia. Regul. Riv. 7: 137–151.

Bowen, S.H. 1984. Detritivory in neotropical fish communities. pp. 59–66. *In*: T. Zaret (ed.) Evolutionary Ecology of Neotropical Freshwater Fishes, Dr W. Junk Publishers, The Hague.

Bremigan, M.T. & R.A. Stein. 1994. Gape-dependent larval foraging and zooplankton size: implications for fish recruitment across systems. Can. J. Fish. Aquat. Sci. 51: 913–922.

Butler, J.L. & D. Pickett. 1988. Age-specific vulnerability of Pacific sardine, *Sardinops sagax*, larvae to predation by northern anchovy, *Engraulis mordax*. U.S. Fish. Bull. 86: 163.

Bye, V.C. 1984. The role of environmental factors in the timing of reproductive cycles. pp. 187–206. *In*: G.W. Potts & R.J. Wootton (ed.) Fish Reproduction: Strategies and Tactics, Academic Press, London.

Cadwallader, P.L. 1978. Some causes of the decline in range and abundance of native fish in the Murray-Darling river system. Proc. Roy. Soc. Vict. 90: 211–224.

Cadwallader, P.L. & G.N. Backhouse. 1983. A guide to the freshwater fish of Victoria. Victorian Government Printing Office, Melbourne. 249 pp.

Cadwallader, P.L., G.N. Backhouse, G.J. Gooley & J.A. Turner. 1979. New techniques for breeding and raising Murray cod. Aust. Fish. 38: 9–16.

Cadwallader, P.L., A.K. Eden & R.A. Hook. 1980. Role of streamside vegetation as a food source for *Galaxias olidus* Gunther (Pisces: Galaxiidae). Aust. J. Mar. Freshw. Res. 31: 257–262.

Cadwallader, P.L. & B. Lawrence. 1990. Chapter 22: Fish. pp. 317–363. *In*: N. Mackay & D. Eastburn (ed.) The Murray, Murray Darling Basin Commission, Canberra.

Cadwallader, P.L. & P.L. Rogan. 1977. The Macquarie perch, *Macquaria australasica* (Pisces: Percichthyidae) of Lake Eildon, Victoria. Aust. J. Ecol. 2: 409–418.

Copp, G.H. 1992. Comparative microhabitat use of cyprinid larvae and juveniles in a lotic floodplain channel. Env. Biol. Fish. 33: 181–193.

Crome, F.H.J. 1986. Australian waterfowl do not necessarily breed on a rising water level. Aust. Wildl. Res. 13: 461–480.

Crome, F.H.J. & S.M. Carpenter. 1988. Plankton community cycling and recovery after drought – dynamics in a basin on a flood plain. Hydrobiologia 164: 193–211.

Cushing, D.H. 1990. Plankton production and year-class strength in fish populaions: an update of the match/mismatch hypothesis. Adv. Mar. Biol. 26: 249–293.

Davies, BR. & K.F. Walker. 1986. The ecology of river systems, Dr W. Junk Publishers, Dordrecht. 793 pp.

Davis, T.L.O. 1977. Reproductive biology of the freshwater catfish, *Tandanus tandanus* Mitchell, in the Gwydir River, Australia. II. Gonadal cycle and fecundity. Aust. J. Mar. Freshw. Res. 28: 159–169.

de Vlaming, V.L. 1972. Environmental control of teleost reproductive cycles: a brief review. J. Fish Biol. 4: 131–140.

Ferrari, I., A. Farabegoli & R. Mazzoni. 1989. Abundance and diversity of planktonic rotifers in the Po River. Hydrobiologia 186/187: 201–208.

Finger, T.R. & E.M. Stewart. 1987. Response of fishes to flooding regime in lowland hardwood wetlands. pp. 86–92. *In*: W.J. Matthews & D.C. Heins (ed.) Community and Evolutionary Ecology of North American Stream Fishes, University of Oklahoma Press, Norman.

Frank, K.T. 1988. Independent distributions of fish larvae and their prey: natural paradox or sampling artifact? Can. J. Fish. Aquat. Sci. 45: 48–59.

Fuiman, L.A. 1989. Vulnerability of Atlantic herring larvae to predation by yearling herring. Mar. Ecol. Prog. Ser. 51: 291–299.

Gehrke, P.C. 1990a. Spatial and temporal dispersion patterns of golden perch, *Macquaria ambigua*, larvae in an artificial floodplain environment. J. Fish Biol. 37: 225–236.

Gehrke, P.C. 1990b. Clinotactic responses of larval silver perch (*Bidyanus bidyanus*) and golden perch (*Macquaria ambigua*) to simulated environmental gradients. Aust. J. Mar. Freshw. Res. 41: 523–528.

Gehrke, P.C. 1991. Avoidance of inundated floodplain habitat by larvae of golden perch (*Macquaria ambigua* Richardson): influence of water quality or food distribution? Aust. J. Mar. Freshw. Res. 42: 707–719.

Gehrke, P.C. 1992. Diel abundance, migration and feeding of fish larvae (Eleotridae) in a floodplain billabong. J. Fish Biol. 40: 695–707.

Gehrke, P.C. 1994. Influence of light intensity and wavelength on phototactic behaviour of larval silver perch *Bidyanus bidyanus* and golden perch *Macquaria ambigua* and the effectiveness of light traps. J. Fish Biol. 44, 741–751.

Gehrke, P.C., M.B. Revell & A.W. Philby. 1993. Effects of river red gum, *Eucalyptus camaldulensis*, litter on golden perch, *Macquaria ambigua*. J. Fish Biol. 43: 265–279.

Gotceitas, V., V. Puvanendran, L.L. Leader & J.A. Brown. 1996. An experimental investigation of the 'match/mismatch' hypothesis using larval Atlantic cod. Mar. Ecol. Prog. Ser. 130: 29–37.

Goulding, M. 1980. The fishes and the forest. Explorations in Amazonian natural history. University of California Press, Berkeley. 280 pp.

Goulding, M. 1981. Man and fisheries on an Amazonian frontier. Dr W. Junk Publishers, The Hague. 137 pp.

Harris, J.H. 1984. Zoogeography of the Australian freshwater fish fauna. pp. 211–222. *In*: M. Archer & G. Clayton (ed.) Vertebrate Zoogeography and Evolution in Australasia, Hesperian Press, Perth.

Harris, J.H. & P.C. Gehrke. 1994. Modelling the relationship between streamflow and population recruitment to manage freshwater fisheries. Aust. Fish. 6: 28–30.

Hildrew, A.G., C.R. Townsend & A. Hasham. 1985. The predatory Chironomidae of an iron-rich stream: feeding ecology and food web structure. Ecol. Entomol. 10: 403–413.

Holden, M.J. & J. Green. 1960. The hydrology and plankton of the River Sokoto. J. Anim. Ecol. 29: 65–84.

Holland, L.E., C.F. Bryan & J.P. Newman, Jr. 1983. Water quality and the rotifer population in the Atchafalaya River Basin. Hydrobiologia 98: 55–69.

Houde, E.D. 1987. Fish early life dynamics and recruitment variability. American Fisheries Society Symposium 2: 17–29.

Humphries, P. 1989. Variation in the life history of diadromus and landlocked populations of the spotted galaxias, *Galaxias truttaceus* Valenciennes, in Tasmania. Aust. J. Mar. Freshw. Res. 40: 501–518.

Humphries, P. 1995. Life history, food and habitat of southern pygmy perch, *Nannoperca australis*, in the Macquarie River, Tasmania. Mar. Freshw. Res. 46: 1159–1169.

Ingram, B.A. & J.W. Douglas. 1995. Threatened fishes of the world: *Maccullochella macquariensis* (Cuvier, 1829) (Percichthyidae). Env. Biol. Fish. 43: 38.

Ingram, B.A. & M.A. Rimmer. 1992. Induced breeding and larval rearing of the endangered Australian freshwater fish trout cod, *Maccullochella macquariensis* (Cuvier) (Percichthyidae). Aquat. Fish. Man. 24: 7–17.

Jackson, P.D. 1978. Spawning and early development of the river blackfish, *Gadopsis marmoratus* Richardson (Gadopsiformes: Gadopsidae), in the McKenzie River, Victoria. Aust. J. Mar. Freshw. Res. 29: 293–298.

Jobling, M. 1995. Environmental biology of fishes. Chapman & Hall Publishers, London. 455 pp.

Junk, W.J., P.B. Bayley & R.E. Sparks. 1989. The flood pulse concept in river-floodplain systems. pp. 110–127. *In*: D.P. Dodge (ed.) Proceedings of the International Large River Symposium, Can. Sp. Publ. Fish. Aquat. Sci. 106.

Junk, W.J., M.G.M. Soares & U. Saint-Paul. 1997. The fish. pp. 385–408. *In*: W.J. Junk (ed.) The Central Amazon Floodplain: Ecology of a Pulsing System, Springer-Verlag, Berlin.

Junk, W.J. & G.H. Weber. 1996. Amazonian floodplains: a limnological perspective. Verh. internat. Verein. Limnol. 26: 149–157.

Koehn, J. 1986. Approaches to determining flow and habitat requirements for native freshwater fish in Victoria. pp. 95–115. *In*: I.C. Campbell (ed.) Stream Protection: the Management of Rivers for Instream Uses, Water Studies Centre, Chisolm Institute Technology, Melbourne.

Koehn, J.D. & W.G. O'Connor. 1990. Biological Information for Management of Native Freshwater Fish in Victoria. Government Printer, Melbourne. 165 pp.

Lake, J.S. 1967a. Rearing experiments with five species of Australian freshwater fishes. I. Inducement to spawning. Aust. J. Mar. Freshw. Res. 18: 137–153.

Lake, J.S. 1967b. Rearing experiments with five species of Australian freshwater fishes. II. Morphogenesis and ontogeny. Aust. J. Mar. Freshw. Res. 18: 155–173.

Lake, J.S. 1971. Freshwater fishes and rivers of Australia. Thomas Nelson & Sons Ltd, Melbourne. 61 pp.

Lake, P.S. 1995. Of floods and droughts: river and stream ecosystems of Australia. pp. 659–694. *In*. C.E. Cushing, K.W. Cummins & G.W. Minshall (ed.) Ecosystems of the World 22, River and Stream Ecosystems, Elsevier Press, London.

Lancaster, J. & A.L. Robertson. 1995. Microcrustacean prey and macroinvertebrate predators in a stream food web. Freshw. Biol. 34: 123–134.

Lawrence, R. 1971. Habitat and economy: a historical perspective. pp. 249–261. *In*: D.J. Mulvaney & J. Golson (ed.) Aboriginal Man and Environment in Australia, Australian National University, Canberra.

Lightfoot, G.W. & N.V. Jones. 1996. The relationship between the size of 0+ roach, *Rutilus rutilus*, their swimming capabilities, and distribution in an English river. Folia Zool. 45: 355–360.

Litvak, M.K. & W.C. Leggett. 1992. Age and size-selective predation on larval fishes: the bigger-is-better hypothesis revisited. Mar. Ecol. Prog. Ser. 81: 13–24.

Llewellyn, L.C. 1968. Tagging gives answers to fish growth queries. The Fisherman, June: 1–5.

Llewellyn, L.C. 1971. Breeding studies on the freshwater forage fish of the Murray-Darling river system. The Fisherman, June: 1–12.

Llewellyn, L.C. 1973. Spawning, development and temperature tolerance of the spangled perch *Madigania unicolor* (Gunther) from inland waters in Australia. Aust. J. Mar. Freshw. Res. 24: 73–94.

Llewellyn, L.C. 1974. Spawning, development and distribution of the southern pygmy perch *Nannoperca australis australis* Gunther from inland waters in Eastern Australia. Aust. J. Mar. Freshw. Res. 25: 121–149.

Llewellyn, L.C. & M.C. MacDonald. 1980. Family Percichthyidae. Australian freshwater basses and cods. pp. 142–149. *In*: R.M. McDowall (ed.) Freshwater Fishes of South-eastern Australia, Reed, Sydney.

Lowe-McConnell, R.H. 1975. Fish communities in tropical freshwaters. Longman, London. 337 pp.

Mackay, N.J. 1973. Histological changes in the ovaries of the golden perch, *Plectroplites ambiguus*, associated with the reproductive cycle. Aust. J. Mar. Freshw. Res. 24: 95–101.

Mann, R.H.K. & J.A.B. Bass. 1997. The critical water velocities of larval roach (*Rutilus rutilus*) and dace (*Leuciscus leuciscus*) and implications for river management. Regul. Riv. 13: 295–301.

Margulies, D. 1990. Vulnerability of larval white perch, *Morone americana*, to fish predation. Env. Biol. Fish. 27: 187–200.

McComb, A.J. & P.S. Lake. 1990. Australian wetlands, Angus & Robertson Publishers, Sydney. 258 pp.

McDowall, R.M. & W. Fulton. 1996. Galaxiids. pp. 52–77. *In*: R.M. McDowall (ed.) Freshwater Fishes of South-eastern Australia, Reed, Sydney.

McKinnon, L.J. 1995. Emergence of Murray crayfish *Euastacus armartus* (Decapoda: Parastacidae), from the Murray River due to post flooding hypoxia. Proc. Roy. Soc. Vict. 107: 31–38.

McKinnon, L.J. & N. Shepheard. 1995. Factors contributing to a fish kill in the Broken Creek. Vic. Nat. 112: 93–99.

Merrick, J.R. & G.E. Schmida. 1984. Australian freshwater fishes: biology and management. Griffin, Adelaide. 409 pp.

Miller, T.J., L.B. Crowder, J.A. Rice & E.A. Marschall. 1988. Larval size and recruitment mechanisms in fishes: towards a conceptual framework. Can. J. Fish. Aquat. Sci. 45: 1657–1670.

Milton, D.A. & A.H. Arthington. 1984. Reproductive strategy and growth of the crimson-spotted rainbowfish, *Melanotaenia splendida fluviatilis* (Castelnau) (Pisces: Melanotaeniidae) in South-eastern Queensland. Aust. J. Mar. Freshw. Res. 35: 75–83.

Milton, D.A. & A.H. Arthington. 1985. Reproductive strategy and growth of the Australian smelt, *Retropinna semoni* (Weber) (Pisces: Retropinnidae), and the olive perchlet, *Ambassis nigripinnis* (De Vis) (Pisces: Ambassidae), in Brisbane, South-eastern Queensland. Aust. J. Mar. Freshw. Res. 36: 329–341.

Milward, N.E. 1969. Development of the eggs and early larvae of the Australian smelt, *Retropinna semoni* (Weber). Proc. Linn. Soc. NSW 90: 1–7.

Neiff, J.J. 1996. Large rivers of South America: toward the new approach. Verh. internat. Verein. Limnol. 26: 167–180.

O'Connor, W.G. & J.D. Koehn. 1991. Spawning of the mountain galaxias *Galaxius olidus* Gunther, in Bruce's Creek, Victoria. Proc. Roy. Soc. Vic. 103: 113–123.

Pace, M.L., S.E.G. Findlay & D. Lints. 1992. Zooplankton in advective environments: the Hudson River community and a comparative analysis. Can. J. Fish. Aquat. Sci. 49: 1060–1069.

Pepin, P. & T.H. Shears. 1995. Influence of body size and alternate prey abundance on the risk of predation to fish larvae. Mar. Ecol. Prog. Ser. 128: 279–285.

Perlmutter, D.G. & J.L. Meyer. 1991. The impact of a stream-dwelling harpacticoid copepod upon detritally associated bacteria. Ecology 71: 2170–2180.

Poff, N.L. & J.V. Ward. 1989. Implications of streamflow variability and predictability for lotic community structure: a regional analysis of streamflow patterns. Can. J. Fish. Aquat. Sci. 46: 1805–1818.

Puckridge, J.T. & K.F. Walker. 1990. Reproductive biology and larval development of a gizzard shad, *Nematalosa erebi* (Gunther) (Dorosomatinae: Teleostei), in the River Murray, South Australia. Aust. J. Mar. Freshw. Res. 41: 695–712.

Puckridge, J.T., F. Sheldon, K.F. Walker & A.J. Boulton. 1998. Flow variability and the ecology of large rivers. Mar. Freshw. Res. 49: 55–72.

Reid, H.P. & D.A. Holdway. 1995. Early development of the Australian crimson-spotted rainbowfish, *Melanotaenia fluviatilis* (Pisces: Melanotaeniidae). Mar. Freshw. Res. 46: 475–480.

Reynolds, L.F. 1983. Migration patterns of five fish species in the Murray-Darling river system. Aust. J. Mar. Freshw. Res. 34: 857–871.

Richter, B.D., J.V. Baumgartner, J. Powell & D.P. Braun. 1996. A method for assessing hydrologic alteration within ecosystems. Cons. Biol. 10: 1163–1174.

Roberts, J. & G. Sainty. 1996. Listening to the Lachlan. Sainty & Associates, Potts Point. 101 pp.

Roberts, T.R. 1973. Ecology of fishes in the Amazon and Congo basins. pp. 239–254. *In*: B.J. Meggers & E.S. Ayers (ed.) Tropical Forest Ecosystems in Africa and South America: a Comparative Review, Smithsonian Institution Press, Washington, D.C.

Robertson, A.L. 1990. The population dynamics of Chydoridae and Macrothicidae (Cladocera: Crustacea) from the River Thames, UK. Freshw. Biol. 24: 375–389.

Robertson, A.L. 1995. Secondary production of a community of benthic Chydoridae (Cladocera: Crustacea) in a large river, UK. Arch. Hydrobiol. 134: 425–440.

Rowland, S.J. 1983 Spawning of the Australian freshwater fish Murray cod, *Maccullochella peeli* (Mitchell), in earthen ponds. J. Fish Biol. 23: 525–534.

Rowland, S.J. 1989. Aspects of the history and fishery of the Murray cod, *Maccullochella peeli* (Mitchell) (Percichthyidae). Proc. Linn. Soc. NSW 111: 201–203.

Rowland, S.J. 1992. Diet and feeding of Murray cod (*Maccullochella peeli*) larvae. Proc. Linn. Soc. NSW 113: 193–201.

Rundle, S.D. & A.G. Hildrew. 1992. Small fish and small prey in the food webs of some southern English streams. Arch. Hydrobiol. 125: 25–35.

Sager, D.R. 1987. Distribution of larval gizzard shad in the Upper Cape Fear River, North Carolina. American Fisheries Society Symposium 2: 174–178.

Schiemer, F. & T. Spindler. 1989. Endangered fish species of the Danube River in Austria. Regul. Riv. 4: 397–407.

Schiemer, F., T. Spindler, H. Wintersberger, A. Schneider & A. Chovanec. 1991. Fish fry associations: important indicators for the ecological status of large rivers. Verh. internat. Verein. Limnol. 24: 2497–2500.

Sempeski, P. & P. Gaudin. 1995. Habitat selection by grayling – II. Preliminary results on larval and juvenile daytime habitats. J. Fish Biol. 47: 345–349.

Sempeski, P., P. Gaudin & E. Heroum. 1998. Experimental study of young grayling (*Thymallus thymallus*) physical habitat selection factors in an artificial stream. Arch. Hydrobiol. 141: 321–332.

Tan, L.W. & R.S. Shiel. 1993. Responses of billabong rotifer communities to inundation. Hydrobiologia 255/256: 361–369.

Thorp, J.H., A.R. Black, K.H. Haag & J.D. Wehr. 1994. Zooplankton assemblages in the Ohio River: seasonal, tributary and navigation dam effects. Can. J. Fish. Aquat. Sci. 51: 1634–1643.

Thorp, J.H. & M.D. Delong. 1994. The riverine productivity model: an heuristic view of carbon sources and organic processing in large river ecosystems. Oikos 70: 305–308.

Vannote, R.L., G. Minshall, K.W. Cummins, J.R. Sedell & C.E. Cushing. 1980. The river continuum concept. Can. J. Fish. Aquat. Sci. 37: 130–137.

Walker, K.F. 1986. The Murray-Darling river system. pp. 631–659. *In*: B.R. Davies & K.F. Walker (ed.) The Ecology of River Systems, Dr W. Junk Publishers, Dordrecht.

Walker K.F. 1992. The River Murray, Australia: a semiarid lowland river. pp. 472–492. *In*: P. Calow & G.E. Petts (ed.) The Rivers Handbook, Vol.1, Blackwell Scientific Publications, Oxford.

Walker, K.F., F. Sheldon & J.T. Puckridge. 1995. A perspective on dryland river ecosystems. Regul. Riv. 11: 85–104.

Ward, J.V. & J.A. Stanford. 1995. The serial discontuity concept: extending the model to floodplain rivers. Regul. Riv. 10: 159–168.

Watkins, MS., S. Doherty & G.H. Copp. 1997. Microhabitat use by 0+ and older fishes in a small English chalk stream. J. Fish Biol. 50: 1010–1024.

Wehr, J.D. & J.H. Thorp. 1997. Effects of navigation dams, tributaries, and littoral zones on phytoplankton communities in the Ohio River. Can. J. Fish. Aquat. Sci. 54: 378–395.

Welcomme, R.L. 1979. The fisheries ecology of floodplain rivers. Longman, London. 317 pp.

Welcomme, R.L. 1985. River fisheries. Food and Agriculture Organization of the United Nations, FAO Fisheries Technical Paper 262. 303 pp.

Wharton, J.C. 1968. Spawning areas of the Macquarie perch, *Macquaria australasica* above the Eildon lake (Victoria). Australian Society for Limnology Newsletter, 6: 11–13.

Whitley, G.P. 1937. Further ichthyological miscellanea. Mem. Qld. Mus. 11: 113–148.

Williams, J.B. 1982. Temporal and spatial patterns of abundance of the Chydoridae (Cladocera) in Lake Itasca, Minnesota. Ecology 63: 345–353.

Environmental Biology of Fishes **56:** 153–165, 1999.

Comparison of growth plasticity in the laboratory and field, and implications for the onset of juvenile development in sofie, *Chondrostoma toxostoma*

Rodolphe E. Gozlan[a,b], Gordon H. Copp[a] & Jean-Noel Tourenq[b]
[a]*Landscape & Ecology Research Group, Department of Environmental Sciences, University of Hertfordshire, College Lane, Hatfield, Herts AL10 9AB, U.K. (e-mail: g.h.copp@herts.ac.uk)*
[b]*Centre d'Écologie des Systèmes Aquatiques Continentaux, UMR C5576 – CNRS/UPS, Université Paul Sabatier, 118 route de Narbonne, Toulouse, 31062 France*

Received 3 February 1998 Accepted 2 December 1998

Key words: ontogeny, metamorphosis, metabiosis, cyprinids, River Garonne

Synopsis

The aim of the present study was to determine the effect of environmental conditions (controlled and natural) on the duration and sequence of developmental steps in sofie, *Chondrostoma toxostoma*, early ontogeny. Few previous studies on the early development of fishes have included relative growth and none have compared relative growth in the laboratory and the field. Such comparison is important to quantify the morphological development of different parts of fish during their early ontogeny, to determine potential variations in growth that may occur under laboratory conditions and to understand better the plastic nature of relative growth. Early development and relative growth of 23 characters were examined in specimens of sofie reared under both laboratory and natural conditions in tributaries of the River Garonne basin (France). The sofie is still present in this basin despite progressive localised extinction in the rivers of south western Europe over the last 30 years. Growth of field and laboratory embryos (in degree days, °days) was the same up to larva step 1 (9 mm SL), but thereafter was markedly slower in the laboratory than in the field. Ontogenetic rate in the field was twice that in the laboratory, suggesting a precocial (specialist) form under natural conditions and an altricial (generalist) form under laboratory conditions. Stabilisation of relative growth, i.e. end of the remodelling process (metamorphosis), occurred well after all larval characteristics (remnants of finfold, rapid allometric growth) had disappeared and all the juvenile structures had appeared (nasal septa, complete scale cover). In the field, this stabilisation occurred in specimens of approximately 50 mm SL, suggesting that metamorphosis ends and the juvenile period begins at the end of the sofie's first (0+) year of life.

Introduction

The relative growth of morphological characters in fishes is related to overall growth in response to ecological conditions (Balon 1984, Kawamura & Washiyama 1989, Norton et al. 1995). Thus, information on a fish species' developmental biology can be obtained through quantification of that species' relative growth during early life history under experimental and natural conditions. The theory of saltatory ontogeny (Balon 1986, 1990) suggests that during development a fish goes through a succession of extended, stabilised steps separated by periods of rapid changes, i.e. thresholds (Krupka 1988, Kováč 1992, Fuiman 1994, Holden & Bruton 1995). Metamorphosis separates the larva and juvenile periods (Balon 1990). However, defining the end of metamorphosis and the beginning of the juvenile period in fishes that undergo indirect development is difficult. Combining relative growth measurements with the external and functional descriptions should highlight other aspects of metamorphosis and metabiosis, providing a quantitative answer to define when they have ended.

If environmental conditions can trigger adaptation in early ontogeny (Balon 1988, 1989, 1990), then this link between organism–environment can be used to assess the impact of controlled (invariable) laboratory conditions on the plasticity of developmental thresholds, and the rates of metabiosis and metamorphosis. Little attention has been directed towards the differences due to rearing conditions in establishing the definite phenotype of an altricial form. However, it is of great importance to understand how a species form has the ability to generate different phenotypes (altricial and precocial forms, Balon 1990) under different environmental signals, via slow differentiation and metamorphosis. Few previous studies on the early development of fishes have included relative growth and none have compared relative growth in the laboratory and the field. The aim of the present study was to determine the influence of environmental conditions on the patterns and rates of early development in a species that undergoes indirect development. To achieve this aim, we compared, within a saltatory framework, the early ontogeny of the sofie, *Chondrostoma toxostoma* (Vallot, 1836), reared under laboratory and natural conditions. A greater understanding of organism–environment relationships is of particular importance to the conservation of endangered or locally-extinct fish species such as the sofie, which has either been replaced in much of its original range in France by Danubian nase, *C. nasus*, or declined due to river regulation and/or water pollution. However, the management of ecosystems for endangered fishes is new and still in its early development. An understanding of early fish ontogeny and its plastic capacity will provide new perspectives that can lead to better adapted, reared larvae for reintroduction into rivers where reintroduction is the only possible option for re-establishing a species' original range. Effectively, Meffe (1986) highlighted that 'the central problem in conservation genetics is loss of genetic variation resulting in erosion of evolutionary flexibility'.

Material and methods

Sofie larvae, juveniles and adults were sampled twice a week in the River Garonne and its tributaries during the spring and summer of 1996 as soon as reproduction had ended. The larvae were collected by point abundance sampling (Copp & Peňáz 1988, Copp & Garner 1995), using a hand net (0.5 μm mesh net) for young larvae and by electrofishing for older larvae and juveniles. Each sampling location was approached as quietly as possible by foot. The net was quickly plunged under and raised through at the sampling point, the net was preceded by the anode in the case of electrofishing. The specimens were preserved in 4% formalin immediately after capture.

The ontogeny of larvae reared in the laboratory was followed using progeny of ripe brood stock (four females: total length, TL = 12–14 cm) taken from the field by electrofishing and incited to yield eggs in the laboratory by gently stroking their abdomen; the gametes were mixed directly with the sperm of eight males (TL = 12–16 cm) in a bucket. The fertilised eggs were spread on an artificial substrate in a tank (50 cm × 30 cm × 30 cm). After hatching, ten free-embryos and larvae were sampled randomly every day and preserved in 4% formalin. When the finformed phase was over, the sampling frequency was decreased to one collection per week. See Gozlan et al. 1999 (this volume) for details. In order to express standard length (SL) as degree days (°days), water temperatures were recorded every hour in both the laboratory and field for the entire study period.

Twenty-three mensural characters related to swimming capacity, feeding behaviour and habitat were measured on 249 sofie from the field and 300 laboratory-reared specimens as per Holčík (1989) and Kováč (1992): A–C = anal–caudal distance; Gape; H = maximum height; Hl = head length; hpC = caudal peduncle depth; ina = inter-nasal distance; iO = inter-orbital distance; lac = head width; laco = body width; lapc = peduncle width; lC1 = caudal length of inferior lobe; lC2 = middle caudal length; lC3 = caudal length of superior lobe; lP = pectoral length, lpA = pre-anal length; lPb = length of pectoral base; lpD = pre-dorsal length; lpP = pre-pectoral length; mh = minimum height; Oh = orbital horizontal distance; Ov = orbital vertical distance; poO = post-orbital distance; prO = pre-orbital distance. Measurements were to the nearest 0.1 mm using a binocular microscope fitted with an ocular micrometer for small larvae and using a pair of callipers for larger specimens, with notochord length (NL) for preflexion of the urostyle and SL for postflexion taken for each individual. The different developmental intervals are defined in Gozlan et al. 1999 (this volume).

For bivariate analysis, the character measurements were first expressed without any transformation with respect to SL. Because the use of proportional data is inadequate for regression analysis and invalidates statistical or biological analysis of data (Marr 1955, Atchley et al. 1976), residual analysis of linear regressions between each character and SL was used to identify growth irregularities (Sagnes et al. 1997). Characters that develop in shifts are characterised by residuals distributed according to a specific structure made by different line segments (Yoccoz 1988).

For multivariate analysis to reveal morphological groups based on all 23 characters, double centred principal component analysis, PCA (Okamoto 1972), using the ADE programme library (Chessel & Doledec 1993), was applied to the combined field and laboratory measurements. Ordinations of individual specimens were regrouped by developmental status and rearing conditions with inertia ellipses based on uniform weightings: LL < 15 = laboratory larvae < 15 mm SL; LL > 15 = laboratory larvae > 15 mm SL; FS < 23 = field larvae < 23 mm SL; FS23–49 = field specimens from 23 to 50 mm SL; FS50–135 = field individuals from 50 to 135 mm SL; Adult = young adults.

Results

Growth in °days was similar for both field and laboratory embryos (larvae) up to 9 mm SL (larva step 1). Thereafter, growth of larvae in the laboratory was markedly lower than that in the field for the same number of °days (Figure 1). Field specimens needed half the °days (400) to reach larva step 8 as did those in the laboratory, indicating an ontogenetic rate in the laboratory that was twice as slow as that under natural conditions.

Growth was generally not linear, using three body and three head characters as representative examples of the other characters (Figures 2, 3), and levelled off briefly in some cases. In laboratory specimens, growth of head characters and minimum height slowed down twice, once at approximately 8 mm SL and once at 15 mm SL. In laboratory specimens, no specimens <11 mm SL were captured, which may explain why only one case of levelling off was observed, at approximately 23 mm SL. In general, levelling off was more pronounced in the laboratory than in the field, particularly for certain characters such as minimum height. At a given SL, the values of some mensural characters in the field were different to those found in the laboratory. These morphological changes (break points) were

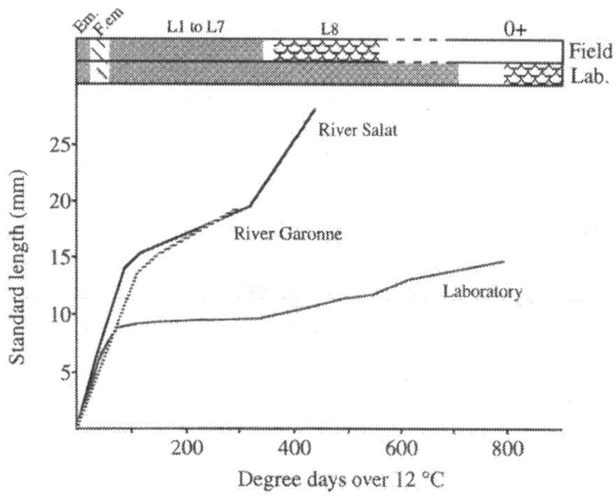

Figure 1. Growth in standard length (mm) of 0+ sofie, *Chondrostoma toxostoma*, over time with respect to water temperature (degree days over 12°) in the laboratory and in two tributaries (R. Salat, R. Touch) of the River Garonne (Em = embryo; F.em = free embryo; L1 to L7 = larva steps 1 to 7; L8 = larva step 8).

apparent in the residuals of the six example characters (Figures 4, 5) and many other characters (Figures 6, 7). In general, the residuals shifted in value at approximately 8 and 15 mm SL in laboratory specimens and at approximately 23 mm SL in field specimens.

The composite analysis of all characters provided by PCA accounted for 66% of the variation (Figure 8a), with sub-ellipses of the six groupings of specimens (by rearing conditions and size class) revealing two morphotypes (Figure 8b): morphotype 1 = LL < 15, LL > 15, FS < 23 and FS23–49; morphotype 2 = ellipses FS50–135 and adult. The principal mensural characters distinguishing these two groups (Figure 8c, longest vectors) were: vertical and horizontal eye diameter (Ov, Oh), anal–caudal distance (A–C), gape (G), body width (laco), peduncle width (lapC), maximum height (H), caudal peduncle depth (hpC) and to a minor degree, minimum height (mh), pre-orbital distance (prO), inter-orbital distance (iO) and middle caudal length (lC2).

Morphotype 1 consisted of individuals with eyes bulged and close together as well as a very large mouth. Ellipses LL<15 and LL>15 overlap but are orientated differently along the axes (Figure 8b), highlighting the differences in morphology revealed in the residuals (Figures 5, 6). Ellipses for young larvae from the laboratory (LL<15) and field (FS<23)

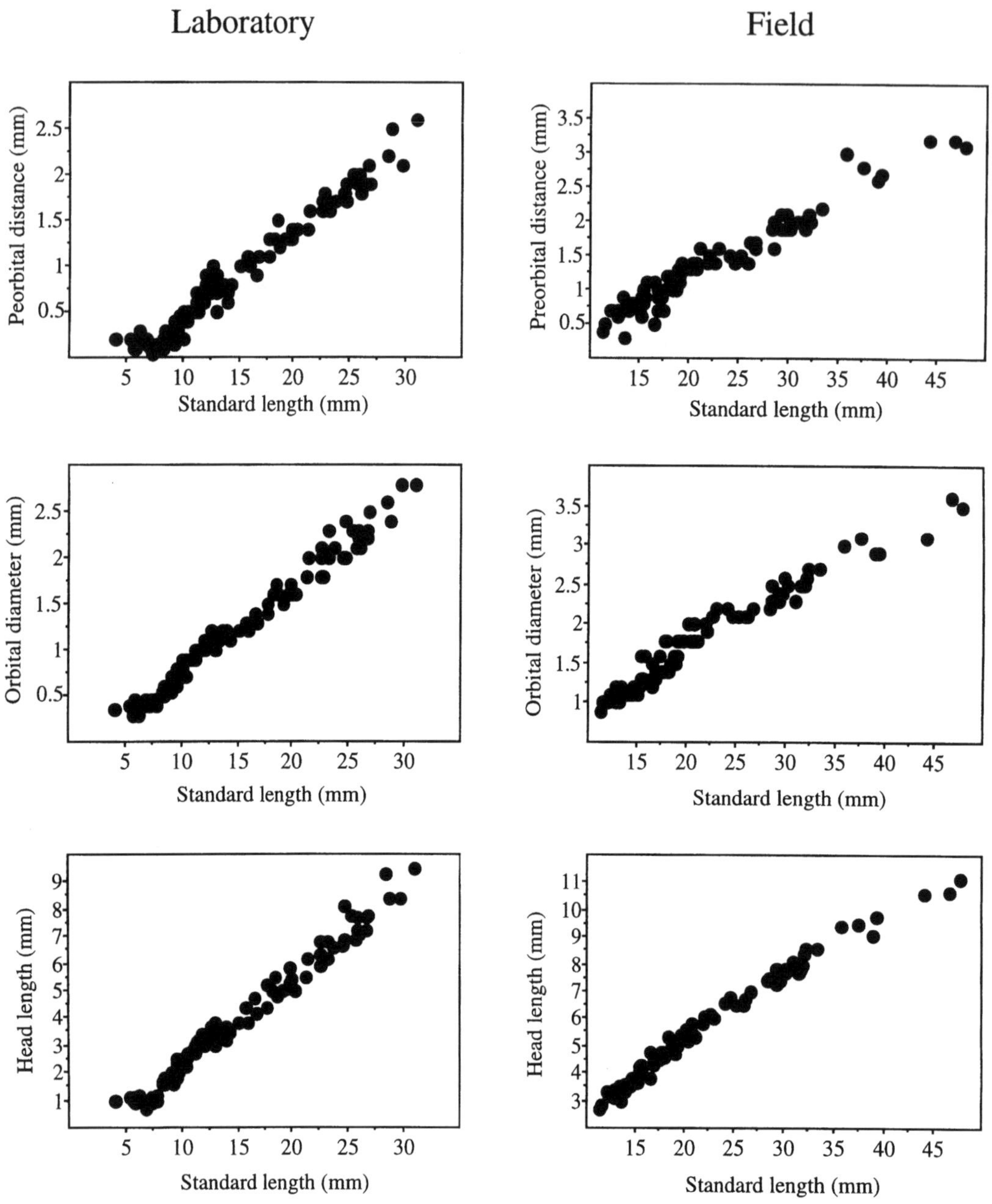

Figure 2. Growth of three morphological head characters (pre-orbital distance, orbital diameter, head length) in sofie, *Chondrostoma toxostoma*, reared under laboratory and natural conditions.

showed considerable overlap, but the deviation in morphology between the two environmental conditions became more apparent in the older larvae, with much less overlap between LL>15 and FS23–49. These differences exist both before and after the first changes in relative growth, but are more pronounced in specimens >15 mm SL (laboratory) and >23 mm SL (field). Morphotype 2 consisted of ellipses for specimens 50–135 mm SL (juveniles) and young adults, which were individuals with a smaller head, a very large body, a greater peduncle width and a large body depth.

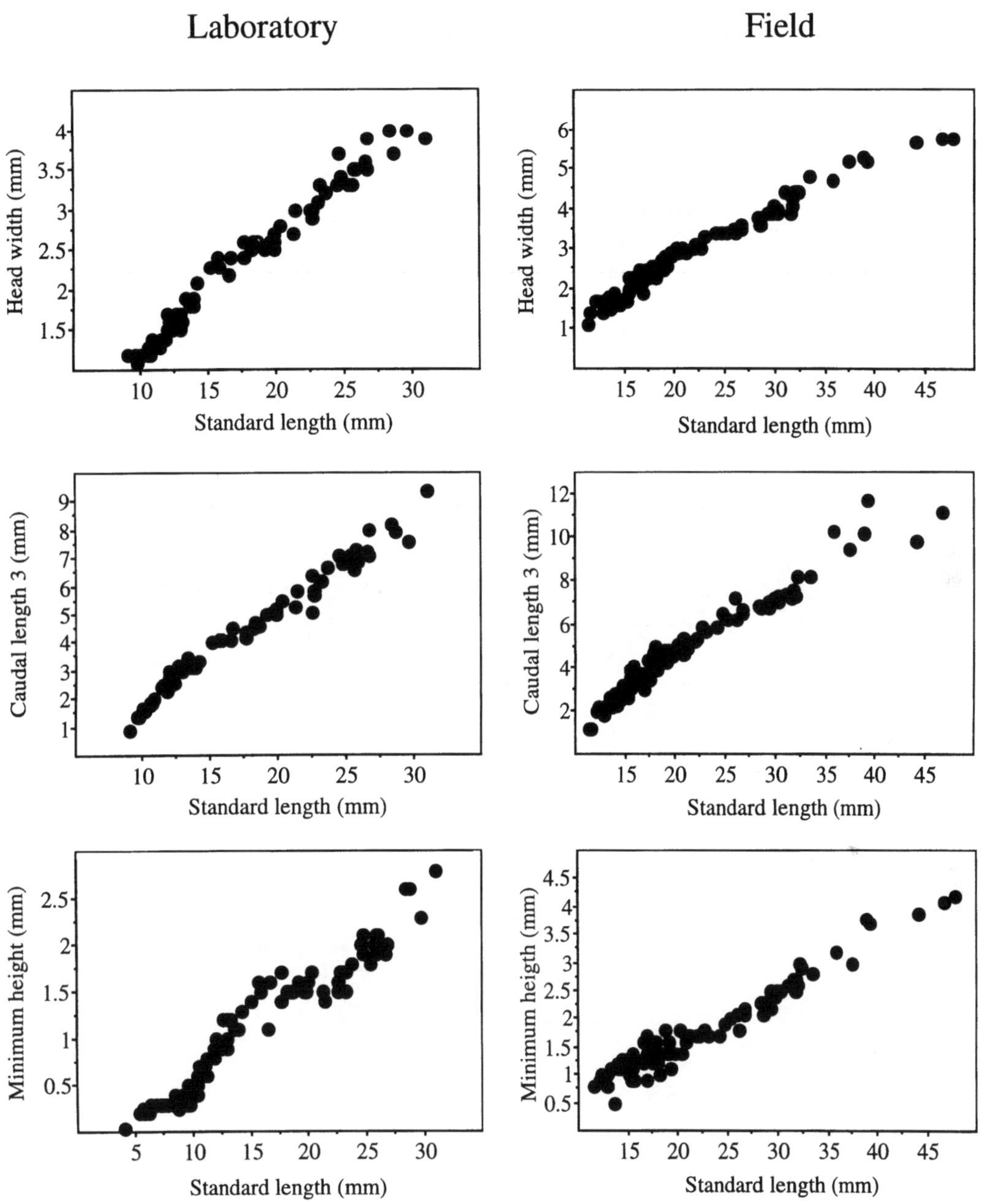

Figure 3. Growth of three morphological body characters (head width, caudal length, minimum height) in sofie, *Chondrostoma toxostoma*, reared under laboratory and natural conditions.

Discussion

Every species, as a result of its complex life history, has adapted to fit as best as possible to its niche (Luczkovich et al. 1995). At the same time, each species' morphology reflects its performance (Winemiller et al. 1995) and behavioural adaptation, e.g. prey capture, predator avoidance, microhabitat use (Roussel & Bardonnet 1999 this volume). This adaptation is continuous, with the organism responding to changes in environmental conditions through the rate and timing of ontogenetic processes. Such adaptation was apparent in our study of the sofie (Figure 1). The onset of adult structures (e.g. scale cover, nasal septa) occurred at about 15 mm

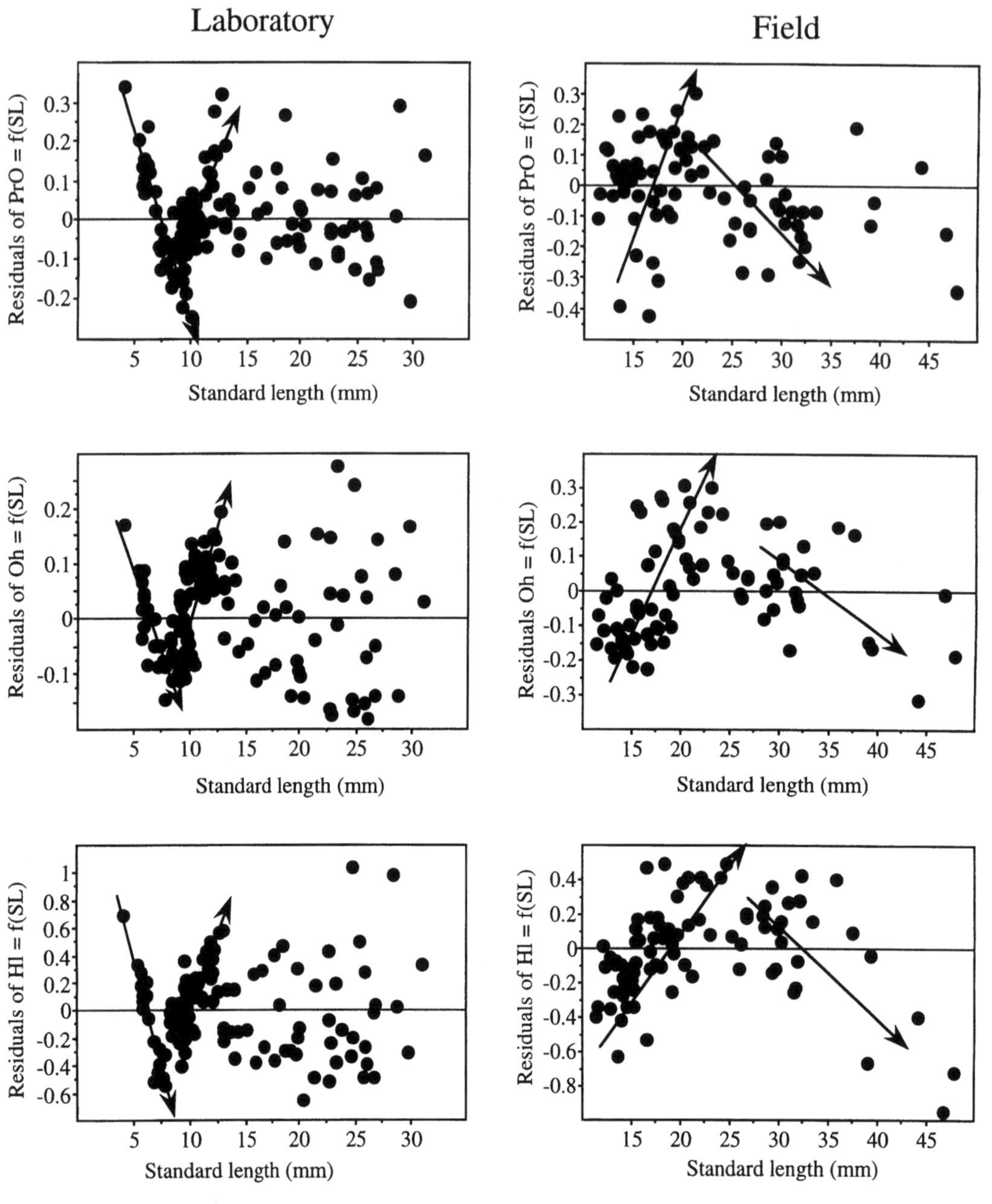

Figure 4. Residuals from linear regressions of head length (Hl), orbital diameter (Oh) and preorbital distance (PrO) for sofie, *Chondrostoma toxostoma*, reared under laboratory and natural conditions. Arrows provided to aid interpretation.

SL under laboratory conditions (Gozlan et al. 1999, this volume) and at 25 mm SL in the field. Overall, two shifts observed in the relative growth of mensural characters in young sofie (Figures 3–6) corresponded to ontogenetic thresholds. The first shift coincided with transition from larva steps 1 and 2, whereas the second shift was either the transition between larva steps 7 and 8 or between larva step 7 and the juvenile period. The fact that the second shift occurred at different lengths under field and laboratory conditions (Figures 4–8), combined with the differences in ontogenetic rate (Figure 1), reveals the plasticity of thresholds and reflects

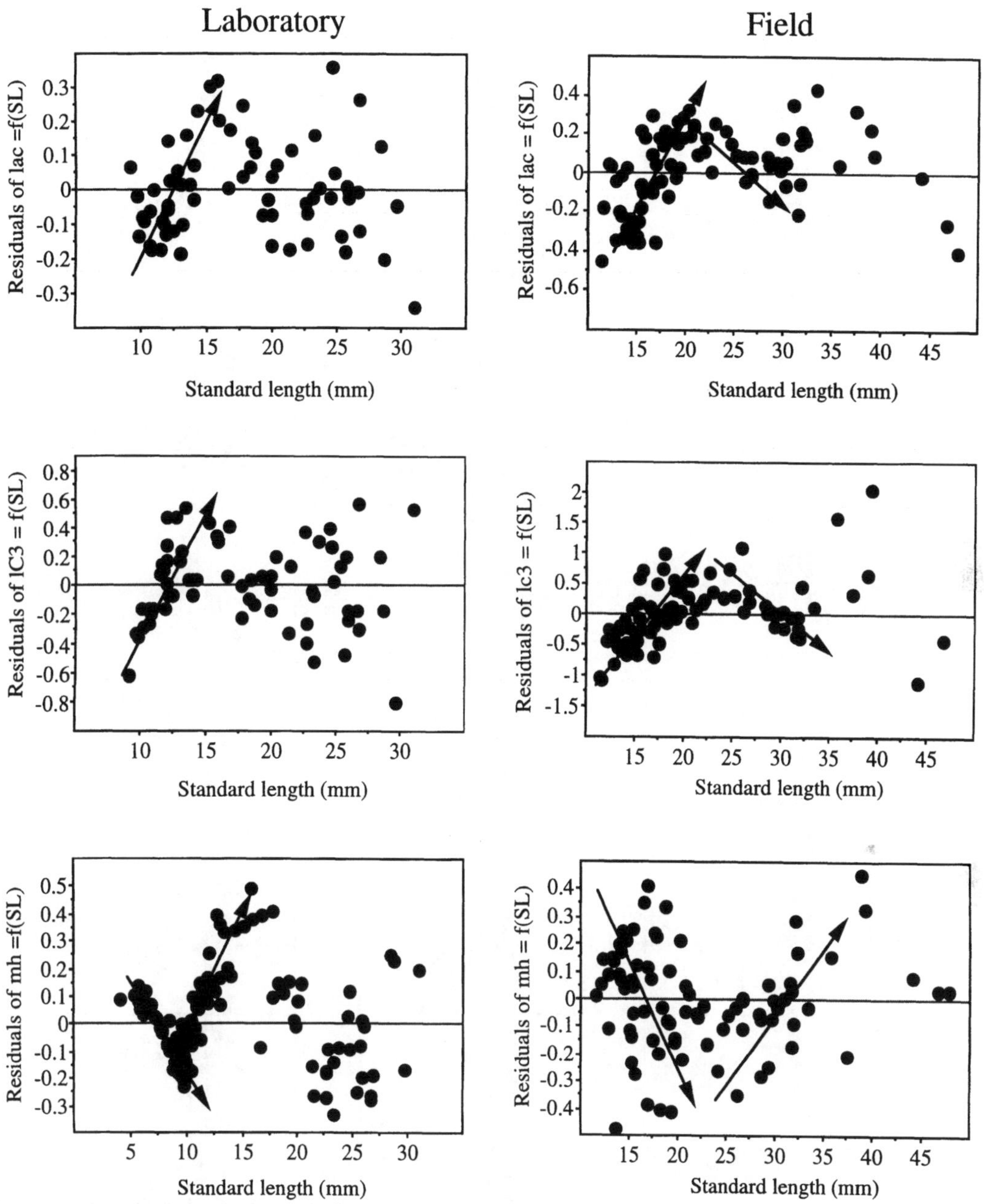

Figure 5. Residuals from linear regressions of minimum body height (mh), caudal length (lC3) and head width (lac) for sofie, *Chondrostoma toxostoma*, reared under laboratory and natural conditions. Arrows provided to aid interpretation.

the quality of the habitat as well as variations in natural resource use (Motta & Kotrschal 1992, Motta et al. 1995, Wainwright & Richard 1995).

One of the issues addressed here is the definition of the larva-to-juvenile transition. The search for natural boundaries to define stabilised steps (as a result of rapid changes in integrative actions) is still controversial. The start of the larva period is increasingly accepted to be the onset of exogenous feeding, though the debate continues (Kováč & Copp 1999 this volume), including discussion of a 'larva state', which addresses the issue of 'ability to feed exogenously' as opposed to the 'onset

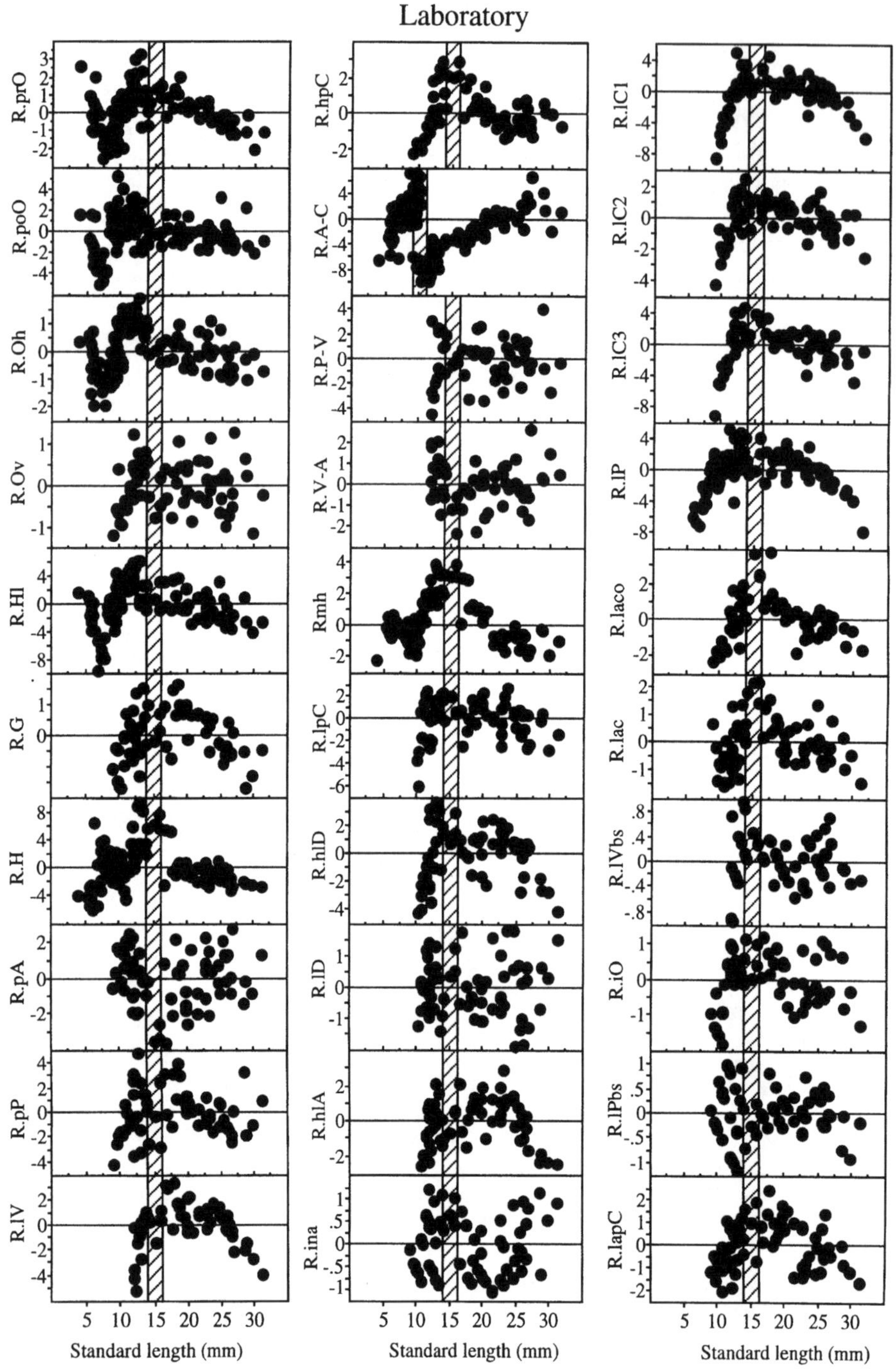

Figure 6. Residuals from linear regressions of mensural characters for sofie, *Chondrostoma toxostoma*, reared under laboratory conditions. A–C = anal–caudal distance; G = gape; H = maximum height; Hl = head length; hpC = caudal peduncle depth; ina = inter-nasal distance; iO = inter-orbital distance; lac = head width; laco = body width; lapC = peduncle width; lC1 = caudal length of inferior lobe; lC2 = middle caudal length; lC3 = caudal length of superior lobe; lP = pectoral length, lpA = pre-anal length; lPb = length of pectoral base; lpD = pre-dorsal length; lpP = pre-pectoral length; mh = minimum height; Oh = orbital horizontal distance; Ov = orbital vertical distance; poO = post-orbital distance; prO = pre-orbital distance. Grey column indicates SL at which residuals are distributed according to a specific structure (Yoccoz 1988).

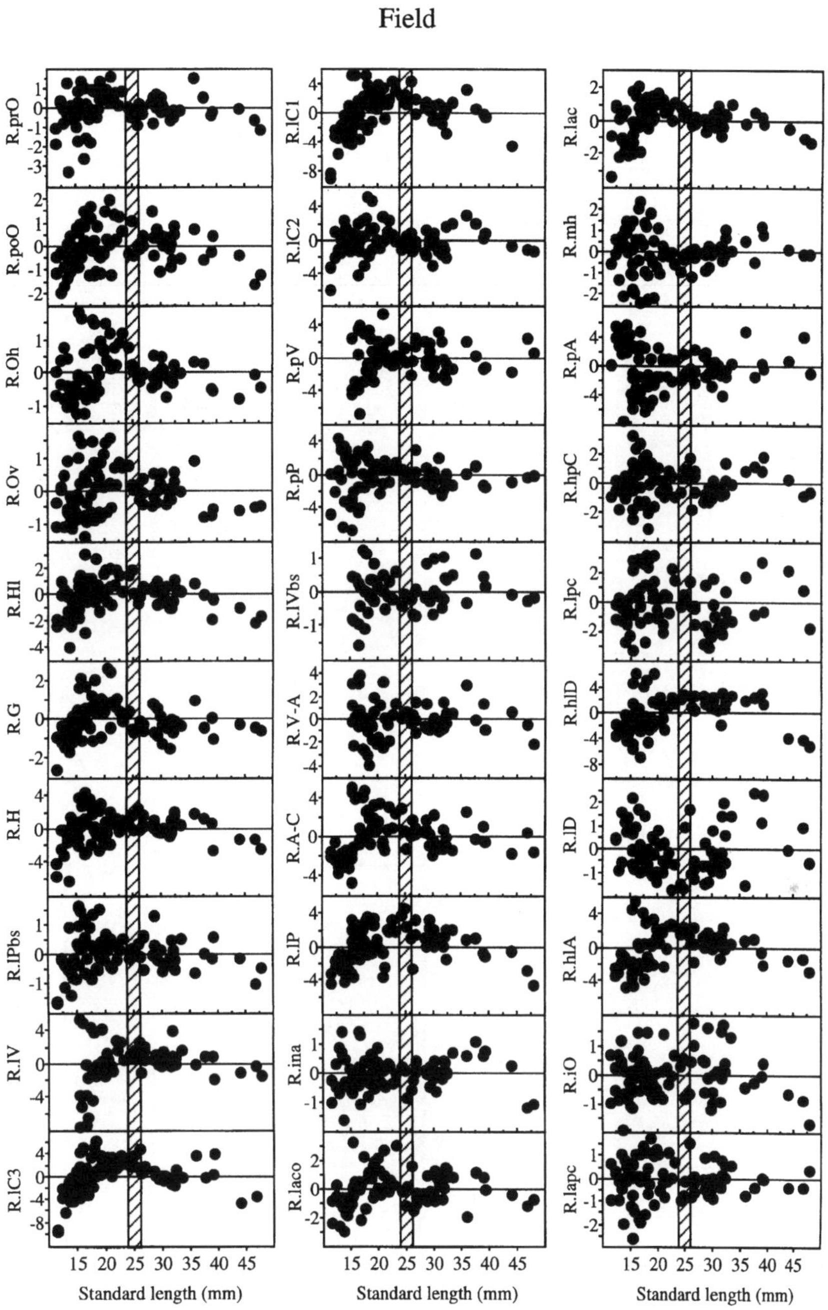

Figure 7. Residuals from linear regressions of mensural characters for sofie, *Chondrostoma toxostoma*, collected in the field. See Figure 6 for codes and explanation.

of exogenous feeding' (Pavlov 1999 this volume). The larva is a dynamic form with many temporary structures (e.g. finfold, notochord) and a body shape that is undergoing change, however the end of larval development and the start of the juvenile period remains unresolved (Balon 1990, 1999 this volume, Kováč et al. 1999 this volume). In the sofie, field specimens of 23–49 mm SL and laboratory specimens of about 15 mm SL possessed the array of adult structures (olfactory pits divided by membranous septum, scale cover, mucus, inferior

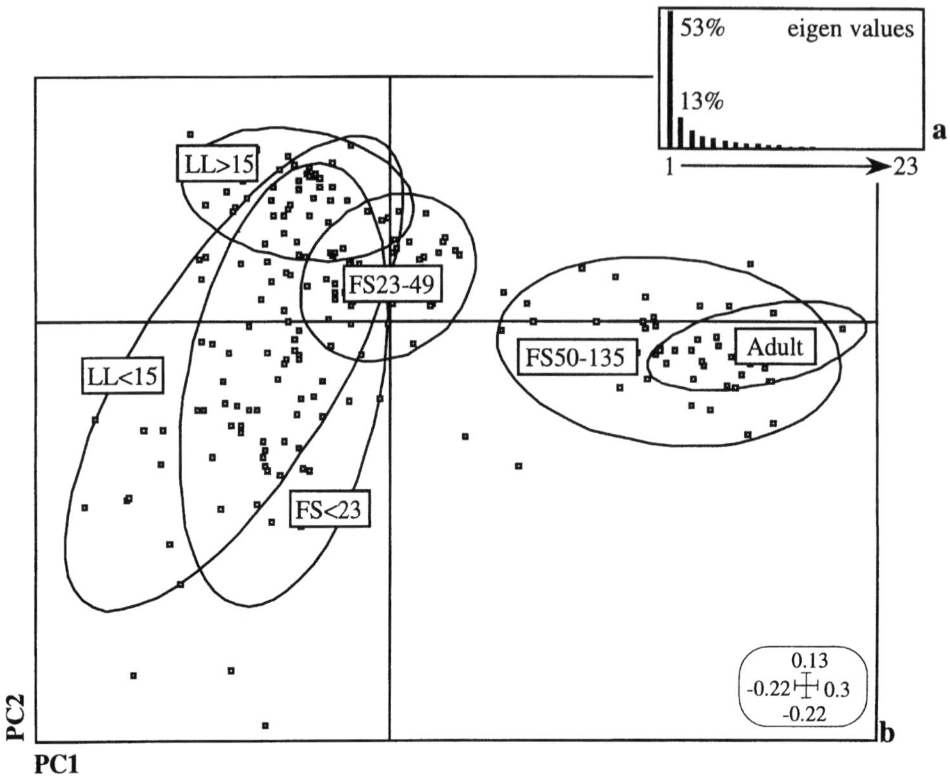

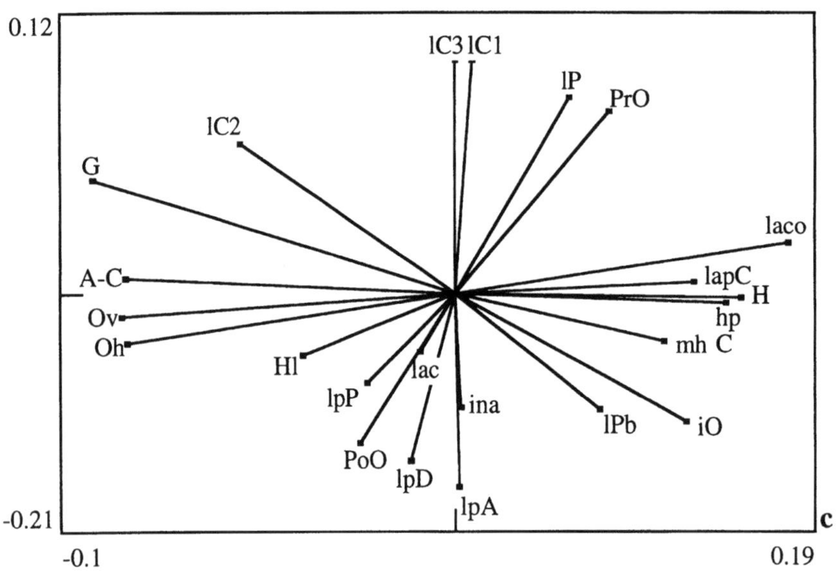

Figure 8. Doubled-centred principal component (PC) analysis of the individuals-by-morphological variables matrix for sofie from the laboratory and field: a – eigen values, b – PC1 vs. PC2 ordination with ellipses based on uniform weightings, c – correlation circle for 23 morphological characters (LL<15 = laboratory < 15 mm SL; LL>15 = laboratory larvae > 15 mm SL; FS<23=field larvae <23 mm SL; FS23–49 = field specimens from 23 to 50 mm SL; FS50–135 = field specimens from 50 to 135 mm SL; adult = young adults).

mouth), and thus could be viewed as having achieved the juvenile period as described in numerous previous studies of other species. The scales, like an armour, and the mucus, like a barrier against infections and parasites, are mainly involved in drag reduction (Videler 1994). At this stage, swimming capacity is increasing, with some dispersion of young sofie out into the main channel of the Garonne. Similarly, by these sizes, the mouth changed from terminal to the typical *Chondrostoma* inferior (Gozlan et al. 1999 this volume), and sofie in the wild had begun to graze on the surface of rocks.

However, the appearance of adult structures in the sofie did not coincide with 'stabilised' relative growth; specimens 23–49 mm SL were still morphologically different to those >50 mm SL (Figure 8), being morphologically closer to the larva morphotype. And, the larger and older sofie (50–135 mm SL) were closer to the adult morphotype, which should not be viewed as a 'definitive phenotype' being that this state does not exist in practice (Kacem et al. 1998, Kováč et al. 1999 this volume). Hence, the shift in relative growth observed in sofie at about 23 mm SL could be interpreted, not as the end of the metamorphosis, but rather as the threshold that initiates the last interval of the remodelling process (e.g. Copp & Kováč 1996), a 'pre-juvenile' interval that is principally metamorphic and completes the 'end-of-larva-period' metabiosis.

The threshold ending this 'pre-juvenile' interval was characterised by the stabilisation of relative growth, which occurred in sofie at approximately 50 mm SL, the mean length of sofie at the end of their first winter. The fact that shifts in habitat use and swimming capacity (Gozlan 1998) coincided with this 'stabilisation' of relative growth, rather than with the shifts observed at about 23 mm SL, suggests a more 'decisive' type of change in organism-to-environment interactions (metabiotic) than one purely of form (metamorphic) (Balon 1999 this volume). This conforms with Balon's definition of the start of juvenile development 'With the completion of metamorphosis the fish acquire the definitive phenotype and look like small adults; but their gonads are not mature – they are juveniles. Most of the time the end of metamorphosis and the beginning of the juvenile period coincide with a change of habitat' (Balon 1999 this volume). Thus, sofie of 23–50 mm SL, despite possessing most adult structures, are not juveniles; they should be viewed as larvae that are in the last 'metamorphic' (transitional) step of the metabiotic process. Perhaps 'metamorphic larvae' would be an appropriate term. This is not a semantic remark only, but more an aid for interpreting the ecology of 0+ sofie within an ontogenetic context, as well as the ultimate consequences for 0+ mortality. It also suggests that some more functional aspects of fish development (e.g. fin-area relationships to propulsion) are probably being overlooked or under-emphasised in developmental studies.

The fundamental assumption that the interaction of a fish with its environment will determine its functional morphology (Long 1995), was indeed evident in the present study (Figure 8). Ontogenetic thresholds are the more probable time for changes in development to occur, altering the 'entire organisation of steps which follow' (Balon 1985, 1989, 1990). The morphological differences in sofie larvae reared in the laboratory and field (Figure 8) were probably the result of morphodynamic changes that occurred during the larva 1 to larva 2 threshold (Figures 4–7). Up to this threshold, development of the laboratory and field specimens was similar (Figure 1). However, the entire organisation of subsequent steps appears to have modified in response to the different environmental conditions of the laboratory and field. In the 'stable' laboratory conditions, extension of the larva period and a decreased ontogenetic rate seem to have generated an altricial (generalist) form of sofie compared to the precocial form of the wild (specialist) specimens. Establishment of this dwarf (altricial) form of sofie larvae is characterised by their different morphotype (smaller) and their slow differentiation. The fact that these changes occurred in all of the laboratory larvae (Figures 4–8) suggests synchronisation (see Balon 1989, 1990). However, the environmental signal that sets these changes into place remains unclear.

One possible explanation for the differences in development and growth under field and laboratory conditions lies in the reduced diel variability of the laboratory environment, which has implications for metabolism, stress responses and physiology in general. Habitat use, food availability and foraging activity/patterns in young and small riverine fishes in general follow distinct diel patterns (e.g. Copp & Jurajda 1998, Baras & Nindaba 1999 this volume, Bischoff & Freyhof 1999 this volume), with habitat shift-related responses to water temperature gradients (an estimated 0.92 °days day^{-1} advantage in warmer river microhabitats) believed to enhance metabolism in minnow *Phoxinus phoxinus* (Garner et al. 1998). In the sofie, the higher quality of food and greater variability of the natural environment may provide the necessary environmental stimulus to maintain the precocial form. Growth variability due to environmental conditions is receiving increasing study, and the available evidence suggests that juvenile fish can compensate for variations in food availability (DeAngelis et al. 1993, Ali & Wootton 1998). And, environmental variations in 'growth opportunity' have recently been linked to endocrine differences (growth hormone stimulation of growth factor) associated with smolting plasticity in salmonids (Beckman & Dickhoff 1998), suggesting a physiological paradigm in which the endocrine system provides the mechanism through which phenotype plasticity and life history diversity develop (Finch & Rose 1995).

In conclusion, complete acquisition of adult structures and the stabilisation of relative growth did not coincide in the sofie. The interval between these two events could be viewed as the last 'metamorphic' (transitional) step of the metabiotic process, either as the last larva step (metamorphic larvae) or as a separate period (metamorphosis period). Morphology was found to be an efficient tool both in emphasising the plastic character of thresholds and in addressing the question of larva-to-juvenile transition, however other functional and physiological aspects require greater attention in conjunction with morphological studies. The response of ontogenetic rate, and thus the onset of juvenile development, in the sofie to different environmental conditions is evinced, though less emphatically, in other cyprinids such as the roach, *Rutilus rutilus* (Copp & Kováč 1996), suggesting that the comparison of wild and laboratory-reared roach requires re-examination within the precocial–altricial context.

References cited

Ali, M. & R.J. Wootton. 1998. Do random fluctuations in the intervals between feeding affect growth rate in juvenile three-spined sticklebacks. J. Fish Biol. 53: 1006–1014.

Atchley, W.R., C.T. Gaskins & D. Anderson. 1976. Statistical properties of ratios. I. Empirical results. Systemat. Zool. 25: 137–148.

Balon, E.K. 1984. Reflections on some decisive events in the early life of fishes. Trans. Amer. Fish. Soc. 113: 178–185.

Balon, E.K. 1985. The theory of saltatory ontogeny and life history models revisited. pp. 13–30. *In*: E.K. Balon (ed.) Early Life Histories of Fishes, Dr W. Junk Publishers, Dordrecht.

Balon, E.K. 1986. Saltatory ontogeny and evolution. Rivista Biol.-Biol. Forum 79: 151–190.

Balon, E.K.1988. Tao of life: universality of dichotomy in biology. The epigenetic mechanisms. Rivista Biol.-Biol. Forum 81: 339–380.

Balon, E.K. 1989. The Tao of life: from the dynamic unity of polar opposites to self organisation. pp. 7–40. *In*: M.N. Bruton (ed.) Alternative Life-history styles of Animals, Kluwer Academic Publishers, Dordrecht.

Balon, E.K. 1990. Epigenesis of an epigeneticist: the development of some alternative concepts on early ontogeny and evolution of fishes. Guelph Ichthyol. Rev. 1: 1–48.

Balon, E.K. 1999. Alternative ways to become a juvenile or a definitive phenotype (and on some persisting linguistic offenses). Env. Biol. Fish. 56: 17–38 (this volume).

Baras, E. & J. Nindaba. 1999. Seasonal and diel utilisation of inshore microhabitats by larvae and juveniles of *Leuciscus cephalus* and *Leuciscus leuciscus*. Env. Biol. Fish. 56: 183–197 (this volume).

Beckman, B.R. & W.W. Dickhoff. 1998. Plasticity of smolting in spring chinook salmon: relation to growth and insulin-like growth factor-I. J. Fish Biol. 53: 808–826.

Bischoff, A. & J. Freyhof. 1999. Ontogenetic shifts in resource use of barbel, *Barbus barbus*, during the first year of life. Env Biol. Fish. (this volume).

Chessel, D. & S. Dolédec. 1993. ADE Version 3.1 Hypercard© stacks and Quick Basic Microsoft© programme library for the analysis of environmental data. URA CNRS 1451, Universite de Lyon I, France.

Copp, G.H. & M. Peňáz. 1988. Ecology of fish spawning and nursery zones in the flood plain, using a new sampling approach. Hydrobiologia 169: 209–224.

Copp, G.H. & P. Garner. 1995. Evaluating microhabitat use of fish larvae and juveniles with point abundance sampling. Folia Zool. 44: 145–158.

Copp, G.H. & P. Jurajda. 1998. Size-structured diel use of banks by riverine fishes. Aquat. Sci. 60: 000-000 (in press).

Copp, G.H. & V. Kováč. 1996. When do fish with indirect development become juveniles? Can. J. Fish. Aquat. Sci. 53: 746–752.

DeAngelis, D.L., K.A. Rose, L.B. Crowder, E.A. Marschall & D. Lika. 1993. Fish cohort dynamics: application of complementary modeling approaches. Amer. Natur. 142: 604–622.

Finch, C.E. & M.R. Rose. 1995. Hormones and the physiological architecture of life history evolution. Quat. Rev. Biol. 70: 1–52.

Fuiman, L.A. 1994. The interplay of ontogeny and scaling in the interactions of fish larvae and their predators. J. Fish Biol. 45: 55–79.

Garner, P., S. Clough, S.W. Griffiths, D. Deans & A. Ibbotson. 1998. Use of shallow marginal habitat by *Phoxinus phoxinus*: a trade-off between temperature and food? J. Fish Biol. 52: 600–609.

Gozlan, R.E. 1998. Environmental biology of the sofie *Chondrostoma toxostoma* (Cypinidae), with emphasis on early development. Ph.D. Thesis, University of Hertfordshire, Hatfield. 210 pp.

Gozlan, R.E., G.H. Copp & J.N. Tourenq. 1999. Early development of the sofie, *Chondrostoma toxostoma*. Env. Biol. Fish. 56: 67–77 (this volume).

Holčík, J. 1989. General introductions to fishes. Acipenseriformes. Freshwater fishes of Europe, Volume I/II, AULA – Verlag, Wiesbaden. 469 pp.

Holden, K.K. & M.N. Bruton. 1995. The early ontogeny of the southern mouth-brooder, *Pseudocrenilabrus philander* (Pisces, Cichlidae). Env. Biol. Fish. 41: 311–329.

Kacem, A., F.J. Meunier & J.L. Baglinière. 1998. A quantitative study of morphological and histological changes in the skeleton of *Salmo salar* during its anadromous migration. J. Fish Biol. 53: 1096–1109.

Kawamura, G. & N. Washiyama. 1989. Ontogenic changes in behavior and sense organ morphogenesis in largemouth bass and *Tilapia nilotica*. Trans. Amer. Fish. Soc. 118: 203–213.

Kováč. V. 1992. Early development of the yellow pope, *Gymnocephalus schraetser*. Folia Zool. 41: 365–377.

Kováč. V., G.H. Copp & M.P. Francis. 1999. Morphometry of the stone loach, *Barbatula barbatula*: do mensural characters reflect the species' life history thresholds? Env. Biol. Fish. 56: 105–115 (this volume).

Kováč. V. & G.H. Copp. 1999. Prelude: looking at early development in fishes. Env. Biol. Fish. 56: 7–14 (this volume).

Krupka, I. 1988. Early development of the barbel (*Barbus barbus* (L, 1758)). Práce Ustavu Rybárstva Hydrobiol. 6: 115–138.

Long, J.H. 1995. Morphology, mechanics, and locomotion: the relation between the notochord and swimming motions in sturgeon. Env. Biol. Fish. 44: 199–211.

Luczkovich, J.J., S.F. Norton & R.G. Gilmore. 1995. The influence of oral anatomy on prey selection during the ontogeny of two percoid fishes, *Lagodon rhomboides* and *Centropomus undecimalis*. Env. Biol. Fish. 44: 79–95.

Marr, J.C. 1955. The use of morphometric data in systematic, racial and relative growth studies in fishes. Copeia 1955: 23–31.

Meffe, G.K. 1986. Conservation genetics and the management of endangered fishes. Fisheries 11: 14–22.

Motta, P.J. & K.M. Kotrschal. 1992. Correlative, experimental, and comparative evolutionary approaches in ecomorphology. Netherl. J. Zool. 42: 400–415.

Motta, P.J., S.F. Norton & J.J. Luczkovich. 1995. Perspectives on ecomorphology of bony fishes. Env. Biol. Fish. 44: 11–20.

Norton, F., J.J. Luczkovich & P.J. Motta. 1995. The role of ecomorphological studies in the comparative biology of fishes. Env. Biol. Fish. 44: 287–304.

Okamoto, M. 1972. Four techniques of principal component analysis. J. Japan. Stat. Soc. 2: 63–69.

Pavlov, D.A. 1999. Features of transition from larva to juvenile in fishes with different types of early ontogeny. Env. Biol. Fish. 56: 41–52 (this volume).

Roussel, J.-M. & A. Bardonnet. 1999. Ontogeny of diel pattern of stream-margin habitat use by emerging brown trout, *Salmo trutta*, in experimental channels: influence of food and predator presence. Env. Biol. Fish. 56: 253–262 (this volume).

Sagnes, P., P. Gaudin & B. Statzner. 1997. Shifts in morphometrics and their relation to hydrodynamic potential and habitat use during grayling ontogenies. J. Fish Biol. 50: 846–858.

Videler, J.J. 1994. Fish swimming. Fish & Fisheries, Chapman & Hall, London. 241 pp.

Wainwright, P.C. & B.A. Richard. 1995. Predicting patterns of prey use from morphology of fishes. Env. Biol. Fish. 44: 97–113.

Winemiller, K.O., L.C. Kelso-Winemiller & A.L. Brenkert. 1995. Ecomorphological diversification and convergence in fluvial cichlid fishes. Env. Biol. Fish. 44: 235–261.

Yoccoz, N.G. 1988. Le rôle du modèle euclidien d'analyse des données en biologie évolutive. Thèse de doctorat, Université Lyon I. 254 pp.

Part 3. Ontogeny of predator-prey interactions

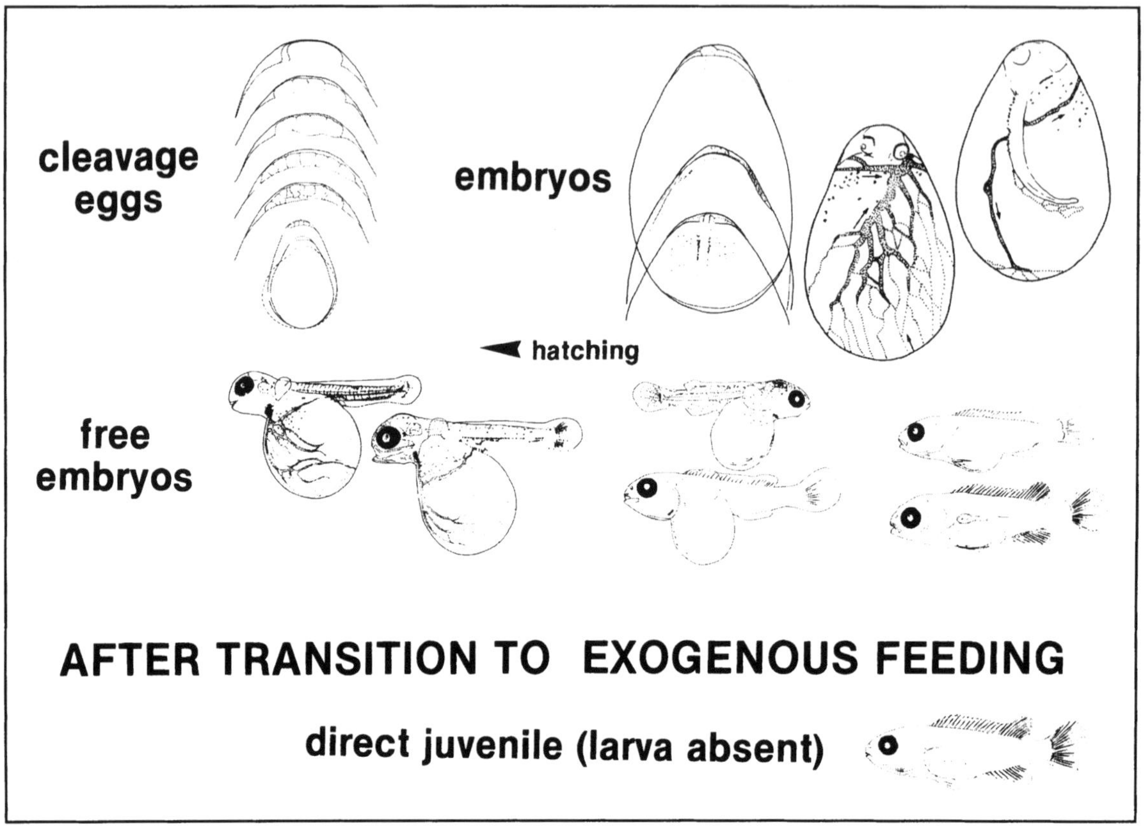

Labeotropheus fuelleborni

Environmental Biology of Fishes **56:** 169–181, 1999.

A review of predation impact by 0+ fish on zooplankton in fresh and brackish waters of the temperate northern hemisphere

Thomas Mehner[a] & Ralf Thiel[b]
[a] *Institute of Freshwater Ecology and Inland Fisheries, Department of Biology and Ecology of Fishes, P.O.B. 850119, D-12561 Berlin, Germany (e-mail: mehner@igb-berlin.de)*
[b] *Institute of Hydrobiology und Fisheries Sciences, Elbelabor, Hamburg University, Große Elbstr. 268, D-22767 Hamburg, Germany*

Received 17 March 1997 Accepted 7 August 1998

Key words: 0+ fish feeding, zooplankton dynamics, ontogenetic development, trophic interactions, morphology, physiology

Synopsis

To assess potential differences in predation impact on zooplankton communities by small (larva) and larger 0+ juvenile fish, 18 studies were reviewed from fresh water and the brackish Baltic Sea of the northern hemisphere temperate region. These case studies were performed either in the field or in mesocosm experiments. Larva stocks were found to exert only minor impact on small zooplankton species such as rotifers, copepodids and small cladocerans. In contrast, stocks of 0+ juveniles were found to have the potential to depress populations of large cladocerans and copepods, especially during late summer and autumn. However, studies where both 0+ juvenile fish consumption and zooplankton dynamics and production were exactly quantified are still very rare, and therefore final evaluation of this interaction cannot be made. In addition, papers were summarized that describe differences in morphological and physiological performance between larva and 0+ juvenile fish. The greater impact of 0+ juvenile fish on large zooplankton may be explained by their larger mouth gape and by their better developed abilities to detect and consume their prey items. However, this partly is lessened by the lower energy requirements of juvenile fish compared with identical biomasses of fish larvae, although larva bioenergetics remains only fragmentarily understood. Consequently, selective predation by fish larvae on particular small zooplankton prey may be more important than has been detected so far.

Introduction

That many fish in their first year of life (0+) mainly feed on zooplankton prey has been well known for more than 30 years (e.g. Rogowski & Tesch 1960, Applegate & Mullan 1967, Wong & Ward 1972). However, an ongoing controversy exists whether larva and 0+ juvenile fish may drive zooplankton succession by predation (reviewed by Gliwicz & Pijanowska 1989, see also papers in Mehner & Winfield 1997) or whether these fish are 'too dilute' to exert significant predation impact on dynamics of their prey (Cushing 1983). Succession of different zooplankton taxa during the season may also be influenced by resource availability and competition (Threlkeld 1985, Lampert et al. 1986, DeMott 1989).

Populations of all fish species are strongly size-structured (Werner & Gilliam 1984). Although the transition between larva and juvenile development remains ill defined (Balon 1975, Copp & Kováč 1996), interval-specific differences in feeding influence on zooplankton species of either fish larvae or larger 0+ juvenile fish may be expected. For this purpose, we have reviewed the literature for the period since 1980. In the first part, we summarize selected case studies from temperate fresh waters and from the brackish Baltic Sea, which describe either positive or no correlation between predation by 0+ fish and zooplankton succession. Here we address the question whether the period of ontogenetic development (larva versus 0+ juveniles) is a good predictor of degree of predation impact. Since in the limnological literature the ontogenetic

Table 1. Summary of case studies from fresh and brackish waters of the northern hemisphere temperate region where predation of 0+ fish on zooplankton communities was analysed. Fish length, mass and population density or biomass is always given for the time before the zooplankton decline. All biomass values are wet masses.

Lake	Surface area (ha) Depth (m) Trophic state	Zooplankton species	Time of decline	Fish species	Fish length (mm)	Fish population density and/or biomass	Author(s)
Lake Opinicon (Canada)	222 ha max. depth 10 m eutrophic	*Bosmina longirostris* Cyclopoid copepods	May/June	*Perca flavescens* *Percina caprodes* *Pomoxis nigromaculatus* *Ambloplites rupestris* *Lepomis gibbosus* *Lepomis macrochirus*	4–23 mm	about 7 ind m^{-3} in June, 14 ind m^{-3} in July	Keast (1980)
Lake Itasca (U.S.A.)	436 ha mean depth 5.2 m mesotrophic	Cladocera Copepoda	June/July	*Perca flavescens* Other species not given	25 mm larvae	not determined	Whiteside et al. (1985), Whiteside (1988)
Alderfen Broad (U.K.) and enclosures	4.7 ha mean depth 0.8 m trophic state not given	*Daphnia hyalina* *Bosmina longirostris* *Ceriodaphnia quadrangula*	June/July	*Rutilus rutilus*	not given 20 mm in enclosures	max 0.8 ind m^{-2}; 5.9 ind m^{-2} in enclosures	Cryer et al. (1986)
Lake Pääjärvi (Finland) Experimental tanks	1340 ha mean depth 14.4 m oligo-mesotrophic	*Bosmina longirostris*	July	*Rutilus rutilus*	15–18 mm	208 ind m^{-2} max: 900 ind m^{-2}	Kairesalo & Seppälä (1987)
Loch Kinord Loch Davan (U.K.)	82.2/41.9 ha mean depth 1.2 m oligo-mesotrophic	*Cyclops strenuus abyssorum* *Daphnia longispina*	June	*Perca fluviatilis*	8–20 mm	max: 40.4 mg dry weight at the 25–75 cm depth stratum	Treasurer (1992)
Lake Shelbyville (U.S.A.), and enclosures	4500 ha max. depth 18 m trophic state not given	*Acanthocyclops vernalis*	June	*Dorosoma cepedianum* *Lepomis macrochirus*	not given	70–105 ind m^{-3}, in the lake max. 170 ind m^{-3}	Welker et al. (1994)
Barther Strom (Germany)	210 ha mean depth 1.6 m eutrophic	*Eurytemora affinis*	May	*Perca fluviatilis* *Clupea harengus*	6–18 mm 12–27 mm	max: 0.6 g m^{-3} max: 1 g m^{-3}	Mehner (1996)
Kokosing Lake (U.S.A.) and enclosures	65 ha mean depth 2.0 m trophic state not given	*Daphnia sp.* *Bosmina sp.*	June	*Dorosoma cepedianum*	4–15 mm in the lake 18–58 mm in enclosures	critical: 6 ind m^{-3}	DeVries & Stein (1992)

Knox Lake Kokosing Lake (U.S.A.)	225 ha max. depth 9.6 m trophic state not given	*Daphnia sp.* Copepods	June June	*Dorosoma cepedianum*	5–30 mm	38–210 ind m^{-3} 1– 30 ind m^{-3}	Dettmers & Stein (1992)
Oneida Lake (U.S.A.)	20700 ha mean depth 6.8 m eutrophic	*Daphnia pulex* *Daphnia galeata*	July	*Perca flavescens*	not given	critical: 20 kg ha^{-1}	Mills & Forney (1983)
Lake Erie (U.S.A.)	mean depth 6–10 m eutrophic	*Daphnia galeata mendotae* *Daphnia retrocurva*	July June	*Perca flavescens* *Morone americana*	21–48 mm	max: 0.25 ind m^{-3} max: 1.25 ind m^{-3}	Wu & Culver (1994)
Lake St. George (Canada) Enclosures	10.3 ha max. depth 15 m mesotrophic	*Daphnia galeata*	July	*Perca flavescens*	20 mm	critical: 30–50 kg ha^{-1}	Post & McQueen (1987)
Bautzen reservoir (Germany), and enclosures	533 ha mean depth 7.4 m hypertrophic	*Daphnia galeata*	June/July	*Perca fluviatilis* *Stizostedion lucioperca*	20–30 mm	critical: 60 kg ha^{-1}	Hülsmann & Mehner (1997), Mehner et al. (1997)
Tjeukemeer (The Netherlands)	2150 ha mean depth 1.5 m eutrophic	*Daphnia hyalina* *Daphnia cucullata* *Daphnia galeata*	June/July	*Osmerus eperlanus* *Stizostedion lucioperca* *Perca fluviatilis*	7–30 mm	0.7–2.4 ind m^{-3} in September	Vijverberg et al. (1990), Boersma et al. (1996)
Baltic Sea (Sweden)	max. depth 27–40 m meso-eutrophic	*Bosmina longispina maritima*	August	*Clupea harengus* *Sprattus sprattus*	8–82 mm	larvae: max 9 ind m^{-2} juveniles: max 0.5 ind m^{-2}	Rudstam et al. (1992)
Barther Bodden (Germany)	2730 ha mean depth 1.5 m eutrophic	*Eurytemora affinis* *Bosmina longirostris*	May/June	*Perca fluviatilis* *Clupea harengus* *Rutilus rutilus* *Gasterosteus aculeatus* and 9 less abundant species	22 mm 35 mm 12 mm 12 mm	11.8 ind m^{-2} (littoral) 2.5 ind m^{-2} (pelagial) 0.64 g m^{-2} (littoral) 0.17 g m^{-2} (pelagial)	Thiel (1996)
Fish ponds, Ohio (U.S.A.)	0.4 ha mean depth 1 m	*Daphnia galeata mendotae*, *Cyclops vernalis*	after 4 weeks	*Stizostedion vitreum*	9–40 mm	stocking density: 10 ind m^{-3} and 50 ind m^{-3}	Qin & Culver (1995)

intervals of fish development are frequently not clearly identified (e.g. fry, fingerlings, 0+, age-0, young-of-year, juvenile, underyearling), we roughly separated the studies according to the fish lengths reported. All 0+ fish that were smaller than 20 mm total length (TL) were referred to as 'larvae', whereas larger 0+ fish were referred to as '0+ juveniles'. This grouping is rather pragmatic, and within a few studies fish lengths below and above 20 mm were simultaneously observed. Although there is evidence that, at least in Eurasian fish species, the transition to juveniles may be completed at 20–25 mm TL (e.g. Koblitskaya 1981), some weakness remains about the ontogenetic interval of fish about 20–30 mm long (see Copp & Kováč 1996). We paid further attention to season and species involved in zooplankton declines, and to the method by which predation impact was calculated. The basic data on the lakes studied, their zooplankton communities and 0+ fish composition, and observed 0+ fish densities are given in Table 1. Although some more studies exist where influence of predation by 0+ fish on zooplankton succession was assessed (e.g. Murtaugh 1985, Larsson et al. 1985, Luecke et al. 1990, Zalewski et al. 1990, R.G. Werner et al. 1996), they were excluded because no data on 0+ fish lengths were reported. In the second part of our review, we analyse potential differences in feeding mechanisms and metabolic performance between fish larvae and juveniles and assess their influence on the predation impact of larva and 0+ juveniles on zooplankton communities.

Case studies

Predation impact of fish larvae on zooplankton

In one of the earliest studies, Keast (1980) demonstrated that the density of cyclopoid nauplii and copepodids and the small cladoceran species *Bosmina longirostris* declined after the mass hatching of fish embryos in Lake Opinicon (Canada) during spring and remained at low levels in summer (Table 1). Since fish larvae fed mainly on these zooplankters, Keast (1980) assumed that the succession of species in the open-water zooplankton community was controlled by predation between May and July. However, he did not calculate the daily consumption of larva stocks, therefore his assumption could be justified.

Similar conclusions were also drawn for the development of the littoral zooplankton community in the mesotrophic Lake Itasca (Minnesota) by Whiteside et al. (1985) and Whiteside (1988). They found that the inshore migration of 0+ juvenile yellow perch, *Perca flavescens*, about 25 mm long coincided with the beginning of exogeneous feeding of other fish larvae in inshore areas in the first weeks of July. The authors concluded that the decline of littoral zooplankton populations during that exact time could be attributed to the influence of 0+ fish predation. Although the data gave evidence for strong predation impact, we suggest that fish density and consumption have to be quantified to corroborate such a conclusion.

Strong and fast declines in population densities of *Daphnia hyalina* and *B. longirostris* were observed in the small, shallow Alderfen Broad (United Kingdom), during the first weeks of July in two of four years investigated by Cryer et al. (1986). The authors concluded from fish diet analysis and additional mesocosm experiments that feeding by roach, *Rutilus rutilus*, larvae (20 mm TL in mesocosms, not determined in the lake) significantly contributed to that zooplankton decline. Maximum larva densities were determined as 2.5 ind. m^{-2} in early June and 0.7 ind. m^{-2} in early July; and in the mesocosms 5.9 ind. m^{-2} were stocked. By contrast, in those years when zooplankton density fluctuated only slightly and a population decline in early summer did not occur, low reproduction of roach and thus low densities of roach larvae (<0.1 ind. m^{-2}) were found. However, also these authors failed to calculate consumption of roach larvae to corroborate their qualitative analysis.

Kairesalo & Seppälä (1987) described a mesocosm experiment in the littoral zone of the mesotrophic Lake Pääjärvi (Finland) where roach larvae (total length 15–18 mm) eliminated nearly all *Bosmina* within three days. They estimated that the highest natural densities of roach larvae ever found in this area (900 ind. m^{-2}) were able to ingest approximately 400000 *Bosmina* m^{-2} day^{-1}, thus exhaustion of this zooplankton species due to feeding is very likely in this low-productive system. Consequently, they assumed that drastic population declines of *Bosmina* observed in the same lake during preceding years (Lehtovaara & Sarvala 1984) may also be explained by fish larva predation, although this has not been proven by field studies.

For Lake Shelbyville (U.S.A.), Welker et al. (1994) estimated whether larvae of gizzard shad, *Dorosoma cepedianum* (5–28 mm TL), and bluegill, *Lepomis macrochirus* (4–23 mm TL), may impact zooplankton by predation. They found that the decrease of the total zooplankton biomass occurred simultaneously with the peak in larva density (about 170 larvae m^{-3}) during the middle of June. In an accompanying mesocosm experiment lasting 14 days, a density of 70–105 larvae m^{-3} also resulted in a decrease of the macrozooplankton density. However, among individual zooplankton taxa, only the copepod *Acanthocyclops vernalis* was significantly reduced in all mesocosms relative to the fishless controls. Although changes in zooplankton biomass were reasonably well correlated with density of larvae, daily consumption of larvae was not calculated to quantify this relationship.

Predation impact of Eurasian perch, *Perca fluviatilis*, larvae (length range 5–20 mm) was studied in two Scottish lochs (Kinord and Davan) in May and June by Treasurer (1992). He calculated food consumption of fish by the gastric evacuation method during 24-h-cycles, and compared these daily values with zooplankton standing crop. Consumption was always lower than 1.5% of total zooplankton biomass in Loch Davan, but reached maximum daily values of 38.6% in Loch Kinord. However, a drastic decline of *Cyclops* sp. in Loch Kinord during first week of June could not be explained by predation, since the consumption/biomass ratio was less than 1% during this period. Also, the population of *Daphnia longispina* was not influenced by fish larva predation and did not decline in summer, although 12% of its biomass was eaten by perch larvae on one sampling date.

Comparable results were also obtained for a small inlet of the Baltic Sea. Mehner (1996) compared feeding of perch larvae (6–22 mm TL) and herring, *Clupea harengus*, larvae and 0+ juveniles (12–28 mm TL) with production of the calanoid copepod *Eurytemora affinis* in the Barther Strom (Southern Baltic Sea, Germany). Consumption estimates made with the aid of two bioenergetics models were independently corroborated by direct consumption calculations in the field (Mehner & Heerkloss 1994). The studies demonstrated clearly that the decline in copepod density observed at the end of May in each year could not be attributed to the predation impact by fish larvae, since consumption was always about one magnitude lower than zooplankton production.

A combination of field study and mesocosm experiments to test the feeding influence of gizzard shad larvae on zooplankton was performed in Kokosing Lake (U.S.A.) by DeVries & Stein (1992). The abundance peak of gizzard shad larvae (4–15 mm long) was associated with a strong decline of crustacean zooplankton in the lake during June of two years. Natural peak densities of larvae were estimated between 14 and 84 ind. m^{-3}, but even much lower densities (about 6 ind. m^{-3}) were determined as being responsible for zooplankton declines in the mesocosms. However, the shad stocked into the mesoscosms where larger (18 mm TL at the beginning of the experiment, 38–55 mm TL at the end) than the fish that had been caught in the lake during June. Consequently, the fish biomasses in the mesocosms did not mimick the situation in the lake during the zooplankton decline. By determining food consumption of shad larvae in the lake with the aid of gastric evacuation rates, Dettmers & Stein (1992) estimated that the zooplankton decline observed in Kokosing Lake in June could not be attributed to the predation impact of fish. Compared with zooplankton production, only unreasonably high larva densities (maximum 212 ind. m^{-3}) could be expected to consume enough zooplankton to contribute substantially to the decline. However, in the adjacent Knox Lake, the predation impact of only 7 shad larvae m^{-3} may have caused the spring zooplankton decline due to the substantially lower zooplankton biomass and production in that lake (Dettmers & Stein 1992).

Predation impact by 0+ juvenile fish on zooplankton

One of the best investigated case studies appears to be the 0+ juvenile yellow perch – *Daphnia* interactions in Oneida Lake (U.S.A.) that has been analysed for more than 10 years (Noble 1975, Mills & Forney 1983, Mills et al. 1987). *Daphnia pulex* or *D. galeata* were found to disappear completely from the lake if the 0+ juvenile yellow perch biomass exceeded about 20 kg ha^{-1} by the end of July. In all years when the 0+ yellow perch biomass was lower, the density of daphnids remained nearly constant during the entire summer. 0+ yellow perch biomass in July was the consequence of the combined, but antagonistic effects of yellow perch reproduction success and predation impact of walleye, *Stizostedion vitreum*, on 0+ yellow perch. Unfortunately, consumption rates of 0+ juvenile yellow perch were not determined in any of the years, therefore the influence of predatory mortality on *Daphnia* population dynamics cannot be evaluated.

Zooplankton mortality was compared with zooplankton consumption by 0+ juvenile yellow perch (17–62 mm TL, consumption estimated by a bioenergetics model) in Lake Erie by Wu & Culver (1994). Dominant zooplankton species were *Daphnia galeata mendotae* and *Daphnia retrocurva*. The latter species declined drastically in June, whereas *D. g. mendotae* peaked during that time, but declined in July. However, consumption estimates for 0+ juvenile yellow perch accounted for only 50% of the observed *Daphnia* mortality during that time. It was concluded that predation by yellow perch contributed little to the dynamics of daphnids during the entire season. Besides the impact of low food resources on the *Daphnia* populations, the feeding influence of 0+ white perch, *Morone americana*, may also be important, but could not be estimated due to the lack of appropriate parameters for a bioenergetics model for this species.

A 'critical' 0+ juvenile yellow perch biomass between 30 and 50 kg ha^{-1} was determined for Lake St. George, Canada, by Post & McQueen (1987) and McQueen & Post (1988). These fish biomasses were responsible for the collapse of *Daphnia galeata mendotae* in mesocosms during a 4 month experimental period. Initial size of yellow perch was about 20 mm TL and 0.11 g wet body mass. However, since the fish were stocked sequentially into the four mesocosms, there were differences in population biomasses between the mesocosms at the same date. Nevertheless, the *Daphnia* decline was also observed to occur sequentially and was triggered at the point when yellow perch biomass had exceeded the critical value. However, yellow perch consumption was not estimated, therefore this correlation cannot be justified.

In a similar mesocosm experiment, Hülsmann & Mehner (1997) determined the critical 0+ juvenile Eurasian perch biomass (20–30 mm TL) for the long-term biomanipulated Bautzen Reservoir (Germany). Fish consumption was calculated with the aid of the bioenergetics model of Kitchell et al. (1977) and was compared with the total daily *Daphnia galeata* mortality rate. In Bautzen Reservoir, 0+ Eurasian perch biomass must exceed 60 kg ha^{-1} if daily *Daphnia* mortality during late June and July was to be accounted for exclusively by predation. However, during additional field studies over two years, this 0+ fish biomass could never be observed in Bautzen Reservoir, and predation accounted for a maximum of 2% of total daily *Daphnia* mortality in pelagic areas (Mehner et al. 1998b). We concluded that direct predation effects by 0+ juvenile fish may be of minor importance for the midsummer decline of daphnids, whereas indirect demographic effects due to selective feeding of fish may be partially responsible for the decrease in *Daphnia* biomass (Mehner et al. 1998a).

Seasonal variability of predation impact by fish larvae and 0+ juveniles on zooplankton

As a result of a two-year field study, the 0+ fish of several species (Eurasian perch, smelt, *Osmerus eperlanus*, and zander, *Stizostedion lucioperca*) were determined as playing the key role in the food web of Lake Tjeukemeer (The Netherlands, Vijverberg et al. 1990). The authors calculated that consumption by 0+ fish on large zooplankton species was more than three times higher than was consumption by adult zooplanktivorous fish. The most intense predation was found in June and July. Consequently, the decline in *Daphnia* abundance during that period was concluded to be driven mainly by 0+ juvenile fish predation. However, 0+ juvenile fish densities also declined during the season, since these fish were the main food component of the dominant piscivorous predator in the lake, adult zander. Therefore, the reduced number of 0+ fish allowed the *Daphnia* population to rebound in August and September.

Continuing the studies of Vijverberg et al. (1990) and Boersma et al. (1996) assessed the predation impact of 0+ fish (7–100 mm TL) on the population dynamics of *Daphnia galeata* and *D. cucullata* in Tjeukemeer. Mortality of daphnids was calculated separately for four size classes thereof, whereas estimates of fish consumption were made by using fixed values of food conversion efficiency combined with growth rates of fish. Although the smallest size class of daphnids (<0.5 mm) showed high mortality during the entire season, this mortality could not be explained by 0+ fish predation. Feeding selectivity of the dominant 0+ smelt, zander, and Eurasian perch was strongly negative with respect to the smallest daphnids, but positive for the larger *Daphnia* groups (>0.5 mm). However, the calculated uptake of larger daphnids by the fish was also significantly lower than the observed *Daphnia* mortality in these size classes until July, and the midsummer decline of daphnids thus could not have been triggered by 0+ fish predation. Nevertheless, regulation of *Daphnia* numbers by 0+ juvenile fish predation may have been possible later, since consumption could explain all mortality in the second half of the year. From the first impression, these results appear to contrast with the investigations made in the same water by Vijverberg et al. (1990, see above), but the latter study involved calculations using mean annual values of *Daphnia* abundance such that the greater impact of 0+ juvenile fish was averaged with the lower impact of fish larvae. This illustrates the importance of sampling on a fine temporal scale in such studies (Wanzenböck et al. 1997).

Thiel (1996) calculated the predation impact by 0+, 1+ and small adult zooplanktivorous fish on the zooplankton community in a shallow brackish bay of the Southern Baltic Sea (Barther Bodden, Germany). By using a bioenergetics model of Vinberg (1956) for fish consumption estimates, he showed that fish predation led to a total collapse of the copepod and cladoceran populations in May and June in one of two investigated years, since consumption exceeded zooplankton production. During that time, consumption by fish larvae (roach and threespine stickleback, *Gasterosteus aculeatus*) and 0+ juveniles (Eurasian perch and herring) accounted for about 85% of total zooplankton consumption. Also, in the other year of the study, population density of the dominating calanoid copepod *Eurytemora affinis* declined at the end of May. However, in this year consumption by fish could not explain this decline, since zooplankton biomass and production were substantially higher than in the year before and exceeded consumption substantially. We have to conclude, therefore, that ressource availability may also be responsible for the yearly collapse of copepods in spring (see Mehner 1996).

A similar study was conducted in coastal areas of the Baltic Sea by Rudstam et al. (1992) and Arrhenius & Hansson (1993). Compared with zooplankton production, plankton consumption rates of adult and 0+ herring and sprat, *Sprattus sprattus*, were low in July, but increased to a peak in August coinciding with a late summer decline in crustacean zooplankton biomass. From the end of August until October, consumption was similar to or even higher than zooplankton production. During that time, 0+ juvenile fish contributed to about 50% of total consumption of zooplanktivorous fish.

Finally, Qin & Culver (1995) described an experiment with larvae and 0+ juveniles of walleye, which were stocked into six small ponds at about 9 mm TL. They found that direct predation by 0+ walleye resulted in a much smaller increase of cladoceran and copepod biomasses (*Daphnia galeata mendotae*, *Cyclops vernalis*, and others) in the high density (50 ind. m^{-3}) fish ponds compared with ponds with lower (10 ind. m^{-3}) fish density. However, during the first week, when walleye larvae were still relatively small (9–17 mm TL), no difference between both treatments could be observed. Simultaneous calculations of consumption using a bioenergetics model indicated that predation impact increased linearly with increasing fish size, but only from the second week after stocking (Madon & Culver 1993).

Summary of case studies

Summarizing the results of the 18 case studies described above and in Table 1, the predation impact on zooplankton differs between fish larvae and 0+ juvenile fish. Four conclusions can be drawn:

(1) According to several descriptive studies (Keast 1980, Cryer et al. 1986, Whiteside 1988, DeVries & Stein 1992, Welker et al. 1994), fish larvae (size range 4 to about 20 mm TL) may exert significant predation pressure on rotifers, small cladocerans (*Bosmina* spp.) and small developmental stages of copepods (nauplii and copepodids) due to very high fish larva densities. There is a regular decline of these small zooplankters in many waters during May/June, which has been explained by the superior competitive abilities of the coincidingly increasing *Daphnia* populations (Sommer et al. 1986). Perhaps, the influence of fish larva predation on this decline has been neglected so far. However, quantitative approaches investigating predation by fish larvae resulted in a low impact of larva on small zooplankton (Dettmers & Stein 1992, Treasurer 1992, Mehner 1996).

(2) 0+ juvenile fish (>20 mm total length) may be able to control the dynamics of larger zooplankton species (*Daphnia* spp., adult copepods) by predation. Clear evidence for strong predation impact came presumably from mesocosm studies (Cryer et al. 1986, Post & McQueen 1987, DeVries & Stein 1992, Hülsmann & Mehner 1997), but fish densities in the mesocosms were sometimes higher than in the surrounding lakes. In all quantitative field studies where zooplankton consumption by 0+ juvenile fish was compared with zooplankton production or mortality, a minor predation impact by 0+ juvenile fish on midsummer zooplankton declines was calculated (Rudstam et al. 1992, Wu & Culver 1994, Boersma et al. 1996, Thiel 1996, Mehner et al. 1998b). These results indicate that factors other than predation may also influence zooplankton dynamics (resource availability, lake productivity, competition; reviewed by DeMott 1989). However, a stronger predation impact by 0+ juvenile fish was found during late summer and autumn (Rudstam et al. 1992, Wu & Culver 1994, Boersma

et al. 1996), and a linear increase of predation impact with increasing fish length was found in the pond experiments (Qin & Culver 1995).

(3) 'Critical' densities or biomasses of dominant fish larvae or 0+ juvenile fish that are sufficient to induce declines of certain zooplankton species are difficult to compare (Mills & Forney 1983, Post & McQueen 1987, Hülsmann & Mehner 1997). They must be regarded as lake-specific values since they also depend on the density and actual reproductive capacity of the zooplankton species under consideration and on lake morphometry, productivity and presence of refuges for fish and zooplankton. Therefore, these values may fluctuate within one water body or even within one season.

(4) Although the diversity of lakes included into this review was large with respect to surface area (4.7–20 700 ha), depth (0.8–40 m) and trophic state (oligotrophic–hypertrophic), quantitative evidence of strong predation impact by 0+ fish on zooplankton is surprisingly rare. We have to conclude that the sampling technique to estimate 0+ fish abundance has not yet adequately developed (Wanzenböck et al. 1997) such that predation impact was generally underestimated. Alternatively, this fact may indicate that bottom-up forces (resource availability) have a much greater impact on zooplankton dynamics than top-down forces.

Potential mechanisms that may explain the differences in predation impact

In this and the following section, we review the literature to find patterns in ontogenetic development of fish that may explain differences in feeding of fish larvae compared with 0+ juvenile fish. Besides morphological constraints of the feeding process (first section), metabolic performance and digestive abilities (second section) were also compared.

Morphological differences between larvae and 0+ juveniles

The most striking difference between fish larvae and juveniles with respect to feeding process is the larger mouth of 0+ juveniles. Gape width or diameter influences foraging of fish by constraining the size of zooplankton that fish larvae can capture and consume (Wong & Ward 1972, Schael et al. 1991). The mouth diameter increases, for example in 0+ yellow perch, from about 0.5 mm in fish 7 mm long to about 4 mm in a 30 mm long individual (Arts & Evans 1987). This increase has strong implications for the range of zooplankton species that can be eaten by the fish (Furnass 1979, Keast 1980, Mills et al. 1984). For many fish larvae, only rotifers and small copepod nauplii are available as food of appropriate size. With increasing length and mouth diameter of fish, the sequential inclusion of small cladocerans, copepods, large cladocerans and finally invertebrate predators (*Leptodora*, *Bythotrephes*, *Chaoborus* larvae, Mysidacea) in the diet has been observed (Hartmann 1983, Hammer 1985, Thiel et al. 1996). Consequently, mouth diameter was found to be a good predictor for the maximum prey size ingested by fishes (Hartmann 1986, Ghan & Sprules 1993).

Owing to their increased mouth diameter, larger 0+ juvenile fish basically are able to ingest all zooplankton species. It has been assumed that zooplanktivorous fish should always select the largest plankton specimens available (Brooks & Dodson 1965); this has been corroborated by the 'optimal foraging theory' (Werner & Hall 1974). In addition, most 0+ juvenile fish are visual particulate predators and should prefer the more conspicuous, egg-bearing females of zooplankton species as shown also for particulate-feeding adult fishes (Winfield & Townsend 1983, reviewed by Lazzaro 1987). Hence, one would expect strong predation impact at even moderate 0+ fish densities due to the demographic consequences of selective predation on zooplankton populations (Gliwicz 1994). However, the mean size of ingested prey was substantially smaller than would be predicted by the maximum mouth gape (below 60% in *Lota lota*, Ghan & Sprules 1993), or was even constantly below 0.6 mm in gizzard shad independently of fish size (Bremigan & Stein 1994). Similarly, poor correlations between mean prey size and mouth diameter were found in other species (black crappie, *Pomoxis nigromaculatus* and freshwater drum, *Aplodinotus grunniens*, Schael et al. 1991). Also, the mean size of daphnids in fish stomachs was often smaller than in the study lakes and thus below the size when daphnids mature and start carrying eggs (Hansen & Wahl 1981, Mills et al. 1984, Mehner et al. 1998a). This may be due to motivational constraints in handling relatively large prey (Wanzenböck 1995). Consequently, demographic effects of fish predation on *Daphnia* dynamics are more likely in adult or larger 0+ juvenile fish than in smaller juveniles immediately after the transition from larvae to juveniles. This partly could account for the increasing predation impact of 0+ fish during the season found in some studies (Rudstam et al. 1992, Boersma et al. 1996). In addition, the preference for copepods over daphnids appears to be common at least in 0+ juvenile percids up to 30–35 mm TL (Guma'a 1978, Mills et al. 1984, Post et al. 1992, Wu & Culver 1994). This fact additionally reduces the potential feeding impact of 0+ fish on daphnids until summer, but makes predation control of copepod populations more likely (compare to Keast 1980, Dettmers & Stein 1992, Treasurer 1992).

Fish larvae are less effective swimmers than the 0+ juveniles due to the not fully differentiated system of muscles (Forstner et al. 1983, Batty 1984, El-Fiky et al. 1987), and their incompletely developed sensory abilities to detect prey (Blaxter 1986, Wanzenböck & Schiemer 1989, Wahl et al. 1993, Wanzenböck et al. 1996). Moreover, the time costs of prey attacks were higher in larvae than in juvenile cyprinids (Wanzenböck 1992). Hence it can be assumed that fish larvae are less able than 0+ juveniles to select particular prey species from the zooplankton community.

Fish larvae experience higher predation mortality by piscivorous fish than do larger 0+ juveniles, since the interaction between piscivorous predators and their prey is also gape-limited (Rice et al. 1987, Post & Evans 1989, Miller et al. 1992, Hambright 1994, Mehner et al. 1996). Therefore, control of 0+ fish densities by fish predation is more likely during the larva and early juvenile periods than during the end of the first growing season. High water temperatures thus may have a marked influence on the development of 0+ fish density in spring and early summer due to the reduced vulnerability of faster grown and hence larger 0+ fish to their predators (Mehner et al. 1997).

Owing to the data presented above, the impact of 0+ fish should depend upon the fish species involved. Some fish species, which are overwhelmingly benthivorous as juveniles and adults, may start feeding on benthos relatively early in life and will perhaps not influence zooplankton dynamics (e.g. ruffe, *Gymnocephalus cernuus*, M.-G. Werner et al. 1996). Species that are piscivorous as adults tend to have larger mouth gapes than species that remain planktivorous throughout their life (Schael et al. 1991, Bremigan & Stein 1994). Thus they should impact larger zooplankton, especially daphnids, earlier than do 0+ fish with small mouth gapes (see Kurmayer & Wanzenböck 1996). Indeed, the uptake of daphnids by, for example, walleye larvae has been described (Mathias & Li 1982, Qin & Culver 1995). On the other hand, this may be masked by lower densities of ultimately piscivorous species compared with densities of 0+ fish from other trophic guilds due to the lower number of deposited eggs per individual female (Post & Kitchell 1997). However, 11 out of the 18 case studies cited above included the facultative piscivorous yellow and Eurasian perch. Obviously, these species combine a relatively large mouth (Schael et al. 1991) with high fecundities and hence high larva densities, and may have good prerequisites to impact zooplankton by predation even early in the season.

Differences in daily food rations between fish larvae and 0+ *juveniles*

Average oxygen consumption per individual (here used as a parameter to describe metabolic requirements) increases exponentially with increasing fish size, with the mass-specific exponent being about 0.8 in adult fish (Vinberg 1956). If growth and egestion and excretion are constant, then fish larvae consume more food per unit weight and day than 0+ juvenile fish. However, the mass-specific exponents in fish larvae and 0+ juveniles were found to be highly variable (0.45–1.33) and thus differ from those estimated in 1+ juveniles or adults (Rombough 1988, Post & Lee 1996). Hence, mass-specific oxygen consumption in 0+ fish could either decrease (exponent <1.0) or increase with increasing fish length, especially during the larva period (Rombough 1988, Post & Lee 1996). This may be attributed to the dominance of cutaneous respiration found in fish larvae, in which gill differentiation is not yet finished (Rombough 1988). This was experimentally corroborated by an increasing daily food ration with increasing length for *Coregonus* sp. larvae up to an age of 23 days (Hofer & Bürkle 1986). By contrast, most of the extraordinarily high daily rations ($>200\%$, dry weight basis) were determined in fish larvae fed zooplankton *ad libitum* in laboratory experiments (Marmulla & Rösch 1990, Troschel & Rösch 1991).

Direct food consumption estimates in fish larvae in the field during 24 h-cycles tend to be lower than larva consumption values obtained using bioenergetics models with parameters for 0+ juvenile fish (Arrhenius & Hansson 1994a,b, Mehner 1996). Therefore, extrapolation of bioenergetics models developed for 0+ or 1+ juvenile fish to larvae may result in erroneous consumption estimates, and the use of bioenergetics models can be recommended only in those cases where appropriate parameters for the energy budgets were also directly measured in fish larvae (e.g. Karjalainen et al. 1997, Worischka & Mehner 1998).

Daily consumption rates may also depend on daily growth rates with higher growth requiring a proportionally higher food uptake. Wieser (1991) showed that this assumption does not hold true in fish larvae growing with rates higher than $0.06\,d^{-1}$ (at 15°C) and $0.08\,d^{-1}$ (at 20°C). In these cases, measured oxygen consumption did not fully represent the assumed metabolic costs of protein synthesis, and it was concluded that the efficiency of growth was higher in fast-growing larvae than in slow-growing larvae or 0+ juvenile fish.

The ingestion difference may be outweighed by the fact that fish larvae have poorly developed digestive tracts (Dabrowski 1984). For example, length, differentiation and proteolytic activity of the gut increase with greater length of the stomachless roach (Hofer & Nasir-Uddin 1985, Mark et al. 1989). Similarly, the stomach is not yet fully differentiated in specimens smaller than 20 mm TL of many other species (for example in Eurasian perch, Treasurer 1992). Consequently, larvae must presumably eat more prey biomass to reach an energy intake similar to that of 0+ juvenile fish. In addition, the suitability of prey should be ranked rather by nutritional value, digestibility, and assimilability than by biomass (or size, Mills et al. 1986).

Summarizing from the view of basic physiology, a given biomass of fish larvae should consume more food than an identical amount of 0+ juvenile fishes. However, this general rule is biased by several specific peculiarities attributed to the incomplete development of fish larvae. Thus, more work on larva bioenergetics is required, and only very tentative conclusions may be drawn from the existing, limited information.

Conclusion

The case studies reviewed here have shown basically that 0+ fish may be able to influence zooplankton successions by selective predation in fresh waters or in brackish regions of the northern temperate area. However, the exclusive control of zooplankton dynamics by 0+ juvenile fish predation in late spring or early summer (for example the induction of a midsummer decline of large cladocerans) could not be proven in either case studies from the field or by the results of temporally restricted mesocosm experiments, although the latter provided some corroboration. In contrast, control of zooplankton dynamics by 0+ juvenile fish predation was found in field studies during late summer and autumn. This seasonal pattern may be explained by the morphological constraints of feeding and swimming in fish larvae compared with the improved abilities of larger 0+ juveniles at the end of their first year of life. Nevertheless, it may be possible that we have overlooked the consequences that predation by first-feeding larvae may have on the development of zooplankton communities in fresh waters, which result from high densities of fish larvae and their narrow range of potential prey of suitable size.

In general, it may be concluded that predation impact by 0+ fish on their prey is mainly triggered by length of fish. However, since most of the studies failed to determine the ontogenetic interval of 0+ fish development, this argument cannot be proved. It may be possible that ontogenetic diet and habitat shifts in 0+ fish clearly correspond to the ontogenetic interval if the studies would have focused on fish individuals rather than on average population characteristics. Therefore, combining fish ecology and ontogeny more strongly may improve considerably our understanding of fish–zooplankton interactions.

Acknowledgements

We would like to thank Ian J. Winfield, David A. Culver, Peter C. Gehrke, Josef Wanzenböck, Gordon Copp and an anonymous reviewer for helpful criticisms on earlier versions of this manuscript. This review was initiated by discussions with Jürgen Benndorf and many other colleagues during the studies of T.M. at the Institute of Hydrobiology (Dresden University of Technology, Germany).

References cited

Applegate, R.L. & J.W. Mullan. 1967. Food of young largemouth bass, *Micropterus salmoides*, in a new and old reservoir. Trans. Amer. Fish. Soc. 96: 74–77.

Arrhenius, F. & S. Hansson. 1993. Food consumption of larval, young and adult herring and sprat in the Baltic Sea. Mar. Ecol. Progr. Ser. 96: 125–137.

Arrhenius, F. & S. Hansson. 1994a. In situ food consumption by young-of-the-year Baltic Sea herring *Clupea harengus*: a test of predictions from a bioenergetic model. Mar. Ecol. Progr. Ser. 110: 145–149.

Arrhenius, F. & S. Hansson. 1994b. Erratum. Mar. Ecol. Progr. Ser. 114: 314.

Arts, M.T. & D.O. Evans. 1987. Precision micrometer measurement of mouth gape of larval fish. Can. J. Fish. Aquat. Sci. 44: 1786–1791.

Balon, E.K. 1975. Terminology of intervals in fish development. J. Fish. Res. Board Can. 32: 1663–1670.
Batty, R.S. 1984. Development of swimming movements and musculature of larval herring (*Clupea harengus*). J. Exp. Biol. 110: 217–229.
Blaxter, J.H.S. 1986. Development of sense organs and behaviour of teleost larvae with special reference to feeding and predator avoidance. Trans. Amer. Fish. Soc. 115: 98–114.
Boersma, M., O.F.R. van Tongeren & W.M. Mooij. 1996. Seasonal patterns in the mortality of *Daphnia* species in a shallow lake. Can. J. Fish. Aquat. Sci. 53: 18–28.
Bremigan, M.T. & R.A. Stein. 1994. Gape-dependent larval foraging and zooplankton size: implications for fish recruitment across systems. Can. J. Fish. Aquat. Sci. 51: 913–922.
Brooks, J.L. & S.I. Dodson. 1965. Predation, body size, and the composition of the plankton. Science 150: 28–35.
Copp, G.H. & V. Kováč. 1996. When do fish with indirect development become juveniles? Can. J. Fish. Aquat. Sci. 53: 746–752.
Cryer, M., G. Peirson & C.R. Townsend. 1986. Reciprocal interactions between roach, *Rutilus rutilus*, and zooplankton in a small lake: prey dynamics and fish growth and recruitment. Limnol. Oceanogr. 31: 1022–1038.
Cushing, D.H. 1983. Are fish larvae too dilute to affect the density of their food organisms? J. Plankt. Res. 5: 847–854.
Dabrowski, K. 1984. The feeding of fish larvae: present, a state of the art and perspective. Repr. Nutr. Develop. 20: 807–833.
DeMott, W.R. 1989. The role of competition in zooplankton succession. pp. 195–252. *In*: U. Sommer (ed.) Plankton Ecology: Succession in Plankton Communities, Springer Verlag, Berlin.
Dettmers, J.M. & R.A. Stein. 1992. Food consumption by larval gizzard shad: zooplankton effects and implications for reservoir communities. Trans. Amer. Fish. Soc. 121: 494–507.
DeVries, D.R. & R.A. Stein. 1992. Complex interactions between fish and zooplankton: quantifying the role of an open-water planktivore. Can. J. Fish. Aquat. Sci. 49: 1216–1227.
El-Fiky, N., S. Hinterleitner & W. Wieser. 1987. Differentiation of swimming muscles and gills, and development of anaerobic power in the larvae of cyprinid fish (Pisces, Teleostei). Zoomorphology 107: 126–132.
Forstner, H., S. Hinterleitner, K. Mähr & W. Wieser. 1983. Towards a better definition of "metamorphosis" in *Coregonus* sp.: biochemical, histological, and physiological data. Can. J. Fish. Aquat. Sci. 40: 1224–1232.
Furnass, T.I. 1979. Laboratory experiments on prey selection by perch fry (*Perca fluviatilis*). Freshwat. Biol. 9: 33–43.
Ghan, D. & W.G. Sprules. 1993. Diet, prey selection, and growth of larval and juvenile burbot *Lota lota* (L.). J. Fish Biol. 39: 47–64.
Gliwicz, Z.M. 1994. Relative significance of direct and indirect effects of predation by planktivorous fish on zooplankton. Hydrobiologia 272: 201–210.
Gliwicz, Z.M. & J. Pijanowska. 1989. The role of predation in zooplankton succession. pp. 253–295. *In*: U. Sommer (ed.) Plankton Ecology: Succession in Plankton Communities, Springer Verlag, Berlin.
Guma'a, S.A. 1978. The food and feeding habits of young perch, *Perca fluviatilis*, in Windermere. Freshwat. Biol. 8: 177–187.
Hambright, D. 1994. Morphological constraints in the piscivore-planktivore interaction: implications for the trophic cascade hypothesis. Limnol. Oceanogr. 39: 897–912.
Hammer, C. 1985. Feeding behaviour of roach (*Rutilus rutilus*) larvae and fry of perch (*Perca fluviatilis*) in Lake Lankau. Arch. Hydrobiol. 103: 61–74.
Hansen, M.J. & D.H. Wahl. 1981. Selection of small *Daphnia pulex* by yellow perch fry in Oneida lake, New York. Trans. Amer. Fish. Soc. 110: 64–71.
Hartmann, J. 1983. Two feeding strategies of young fishes. Arch. Hydrobiol. 96: 496–509.
Hartmann, J. 1986. Interspecific predictors of selected prey of young fishes. Arch. Hydrobiol. Beih. Ergebn. Limnol. 22: 373–386.
Hofer, R. & A. Nasir-Uddin. 1985. Digestive processes during the development of the roach, *Rutilus rutilus* (L.). Hydrobiologia 26: 683–689.
Hofer, R. & O. Bürkle. 1986. Daily food consumption, gut passage rate and protein utilization in whitefish larvae (*Coregonus* sp.). Arch. Hydrobiol. Beih. Ergebn. Limnol. 22: 189–196.
Hülsmann, S. & T. Mehner. 1997. Predation by underyearling perch (*Perca fluviatilis*) on a *Daphnia galeata* population in a short-term enclosure experiment. Freshwat. Biol. 38: 209–219.
Kairesalo, T. & T. Seppälä. 1987. Phosphorus flux through a littoral ecosystem: the importance of cladoceran zooplankton and young fish. Int. Revue ges. Hydrobiol. 72: 385–403.
Karjalainen, J., D. Miserque & H. Huuskonen. 1997. The estimation of food consumption in larval and juvenile fish: experimental evaluation of bioenergetics model. J. Fish Biol. 51(Suppl. A): 39–51.
Keast, A. 1980. Food and feeding relationships of young fish in the first weeks after the beginning of exogenous feeding in Lake Opinicon, Ontario. Env. Biol. Fish. 5: 305–314.
Kitchell, J.F., D.J. Stewart & D. Weininger. 1977. Applications of a bioenergetics model to yellow perch (*Perca flavescens*) and walleye (*Stizostedion vitreum vitreum*). J. Fish. Res. Board Can. 34: 1922–1935.
Koblitskaya, A.A. 1981. Identification key of yearlings of fish in the Volga delta. Izd. Nauka, Moskva. 208 pp. (in Russian).
Kurmayer, R. & J. Wanzenböck. 1996. Top-down effects of underyearling fish on a phytoplankton community. Freshwat. Biol. 36: 599–609.
Lampert, W., W. Fleckner, H. Rai & B.E. Taylor. 1986. Phytoplankton control by grazing zooplankton: a study on the spring clear-water phase. Limnol. Oceanogr. 31: 478–490.
Larsson, P., G. Johnsen & A.L. Steigen. 1985. An experimental study of the summer decline in a *Daphnia* population. Verh. internat. Verein. Limnol. 22: 3131–3136.
Lazzaro, X. 1987. A review of planktivorous fishes: their evolution, feeding behaviours, selectivities, and impacts. Hydrobiologia 146: 97–167.
Lehtovaara, A. & J. Sarvala. 1984. Seasonal dynamics of total biomass and species composition of zooplankton in the littoral of an oligotrophic lake. Verh. internat. Verein. Limnol. 22: 805–810.
Luecke, C., M.J. Vanni, J.J. Magnuson, J.F. Kitchell & P.T. Jacobson. 1990. Seasonal regulation of *Daphnia* populations by planktivorous fish: implications for the spring clear-water phase. Limnol. Oceanogr. 35: 1718–1733.
Madon, S.P. & D.A. Culver. 1993. Bioenergetics model for larval and juvenile walleyes: an *in situ* approach with experimental ponds. Trans. Amer. Fish. Soc. 122: 797–813.

Mark, W., W. Wieser & C. Hohenauer. 1989. Interactions between developmental processes, growth, and food selection in the larvae and juveniles of *Rutilus rutilus* (L.) (Cyprinidae). Oecologia 78: 330–337.

Marmulla, G. & R. Rösch. 1990. Maximum daily ration of juvenile fish fed on living natural zooplankton. J. Fish Biol. 36: 789–801.

Mathias, J. & S. Li. 1982. Feeding habits of walleye larvae and juveniles: comparative laboratory and field studies. Trans. Amer. Fish. Soc. 111: 722–735.

McQueen, D.J. & J.R. Post. 1988. Cascading trophic interactions: uncoupling at the zooplankton–phytoplankton link. Hydrobiologia 159: 277–296.

Mehner, T. 1996. Predation impact of age-0 fish on a copepod population in a Baltic Sea inlet as estimated by two bioenergetics models. J. Plankt. Res. 18: 1323–1340.

Mehner, T. & R. Heerkloss. 1994. Direct estimation of food consumption of juvenile fish in a shallow inlet of the Southern Baltic. Int. Revue ges. Hydrobiol. 79: 295–304.

Mehner, T. & I.J. Winfield (ed.). 1997. Trophic interactions of age-0 fish and zooplankton in temperate waters. Proceedings of a Plankton Ecology Group (PEG) workshop, Dresden, February 1996. Arch. Hydrobiol. Beih. Ergebn. Limnol. 49: 1–152.

Mehner, T., H. Schultz, D. Bauer, R. Herbst, H. Voigt & J. Benndorf. 1996. Intraguild predation and cannibalism in age-0 perch (*Perca fluviatilis*) and age-0 zander (*Stizostedion lucioperca*): interactions with zooplankton succession, prey fish availability and temperature. Ann. Zool. Fennici 33: 353–361.

Mehner, T., M. Plewa, S. Hülsmann, H. Voigt & J. Benndorf. 1997. Age-0 fish predation on daphnids – spatial and temporal variability in the top-down manipulated Bautzen reservoir, Germany. Arch. Hydrobiol. Beih. Ergebn. Limnol. 49: 13–25.

Mehner, T., M. Plewa, S. Hülsmann & S. Worischka. 1998a. Gape-size dependent feeding of age-0 perch (*Perca fluviatilis*) and age-0 zander (*Stizostedion lucioperca*) on *Daphnia galeata*. Arch. Hydrobiol. 142: 191–207.

Mehner, T., S. Hülsmann, S. Worischka, M. Plewa & J. Benndorf. 1998b. Is the midsummer decline of *Daphnia* really induced by age-0 fish predation? Comparison of fish consumption and *Daphnia* mortality and life history parameters in a biomanipulated reservoir. J. Plankt. Res. (in press).

Miller, T.J., L.B. Crowder, J.A. Rice & F.P. Binkowski. 1992. Body size and the ontogeny of the functional response in fishes. Can. J. Fish. Aquat. Sci. 49: 805–812.

Mills, E.L. & J.L. Forney. 1983. Impact on *Daphnia pulex* of predation by young yellow perch in Oneida Lake, New York. Trans. Amer. Fish. Soc. 112: 151–161.

Mills, E.L., J.L. Confer & R.C. Ready. 1984. Prey selection by young yellow perch: the influence of capture success, visual acuity, and prey choice. Trans. Amer. Fish. Soc. 113: 579–587.

Mills, E.L., J.L. Confer & D.W. Kretchmer. 1986. Zooplankton selection by young yellow perch: the influence of light, prey density, and predator size. Trans. Amer. Fish. Soc. 115: 716–725.

Mills, E.L., J.L. Forney & K.J. Wagner. 1987. Fish predation and its cascading effect on the Oneida Lake food chain. pp. 118–131. *In*: W.C. Kerfoot & A. Sih (ed.) Predation – Direct and Indirect Impacts on Aquatic Communities, University Press, Hanover.

Murtaugh, P.A. 1985. Vertical distribution of zooplankton and population dynamics of *Daphnia* in a meromictic lake. Hydrobiologia 123: 47–57.

Noble, R.L. 1975. Growth of young yellow perch (*Perca flavescens*) in relation to zooplankton populations. Trans. Amer. Fish. Soc. 104: 731–741.

Post, D.M. & J.F. Kitchell. 1997. Trophic ontogeny and life history effects on interactions between age-0 fishes and zooplankton. Arch. Hydrobiol. Beih. Ergebn. Limnol. 49: 1–12.

Post, J.R. & D.J. McQueen. 1987. The impact of planktivorous fish on the structure of a plankton community. Freshwat. Biol. 17: 79–89.

Post, J.R. & D.O. Evans. 1989. Size-dependent overwinter mortality of young-of-the-year yellow perch (*Perca flavescens*): laboratory, *in situ* enclosure, and field experiments. Can. J. Fish. Aquat. Sci. 46: 1958–1968.

Post, J.R. & J.A. Lee. 1996. Metabolic ontogeny of teleost fishes. Can. J. Fish. Aquat. Sci 53: 910–923.

Post, J.R., L.G. Rudstam, D.M. Schael & C. Luecke. 1992. Pelagic planktivory by larval fishes in Lake Mendota. pp. 303–317. *In*: J.F. Kitchell (ed.) Food Web Management – a Case Study of Lake Mendota, Springer Verlag, New York.

Qin, J. & D.A. Culver. 1995. Effect of young-of-the-year walleye (Percidae: *Stizostedion vitreum*) on plankton dynamics and water quality in ponds. Hydrobiologia 297: 217–227.

Rice, J.A., L.B. Crowder & F.P. Binkowski. 1987. Evaluating potential sources of mortality for larval bloater (*Coregonus hoyi*): starvation and vulnerability to predation. Can. J. Fish. Aquat. Sci. 44: 467–472.

Rogowski, U. & F.W. Tesch. 1960. Erste Nahrung freßfähig gewordener Fischbrut. Zeitschr. f. Fisch. N.F. 9: 735–747.

Rombough, P.J. 1988. Respiratory gas exchange and aerobic metabolism. pp. 59–161. *In*: W.S. Hoar & D.J. Randall (ed.) Fish Physiology 9, The Physiology of Developing Fish, A: Eggs and Larvae, Academic Press, San Diego.

Rudstam, L., S. Hansson, S. Johansson & U. Larsson. 1992. Dynamics of planktivory in a coastal area of the Northern Baltic Sea. Mar. Ecol. Progr. Ser. 80: 159–173.

Schael, D.M., L.G. Rudstam & J.R. Post. 1991. Gape limitation and prey selection in larval yellow perch (*Perca flavescens*), freshwater drum (*Aplodinotus grunniens*), and black crappie (*Pomoxis nigromaculatus*). Can. J. Fish. Aquat. Sci. 48: 1919–1925.

Sommer, U., Z.M. Gliwicz, W. Lampert & A. Duncan. 1986. The PEG-model of seasonal succession of planktonic events in fresh waters. Arch. Hydrobiol. 106: 433–471.

Thiel, R. 1996. The impact of fish predation on the zooplankton community in a southern Baltic bay. Limnologica (Berlin) 26: 123–137.

Thiel, R., T. Mehner, B. Köpcke & R. Kafemann. 1996 Diet niche relationships among early life stages of fish in German estuaries. Mar. Freshwat. Res. 47: 123–136.

Threlkeld, S.T. 1985. Resource variation and the initiation of midsummer declines of cladoceran populations. Arch. Hydrobiol. Beih. Ergebn. Limnol. 21: 333–340.

Treasurer, J.W. 1992. The predator-prey relationship of perch, *Perca fluviatilis*, larvae and zooplankton in two Scottish lochs. Env. Biol. Fish. 35: 63–74.

Troschel, H.J. & R. Rösch. 1991. Daily ration of juvenile *Coregonus lavaretus* (L.) fed on living zooplankton. J. Fish Biol. 38: 95–104.

Vijverberg, J., M. Boersma, W.L.T. VanDensen, W. Hoogenboezem, E.H.R.R. Lammens & W.M. Mooij. 1990. Seasonal variation in the interactions between piscivorous fish, planktivorous fish and zooplankton in a shallow eutrophic lake. Hydrobiologia 207: 279–286.

Vinberg, G.G. 1956. Rate of metabolism and food requirements of fish. Isd. Belrusuniversiteta, Minsk. 253 pp. (in Russian).

Wahl, C.M., E.L. Mills, W.N. McFarland & J.S. DeGisi. 1993. Ontogenetic changes in prey selection and visual acuity of the yellow perch, *Perca flavescens*. Can. J. Fish. Aquat. Sci. 50: 743–749.

Wanzenböck, J. 1992. Ontogeny of prey attack behaviour in larvae and juveniles of three European cyprinids. Env. Biol. Fish. 33: 23–32.

Wanzenböck, J. 1995. Changing handling times during feeding and consequences for prey size selection of 0+ zooplanktivorous fish. Oecologia 104: 372–378.

Wanzenböck, J. & F. Schiemer. 1989. Prey detection in cyprinids during early development. Can. J. Fish. Aquat. Sci. 46: 995–1001.

Wanzenböck, J., M. Zaunreiter, C.M. Wahl & D.L.G. Noakes. 1996. Comparison of behavioural and morphological measures of visual resolution during ontogeny of roach (*Rutilus rutilus*) and yellow perch (*Perca flavescens*). Can. J. Fish. Aquat. Sci. 53: 1506–1512.

Wanzenböck, J., M.C. Whiteside & T. Mehner. 1997. Defining a desirable sampling strategy for studies of age-0 fish – zooplankton interactions – a preliminary approach. Arch. Hydrobiol. Beih. Ergebn. Limnol. 49: 137–138.

Welker, M.T., C.L. Pierce & D.H. Wahl. 1994. Growth and survival of larval fishes: roles of competition and zooplankton abundance. Trans. Amer. Fish. Soc. 123: 703–717.

Werner, E.E. & D.J. Hall. 1974. Optimal foraging and the size selection of prey by the bluegill sunfish (*Lepomis macrochirus*). Ecology 55: 1042–1052.

Werner, E.E. & J.F. Gilliam. 1984. The ontogenetic niche and species interactions in size-structured populations. Ann. Rev. Ecol. Syst. 15: 393–425.

Werner, M.-G., T. Mehner & H. Schultz. 1996. Which factors influence the diet composition of age-0 ruffe (*Gymnocephalus cernuus* [L.]) in the Bautzen reservoir (Saxony, Germany)? Limnologica 26: 145–151.

Werner, R.G., B.V. Jockheere, M.D. Clapsadl & J.M. Farrell. 1996. A bioenergetic exploration of piscivory and planktivory during the early life history of two species of freshwater fishes. Mar. Freshwat. Res. 47: 113–121.

Whiteside, M.C. 1988. 0+ fish as major factors affecting abundance patterns of littoral zooplankton. Verh. internat. Verein. Limnol. 23: 1710–1714.

Whiteside, M.C., C.M. Swindoll & W.L. Doolittle. 1985. Factors affecting the early life history of yellow perch, *Perca flavescens*. Env. Biol. Fish. 12: 47–56.

Wieser, W. 1991. Limitations of energy acquisition and energy use in small poikilotherms: evolutionary implications. Funct. Ecol. 5: 234–240.

Winfield, I.J. & C.R. Townsend. 1983. The cost of copepod reproduction: increased susceptibility to fish predation. Oecologia 60: 406–411.

Wong, B. & F.J. Ward. 1972. Size selection of *Daphnia pulicaria* by yellow perch (*Perca flavescens*) fry in West Blue Lake, Manitoba. J. Fish. Res. Board Can. 29: 1761–1764.

Worischka, S. & T. Mehner. 1998. Comparison of field-based and indirect estimates of daily food consumption in larval perch and zander. J. Fish Biol. 53: 1050–1059.

Wu, L. & D.A. Culver. 1994. *Daphnia* population dynamics in western Lake Erie: regulation by food limitation and yellow perch predation. J. Great Lakes Res. 20: 537–545.

Zalewski, M., B. Brewinska-Zaras, P. Frankiewicz & S. Kalinowski. 1990. The potential for biomanipulation using fry communities in a lowland reservoir: concordance between water quality and optimal recruitment. Hydrobiologia 200/201: 549–556.

Environmental Biology of Fishes **56:** 183–197, 1999.

Seasonal and diel utilisation of inshore microhabitats by larvae and juveniles of *Leuciscus cephalus* and *Leuciscus leuciscus*

Etienne Baras & Joseph Nindaba
University of Liège, Laboratory of Fish Demography and Aquaculture, 10, Chemin de la Justice, B-4500 Tihange, Belgium (e-mail: e.baras@ulg.ac.be)

Received 25 September 1997 Accepted 17 August 1998

Key words: Pisces, Cyprinidae, electrofishing, lotic stream, diel migrations, seasonal shifts, resource partitioning, predation

Synopsis

From July 1995 to January 1996, we examined the seasonal variations of the diel dynamics of habitat use by young-of-the-year cyprinid fishes (chub *Leuciscus cephalus* and dace *L. leuciscus*) in a lotic stream (River Ourthe, Southern Belgium). Inshore bays and neighbouring habitats (riparian shelters, entrance of the bay and shallow riffles) were sampled every three hours from 6:00 to 22:00 h, using DC electrofishing with prepositioned frames. In early summer, chub larvae moved exclusively in between the middle of the bay and riparian shelters inside the bay. Juvenile dace and, later in the season, juvenile chub showed diel dynamics of which the amplitude was dependent on temperature and illumination: they moved into the bay in the morning, gathered in greater numbers at mid-day (up to 586 chub and 387 dace m^{-2}), then progressively left the bay and entered neighbouring riffles in the late afternoon or evening. Small fish immigrated earlier into the bay and emigrated later than fish of larger size. By late September, most dace had left the bays, but returned there when water temperatures were $<7°C$. During autumn and winter, juvenile dace and chub of all sizes occupied exclusively inshore shelters with submerged riparian macrophytes or fallen tree leaves (corresponding densities of 0.7, 9.5 and 15.8 dace m^{-2}, and of 6.8, 20.2 and 94.2 chub m^{-2}). These results support the idea that young-of-the-year dace and chub shift from a restricted use of inshore shallow bays to a diel dynamics with alternate inshore–offshore movements at the time when they become juveniles, although the precise timings of these movements are still influenced by fish size and water temperature afterwards. The significance of these dynamics is discussed within a context of trade-off between the use of food resources and avoidance of predators.

Introduction

Lakes, rivers and streams offer a mosaic of microhabitats that determine the diversity and abundance of fish communities. Over the diel cycle, fish move in between different habitats, depending on their needs, preferences and activity (resting, hiding, feeding, and occasionally spawning). These diel patterns can be regarded as an attempt to achieve optimal foraging behaviour, or as a trade-off between predator avoidance and resource use (Brett 1971, Hall et al. 1979, Gliwicz & Jachner 1992). Hence, diel patterns vary substantially over an annual cycle, depending on fish species, size and environmental factors (e.g. temperature, water level, turbidity). Seasonal and diel shifts in habitat use have been widely documented in adult fish (e.g. Baras 1995b), which could be observed, tagged or tracked by remote sensing techniques. By contrast, fewer studies have been dedicated to fish larvae and juveniles, despite their potential role as functional describers of ecosystem integrity and their crucial role in the recruitment of fish populations (Copp et al. 1991, Schiemer et al. 1991, Houde 1994). Dussart (1970), Keast (1978) and Polis (1984) have proposed that these intervals can be regarded as different 'ecological species', and thus treated separately from older conspecifics.

Habitat and resource partitioning among young-of-the-year fishes (YOY) in lakes, at the seasonal or diel level, has been documented by several authors (e.g., Rheinberger et al. 1987, Jessop 1990, Jachner 1991, Persson 1991, Gerkhe 1992). Except for salmonids, riverine fishes have been less investigated (e.g. Lightfood & Jones 1979) until the late 1980s. Studies on microhabitat use in lowland rivers of the 'bream' zone (Huet 1949) were provided by Scott & Nielsen (1989), Copp (1993, 1997a,b), and Garner (1996a, 1997). Scott (1987), Copp (1990a, 1992), Rincón et al. (1992), Baras et al. (1995), and Watkins et al. (1997) conducted similar investigations in lotic rivers and streams of the 'barbel' zone. With few exceptions, these studies concluded that microhabitat use by YOY is strongly influenced by fish reproductive guilds (Balon 1975, 1985), and that their partitioning is determined by the degree of habitat differentiation. They further provided evidence that littoral, shallow and lentic zones (weakly sloped banks, backwaters, inshore bays inside gravel bars) are colonised by large mixed shoals of YOY fish (especially cyprinids) and that the availability of these key habitats influences the abundance of rheophilic species (Copp 1990b, Baras et al. 1996, Watkins et al. 1997).

Despite the obvious importance of these habitats for YOY fish, few studies have addressed the diel dynamics of microhabitat use (Sanders 1992, Copp & Jurajda 1993, Baras 1995a, Bischoff & Scholten 1996, Garner 1996b, Przybylski 1996) and these did not consider the evolution of the diel dynamics throughout the annual cycle, nor its size-related structure (except for Bischoff & Scholten 1996, Copp & Jurajda 1999). The aim of the present work was to examine the use by YOY cyprinid fishes of shallow inshore bays inside gravel bars of the barbel zone (River Ourthe, Belgium), with respect to season, time of the day, fish size and ontogenetic interval. The lithophilous, late spring spawner, chub *Leuciscus cephalus* and phytolithophilous, early spring spawner, dace *L. leuciscus* were selected as ecological models, based on their representativity in this ecosystem (Philippart & Vranken 1983), their frequent presence in bays (Baras 1995a), and the contrasting succession of their habitat use throughout the annual (Copp 1992, Baras et al. 1995) and diel (Copp & Jurajda 1999) cycles.

Study site

The study was conducted in the lower River Ourthe, which is the main tributary of the River Meuse in Belgium. Prior to regulation, this part of the river was typical of the lower barbel zone (Huet 1949). The current fish community is dominated by three categories of cyprinids (Philippart & Vranken 1983): rheophils (essentially barbel *Barbus barbus*, chub, dace and nase *Chondrostoma nasus*), accompanying species (gudgeon *Gobio gobio* and minnow *Phoxinus phoxinus*), and generalists (mainly roach *Rutilus rutilus* and bleak *Alburnus alburnus*).

The study site was located at Hony-sur-Ourthe (50°32′30″ N, 5°35′00″), 14 km upstream of the confluence with the River Meuse (Figure 1a). This site still offers habitat profiles that closely resemble those of intact parts of the river. The river width ranges from 30 to 50 m, depending on water level. Its slope averages 1.3% and its depth does not exceed 1.5 m. The river contains a large gravel bar which represents a major spawning ground for lithophilous cyprinids in spring (Baras et al. 1996). As water level decreases throughout summer and autumn, this gravel bar creates a lateral backwater (hereby named bay due to its shallow depth and small surface; Figure 1b, Table 1), which has been reported as a major nursery for cyprinid larvae and juveniles (Baras et al. 1995, 1996). Weather conditions in 1995 were quite uncommon, as the mean air temperature in summer was amongst the highests since 1900 (source: Belgian Royal Institute of Meteorology), and

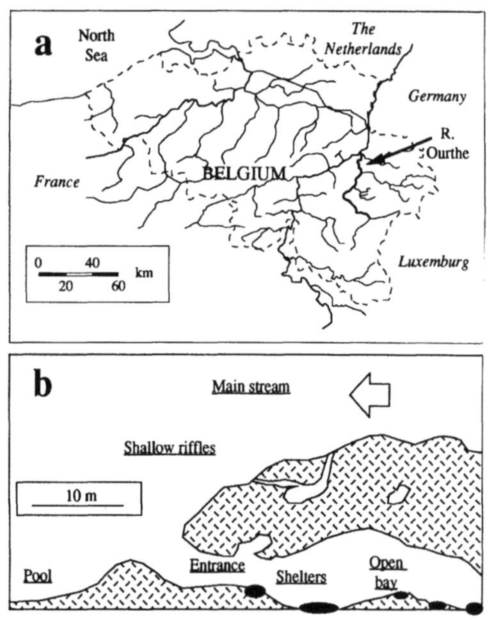

Figure 1. Geographical location (a) and schematic illustration (b) of the bay where larvae and juvenile fish were sampled in the River Ourthe, from July 1995 to January 1996.

Table 1. Characteristics of the main habitat types sampled inside and around the bay in the River Ourthe. Values of depth and velocity (mean ± standard error of measure [SEM]) were obtained from actual measures in the field. For depth and water velocity, the categories sharing at least one common superscript are not significantly different, whereas the other comparisons differ at $p < 0.05$ (Fisher PLSD comparison of means).

Code	Habitat type	Depth mean ± SEM (range)(cm)	Velocity mean ± SEM (range)(cm)	Substrate type	Vegetation presence, type
A	Middle of the bay	13.1 ± 0.9^a (7–22)	0.9 ± 0.3^a (0–6)	Silt, gravel cobble	No
B	Entrance of the bay	15.8 ± 1.6^a (10–30)	9.7 ± 1.7^a (0–23)	Gravel cobble	No
C	Shallow vegetation shelter	15.4 ± 1.3^a (6–25)	1.7 ± 0.9^a (0–13)	Silt	Hydrophytes, submerged riparian macrophytes (≥50% surface)
D	Main stream (shallow riffle)	17.7 ± 2.1^a (10–32)	23.0 ± 3.4^c (13–45)	Cobble	No
E	Deep vegetation shelter	32.4 ± 1.4^b (25–45)	1.5 ± 0.5^a (0–11)	Silt	Hydrophytes, submerged riparian macrophytes (≥50% surface)
F	Fallen tree leaves	28.7 ± 2.8^b (11-42)	0.1 ± 0.1^a (0–2)	Variable mainly silt	Presence of falled leaves of tree (≥50% surface)

rare rainfalls maintained flows close to base flow until 25 December.

Material and methods

Fish larvae and juveniles were collected by 240 V DC electrofishing (EPMC, 2.5 kVA) in prepositioned frames, which enables quantitative sampling of discrete habitats (see Baras 1995a). Each frame encompassed a 2 m^2 homogeneous habitat, which was delimited by a closed cathode (steel bars 2.0 × 0.3 cm), with a 0.04 m^2 steel plate anode placed in its centre. Both electrodes were camouflaged in the substratum layer to eliminate the avoidance behaviour by young fish towards their bright, regular surface. A recolonisation delay of 15 min was allowed between the installation of the frame and application of electrical current. During darkness, hand-help lamps were used to illuminate the sample area after electricity was applied. The frame remained energetised until the last fish was collected with dipnets in order to prevent any escape from the sample area.

Eight sampling sessions were conducted from 11 July 1995 to 18 January 1996. During each session, fish were sampled for one hour every three hours beginning at 6:00, 9:00, 12:00, 15:00, 18:00 and 21:00 h. Nocturnal sampling could not be carried out essentially because the noise from the power generator would have caused a major nuisance to neighbouring houses. At each period of the day, three-to-four frame samples were taken in succession so that all samples were collected within one hour. Sampling took place in different habitat types inside and around the bay, depending on their availability at that time of the year and associated water level (Table 2). In summer, the habitats were taken in shallow riffles outside the bay, at the entrance of the bay, in shallow calms in the middle of the bay and in shallow riparian macrophyte shelters. Deep riparian macrophyte shelters and calm bays with fallen leaves were also sampled during winter time.

Immediately after all fish had been captured, the environment within the sample area (frame) was further characterised by five randomly selected measurement points. We measured depth to the nearest 1 cm and water velocity to the nearest 1 cm s^{-1}, using a magnetic current meter (Marsh McBirney # 201). Vegetation was identified and substratum characterised according to the Wentworth scale. Water temperature (nearest 0.1°C) and dissolved oxygen (nearest 0.1 mg l^{-1}) concentration were measured with a WTW 191 oxymeter at all five points within the frame area, and compared to those in running waters (≥50 cm s^{-1}) at the same time of the day. For each sample site, all captured fish were identified, counted and up to 50 fish of each species were measured for fork length (FL), depending on the number captured. All surviving fish (5–10% mortality) were released near their capture site.

Juveniles of chub and dace were distinguished from larvae using the criteria of Economou et al. (1991),

Table 2. Characteristics of the eight sampling sessions in the River Ourthe in 1995 and 1996. Corresponding descriptions of habitats A, B, C, D, E and F in Table 1. Habitat C was not sampled in September, due to low water levels (measured on a limnimetric scale installed at Hamoir-sur-Ourthe). Water temperatures are daily ranges (6:00–22:00 h). For each species and sampling date, the total number of young-of-the-year fish and maximum number captured per 2 m^2 site are given. The mean size (fork length, FL) of fish are compared by Fisher PLSD comparison of means. For each species, the categories sharing at least one common superscript are not significantly different at $p < 0.05$.

Date	Water level (cm)	Water temperature Stream (°C)	Bay (°C)	Number of sites sampled	Types of habitats sampled	*Leuciscus cephalus* N fish (max/site)	FL range (mean, mm)	*Leuciscus leuciscus* N fish (max/site)	FL range (mean, mm)
11 Jul 1995	54	20.2–21.7	20.2–22.9	25	A, B, C, D	353 (119)	15–26 (19.3)[a]	1067 (413)	21–39 (30.8)[a]
3 Aug 1995	54	20.8–22.3	18.9–29.4	19	A, B, C	1407 (626)	11-49 (25.6)[a]	1312 (775)	33–58 (47.7)[b]
1 Sept 1995	50	14.0–15.4	12.8–21.8	26	A, B, D	1484 (1173)	12–49 (29.8)[c]	450 (305)	46–76 (60.6)[e]
28 Sept 1995	49	11.9–12.0	11.5–12.1	21	A, B, D	160 (57)	17–53 (31.7)[d]	119 (44)	42–72 (58.0)[d]
9 Nov 1995	53	5.8–6.6	5.8–6.8	16	A, C, D, E, F	608 (137)	18–55 (36.0)[e]	455 (191)	41–79 (56.9)[c]
30 Nov 1995	58	4.4–5.2	4.6–6.3	17	A, C, D, E, F	604 (304)	19–52 (35.5)[e]	230 (113)	42–74 (56.3)[c]
19 Dec 1995	56	2.8–3.3	2.7–4.1	18	A, C, D, E, F	1854 (404)	18–53 (34.0)[e]	95 (27)	42–75 (56.9)[c,d]
18 Jan 1995	68	2.2–2.4	0.5–2.5	16	A, C, D, E, F	349 (93)	20–55 (35.8)[e]	122 (86)	47–81 (59.7)[e]

when they had complete scale cover, complete fin differentiation and possessed intestinal loops. The latter criterion, which required the dissection of specimens, could not be applied to all fish collected, as this would have interfered with the structure of the population.

Cut-off sizes were determined in the laboratory on a sample of fish (77 chub, 12–30 mm FL, and 37 dace, 16–29 mm FL) collected in June and July 1995, on dates when the diel dynamics was not investigated. Chub < 21 mm FL and dace < 20 mm FL were all larvae. The corresponding size beyond which all examined specimens were juveniles, were 25 and 22 mm FL, respectively. The transitional size ranges (21–25 and 20–22 mm FL, for chub and dace, respectively) contained both larvae and juveniles.

For each species, the numbers of fish captured in different frame samples were compared by Kruskal-Wallis tests, depending on habitat and time of the day. Analysis of variance (ANOVA), in conjunction with the Fisher PLSD comparison of means, was used to test for differences between the mean size of fishes captured in different sites. Mann-Whitney U-tests, Student's *t*-tests and regression analyses (simple, stepwise multiple regression) were also used when appropriate. Null hypotheses were rejected at $p > 0.05$.

Results

Over eight sampling sessions, 158 frame samples were collected and a total 6819 YOY chub and 3850 YOY dace were captured. The size of chub and dace ranged from 11 to 55 mm and from 21 to 81 mm FL, respectively (Table 2). The size of captured chub and dace increased until September, under low water levels and water temperatures higher than 12°C, then remained constant throughout autumn and early winter. Based on these parameters, distinct analyses were carried out over the three first samples (summer: growth period) and over the four last samples (overwinter), with late September being considered as an intermediate period.

Growth period

During summer, the number of captured 0^+ chub and dace varied substantially with respect to habitat and time of sampling (Table 3). Both 0^+ chub and dace were less abundant in riffles (0.0 and 3.1 fish m^{-2}) or at the bay's entrance (0.2 and 5.8 fish m^{-2}) than in vegetation shelters (30.9 and 20.1 fish m^{-2}) and in the middle of the bay (56.2 chub and 46.4 dace m^{-2}), where their density peaked at mid-day. This general trend

Table 3. Statistical comparisons (Kruskal-Wallis H and Mann-Whitney U-tests) of the number of young-of-the-year chub *Leuciscus cephalus* and dace *L. leuciscus* captured in frame samples during summer and winter, depending on habitat and time of the day. No distinction is made between larvae and juveniles for this analysis. Corresponding descriptions of habitats A, B, C, D, E, F and sampling sessions in Table 1 and 2.

	Leuciscus cephalus		*Leuciscus leuciscus*	
Summer (T° > 12°C: July to early September)				
Habitat type (A, B, C, D) vs. number of fish	$H = 43.6$	$p < 0.001$	$H = 6.93$	$p = 0.074$
Time of day vs. number of fish in open bay (12:00 & 15:00 h vs. others)	$u = 18,\ u' = 54$	$p = 0.092$	$u = 15,\ u' = 57$	$p = 0.049$
Time of day vs. number of fish at entrance of bay (9:00 & 18:00 h vs. others)	$u = 30,\ u' = 42$	$p = 0.574$	$u = 13,\ u' = 59$	$p = 0.032$
Winter (T° < 7°C: November to January)				
Habitat type (C, E, F) vs. number of fish	$H = 20.1$	$p < 0.001$	$H = 18.3$	$p < 0.001$
Time of day vs. number of fish in open bay (12:00 & 15:00 h vs. others)	$u = 322,\ u' = 489$	$p = 0.274$	$u = 394,\ u' = 418$	$p = 0.848$

showed some variation throughout summer, depending on ontogenetic interval, fish size, water temperature, water level and associated availability of riparian vegetation shelters (Figure 2, Table 3, summary in Figure 4).

In July, chub larvae were encountered exclusively in the middle of the bay and in vegetation shelters, with no marked change in fish abundance or size at different times of the day, except for the early morning samples, when few specimens were collected (Figure 2a, Table 4). Fish in the transitional size range were captured exclusively in the same habitats, but proportionally were more abundant in the middle of the bay at mid-day compared to other periods of the daily cycle. In August, chub larvae were also found in the bay and their diel habitat use was not size-structured (Table 4). Juvenile chub showed a more contrasted diel pattern (Figure 2b), occurring in small numbers in the bay in the early morning, gathering in much higher numbers around noon in the middle of the bay (up to 323 fish m^{-2}), and then in shelters during the afternoon. Although no specimens were captured at the entrance of the bay, the high density of juvenile chub in the bay at noon compared to the morning and evening values indicated that these moved in between the bay and the stream already in early August. In early September, the water level was extremely low: only a few vegetation shelters were accessible, and these were used by the smallest juveniles and remaining larvae. Juvenile chub in September showed marked diel dynamics, which was more size-structured than in August (Figure 2c, Table 4). At dawn, only small chub were encountered in the middle of the bay. Later on, larger 0^+ juvenile chub were captured both at the entrance and inside the bay, where they gathered in very large shoals (586 fish m^{-2}) in mid-afternoon. In the early evening, they started leaving the bay, which was devoid of chub at night.

In early July, almost all dace were already juveniles (around 30 mm and 0.3 g). Only 17 fish were ranked in the transitional size range: 14 of them were found in a vegetation shelter and three other specimens in the middle of the bay in the early morning. Juvenile dace used a wider range of habitats (Figure 2a, Table 5). At dawn, only a few small 0^+ juveniles used the bay and riparian shelters. Their numbers and size progressively increased during the day. In late afternoon and early evening, most large 0^+ juvenile dace started moving to the entrance of the bay, then into neighbouring riffles, whereas most small juvenile dace remained in the middle of the bay and in vegetation shelters. From August until the end of the growth period, 0^+ juvenile dace were no longer encountered in vegetation shelters and their diel pattern of bay use was more clear-cut (Figures 2b,c) and size-structured (Table 5). Their stock density in the middle of the bay increased from 5 dace m^{-2} at dawn up to 387 dace m^{-2} at noon, as the bay was progressively colonised by dace of increasing size. Throughout the afternoon and evening, both the density and size of 0^+ juvenile dace in the bay decreased progressively, and their values at dusk were similar to those at dawn. A similar situation was observed in early September, except that no dace were captured in the bay in the first and last samples, and that the stock density peaked (only) at 152 dace m^{-2} in the mid-afternoon.

Stepwise multiple regression analyses were used to test for the influence of water temperature, oxygen level and the amplitude of their gradients between the stream and the bay, on the numbers of chub and dace in the bay on different sampling dates. For 0^+ juvenile dace, these analyses retained the significant influence of the thermal gradient: the warmer the bay comparatively to the stream, the higher the density of 0^+ juvenile dace inside the bay at that moment (simple linear regression tests, df = 5; July: $r = +0.940$, $F = 30.2$, $p = 0.005$; August: $r = +0.977$, $F = 86.6$, $p < 0.001$; September: $r = +0.785$, $F = 6.4$, $p = 0.064$). Similar tests on

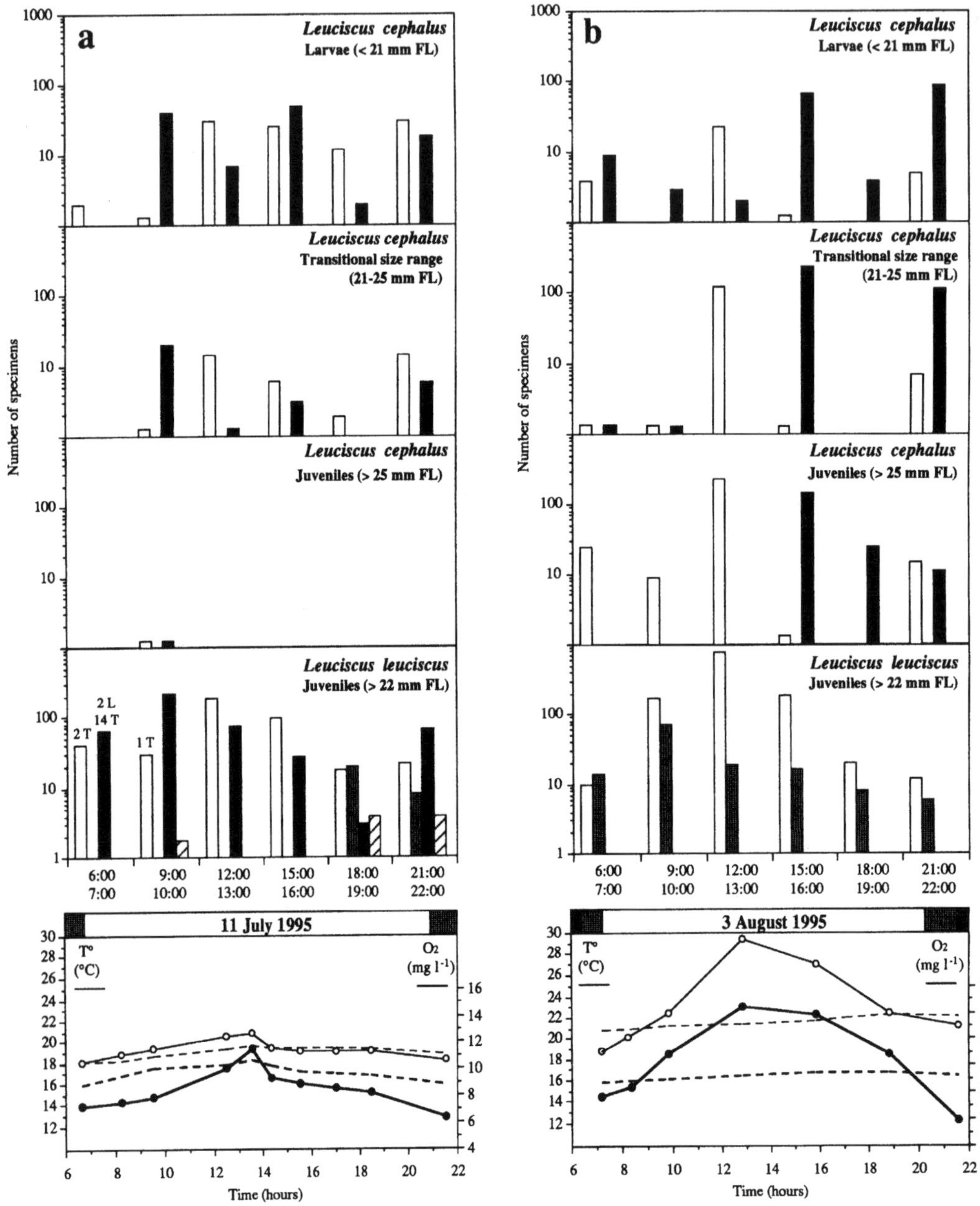

Figure 2a and b

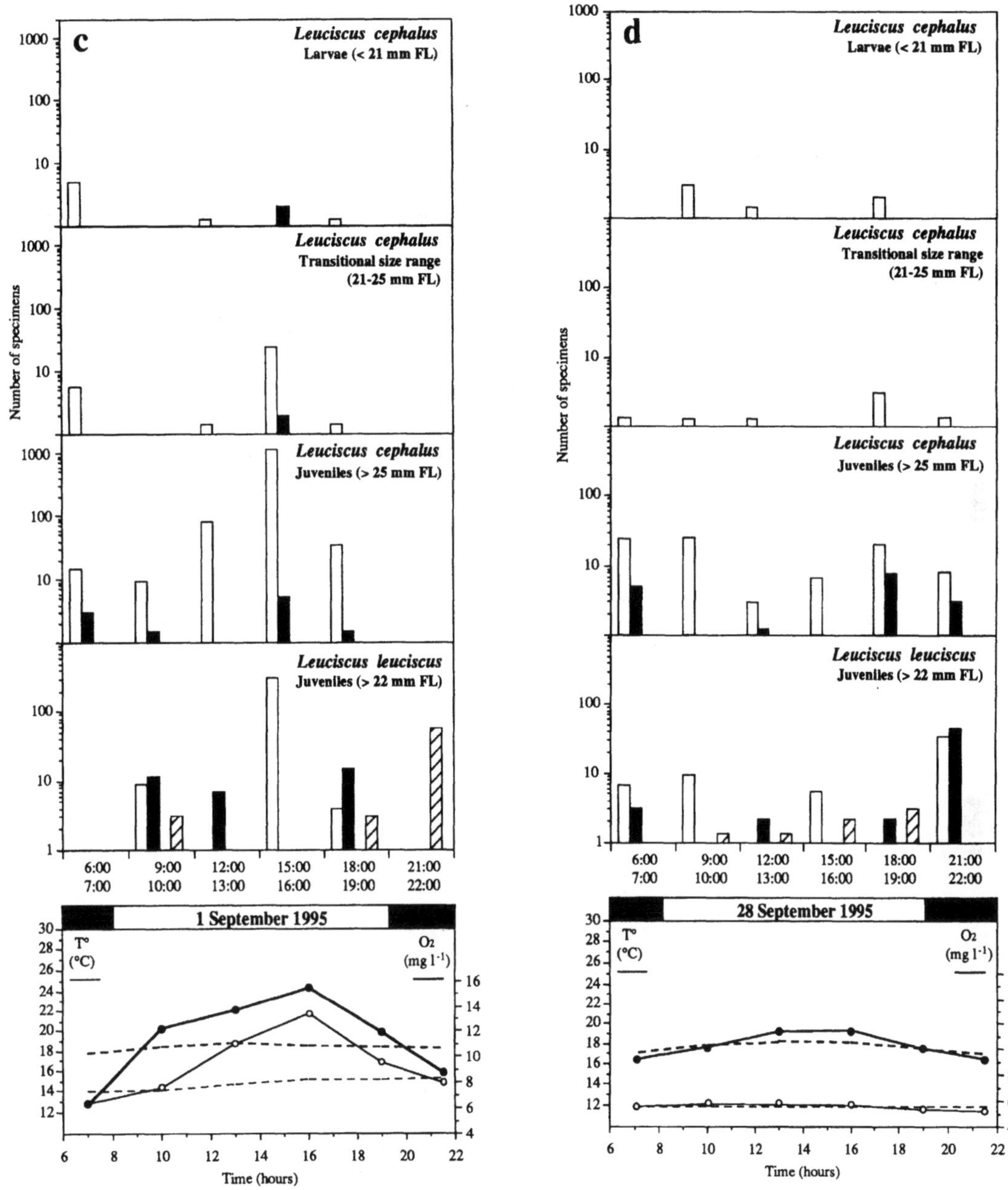

Figure 2c and d

Figure 2. Variations in the numbers of young-of-the-year chub *Leuciscus cephalus* and dace *L. leuciscus* captured in summer, depending on sampling date (a,b,c,d), time of the day and habitat: open and closed bars are the middle of the bay and riparian vegetation shelters. Grey and dashed bars correspond to the entrance of the bay and to shallow riffles in the stream (detailed description in Table 1). Labels L and T on the *L. leuciscus* graph for 11 July 95 refer to the numbers of larvae (L) and of fish in the transitional size range (T), with all other specimens being juveniles. Temperature (thin lines) and oxygen levels (bold lines) in the bay (solid lines with circles) are compared to those in the stream (dotted lines, no symbol).

Table 4. Variation of the mean size (fork length ± standard error of measure, mm) of young-of-the-year chub *Leuciscus cephalus* during summer and early autumn, depending on sampling date, ontogenetic interval, time of the day and habitat. Habitats A, B, C and D as in Table 1. The mean fork lengths of fish are compared by ANOVA and Fisher PLSD comparison of means. Larvae or juveniles were excluded from the analyses when their total numbers were low (see Figure 4), and fish in the transitional size range are not represented. For each sampling date, the categories sharing at least one common superscript are not significantly different whereas the other comparisons differ at $p < 0.05$. No comparison was made when a single fish was captured (*).

Date (1995)	Habitat	6–7:00 h	9–10:00 h	12–13:00 h	15–16:00 h	18–19:00 h	21–22:00 h
11 July (Larvae)	A	16.5 ± 0.5[a]	18.0(*)	18.7 ± 0.3[b]	18.1 ± 0.3[a,b]	18.8 ± 0.3[b]	18.9 ± 0.3[b]
	B	–	–	–	–	–	–
	C	–	18.8 ± 0.2[b]	18.4 ± 0.5[b]	18.2 ± 0.3[a,b]	17.5 ± 1.5[a,b]	18.6 ± 0.2[b]
	D	–	–	–	–	–	–
3 August (Juveniles)	A	19.2 ± 0.5[b,c]	–	18.2 ± 0.3[b]	20.0(*)	–	18.0 ± 1.2[b,c]
	B	–	–	–	–	–	–
	C	–	–	–	31.7 ± 0.9[a]	33.5 ± 0.9[a]	32.2 ± 0.8[a]
	D	no sample	no sample	no sample	no sample	no sample	no sample
1 September (Juveniles)	A	28.1 ± 0.3[a]	34.0 ± 0.5[c,d]	28.9 ± 0.5[a]	35.8 ± 1.1[d]	31.8 ± 0.6[b]	–
	B	29.3 ± 2.4[a,b]	35.7 ± 1.2[c,d]	–	–	33.3 ± 1.3[b,c]	–
	C	no sample	–	31.3 ± 1.9[a,b,c]	–	no sample	no sample
	D	–	–	–	–	–	–
28 September (Juveniles)	A	34.3 ± 1.0[b,c]	33.0 ± 0.7	29.5 ± 1.7[a,b]	32.6 ± 2.3[b,c]	30.9 ± 0.7[a,b]	32.8 ± 2.4[b,c]
	B	32.2 ± 1.2[b,c]	–	33.0(*)	–	31.3 ± 2.6[a,b]	39.3 ± 3.2[c]
	C	no sample	no sample	no sample	no sample	no sample	no sample
	D	–	–	–	–	–	–

Table 5. Variation of the mean size (fork length ± standard error of measure, mm) of young-of-the-year juvenile dace *Leuciscus leuciscus* during summer and early autumn, depending on sampling date, time of the day and habitat. Habitats A, B, C and D as in Table 1. The mean fork lengths of fish are compared by ANOVA and Fisher PLSD comparison of means. For July, dace in the transitional size range are not represented. For each sampling date, the categories sharing at least one common superscript are not significantly different whereas the other comparisons differ at $p < 0.05$. No comparison was made when a single fish was captured (*).

Date (1995)	Habitat	6–7:00 h	9–10:00 h	12–13:00 h	15–16:00 h	18–19:00 h	21–22:00 h
11 July	A	28.6 ± 0.5[a]	29.7 ± 0.6[a]	31.4 ± 0.3[b,c]	31.4 ± 0.3[b,c]	31.5 ± 0.5[b,c]	30.9 ± 0.5[b]
	B	–	–	–	–	32.8 ± 0.4[c,d]	33.0 ± 0.5[c,d]
	C	28.9 ± 0.6[a]	30.9 ± 0.5[b]	32.8 ± 0.2[c]	29.0 ± 0.2[a]	33.0 ± 1.5[b,c]	31.5 ± 0.2[b]
	D	–	32.0 ± 0.0[a,b,c]	–	–	34.5 ± 0.5[c,d]	34.8 ± 0.5[d]
3 August	A	44.5 ± 0.5[a]	46.8 ± 0.3[b,c]	49.1 ± 0.5[d,e]	49.3 ± 0.4[e]	45.3 ± 1.0[a]	45.6 ± 1.3[a,b]
	B	45.6 ± 0.7[a,b]	47.7 ± 0.4[c,d]	48.8 ± 0.6[d,e]	49.4 ± 0.7[e]	47.4 ± 1.1[a,b,c,d]	45.7 ± 0.7[a,b,c]
	C	–	44.0(*)	–	50.0(*)	–	–
	D	no sample	no sample	no sample	no sample	no sample	no sample
1 September	A	–	56.8 ± 0.9[a,b]	–	62.2 ± 0.4	52.3 ± 1.4[a]	–
	B	–	58.5 ± 1.0[b,c]	58.4 ± 1.2[b,c]	–	60.0 ± 0.7[c]	–
	C	no sample	–	no sample	54.7 ± 3.2[a]	no sample	no sample
	D	–	60.3 ± 0.7[c]	–	–	56.0 ± 1.8[a]	62.4 ± 0.7[c]
28 September	A	53.1 ± 2.0[a]	54.7 ± 1.5[a]	–	54.6 ± 2.5[a,b]	–	59.8 ± 1.6[b]
	B	52.7 ± 2.0[a]	–	52.0 ± 0.7[a]	–	50.0 ± 0.7[a]	60.4 ± 0.9[b]
	C	no sample	no sample	no sample	no sample	no sample	no sample
	D	–	72.0(*)	56.0(*)	60.0 ± 1.0[a,b]	55.7 ± 1.5[a,b]	–

the density of juvenile chub in August and September also retained significant relationships with the thermal gradient ($r = +0.945$, $F = 33.3$, $p = 0.004$, and $r = +0.821$, $F = 8.3$, $p = 0.045$, respectively), suggesting that juveniles of both species were influenced in the same way by water temperature. By contrast with juveniles, the density of chub larvae in the bay was independent from the thermal gradient between the stream and the bay (July: $r = +0.007$, $F < 0.1$, $p = 0.990$; August: $r = +0.398$, $F = 0.8$, $p = 0.435$), suggesting that they remained all day long in the bay.

Autumn

In contrast to the three first sampling sessions, sampling in late September took place on a cloudy day. The water temperature in the bay was *circa* 12°C, remained steady all day long and never differed from the thermal regime in the stream by more than 0.4°C (Figure 2d). Very few 0^+ juvenile chub and dace were captured on this occasion ($n = 160$ and 119, respectively). Both species were found in low and variable numbers in the middle of the bay and at its entrance, with dace occasionally found in riffles (Figure 2d). Beyond low stock density, the major difference between this and the summer samples was an (almost) opposite dynamics of habitat use. Both species used the bay less intensively at midday than at others periods of the day or night, and fish captured by night were significantly larger than those captured by day (Tables 4, 5).

Overwinter period

During November to January, 0^+ juveniles of both chub and dace occurred in much smaller numbers than in summer, especially during the January sampling, which took place after the first winter flood (Table 2). All fish were captured exclusively in calm shelters, where the water velocity averaged $< 2.0\,\mathrm{cm\,s^{-1}}$ (Table 1) and neither their numbers nor their size significantly varied between night and day (33.0 vs. 27.0 chub and 8.3 vs. 7.4 dace m^{-2}, Table 3). These findings, along with the absence of fish in the middle of the bay and in faster flowing habitats, suggested that 0^+ juveniles of chub and dace remained in sheltered habitats all day long when water temperature was < 7°C. Shallow vegetation shelters were less frequently occupied by juvenile chub and dace than other shelters (Table 3, Figure 3). From late November onwards, when fallen leaves were more abundant (2–5 cm thick layers) than earlier in the season, chub essentially were encountered in this type of shelter whereas dace equally used this habitat and deep vegetation shelters. There was little variation between the mean size of chub or dace captured in different habitats during autumn and winter (Table 6), except that fish in shallow vegetation shelters were generally smaller than in other habitats.

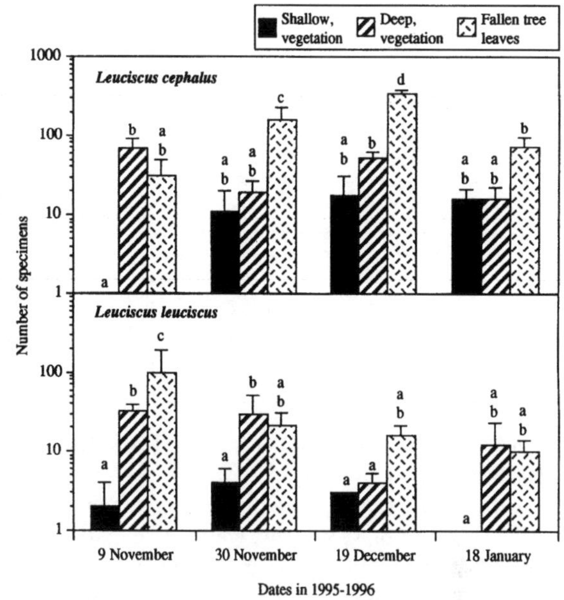

Figure 3. Variations in the numbers of young-of-the-year chub *Leuciscus cephalus* and dace *L. leuciscus* captured during the overwintering period, depending on sampling date and habitat type (environmental conditions as in Table 2). The January sample takes place after the first seasonal flood. No single fish captured in open habitats (Table 1, A, B, D). For each species, the categories sharing at least one common superscript are not significantly different at $p < 0.05$ (Mann-Whitney U-tests).

Discussion

This study demonstrated that inshore bays are used by chub and dace throughout summer, autumn and winter, and that their diel dynamics of habitat use progressively amplifies throughout summer, along with their ontogeny and size, until it becomes most restricted in winter (see idealised picture in Figure 4). The contrast between chub, which were exclusively encountered in the middle of the bay and in riparian vegetation shelters in early summer, and dace, which also used faster flowing habitats at this time of the year, supports the findings of Rheinberger et al. (1987), Scott (1987), Mills (1991), Copp (1992) and Baras et al. (1995). It also

Table 6. Variation of the mean size (fork length ± standard error of measure, mm) of young-of-the-year chub *Leuciscus cephalus* and dace *L. leuciscus* during autumn and winter time, depending on sampling date and habitat. Habitats C, E and F as in Table 1. The mean lengths of fish are compared by ANOVA and Fisher PLSD comparison of means. For each species and sampling date, the categories sharing at least one common superscript are not significantly different whereas the other comparisons differ at $p < 0.05$.

Date	*Leuciscus cephalus*			*Leuciscus leuciscus*		
	C	E	F	C	E	F
9 Nov 1995	36.5 ± 5.5^{b}	35.9 ± 0.4^{b}	36.7 ± 1.0^{b}	51.0 ± 1.6^{a}	$56.0 \pm 0.4^{a,b}$	61.0 ± 0.5^{d}
30 Nov 1995	33.8 ± 0.9^{a}	36.8 ± 0.6^{b}	35.2 ± 0.4^{b}	$54.3 \pm 1.4^{a,b,c}$	$57.4 \pm 0.5^{b,c}$	$53.5 \pm 0.8^{a,b}$
19 Dec 1995	$35.0 \pm 0.7^{a,b}$	33.4 ± 0.4^{a}	$34.8 \pm 0.4^{a,b}$	$53.5 \pm 1.8^{a,b}$	$55.8 \pm 1.3^{a,b}$	57.7 ± 0.7^{c}
18 Jan 1996	$34.2 \pm 1.2^{a,b}$	36.2 ± 0.6^{b}	35.9 ± 0.5^{b}	$56.5 \pm 5.5^{a,b,c}$	60.7 ± 0.9^{d}	$56.6 \pm 1.1^{a,b,c}$

substantiates the importance of shallow banks for small fish (Schiemer & Zalewski 1992).

Although no sample could be collected from 22:00 to 6:00 h in the present study, we suspect that the larvae and small 0^+ juveniles do not leave the bay at night. This permanent use of bays by these fishes can be accounted for by several factors. Young fishes have restricted swimming capacities and would presumably have been displaced by the water velocities in the riffles around the bays. Even if fish larvae swam near their critical swimming speeds (ca. 10 body lengths per second [BL s^{-1}]) and eventually held their position in riffles, these high velocities would considerably reduce their chances to capture drifting prey. Flore & Keckeis (1996) indeed provided evidence that the probability of prey capture by nase larvae sharply decreases at water velocities faster than 4 BL s^{-1}, and is zero at 5 BL s^{-1}. As larvae, cyprinid fishes still have a short intestine relative to their body length (Stroband & Dabrowski 1979). Hence, they can only cope with easily digestible prey, such as Rotifera and other microinvertebrates (Mark et al. 1987), or with small chironomid larvae. These small prey, which indeed were the main food items consumed by larvae of chub and dace during early summer (Nindaba & Baras unpublished), generally occur in much greater numbers in bays compared to other habitats on the margins of the stream (up to six times as high, Reckendorfer et al. 1996). Because of their lentic nature, bays warm up considerably on sunny summer days and offer thermal regimes that favour fast growth. Finally, the shallow depth and presence of vegetation shelters in bays both grant natural refuges to limit or escape fish predator attacks.

Because all factors (limited swimming capacities, high food availability, thermal regime and refuge against predation) concur with the selection of inshore bays by larvae of chub and dace in early summer, it is difficult to determine which ones are prevalent. In any case, their combination accounts for the restricted diel habitat use by these 0^+ fish during early summer.

Juvenile dace and, later in the season, chub showed a much more pronounced and size-structured diel dynamics of habitat use, which can reflect a trade-off between the use of food resources and avoidance of predators (Hall et al. 1979, Copp 1992, Gliwicz & Jachner 1992, Copp & Jurajda 1999). Compared to the situation in July, 0^+ juvenile chub and dace in August and September were much larger fish, with higher absolute needs for food, which probably could no longer be fulfilled through the exclusive use of bays. Their larger size facilitated their access to and exploitation of faster flowing sites, which covered a much larger surface of the stream, and where other prey (larvae of *Diptera* and *Trichoptera*, Nindaba & Baras unpublished data) were accessible during night-time. Yet, it is still uncertain whether 0^+ juvenile rheophils remained offshore all night long in summer, as no sample could be collected in the middle of the night during this study. Recent observations in an English river (River Lee, G.H. Copp personal communication) suggest that bays may also be colonised at night. Under increasing light intensity, the activity and accessibility of macroinvertebrates is much lower (Neveu 1974, Neveu & Échaubard 1975), whereas 0^+ juvenile fish offer more visible targets for diurnal predators. 0^+ juvenile dace and chub would thus start moving to inshore bay refuges at the time when the risk of being eaten would outweigh the benefit granted by foraging. As juvenile dace and chub are both gregarious fish and apparently do not defend a territory, there would be no loss of strategic advantage as a result of their emigration from the stream into the bay.

This interpretation is supported by the observation that the bay was colonised over longer periods of

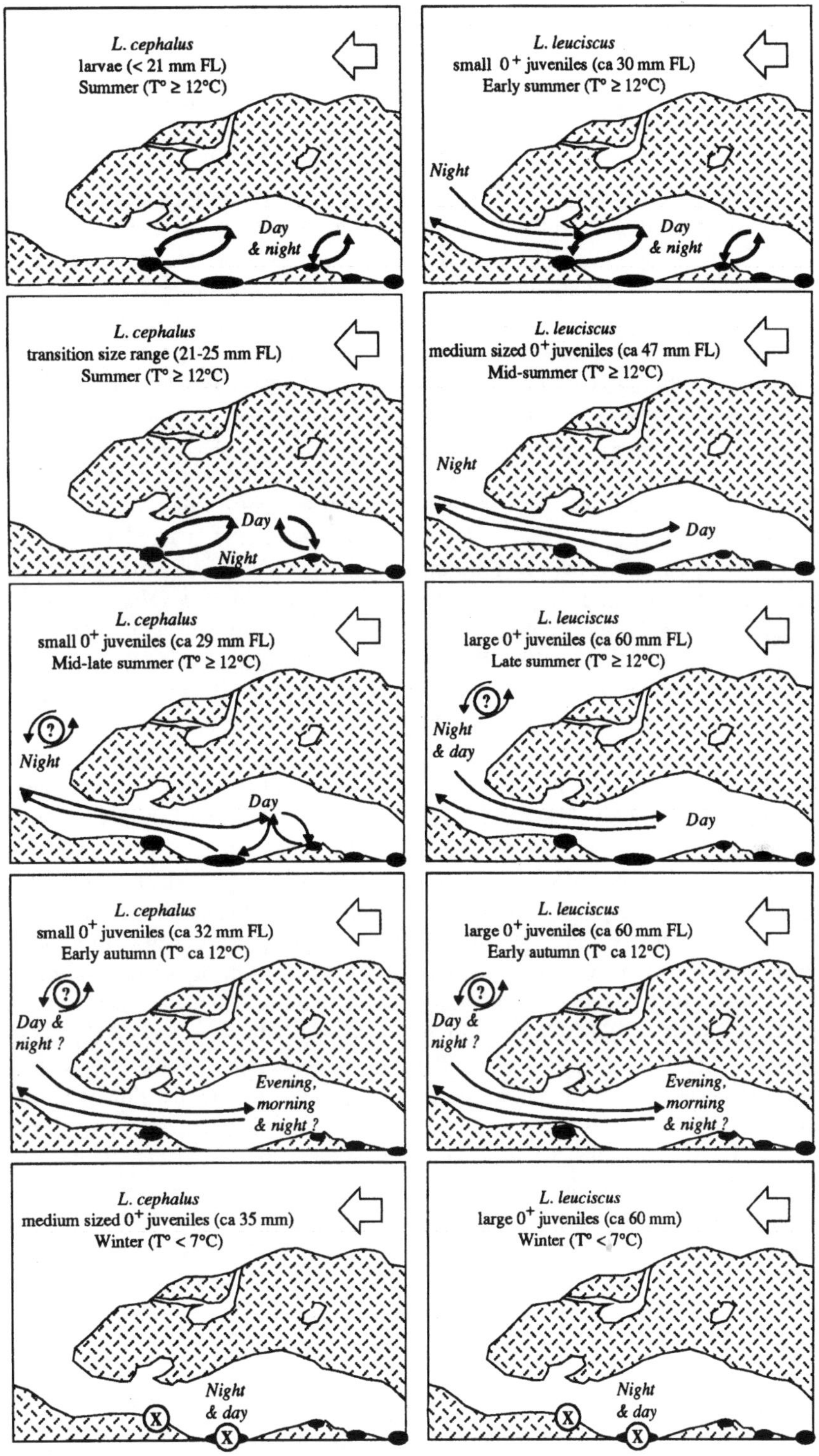

Figure 4. Idealised picture of diel dynamics of inshore bays use by young-of-the-year chub *Leuciscus cephalus* and dace *L. leuciscus*, depending on season, fish size (fork length, FL) and ontogenetic interval. The transitional size range for chub groups larvae and juveniles, in between the cut-off size limits for these stages (21 and 25 mm, respectively).

time by the smallest juveniles, of which the risk of being eaten was highest, as well as by the relationships between the abundance of fish in bays and the thermal gradient from the stream to the bay. Higher temperatures in the bay compared to the stream reflected increasing levels of illumination, and these normally make prey more contrasted to predators. The much more intense colonisation of bays by juvenile dace on hot sunny days, compared to colder cloudy days, may thus correspond to increased predation risk under brighter illumination. The search for high temperature, which enables faster digestion and growth rates, may yet be a complementary driving force in the colonisation of inshore bays by juvenile fishes in summer (Scott 1987, Baras 1995a). However, the size-structure of this colonisation suggests that it is of secondary importance compared to predation, and that it may be a consequence rather than a driving force.

Studies in lakes (Tonn & Paszkowski 1987, Persson 1991, Gliwicz & Jachner 1992) and in deep rivers (Sanders 1992, Copp & Jurajda 1993) reported that YOY fish occurred in greater numbers in littoral areas during darkness than during daylight. This pattern of habitat use was diametrically opposed to those of 0^+ juvenile chub and dace in the present study, and to those of other 0^+ juvenile rheophils in the same environment (barbel and nase, Baras unpublished). As predation pressure was suggested as one of the driving forces of the diel dynamics in this study, as in other studies (Gliwicz & Jachner 1992, Copp & Jurajda 1993, 1999), the difference between this and other studies may originate from different dynamics of predation in their main environment. In deep and calm habitats, piscivorous fish like perch, *Perca fluviatilis*, disperse at night (Thorpe 1977, Craig 1987, Copp & Jurajda 1998) and may indeed cause the nocturnal migrations of juvenile fish into nearshore shallow refuges. In the present study, $\geq 1+$ chub (≥ 90 mm) were the primary potential predators observed in shallow streams with moderate water velocity (10–25 cm s^{-1}), and these obviously exert a diurnal predation (e.g. Baras 1995a, Copp & Jurajda 1999, G.H. Copp personal communication). Little is known about chub nocturnal activity and use of habitat, although Copp & Jurajda (1993) provided evidence that they occur in greater numbers in deeper waters at night.

In late September, both dace and chub shifted to a nocturnal use of inshore bays. Seasonal shifts of daily activity rhythms have been documented in several species (review in Boujard & Leatherland 1992), and in some cases, these were more dependent on water temperature than on day length (Heggenes et al. 1993, Baras 1995b). This may be the case here as the thermal regime in late September was close to the lower temperature threshold for the growth of cyprinids (ca. 12°C; Mann 1974, 1976, Mills & Mann 1985). This shift of activity rhythms may also be associated with seasonal changes in habitat use. Juvenile dace and chub move to deeper and calmer places in autumn (Baras et al. 1995), where they may encounter twilight or nocturnal predators like perch, which could be avoided through the nocturnal use of inshore bays.

During overwinter, the diel and seasonal dynamics of habitat use was most restricted. As most of this study took place under very low flows, these restricted movements probably were the consequences of low water temperatures, which depress the swimming capacity of fishes (Brett 1971, Beamish 1978), and increase the latency and contraction time of muscles (Batty et al. 1991, Kaufman & Wieser 1992). Dace and chub did not starve all over autumn and winter (Nindaba et al. unpublished data) but their diet was composed of food items with a low energetic value (essentially algae: Chlorophycea and Bacillariophycea). Because temperature reduced their metabolism, YOY fish could rely on such a subsistence diet during autumn and winter, but only by selecting habitats where they could minimise their energetic expenditures, such as calm bays. Deep habitats with physical structures were predominantly used, presumably as a passive protection against wandering predators (perch, trout *Salmo trutta*), displacement by high floods and/or frost. As in the summer sampling session, all factors concur to the selection of shelter habitats by juveniles, and experimental studies are needed to determine which ones are prevalent. In any case, the observation of numerous dace and chub in lentic bays during overwintering further substantiates the importance of habitats connected to gravel bars to the recruitment of these species (Baras et al. 1996, Mann 1996).

These results support the idea that young-of-the-year dace and chub shift from a restricted use of inshore shallow bays to a diel pattern with alternate inshore–offshore movements at the time when they become juveniles, although the precise timings of these movements are still influenced by fish size and water temperature afterwards. As the diel dynamics of young-of-the-year dace and chub essentially emerge as a trade-off between feeding and hiding, future studies should focus on determining the spatial and temporal availability of food resources in different habitats, as well as their actual chances of being captured by fishes (see Flore & Keckeis 1996). Similarly, more attention should be dedicated to the food spectrum of predators of 0^+ fishes, and to their seasonal and diel dynamics of habitat use. Finally, it is still uncertain whether juvenile fishes randomly select refuge habitats in bays, or show some consistent fidelity and homing towards particular refuges, of which knowledge could facilitate their escape from predators. Tagging fish ($\geq$ 1.5 g) with passive integrated transponders and the permanent remote interrogation of flat antennas arrays (Armstrong et al. 1996) installed at the shallow entrances of the bays could provide a non invasive way of answering this fundamental question.

Acknowledgements

The authors wish to thank J.C. Philippart (research associate of FNRS, head of LFDA-ULg) for fruitful discussions and support throughout this study. G. Rimbaud, A. Champagne, J. Foriez and L. Hill (LFDA-ULg) assisted in field sampling. We also are indebted to the Burundese Government (Ministry of Scientific Research) who provided a Ph.D. studentship to J. Nindaba.

References cited

Armstrong, J.D., V.A. Braithwaite & P. Rycroft. 1996. A flat-bed passive integrated transponder antenna array for monitoring behaviour of Atlantic salmon parr and other fish. J. Fish Biol. 48: 539–541.

Bain, M.B., J.T. Finn & H.E. Booke. 1985. A quantitative method for sampling riverine microhabitats by electrofishing. N. Amer. J. Fish. Manage. 5: 489–493.

Balon, E.K. 1975. Reproductive guilds of fishes: a proposal and definition. J. Fish. Res. Board Can. 32: 821–864.

Balon, E.K. 1985. Early life histories of fishes: new developmental, ecological and evolutionary perspectives. Developments in Environmental Biology of Fishes 5, Dr Junk Publishers, Dordrecht. 280 pp.

Baras, E. 1995a. An improved electrofishing methodology for the assessment of habitat use by young-of-the-year fishes. Arch. Hydrobiol. 134: 403–415.

Baras, E. 1995b. Seasonal activities of Barbus barbus (L.) – Effect of temperature on time-budgeting. J. Fish Biol. 46: 816–828.

Baras, E., J. Nindaba & J.C. Philippart. 1995. Microhabitat used in a 0^+ rheophilous cyprinid assemblage: quantitative assessment of community structure and fish density. Bull. Fr. Pêche Pisc. 337-338-339: 241–247.

Baras, E., J.C. Philippart & J. Nindaba. 1996. Importance of gravel bars as spawning grounds and nurseries for European running water cyprinids. pp. 367–378. *In*: M. Leclerc et al. (ed) Proceedings of Ecohydraulics 2000, Vol. B, 2nd IAHR International Symposium on Hydraulics and Habitats, INRS eau, Québec City.

Batty, R.S., J.H.S. Blaxter & K. Fretwell. 1991. The effect of temperature on swimming performance of fish larvae during escape responses. Bull. Zool. Museum (Univ. Amsterdam) August 1991: 9.

Beamish, F.W.H. 1978. Swimming capacity. pp. 101–187. *In*: W.S. Hoar & D.J. Randall (ed) Fish Physiology, Vol. 7, Academic Press, London.

Bischoff, A. & M. Scholten. 1996. Diurnal distribution patterns of 0^+ barbel *Barbus barbus* in two different types of habitats in the River Sieg, Germany. Folia Zool. 46(supp. I): 13–20.

Boujard, T. & J.F. Leatherland. 1992. Circadian rhythms and feeding times in fishes. Env. Biol. Fish. 35: 109–131.

Brett, J.R. 1971. Energetic responses of salmon to temperature. A study of some thermal relations in the physiology and freshwater ecology of the sockeye salmon (*Oncorhynchus nerka*). Amer. Zool. 11: 99–113.

Copp, G.H. 1990a. Shifts in the microhabitat of larval and juvenile roach *Rutilus rutilus* (L.) in a floodplain channel. J. Fish Biol. 36: 683–692.

Copp, G.H. 1990b. Effect of regulation on 0^+ fish recruitment in the Great Ouse, a lowland river. Reg. Riv. Res. Manage. 5: 135–145.

Copp, G.H. 1992. Comparative microhabitat use of cyprinid larvae and juveniles in a lotic floodplain channel. Env. Biol. Fish. 33: 181–193.

Copp, G.H. 1993. Microhabitat use of fish larvae and 0^+ juveniles in a small abandoned channel of the upper River Rhône, France. Folia Zool. 42: 153–164.

Copp, G.H. 1997a. Microhabitat use of fish larvae and juveniles in a highly regulated section of the River Great Ouse. Reg. Riv. Res. Manage. 13: 267–276.

Copp, G.H. 1997b. Importance of marinas and off-channel water bodies as refuges for young fishes in a regulated lowland river. Reg. Riv. Res. Manage. 13: 303–307.

Copp, G.H. & P. Jurajda. 1993. Do small riverine fish move inshore at night ? J. Fish Biol. 43 (supp. A): 229–241.

Copp, G.H. & P. Jurajda. 1999. Size-structured diel use of river banks by fish. Aquat. Sci. 61: 75–91.

Copp, G.H., J.M. Oliver, M. Peňáz & A.L. Roux. 1991. Juvenile fishes as functional describers of fluvial ecosystem dynamics: applications on the River Rhône, France. Reg. Riv. Res. Manage. 5: 251–263.

Craig, J.F. 1987. The biology of perch and related fish. Croom Helm, Beckenham. 333 pp.

Dussart, B. 1970. Quelques problèmes d'écologie. Bull. Assoc. Prof. Biol. Géol. 57: 245–257.

Economou, A.N., Ch. Daoulas & T. Psarras. 1991. Growth and morphological development of chub, *Leuciscus cephalus* (L.), during the first year of life. J. Fish Biol. 39: 393–408.

Flore, L. & H. Keckeis. 1996. The effect of water current on foraging behaviour of a rheophilic cyprinid, *Chondrostoma nasus*, during ontogeny: trade-off energetic benefit-swimming costs.

pp. 487–505. *In*: M. Leclerc et al. (ed.) Proceedings of Ecohydraulics 2000, Vol. B, 2nd IAHR International Symposium on Hydraulics and Habitats, INRS eau, Québec City.

Garner, P. 1996a. Microhabitat use and diet of 0^+ cyprinid fishes in a lentic, regulated reach of the River Great Ouse, UK. J. Fish Biol. 48: 367–382.

Garner, P. 1996b. Diel behaviour of juvenile 0-group fishes in a regulated river: the Great Ouse, England. Ecol. Freshwat. Fish. 5: 175–182.

Garner, P. 1997. Sample sizes for length and density estimation of 0^+ fish when using point sampling by electrofishing. J. Fish Biol. 50: 95–106.

Gerkhe, P.C. 1992. Diel abundance, migration and feeding of fish larvae (*Eleotridae*) in a floodplain billabong. J. Fish Biol. 40: 695–707.

Gliwicz, Z.M. & A. Jachner. 1992. Diel migrations of juvenile fish: a ghost of predation past or present. Arch. Hydrobiol. 124: 385–410.

Hall, D.J., E.E. Werner, J.F. Gilliam, G.G. Mittelbach, D. Howard & C.G. Doner. 1979. Diel foraging behaviour and prey selection in the golden shiner (*Notemigonus chrysoleucas*). J. Fish. Res. Board Can. 36: 1029–1039.

Heggenes, J., O.M.W. Krog, O.R. Lindas, J.G. Dogg & T. Bremmer. 1993. Homeostatic behavioural responses in a changing environment: brown trout (*Salmo trutta*) become nocturnal in winter. J. Anim. Ecol. 62: 295–308.

Houde, E.D. 1994. Differences between marine and freshwater fish larvae: implications for recruitment. ICES J. Mar. Sci. 51: 91–97.

Huet, M. 1949. Aperçu de la relation entre la pente et les populations piscicoles des eaux courantes. Schweiz. Z. Hydrol. 11: 332–351.

Jachner, A. 1991. Food and habitat partitioning among juveniles of three fish species in a pelagial of a mesotrophic lake. Hydrobiologia 226: 81–89.

Jessop, B.M. 1990. Diel variations in density, length composition, and feeding activity of juvenile alewife, *Alosa pseudoharengus* Wilson, and blueback herring, A. *aestivalis* Mitchill, at near-surface depth in a hydroelectric dam impoundment. J. Fish Biol. 37: 813–822.

Kaufmann, R. & W. Wieser. 1992. Influence of temperature and ambient oxygen on the swimming energetics of cyprinid larvae and juveniles. Env. Biol. Fish. 33: 87–95.

Keast, A. 1978. Feeding interrelations between age groups of pumpkinseed (*Lepomis gibbosus*) and comparisons with bluegill (*Lepomis macrochirus*). J. Fish. Res. Board Can. 35: 12–27.

Lightfood, G.W. & N.V. Jones. 1979. The relationship between the size of 0 group roach (*Rutilus rutilus* (L.)), their swimming capabilities, and their distribution in a river. pp. 230–236. *In*: Proceedings of the First Brit. Freshwater Fish. Conf., Liverpool.

Mann, R.H.K. 1974. Observations on the age, growth, reproduction and food of the dace *Leuciscus leuciscus* (L.) in two rivers in Southern England. J. Fish Biol. 6: 237–253.

Mann, R.H.K. 1976. Observations on the age, growth, reproduction and food of the chub *Squalius cephalus* (L.) in the River Stour, Dorset. J. Fish Biol. 8: 265–288.

Mann, R.H.K. 1996. Environmental requirements of European non-salmonid fish in rivers. Hydrobiologia 323: 223–235.

Mark, W., R. Hofer & W. Wieser. 1987. Diet spectra and resource partitioning in the larvae and juveniles of three species and six cohorts of cyprinids from a subalpine lake. Oecologia 71: 388–396.

Mills, C.A. 1991. Reproduction and life history. pp. 483–508. *In*: I. Winfield & J.S. Nelson (ed) Cyprinid Fishes, Systematics, Biology and Exploitation, Fish and Fisheries Series 3, Chapman & Hall, London.

Mills, C.A. & R.H.K. Mann. 1985. Environmentally-induced fluctuations in year class strength and their implications for management. J. Fish Biol. 27(supp. A): 209–226.

Neveu, A. 1974. La dérive des stades aquatiques de quelques familles de diptères torrenticoles. Ann. Hydrobiol. 5: 15–42.

Neveu, A. & M. Échaubard. 1975. La dérive estivale des invertébrés aquatiques et terrestres dans un ruisseau du Massif Central, la Couze-Pavin. Ann. Hydrobiol. 6: 1–26.

Persson, L. 1991. Interspecific interactions. pp. 530–551. *In*: I. Winfield & J.S. Nelson (eds) Cyprinid Fishes, Systematics, Biology and Exploitation, Fish and Fisheries Series 3, Chapman & Hall, London.

Philippart, J.C. & M. Vranken. 1983. Atlas des poissons de Wallonie: distribution, écologie, éthologie, pêche, conservation. Cahiers d'Éthologie appliquée 3: 1–395.

Polis, G.A. 1984. Age structure component of niche width and intraspecific resource partitioning: can age groups function as ecological species? Amer. Nat. 123: 541–564.

Przybylski, M. 1996. The diel feeding pattern of bitterling, *Rhodeus sericeus amarus* (Block) in the Wieprz-Krzna Canal, Poland. Pol. Arch. Hydrobiol. 43: 203–212.

Reckendorfer, W., H. Keckeis, G. Winkler & F. Schiemer. 1996. *Chondrostoma nasus* as an indicator of river functioning: field and experimental studies at the University of Vienna. 2) Food availability for 0^+ fish in the Danube: spatial and seasonal changes in plankton, benthos and drift. pp. 25. *In*: J. Freyhof & A. Bischoff (ed.) Abstracts of the Second International Symposium on the Biology of the Genus *Chondrostoma Agassiz*, 1835, Museum Alexander Koenig, Bonn.

Rheinberger, V., R. Hofer & W. Wieser. 1987. Growth and habitat separation in eight cohorts of three species of cyprinids in a subalpine lake. Env. Biol. Fish. 18: 209–217.

Rincón, P.A., P. Barrachina & Y. Bernat. 1992. Microhabitat use by 0^+ juvenile cyprinids during summer in a Mediterranean river. Arch. Hydrobiol. 125: 323–337.

Sanders, R.E. 1992. Day versus night electrofishing catches from near-shore waters of the Ohio and Muskingum rivers. Ohio J. Sci. 92: 51–59.

Schiemer, F. & M. Zalewski. 1992. The importance of riparian ecotones for diversity and productivity of riverine fish communities. Neth. J. Zool. 42: 323–335.

Schiemer, F., T. Spindler, H. Wintersberger, A. Schneider & A. Chovanec. 1991. Fish fry associations: important indicators for the ecological status of large rivers. Verh. internat. Verein. Theor. Angew. Limnol. 24: 2497–2500.

Scott, A. 1987. Prey selection by juvenile cyprinids from running waters. Freshwat. Biol. 17: 129–142.

Scott, M.T. & L.A. Nielsen. 1989. Young fish distribution in backwaters and main channel borders of the Kanawha River, West Virginia. J. Fish Biol. 35(supp. A): 21–27.

Stroband, H.J.W. & K.R. Dabrowski. 1979. Morphological and physiological aspects of the digestive system and feeding in fresh-water fish larvae. pp. 355–374. *In*: M. Fontaine (ed.) La Nutrition des Poissons, CNRS Editions, Paris.

Thorpe, J.E. 1977. Morphology, physiology, behaviour, and ecology of *Perca fluviatilis* L. and *P. flavescens* Mitchill. J. Fish. Res. Board Can. 34: 1504–1514.

Tonn, W.M. & C. Paszkowski. 1987. Habitat use of the central mudminnow (*Umbra limmi*) and yellow perch (*Perca flavescens*) in *Umbra–Perca* assemblages: the walls of competition, predation, and the abiotic environment. Can. J. Zool. 65: 865–870.

Watkins, M.S., S. Doherty & G.H. Copp. 1997. Microhabitat use by 0^+ and older fishes in a small English chalk stream. J. Fish Biol. 50: 1010–1024.

Weatherley, N.S. 1987. The diet and growth of 0-group dace, *Leuciscus leuciscus* (L.), and roach, *Rutilus rutilus* (L.), in a lowland river. J. Fish Biol. 30: 237–247.

Environmental Biology of Fishes **56:** 199–212, 1999.

Seasonal shifts in day-time resource use of 0+ barbel, *Barbus barbus*

Antje Bischoff[a] & Jörg Freyhof[b]
[a] *Institute of Freshwater Ecology and Inland Fisheries, Deptartment of Biology and Ecology of Fishes, POB 850 119, 12561 Berlin, Germany (e-mail: bischoff@igb-berlin.de)*
[b] *Zoological Research Institute and Museum Alexander Koenig, Deptartment of Ichthyology, Adenauerallee 160, 53113 Bonn, Germany*

Received 12 September 1997   Accepted 17 July 1998

Key words: larvae, juveniles, body size, mesohabitat, microhabitat, riffles, diet

Synopsis

Habitat preference and diet of 0+ barbel were studied both on a meso- and a microhabitat scale in the River Sieg (Germany) between May 1993 and January 1995. Changes in mesohabitat use were observed. Barbel moved from shallow bays (larvae and step 1 juveniles) to gravel banks, and subsequently, to riffles (step 2 juveniles). Size-dependent shifts in microhabitat-use were observed during the second juvenile step. These juveniles left the shoreline and preferred microhabitats with stronger current velocities. 0+ barbel in riffles fed on Chironomidae and Ephemeroptera. No remarkable shifts in diet were detected between larvae and juveniles. We suggest that the observed ontogenetic shifts to habitats with high food supply and low predation pressure might contribute to the high abundances of barbel in the River Sieg.

Introduction

During their first year of life, 0+ fishes pass through a series of developmental steps characterized by anatomical and corresponding physiological traits that result in important shifts in resource use (Baltz & Moyle 1982, Balon 1984, 1986, Wieser 1991). The microhabitat use of 0+ fishes has received considerable attention over the last decade (Copp 1990, 1992, 1993, Copp et al. 1994, Baras et al. 1995, Jurajda 1995, Freyhof 1996, Garner 1996, Kurmayer et al. 1996, Staas & Neumann 1996, Watkins et al. 1997). However, only a few studies have taken into account the ontogenetic aspects (Copp 1990, 1992, 1997) and the size-dependency of shifts in meso- and microhabitat use (Copp 1990, Lightfoot & Jones 1996, Wintersberger 1996, Garner 1997a). Body size at a respective step of development is often considered to be the most crucial factor determining resource use during the first year of life (Kamler 1992). For example, diet usually becomes more diverse as fish size increases (Mills 1985); this can be attributed to gape-limited feeding processes in larvae and small juveniles. In addition, since maximum sustainable swimming speed is a direct function of length (Webb & Weihs 1986, Mills 1991, Wieser 1991, Sabo & Orth 1994), 0+ fish may be particularly vulnerable to floods (Harvey 1987, Mann & Bass 1997). As a result, the morphological constraints of the respective developmental steps of 0+ fish may be correlated with the environmental variables in their preferred habitat. In turn, ontogenetic niche shifts (sensu Werner & Gilliam 1984) may occur in response to changes in fish morphology during development.

The aim of our study was to determine the dynamics of resource use of both larvae and juveniles of barbel *Barbus barbus* (L.). This species was chosen because the emerging larvae of barbel are reported to leave the substratum at a relatively large body size (Peňáz 1973), and thus 0+ barbel may be able to perform changes in habitat use at an early interval. The question remains whether such shifts occur and if so, how these might contribute to the success of this species in the River

Sieg, where it is one of the most abundant species of the cyprinid community (Freyhof 1996).

Material and methods

The investigations were undertaken between May 1993 and January 1995 at the River Sieg, a tributary of the German stretch of the Lower River Rhine. The Sieg rises in the Rothaargebirge at 603.5 m. a. s. l. It meets the Rhine after 152.7 km at 48 m. a. s. l. close to Bonn-Beuel. We sampled fishes in the middle and lower part of the epipotamal Sieg. The river was approximately 20–50 m wide. The mean discharge in the last 45 years was 33 $m^3 s^{-1}$. The minimum discharge of about 4 $m^3 s^{-1}$ was usually reached in late summer (level recorder Siegburg-Kaldauen). Mean conductivity was 250 $\mu S cm^{-1}$ and water transparency (Secchi disc) was 60 cm during medium flow conditions. Average water temperature during the summer months from June to August was 19.6°C (mean of the years 1993 and 1994). Biological water quality was characterized as slightly polluted 'betamesosaprobic'. The main part of the river has no water retention structures (two dams provided with fish slopes as rocky ramps downstream of weirs and one with a stepped bypass are located in the study area). The predominant part of the shoreline embankment is regulated and fixed with basaltic rocks. The river bottom consists mainly of coarse gravel.

To analyse habitat shifts on a mesohabitat scale, six habitat types along the riverbank were defined (Table 1). To sample 0+ fish in both meso- and microhabitats with water <60 cm, a DEKA 3000 portable electrofishing unit with an anode of 15 cm diameter was activated at points selected in a random-stratified manner (see Copp & Peňáz 1988, Copp 1989) from the bank. Each sampling point was approached discretly to avoid disturbance. Prior to sampling, the activated electrode was swiftly immersed in the water at the sampling point and all fishes immobilized by the electric field were collected immediately with a separate dip net by a second operator. Fish were collected at 10–25 points within a selected homogenous area (with the sample number being based upon the area of these homogenous areas); fish from the 10–25 points were then treated as one sample. Distance between sampling points was at least 2 m. Because of the observed size selectivity of the galvanotaxis, the densities of fishes per point were calculated according to their total length (Table 2). Reaction distances to the electric field were observed in the field and noted together with the environmental variables (temperature, oxygen, depth, velocity) measured at each fishing point. The reported densities are thus relative to the surface area sampled as in Copp et al. (1994), see also Garner (1997b). Further, a small beach-seine (400 × 140 cm; 1.8 × 0.5 mm mesh size) for larvae and a higher beach-seine for juveniles (1000 × 190 cm; mesh size 2.5 × 2.0 mm) were used in all but riffle habitats. The beach seines were implemented from the bank. Samples were collected by moving the net through a quarter of a circle against the current. Normally, only one sample was collected by beach seine per mesohabitat. Beach seine data were standardised to m^2 using: area (m^2) = $(\pi \times a \times b)/4$, where a = distance from shoreline, and b = distance of the two operators at the shoreline.

All fishes were collected in the daylight hours between 10:30 and 16:00. From each sampling, captured fishes were anaesthetised with chlorobutanol (1.1.1-trichloro-2-methyl-2-propanol) and fixed in 4% formaldehyde. The mean number of individuals caught per mesohabitat type was calculated per square meter of habitat structure. For statistical analysis, electrofishing and beach seine data had to be pooled, as habitats deeper than 60 cm could not be sampled by point abundance sampling by electrofishing.

Additionally, data on the macroinvertebrate fauna of the River Sieg were received from A. Kopynske

Table 1. Environmental variables in different mesohabitats of the River Sieg.

Mesohabitat	Current velocity ($cm s^{-1}$)	Water depth (m)	Slope	Substratum
Backwater	0	0–1.2	steep	mud
River embankment (rip-rap)	0–40	0–1.0	very steep	basaltic rocks
Whirlpool	0–40	0–1.0	steep	gravel
Bay	0–10	0–0.7	very gentle	sand/detritus
Gravel bank	0–60	0–0.6	gentle	gravel
Riffle	0–120	0–0.3	very gentle	coarse gravel

Table 2. Distances of proper reaction (r) and calculated covered area for electrofishing points to calculate relative densities of 0+ fishes by size class of total length (TL) in mm.

TL (mm):	8–19	20–29	30–39	40–49	>50
r from anode ring (mm)	100	200	300	400	500
Ind × m^{-2} = catch × X	$X = 10.4$	$X = 4.2$	$X = 2.3$	$X = 1.4$	$X = 1$
Covered area (m^2)	0.1	0.2	0.4	0.7	1.0

(University of Bonn, Institute of Limnology, unpublished data) and H.-J. Weon (University of Bonn, Zoological Research Institute and Museum A. Koenig, unpublished data). Benthic fauna were sampled using a modified surber-sampler (mesh size: 0.7 mm × 0.7 mm, A. Kopynske, unpublished data) within three habitat structures (lotic, lenitic, lentic) combining methods described by Caspers (1972) and Hynes (1961). Invertebrate drift was sampled using a drift net of 1.60 m length (mesh size: 0.3 mm × 0.3 mm). The net was exposed once a month from June to December (exposition time: 60 min every 3 h over a 24 h period) in a lotic habitat. Furthermore, drift measurements were carried out in August 1995 in different river sections ('before riffle', 'in riffle', 'behind riffle', distance between the sections being 30 m). Drift density D was calculated according to Allan & Russek (1985) as:

$$D = \frac{\text{individuals} \cdot h^{-1}}{m^3 \cdot h^{-1}} \cdot 100.$$

In the laboratory, 0+ fish were identified (Koblitskaya 1981) and measured for total length (TL – tip of the snout to the end of the lower caudal lobe). Samples of fish collected from June to August 1994 in 'riffles' were stored for gut content analyses, because previous investigations have indicated that the larger size of that 0+ barbel captured in riffles, relative to other habitats, might be attributed to the greater food supplies in this habitat type (Freyhof 1996). A total of 80 specimens were examined qualitatively and quantitatively. Mean gut fullness was expressed as the ratio of the total gut length and the length of the section filled with food items. Afterwards, guts were dissected out under a light binocular microscope. The contents were embedded in Gelvatol (polyvinylalcohol) and analyzed using the numerical method (Windell & Bowen 1978, Hyslop 1980). Owing to the difficulty of identifying the ingested food, prey items were identified to family or broad taxonomic groups. To account for differences in gut capacity of barbel larvae and juveniles, the mean number of food items per mm gut length was determined for the main prey categories. Preferences for food types were calculated using specimens sampled during summer 1994. Prey preferences were displayed by the Relativized Electivity E_i^* (Vanderploeg & Scavia 1979). Unlike many alternative indices, E_i^* embodies a measure of the feeder's perception of a food's value as a function of both its abundance and the abundance of other food types present (Lechowicz 1982). E_i^* is calculated by

$$E_i^* = \frac{W_i - \left(\frac{1}{n}\right)}{W_i + \left(\frac{1}{n}\right)},$$

where n is the number of prey types available and W_i is estimated by

$$W_i = \frac{\frac{r_i}{p_i}}{\sum \frac{r_i}{p_i}},$$

where r_i and p_i are the percentage of prey type i in the diet and in the environment respectively; p_i was determined using drift and benthon samples collected in summer 1994 (methods as described above) together with the 0+ barbels in riffles. E_i^* ranges from −1 to +1, with 0 representing no preference. Dietary overlap between larvae and juveniles of barbel was compared using the index proposed by Schoener (1970):

$$O = 1 - \frac{0.5 \sum |p_a - p_b|}{100},$$

where p_a = percentage of a food item in species a, and p_b = percentage of a food item in species b. The index produces values between 0 (no overlap) and 1 (complete overlap).

To detect differences in meso- and microhabitat use, 0+ barbel were assigned to different groups based on their developmental steps following the definition of

Table 3. Categories of environmental variables, number of samples containing 0+ barbel of the respective developmental interval [larvae, J1 = first juvenile step, J2 = second juvenile step following Krupka (1988)] and of the size group in the second juvenile step, respectively (J2A–J2G), number of zero samples and total number of samples in each category for each environmental variable in the 1060 point samples from the River Sieg in the period from May 1993 to January 1995.

Interval		Larvae			J1			J2A			J2B		
Variable/Category		Occupied	Null	Total samples	Occupied	Null	Total samples	Occupied	Null	Total samples	Occupied	Null	Total samples
Velocity (cm s^{-1})													
0	1	33	56	89	22	35	57	29	94	123	20	116	136
1–5	2	18	22	40	24	29	53	29	47	76	23	54	77
6–10	3	10	22	32	15	20	35	18	36	54	18	35	53
11–20	4	10	32	42	13	37	50	25	75	100	36	58	94
21–40	5	7	51	58	8	79	87	31	88	119	37	68	105
41–60	6	1	41	42	2	71	73	16	93	109	36	64	100
61–80	7	0	17	17	1	40	41	8	83	91	15	96	111
81–100	8	0	4	4	0	19	19	2	58	60	5	69	74
101–120	9	0	2	2	1	9	10	0	18	18	0	16	16
>120	10	0	1	1	0	3	3	0	6	6	0	8	8
Distance (m)													
<1.0	1	53	78	131	30	43	73	44	93	137	42	107	149
1.0–2.9	2	19	59	78	19	73	92	39	120	159	47	112	159
3.0–4.9	3	2	40	42	11	55	66	19	78	97	22	73	95
5.0–6.9	4	0	27	27	4	50	54	14	77	91	22	67	89
7.0–9.9	5	5	22	27	20	57	77	28	114	142	33	117	150
>9.9	6	0	22	22	2	64	66	14	116	130	24	105	129
Depth (m)													
<0.2	1	55	147	202	57	196	253	100	263	363	117	206	323
0.2–0.6	2	23	86	109	20	119	139	42	261	303	56	286	342
>0.6	3	1	15	16	9	27	36	16	74	90	17	92	109
Slope													
Very low	1	45	160	205	47	287	334	86	436	522	116	381	497
Low	2	14	35	49	7	13	20	21	21	42	20	38	58
Moderate	3	2	8	10	2	3	5	3	7	10	5	12	17
Steep	4	1	13	14	3	17	20	7	32	39	11	41	52
Substratum													
Basaltic rocks	1	10	22	32	9	64	73	25	156	181	44	154	198
Gravel	2	52	204	256	66	263	329	113	393	506	130	365	495
Plants	3	1	1	2	0	0	0	0	0	0	0	1	1
Sand	4	2	9	11	5	3	8	4	17	21	3	25	28
Detritus	5	12	2	14	6	0	6	13	6	19	10	8	18
Mud	6	2	10	12	0	12	12	3	26	29	3	31	34
Exposition													
Shady	1	1	11	12	1	1	2	2	7	9	1	8	9
Half-shady	2	5	10	15	9	21	30	12	30	42	7	33	40
Sunny	3	73	227	300	76	320	396	144	561	705	182	543	725
Mesohabitat													
Backwater	1	1	9	10	1	14	15	2	26	28	2	29	31
Embankment	2	8	27	35	11	25	36	17	56	73	15	67	82
Whirlpool	3	0	2	2	1	6	7	3	8	11	6	9	15
Shallow bay	4	20	36	56	16	7	23	25	35	60	21	46	67
Gravel bank	5	12	22	34	10	9	19	26	31	57	28	37	65
Riffle	6	38	152	190	47	281	328	85	442	527	118	394	512

J2C			J2D			J2E			J2F			J2G		
Occupied	Null	Total samples	Occupied	Null	Total samples	Occupied	Null	Total samples	Occupied	Null	Total samples	Occupied	Null	Total samples
12	118	130	3	102	105	3	83	86	1	71	72	2	70	72
8	59	67	6	43	49	2	33	35	2	28	30	0	30	30
11	31	42	6	25	31	0	25	25	0	18	18	0	18	18
32	58	90	12	44	56	6	36	42	2	36	38	3	35	38
28	68	96	11	47	58	11	15	26	3	23	26	2	24	26
43	50	93	19	54	73	8	38	46	4	38	42	0	42	42
31	74	105	31	59	90	23	48	71	14	57	71	14	57	71
17	56	73	8	56	64	6	46	52	5	43	48	2	46	48
0	15	15	1	14	15	2	6	8	1	7	8	1	7	8
1	7	8	0	7	7	0	5	5	0	5	5	1	4	5
42	107	149	35	110	145	25	95	120	6	73	79	5	74	79
47	112	159	39	106	145	25	88	113	6	68	74	7	67	74
22	73	95	25	60	85	14	52	66	5	38	43	1	42	43
22	67	89	15	65	80	8	53	61	4	39	43	1	42	43
33	117	150	33	104	137	11	97	108	4	61	65	4	61	65
24	105	129	34	90	124	11	66	77	5	46	51	5	46	51
94	186	280	48	132	180	30	90	120	10	100	110	8	102	110
75	255	330	45	222	267	29	175	204	21	168	189	14	175	189
14	95	109	4	97	101	2	70	72	1	58	59	3	56	59
119	332	451	61	251	312	37	171	208	18	190	208	8	200	208
14	44	58	7	51	58	1	37	38	1	37	38	0	38	38
2	15	17	1	16	17	0	12	12	1	11	12	0	12	12
44	52	1	51	52	1	31	32	1	31	32	2	30	32	8
45	153	198	31	158	189	27	102	129	13	103	116	8	108	116
128	314	442	63	231	294	32	176	208	18	178	196	17	179	196
0	1	1	0	1	1	0	1	1	0	1	1	0	1	1
2	26	28	1	24	25	0	22	22	1	15	16	0	16	16
7	9	16	2	9	11	2	9	11	0	6	6	0	6	6
1	33	34	0	28	28	0	25	25	0	23	23	0	23	23
0	9	9	0	7	7	0	7	7	0	5	5	0	5	5
5	35	40	1	32	33	0	18	18	0	18	18	0	18	18
178	492	670	96	412	508	61	310	371	32	303	335	25	310	335
1	30	31	0	25	25	0	19	19	0	17	17	0	17	17
8	72	80	2	70	72	1	48	49	1	37	38	1	37	38
5	10	15	1	13	14	0	8	8	0	8	8	1	7	8
12	53	65	4	54	58	2	44	46	1	34	35	1	34	35
29	34	63	15	39	54	5	39	44	2	36	38	2	36	38
128	335	463	75	248	323	53	175	228	28	192	220	20	200	220

Krupka (1988). Owing to the low number of 0+ barbel available belonging to the 1st–4th larva step, these specimens were pooled with those of the 5th step, resulting in one group of larvae (TL < 20 mm). The 5th larva step ends with the complete disappearance of the embryonic finfold and the completed formation of the fin apparatus (Krupka 1988). Juvenile specimens were divided up in two steps. The first juvenile step (J1, TL ≤ 29 mm) is characterized (inter alia) by the development of the second pair of barbels and the complete disappearance of the embryonic finfold, the second juvenile step begins when the caudal peduncle is covered with scales and scales in the place of the future lateral line and in four rows below are developed all along the body (Krupka 1988). We justified a further distinction of 0+ juveniles into size classes because they are able to move actively to microhabitats where they can most successfully forage and avoid predators (Boehlert & Mundy 1988) and because swimming ability is a function of length (Mann & Bass 1997). Thus, specimens in the second juvenile step were additionally assigned to seven size classes (J2A = 30–39 mm TL, J2B = 40–49 mm TL, J2C = 50–59 mm TL, J2D = 60–69 mm TL, J2E = ≤79 mm TL, J2F = 80–89 mm TL and J2G = 90–100 mm TL).

Although not an arbitrary choice (see above), we recognize that the subdivision in 10 mm classes is a generalisation imposed to facilitate data analysis. The mean number of fish per developmental interval or size class and sample was calculated. Temporal variability in fish density was determined by comparing the mean number of individuals per sample for the months June to October. Abundances in 1993 and 1994 were not significantly different (Jörg Freyhof unpublished data), so these samples were pooled. Furthermore, the index of dispersion was calculated. Quantitative environmental variables were converted to semi-quantitative categories (Table 3), and species-variable associations were assessed by chi-square (χ^2) analysis. Electivity indices were calculated to determine preference/avoidance between steps and environmental variables, where the index was the frequency of a developmental step and size class respectively in the group of samples having a given category of variable and the frequency of that developmental step/size-group in all samples (negative values indicate avoidance, positive preference). Spearman rank correlation coefficients were calculated to determine which environmental variables correlated most with the frequencies of 0+ barbel. Differences in habitat use between steps/size classes were tested using the Mann-Whitney U-test.

Results

A total of 1060 samples was collected of which 472 samples contained 3580 0+ barbel (Table 3). The mean number of barbel per sample was greatest for larvae (TL < 20 mm) and then decreased rapidly (Figure 1a). The index of dispersion indicates that barbel larvae exhibited the greatest tendency to aggregate (Figure 1b). This inclination decreased as 0+ barbel grew in length and a size-related pattern of abundance was observable during the first year of life (Figure 2). Abundance was highest in the second half of June and declined continuously during the summer and autumn.

Shallow riverbays were the most important habitats for barbel larvae (Figure 3), although they were

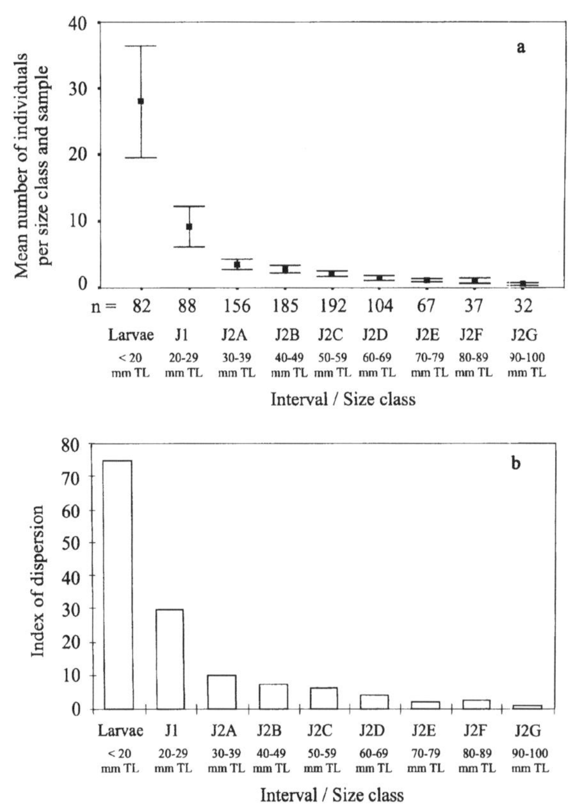

Figure 1. a – Mean number of individuals per size class and sample, and b – index of dispersion for developmental intervals of 0+ barbel [larvae, J1 = first juvenile step, J2 = second juvenile step following Krupka (1988), J2A–J2G = size classes, definition of intervals and size groups given in the text] in the River Sieg during the period May 1993 to October 1994. SE bars given, n = number of samples.

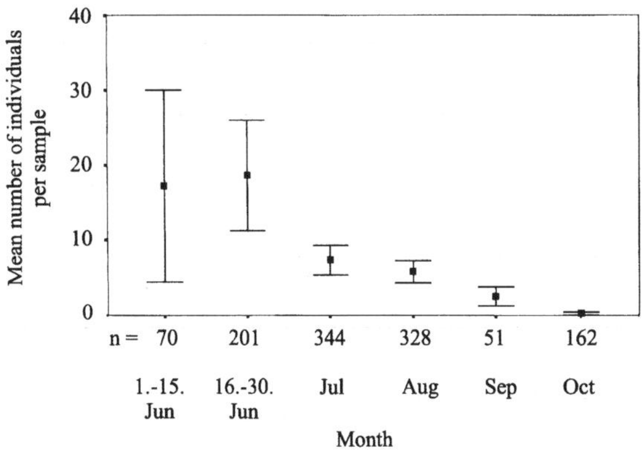

Figure 2. Monthly variation in the mean number of 0+ barbel per sample in the River Sieg during the period May 1993 to October 1994. SE bars given, n = number of samples.

also caught in the marginal zones of other mesohabitats. Barbel avoided all habitats characterized by deep water (backwaters, whirlpools and bank reinforcements). Significant differences with respect to the category 'mesohabitat' were evident between barbel larvae and larger juveniles (J2B–J2F, Table 4). Small juveniles J1 were still more often associated with bay habitats than with shallow gravelbanks, whereas for larger juveniles it was the other way around, with J2C–J2F (>49 mm) being no longer found in bays. As body size increased further, J2 barbel shifted successively to riffles with high current velocities. Almost all juveniles >59 mm TL (J2D–J2F) were captured in riffle habitats.

Differences in microhabitat use were observed with different life history intervals (Figure 3). Barbel larvae (<20 mm TL) were found associated exclusively with the marginal zone, where current velocity was lowest. Microhabitat use by larvae was significantly different from all juvenile intervals with regard to both distance from bank and current velocity (Table 4). J2 juveniles used areas with greater current velocity and this shift is evident in J2C juveniles from the inversion of the correlation coefficient values (Figure 3). Owing to the high number of null-samples, Spearman rank correlation coefficients were rather low. J2E and J2F juveniles (70–89 mm TL) demonstrated preferences for current velocities up to 120 cm s^{-1} (category 9); J2G juveniles preferred even greater current velocity (category 10). Water depth and bank slope influenced the distribution patterns of 0+ barbel. Almost all larvae and juveniles <70 mm were captured in very shallow water (<0.2 m). Larger J2 juveniles (J2F and J2G) were collected in slightly deeper microhabitats (Figure 3), a statistically significant change for 0+ larvae and juveniles up to a size of 49 mm (Table 4). Barbel larvae and smaller juveniles preferred habitats with moderate bank slopes, whereas larger J2 juveniles shifted to very low bank slope, a variable for which size groups J2C–J2G exhibited no significant differences. The change in bank slope concorded with a progressive move away from bankside habitats with fine substrata to the coarser substrata of riffles.

Mean gut fullness of 0+ barbel larvae was significantly higher than that of J2 juveniles (Mann-Whitney U-test, $p < 0.024$). The most important insect orders in the diet of 0+ barbel were Ephemeroptera (barbel larvae: 74.4%, juveniles: 53.4%) and Diptera, especially chironomid larvae (barbel larvae: 25.1%, juveniles: 41.4%). Mean number of prey items per 1 mm gut length of barbel larvae ($n = 38$) was 1.5 for Ephemeroptera, 0.5 for chironomid larvae and 0.01 for all other categories combined, in J2 juveniles ($n = 27$, all sizes combined) these values amounted 1.8, 1.3 and 0.2 respectively. Differences in the mean number of prey items per 1 mm gut length of barbel larvae and J2 juveniles were significant with respect to the prey categories 'chironomid larvae' (Mann-Whitney U-test, $p < 0.0265$) and 'others' (Mann-Whitney U-test, $p < 0.0001$). Barbel larvae used almost exclusively Ephemeroptera and chironomid larvae; avoidance was indicated for all other prey categories. J2 juveniles preferred Ephemeroptera, chironomid larvae and Hydracarina, whereas the latter food component was avoided by barbel larvae. Chironomid larvae were the most preferred food items of both larvae and juveniles, indicated by high values of E_i^* (Table 5). The high dietary overlap (0.78) indicates that diet spectra between larvae and juveniles were similar. The distributions of invertebrates differed in the different river sections of the Sieg (Figure 4). The densities of benthic organisms were significantly lower in both lenitic (t-test, $p < 0.01$) and in lentic than in lotic stretches (Figure 4a). The rate of drifting invertebrates (Figure 4b) displayed significant differences among sections (χ^2-test, $p = 0.011$), and changes in the taxonomic composition of drift were apparent between June and December of 1995 (Figure 5).

Discussion

Based on daytime samples only, 0+ barbel exhibited important niche shifts during their first summer in the River Sieg (Figure 3, Table 4). Food and space may

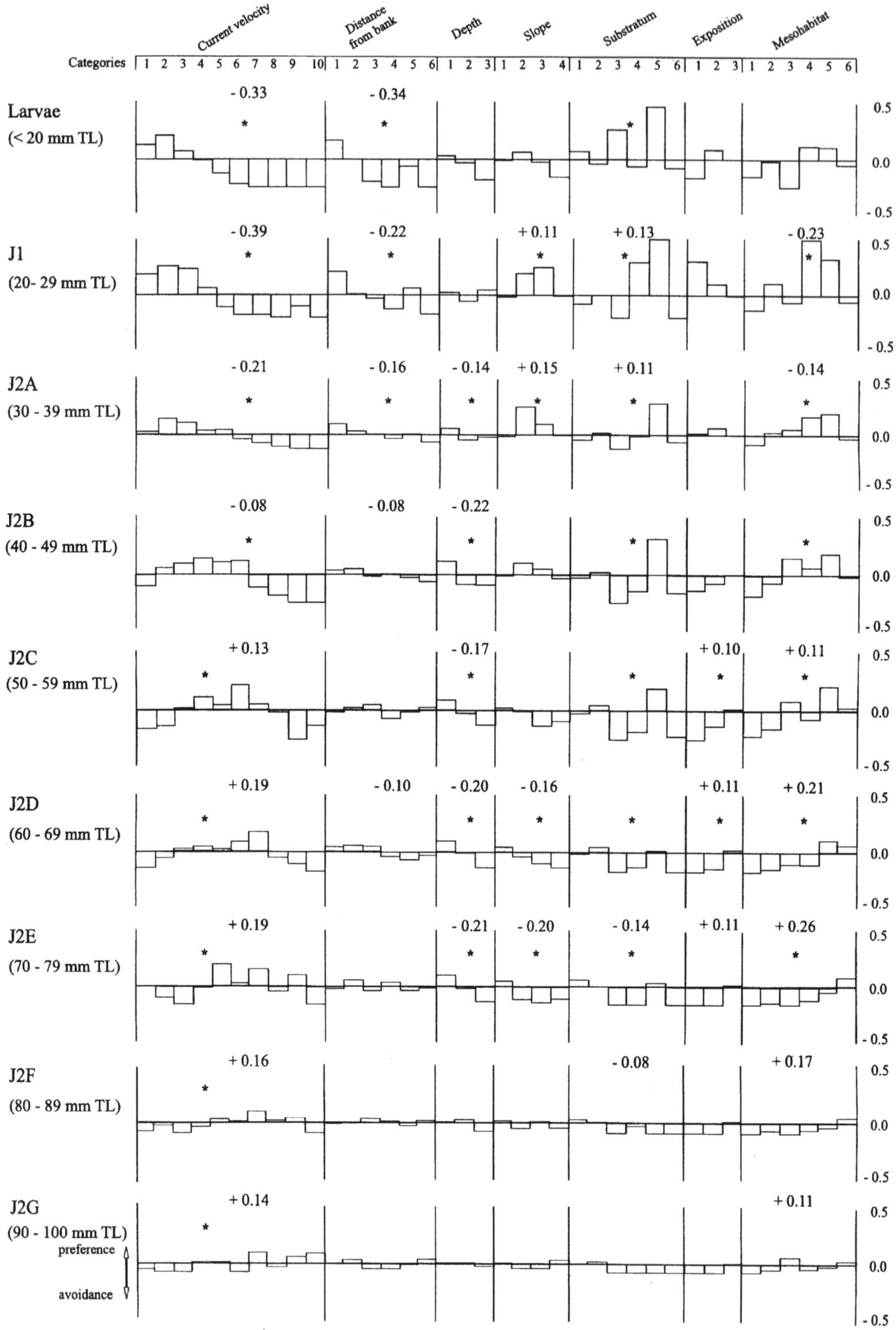

Figure 3. Environmental profiles and χ^2 deviations from expected for developmental steps of 0+ barbel in the River Sieg during the period May 1993 to January 1995. Each histogram represents the difference between the frequency of that developmental interval and size class, respectively, in the group of samples having that category of environmental variable and the frequency of that interval/size class in all samples (abbreviations as in Figure 1). Significant χ^2 associations between intervals/size groups and variables are indicated with an asterisk. Significant Spearman rank correlation coefficients ($p < 0.05$) reflecting the correlation of environmental variables and frequencies of 0+ barbel are given above the histograms. Categories of environmental variables as in Table 3.

Table 4. Between size class comparisons (Mann-Whitney U-test) of microhabitat use for developmental steps of 0+ barbel (abbreviation of steps as in Table 3) of the River Sieg during May 1993 and January 1995.

Interval	Size		J1 (≤29 mm)	J2A (≤39 mm)	J2B (≤49 mm)	J2C (≤59 mm)	J2D (≤69 mm)	J2E (≤79 mm)	J2F (≤89 mm)	J2G (≤100 mm)
Larvae	(<20 mm)	Current velocity	0.050	0.000	0.000	0.000	0.000	0.000	0.000	0.000
		Distance from bank	0.000	0.000	0.000	0.000	0.000	0.000	0.000	0.000
		Depth	n. s.	n. s.	n. s.	0.005	0.025	0.025	0.000	0.000
		Slope	n. s.	n. s.	n. s.	n. s.	0.050	0.025	n. s.	n. s.
		Substratum	n. s.	n. s.	0.002	0.000	0.000	0.000	0.000	0.000
		Exposition	n. s.	n. s.	n. s.	n. s.	0.05	n. s.	n. s.	n. s.
		Mesohabitat	n. s.	n. s.	0.050	0.000	0.000	0.000	0.000	0.025
J1	(≤29 mm)	Current velocity		0.025	0.000	0.000	0.000	0.000	0.000	0.000
		Distance from bank		n. s.	n. s.	0.025	n. s.	n. s.	n. s.	n. s.
		Depth		n. s.	n. s.	n. s.	n. s.	n. s.	0.025	0.025
		Slope		n. s.	n. s.	n. s.	n. s.	0.050	n. s.	n. s.
		Substratum		n. s.	0.025	0.025	0.000	0.000	0.000	0.000
		Exposition		n. s.	0.050	0.025	0.025	0.025	0.050	n. s.
		Mesohabitat		n. s.	n. s.	0.025	0.000	0.000	0.025	0.025
J2A	(≤39 mm)	Current velocity			0.025	0.000	0.000	0.000	0.000	0.000
		Distance from bank			n. s.	0.025	n. s.	n. s.	n. s.	n. s.
		Depth			n. s.	n. s.	n. s.	n. s.	0.025	0.025
		Slope			n. s.	n. s.	0.025	0.025	n. s.	n. s.
		Substratum			0.05	0.025	0.000	0.000	0.025	0.025
		Exposition			n. s.	0.025	0.025	0.025	n. s.	n. s.
		Mesohabitat			n. s.	0.025	0.000	0.000	0.000	0.025
J2B	(≤49 mm)	Current velocity				0.000	0.000	0.000	0.000	0.000
		Distance from bank				n. s.	n. s.	n. s.	n. s.	n. s.
		Depth				n. s.	n. s.	n. s.	0.025	0.025
		Slope				n. s.	0.050	0.025	n. s.	n. s.
		Substratum				n. s.	0.050	0.025	0.050	n. s.
		Exposition				n. s.	n. s.	n. s.	n. s.	n. s.
		Mesohabitat				n. s.	0.025	0.025	0.025	0.050
J2C	(≤59 mm)	Current velocity					n. s.	0.025	0.025	0.025
		Distance from bank					0.025	n. s.	n. s.	n. s.
		Depth					n. s.	n. s.	n. s.	n. s.
		Slope					n. s.	n. s.	n. s.	n. s.
		Substratum					n. s.	0.025	n. s.	n. s.
		Exposition					n. s.	n. s.	n. s.	n. s.
		Mesohabitat					n. s.	0.025	0.050	n. s.

Table 4. (*Continued*)

Interval	Size		J1 (≤29 mm)	J2A (≤39 mm)	J2B (≤49 mm)	J2C (≤59 mm)	J2D (≤69 mm)	J2E (≤79 mm)	J2F (≤89 mm)	J2G (≤100 mm)
J2D	(≤69 mm)	Current velocity						n. s.	0.050	0.050
		Distance from bank						n. s.	n. s.	n. s.
		Depth						n. s.	n. s.	n. s.
		Slope						n. s.	n. s.	n. s.
		Substratum						n. s.	n. s.	n. s.
		Exposition						n. s.	n. s.	n. s.
		Mesohabitat						n. s.	n. s.	n. s.
J2E	(≤79 mm)	Current velocity							n. s.	n. s.
		Distance from bank							n. s.	n. s.
		Depth							n. s.	n. s.
		Slope							n. s.	n. s.
		Substratum							n. s.	n. s.
		Exposition							n. s.	n. s.
		Mesohabitat							n. s.	n. s.
J2F	(≤89 mm)	Current velocity								n.s.
		Distance from bank								n.s.
		Depth								n.s.
		Slope								n.s.
		Substratum								n.s.
		Exposition								n.s.
		Mesohabitat								n.s.

be regarded as resources because they represent quantities that can be reduced by an organism's activities (Begon et al. 1991). The value of 'space' is determined by factors such as current velocity, riparian or instream cover etc. Thus, depending on body size and ontogenetic development, 0+ barbel are able to exploit a variety of these resources.

At the mesohabitat level, barbel larvae and J1 juveniles were mainly restricted to shallow riverbays (Figure 3), which is similar to Wintersberger's (1996) observations for 0+ barbel in the Danube River. Water velocity is a dominant factor in the ecology of cyprinid larvae (Mills 1991), and this is evident in our study in which barbel larvae preferred low or no water velocities. Similarly, Mann & Mills (1986) found that newly hatched dace, *Leuciscus leuciscus*, were confined to riparian habitats in the River Frome where water velocities were $<2.0\,\mathrm{cm\,s^{-1}}$. Also important for barbel larvae are shallow areas (Figure 3), which provide efficient shelter from predators (Power et al. 1985, Schlosser 1987, Copp 1992).

With increasing body size, the ecological requirements of young fish undergo important changes (Miller et al. 1988). The move of 0+ barbel juveniles to gravel bank habitats we observed during the present study coincides with the completion of the fin apparatus (J1, 20 mm TL), which enables the fish to move more swiftly (Krupka 1988). In fact, Rincon et al. (1992) and Lightfoot & Jones (1996) found similar patterns in different river systems (wider use of the stream), which corresponded with increased swimming capacity (Wintersberger 1996). These shifts may also favour predator avoidance and reduce interspecific competition. Riffles for instance offer an environment where piscivores are frequently absent, and most of the accompanying 0+ cyprinids are not able to follow juvenile barbel into areas with very high current velocities (Antje Bischoff unpublished data). In summary, barbel have the potential to go through several mesohabitat shifts during their first summer, when suitable lotic habitats are available (see Freyhof 1996, Wintersberger 1996).

At the microhabitat level, our results for 0+ barbel can be interpreted only partly in light of previous investigations. Some studies (e.g. Watkins et al. 1997) have reported barbel larvae and 0+ juveniles in a small stream to prefer lentic habitats, whereas other studies of 0+ juvenile barbel in larger rivers revealed preferences for lotic habitats (Copp et al. 1994, Baras et al. 1995). Our findings, however, reveal that analyses of microhabitat use must take into account ontogenetic development, because 0+ barbel larvae and juveniles of different stages and sizes differ in their ecological requirements (Figure 3, Table 4). This is corroborated by the results of Wintersberger (1996). Microhabitat use of 0+ barbel is strongly influenced by water velocity (Figure 3, Copp et al. 1994, Wintersberger 1996). A shift in habitat use from low water velocity to areas with elevated water velocity can be attributed to increases not only in swimming ability but also in foraging efficiency; in particular the move by J2 juveniles to riffle habitats.

Macroinvertebrates represent the most important food resource for the cyprinid community in the River Sieg (Freyhof 1996), with densities of both benthic and drifting organisms being highest in lotic habitats (Figure 4). Thus, the probability of encountering macroinvertebrate prey items will be higher in riffles than in other habitats. In such conditions, emigrating to areas with higher current velocities, that is to say a shift to profitable feeding positions, may be an adaptive style for juvenile 0+ barbel. Although juveniles shift to higher currents, they can also use lentic zones within a given microhabitat. The bottom dwelling 0+ barbel may be able to decrease swimming costs by using micro turbulences behind coarse stones (Freyhof 1996). The preference for coarse substrata by larger individuals (Figure 3) agrees with these suggestions. However, clear evidence for a trade-off between a high energy intake (due to high prey densities in riffles) and energy expenditures (due to high swimming costs) cannot be derived from our data. Further investigations, for instance experiments as shown for minnow *Phoxinus phoxinus* by Garner (1998), are needed to provide more detailed information on the factors determining shifts in habitat use of 0+ barbel.

The diet of barbel larvae and J2 juveniles was similar to that of the adults (Adámek & Obrdlík 1977, Losos et al. 1980), being specialized for feeding on Diptera larvae in the sediments or organisms associated with the sediments (Granando-Lorencio & Garcia Novo 1986), and of 0+ barbel in the Danube River (Thomas Spindler unpublished data). The significantly higher gut fullness of larvae can be attributed to the fact that macroinvertebrate densities were highest in June and decreased considerably during summer (Figure 5). Barbel larvae and juveniles captured in riffles of the River Sieg both selected only two main prey categories out of a broad spectrum of food items available (Table 5). Note, however, that this behaviour may be more advantageous in

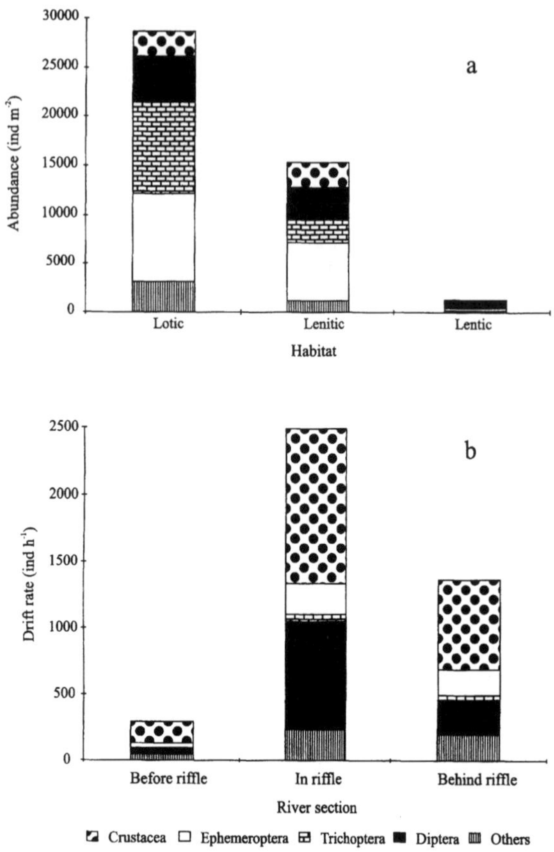

Figure 4. a – Benthic macroinvertebrate abundances (individuals per m^2) in three different river sections (number of samples in each section = 12) of the River Sieg. Each bar represents the sum of samples from the months June and August 1993 (redrawn from unpublished data from A. Kopynske). b – Invertebrate drift rate (individuals per 1 h) before, in, and behind a riffle habitat of the River Sieg in August 1995 (redrawn from unpublished data from H.-Y. Weon).

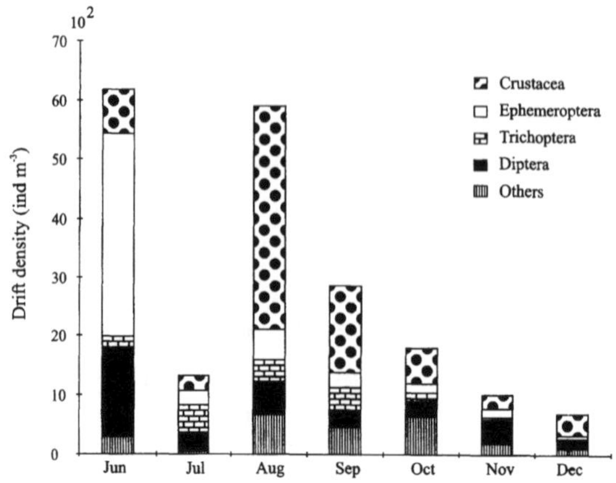

Figure 5. Seasonal development of the drift density in a riffle of the River Sieg from June to December 1995 (redrawn from unpublished data from H.-Y. Weon).

riffles than in other habitats. We presume that foraging in riffles is cost effective for 0+ barbel, because these habitats are highly productive and harbour high densities of bottom dwelling and drifting benthic organisms, especially chironomids and ephemeroptera (Figure 4, Lister & Genoe 1970).

Lobón-Cervía & De Diego (1988) regard barbel as a facultative generalist with the ability to forage on the highest diversity of prey when it is accessible (generalist) and to move to one or very few species only when these become more abundant in relation to the abundance of other species in that particular environment (facultative). Similar considerations could explain the diet spectra observed during our study, and the feeding behaviour of larvae and J2 juveniles should then be attributed to the second case. This might also explain why no differences were found in the diet of barbel larvae and juveniles. However, more detailed data on the prey availability (i.e. horizontal and vertical

Table 5. Food selection (Vanderploeg & Scavia's electivity E_i^*) by 0+ barbel larvae and J2 juveniles (all sizes combined) in a riffle of the River Sieg during summer 1994 (abbreviation of steps as in Table 3).

Category	E_i^*	
	Barbel larvae	Barbel J2
Nematoda	−1.000	–
Bivalvia	−1.000	−1.000
Oligochaeta	−1.000	−1.000
Hirudinea	−1.000	−1.000
Isopoda	−1.000	−1.000
Amphipoda	−1.000	−1.000
Ephemeroptera larvae	0.660	0.360
Plecoptera larvae	−1.000	−1.000
Coleoptera larvae	−1.000	−1.000
Coleoptera imagines	−1.000	−1.000
Trichoptera larvae	−1.000	0.006
Chironomidae larvae	0.840	0.790
Chironomidae pupae	−1.000	−0.250
Simulidae larvae	−1.000	−0.620
Dixidae larvae	−1.000	−1.000
Anflug	−1.000	−0.830
Hydracarina	−0.410	0.740
Ostracoda	−1.000	−0.060
Copepoda	−1.000	−1.000
Cladocera	−1.000	–
Fishes	−1.000	−1.000
Fish eggs	−1.000	−1.000

distribution, density and size composition of prey) as well as diel foraging patterns are needed to confirm these suggestions, with regard not only to size, but also the position of fins, mouth gape, development of the digestive system to feeding habits during ontogeny (Mérigoux & Ponton 1998). We conclude from our results that the exploitation of a range of meso- and microhabitats in the Sieg, in particular riffle habitats with a high food supply, by different stages of 0+ barbel may indeed contribute to the success of the species in this river.

Acknowledgements

The authors would like to thank two anonymous referees, the guest editors and T. Mehner for helpful comments on an earlier version of this manuscript. We are also indebted to G.H. Copp for valuable advice and for support in attending the workshop. The project was funded by the Ministry of Environment, Environmental planning and Agriculture of North Rhine-Westfalia (Germany).

References cited

Adámek, Z. & P. Obrdlík. 1977. Food of important cyprinid species in the warmed barb-zone of the Oslava River. Folia Zool. 26: 171–182.

Allan, J.D. & E. Russek. 1985. The quantification of stream drift. Can. J. Fish. Aquat. Sci. 42: 210–215.

Balon, E.K. 1984. Reflections on some decisive events in the early life of fishes. Trans. Amer. Fish. Soc. 113: 178–185.

Balon, E.K. 1986. Saltatory ontogeny and evolution. Riv. Biol. 79: 151–190.

Baltz, D.M. & P.B. Moyle. 1982. Life history characteristics of tule perch (*Hysterocarpus traski*) populations in contrasting environments. Env. Biol. Fish. 7: 229–242.

Baras, E., J. Nindaba & J.C. Philippart. 1995. Microhabitat use in a 0+ rheophilous cyprinid assemblage: quantitative assessment of community structure and fish density. Bull. fr. Pêche Piscicult. 337/9: 241–247.

Begon, M., J.L. Harper & C.R. Townsend. 1991. Ökologie – Individuen, Populationen, Lebensgemeinschaften. Birkhäuser Verlag, Basel. 1024 pp.

Boehlert, G.W. & B.C. Mundy. 1988. Roles of behavioral and physical factors in larval and juvenile fish recruitment to estuarine nursery areas. Amer. Fish. Soc. Symp. 3: 51–67.

Caspers, N. 1972. Ökologische Untersuchungen der Invertebratenfauna von Waldbächen des Naturparks Kottenforst Ville. Decheniana 112: 189–218.

Copp, G.H. 1989. Electrofishing for fish larvae and 0+ juveniles: equipment modifications for increased efficiency with short fishes. Aqua. Fish. Manag. 20: 453–462.

Copp, G.H. 1990. Shifts in the microhabitat of larval and juvenile roach *Rutilus rutilus* (L.), in a floodplain channel. J. Fish Biol. 36: 683–692.

Copp, G.H. 1992. Comparative microhabitat use of cyprinid larvae and juveniles in a lotic floodplain channel. Env. Biol. Fish. 33: 181–193.

Copp, G.H. 1993. The upper Rhône revisited: an empirical model of microhabitat use by 0+ juvenile fishes. Folia Zool. 42: 329–340.

Copp, G.H. 1997. Microhabitat use of fish larvae and 0+ juveniles in a highly regulated section of the River Great Ouse. Reg. Riv. 13: 267–276.

Copp, G.H. & M. Peňáz. 1988. Ecology of fish spawning and nursery zones in the floodplain, using a new sampling approach. Hydrobiologia 169: 209–224.

Copp, G.H., G. Guti, B. Rovný & J. Černý. 1994. Hierarchical analysis of habitat use by 0+ juvenile fish in Hungarian/Slovak flood plain of the Danube River. Env. Biol. Fish. 40: 329–1994.

Freyhof, J. 1996. Distribution of YOY-barbel *Barbus barbus* (L.) in the River Sieg/Germany. pp. 259–267. *In*: A. Kirchhofer & D. Hefti (ed.) Conservation of Endangered Freshwater Fish in Europe, Birkhäuser Verlag, Basel.

Garner, P. 1996. Microhabitat use and diet of 0+ cyprinid fishes in a lentic, regulated reach of the River Great Ouse, UK. J. Fish Biol. 48: 367–382.

Garner, P. 1997a. Habitat use by 0+ cyprinid fish in the River Great Ouse, East Anglia. Freshwat. Forum 8: 2–27.

Garner, P. 1997b. Sample sizes for length and density estimation of 0+ fish when using point sampling by electrofishing. J. Fish Biol. 50: 95–106.

Garner, P. 1998. Use of shallow marginal habitat by *Phoxinus phoxinus*: a trade-off between temperature and food? J. Fish Biol. 52: 600–609.

Granado-Lorencio, C. & F. García-Novo. 1986. Feeding habits of the fish community in a eutrophic reservoir in Spain. Ekologia Polska 34: 95–110.

Harvey, B.C. 1987. Susceptibility of young-of-the-year fishes to downstream displacement by flooding. Trans. Amer. Fish Soc. 116: 851–855.

Hynes, H.B.N. 1961. The invertebrate fauna of a Welsh mountain stream. Arch. Hydrobiol. 57: 344–388.

Hyslop, E.J. 1980. Stomach contents analysis – a review of methods and their application. J. Fish Biol. 17: 411–429.

Jurajda, P. 1995. Effect of channelization and regulation on fish recruitment in a floodplain river. Reg. Riv. 10: 207–215.

Kamler, E. 1992. Early life history of fish. An energetics approach. Chapman & Hall, London. 267 pp.

Koblitskaya, A.F. 1981. Identification keys for young of freshwater fishes. Consumer and Food Industry Press, Moscow. 208 pp. (in Russian).

Krupka, I. 1988. Early development of the barbel [*Barbus barbus* (Linnaeus, 1758)]. Práce Ústaru Rybárstva a Hydrobiológie (Bratislava) 6: 115–138.

Kurmayer, R., H. Keckeis, S. Schrutka & I. Zweimüller. 1996. Macro- and microhabitats used by 0+ fish in a side-arm of the River Danube. Arch. Hydrobiol. (Suppl. 113, Large Rivers) 10: 425–432.

Lechowicz, M.J. 1982. The sampling characteristics of electivity indices. Oecologia 52: 22–30.

Lightfoot, G.W. & N.V. Jones. 1996. The relationship between the size of 0+ roach, *Rutilus rutilus*, their swimming capabilities, and distribution in an English river. Folia Zool. 45: 355–360.

Lister, D.B. & H.S. Genoe. 1970. Stream habitat utilisation by cohabiting underyearlings of chinook (*Oncorhynchus tshawytscha*) and coho (*O. kisutch*) salmon in the Big Qualicum River, British Columbia. J. Fish Res. Board Can. 27: 1215–1224.

Lobón-Cervía, J. & A. De Diego. 1988. Feeding strategy of the barbel (*Barbus bocagei* Steind.) with relation to benthos composition. Arch. Hydrobiol. 114: 83–95.

Losos, B., M. Peňáz & J. Kuvičková. 1980. Food and growth of fishes of the Jihlava River. Acta Scientarum Naturalium bohemoslovacae, Brno 14: 1–46.

Mann, R.H.K. & C.A. Mills. 1986. Biological and climatic influences on the dace *Leuciscus leuciscus* in a southern chalk stream. F. B. A. Ann. Rep. 54: 123–136.

Mann, R.H.K. & J.A.B. Bass. 1997. The critical water velocities of larval roach (*Rutilus rutilus*) and dace (*Leuciscus leuciscus*) and implications for river management. Reg. Riv. 13: 295–301.

Mérigoux, S. & D. Ponton. 1998. Body shape, diet and ontogenetic diet shifts in young fish of the Sinnamary River, French Guiana, South America. J. Fish Biol. 52: 556–569.

Miller, T., L.B. Crowder, J.A. Rice & E.A. Marschall. 1988. Larval size and recruitment mechanism in fishes: towards a conceptual framework. Can. J. Fish Aquat. Sci. 45: 1657–1670.

Mills, C.A. 1985. Sources of variation in the feeding of larval dace *Leuciscus leuciscus* in an English River. Trans. Amer. Fish Soc. 114: 519–524.

Mills, C.A. 1991. Reproduction and life history. pp. 483–508. *In*: I.J. Winfield & J.S. Nelson (ed.) Cyprinid Fishes – Systematics, Biology and Exploitation, Chapman & Hall, London.

Peňáz, M. 1973. Embryonic development of the barbel *Barbus barbus* (Linnaeus, 1758). Zool. Listy 22: 363–374.

Power, M.E., W. Matthews & A.J. Steward. 1985. Grazing minnows, piscivorous bass, and stream algae: dynamics of a strong interrelation. Ecology 66: 1448–1456.

Rincon, P.A., P. Barrachina & Y. Bernat. 1992. Microhabitat use by 0+ juvenile cyprinids during summer in a Mediterranean river. Arch. Hydrobiol. 125: 323–337.

Sabo, M.J. & D.J. Orth. 1994. Temporal variation in microhabitat use by age-0 smallmouth bass in the North Anna River, Virginia. Trans. Amer. Fish Soc. 123: 733–746.

Schoener, T.W. 1970. Nonsynchronous spatial overlap of lizards in patchy habitats. Ecology 51: 408–418.

Schlosser, I.J. 1987. The role of predation in age- and size-related habitat use by stream fishes. Ecology 33: 651–659.

Staas, S. & D. Neumann. 1996. The occurence of larval and juvenile 0+ fish in the lower River Rhine. Arch. Hydrobiol. (Suppl. 113, Large Rivers) 10: 325–332.

Vanderploeg, H.A. & D. Scavia. 1979. Calculation and use of selectivity coefficients of feeding: zooplankton grazing. Ecol. Modelling 7: 135–149.

Watkins, M.S., S. Doherty & G.H. Copp. 1997. Microhabitat use by 0+ and older fishes in a small English chalk stream. J. Fish Biol. 50: 1010–1024.

Webb, P.W. & D. Weihs. 1986. Functional locomotion morphology of early life history stages of fishes. Trans. Amer. Fish Soc. 115: 115–127.

Werner, E.E. & J.F. Gilliam. 1984. The ontogenetic niche and species interactions in size-structured populations. Ann. Rev. Ecol. Syst. 15: 393–425.

Wieser, W. 1991. Physiological energetics and ecophysiology. pp. 427–455. *In*: I.J. Winfield & J.S. Nelson (ed.) Cyprinid Fishes – Systematics, Biology and Exploitation, Chapman & Hall, London.

Windell, J.T. & S.H. Bowen. 1978. Methods for study of fish diets based on analysis of stomach contents. pp. 219–226. *In*: T. Bagenal (ed.) Methods for Assessment of Fish Production in Fresh Waters, IBP Handbook No. 3, Blackwell Scientific Publications, Oxford.

Wintersberger, H. 1996. Spatial resource utilization and species assemblages of larval and juvenile fishes. Arch. Hydrobiol. (Suppl. 115, Large Rivers) 11: 29–44.

Environmental Biology of Fishes **56:** 213–228, 1999.

Vertical distribution and feeding activity of metamorphosing sole, *Solea solea*, before immigration to the Bay of Vilaine nursery (northern Bay of Biscay, France)

Françoise Lagardère[a], Rachid Amara[b,c] & Lucette Joassard[a]
[a]*Centre de Recherche en Ecologie Marine et Aquaculture de L'Houmeau (CNRS-IFREMER), B.P. 5, F-17137 L'Houmeau, France (e-mail: flagarde@ifremer.fr)*
[b]*IFREMER-Centre de Nantes, B.P. 1049, F-44037 Nantes, France*
[c]*Present address: Université du Littoral, Côte d'Opale - UPRESS-A ELICO 8013, CNRS, Avenue Foch, B.P. 59, F-62930 Wimereux, France*

Received 19 March 1998 Accepted 2 December 1988

Key words: flatfish, metamorphic changes, near-bottom shift, planktonic diet, feeding rhythm, inshore transfer

Synopsis

To evaluate the impact of metamorphosis on the vertical distribution and feeding activity of sole, *Solea solea*, larvae passing from offshore spawning grounds to the Bay of Vilaine, sampling series at fixed stations were carried out in April 1991 and April 1993 at depths from 50 to 30 m. Comparisons between plankton and bottom sampling series indicated differences in vertical distribution of larvae in pre-metamorphic and metamorphic steps. Metamorphosing larvae displayed a tendency to concentrate in the lower part of the water column, mainly during the day. Gut contents, analysed for prey identification, fullness index and carbon content, indicated that metamorphosing larvae fed mostly on plankton. Variations in fullness index were observed not only during the day, but also depended on tide and wind-induced mixing conditions. Larvae sampled in mixed spring-tide waters had highly variable carbon estimates, resulting in unclear diel activity. More larvae fed actively at neap-tide, which allowed the observation of a diurnal feeding activity through hourly changes in carbon estimates. It is concluded that immigrating sole were not yet able to settle but prepared themselves for demersal life (i) without undergoing starvation and (ii) by modifying the patterns of vertical distributions. The presence of a larval swimbladder suggests they can adjust their vertical movements, depending on tidal cycles, which could in turn favour coastal accumulation of metamorphosing larvae and pulses of new settlers entering the nursery grounds.

Introduction

As in other estuarine-dependent species, diel vertical migrations of common sole, *Solea solea* (L.), larvae are presumably among the key processes that govern their transfer to estuarine nurseries. These movements appear to be modulated by changes occurring at metamorphosis, and particularly when the larvae are crossing between the offshore spawning grounds and the near-shore nurseries (Amara et al. 1998). Symonds & Rogers (1995) emphasized that events occurring in this transitional area are not fully understood, and they will be of obvious relevance to recruitment success.

Studies in the northern part of the Bay of Biscay since 1985 on recruitment processes of the common sole (Koutsikopoulos et al. 1993) have identified their spawning grounds on the shelf (Arbault et al. 1986, Koutsikopoulos & Lacroix 1992), as well as nursery areas in the Vilaine estuary, about 70–80 km from the main spawning grounds, exploited by both late larvae and early juveniles (Marchand & Masson 1989, Marchand 1991, 1992). Because newly settled juveniles have never been observed on the spawning grounds, the mechanisms of larva immigration in the common sole remain unknown. Whereas, in the North Sea plaice, *Pleuronectes platessa* (L.), the eggs, embryos and larvae are transported by residual currents until they reach coastal areas where larvae use selective tidal transport to settle in shallow nursery grounds (Creutzberg et al. 1978, Rijnsdorp et al. 1985, Van der Veer 1986). The use of similar mechanisms has been proposed to explain how sole cross the space separating offshore spawning grounds of coastal nurseries (Marchand & Masson 1989, Marchand 1992). In a more general context, the vertical distribution of fish larvae and their behaviour determines their possibilities of horizontal movements and thereafter their degree of dispersal (MacCall 1990). Moreover, vertical migrations, when occurring, are often associated with an ingestion–digestion cycle, predator-avoidance reactions and transport mechanisms (see reviews in Norcross & Shaw 1984, Neilson & Perry 1990).

From extensive surveys in the Bay of Biscay, sole larvae are known to spread around spawning areas by drifting. During this phase, they started in ontogenetic and circadian vertical migrations (Koutsikopoulos et al. 1991). The range of these vertical migrations increased until larvae began to metamorphose ['stage 4', according to the definition of Koutsikopoulos et al. (1991)]. Surprisingly, early and late transforming larvae ('stage 5') immigrating to the estuary maintained this behaviour, reaching the surface at night regardless of the direction of tidal currents (Marchand & Masson 1989, Marchand 1991). Hydrographic and biological results thus have not provided direct evidence for significant onshore advection of larvae, or behavioural selection of onshore currents by larvae. Diffusion was hence assumed to play a prominent part in onshore transfer (Koutsikopoulos et al. 1991). Le Cann et al. (1992) also found that residual currents were weak (2 to 4 $cm\ s^{-1}$) and northwesterly orientated on the shelf, i.e. unfavourable to onshore advection of larvae; however, they suggested a reappraisal of hypotheses for the transfer processes, including vertical migrations in models. As a result, Le Cann et al. (1992) did not exclude interactions between the vertical distribution of larvae and the solar semi-diurnal component of tidal currents, permitting a selective transfer to near-shore nurseries, especially during spring tides.

Diel vertical migrations of sole larvae have been extensively investigated under laboratory conditions (see review in Champalbert et al. 1994), indicating that behavioural changes take place during the transition from larvae to juveniles. These changes were mostly attributed to reactions to light, current and pressure, but also to feeding conditions. Champalbert & Koutsikopoulos (1995) inferred that vertical migrations of larvae on the shelf could be synchronized by an endogeneous diel rhythm, upward and downward movements being triggered off by dusk and dawn, respectively. By simulating horizontal trajectories from diffusive processes and active movements, these authors confirmed the suggestion by Le Cann et al. (1992) that a reduction in diffusive processes leads to onshore movements. These simulations were based on 'idealized' vertical ascents of larvae up to the sea surface and on the assumption that the nutritional status (viz starvation) may increase nocturnal movements. However, from previous studies sole larvae are known to be diurnal predators (Blaxter 1972, Appelbaum et al. 1983), ingesting zooplanktonic prey (Last 1978) during their pelagic phase. It is thus very unlikely that the nocturnal swimming activity, up to the surface, was a feeding migration, although these simulations could explain, to a certain extent, the movements of larvae on the spawning grounds. In addition, the small number of metamorphosing sole caught in plankton samples (Boulhic et al. 1992) led us to suspect that larvae reduce the amplitude of their vertical ascents, as described by Gibson et al. (1978) for plaice, even though larvae settling in the estuarine nurseries still performed diel vertical movements. In these areas, larvae were completing transition to the juvenile period by shifting to flatfish swimming behaviour and initiating feeding on small benthic prey (Marchand & Masson 1989, Marchand 1992). During settlement, changes in seasonal pattern of distribution of steps suggested immigration of successive cohorts (Marchand 1991, 1992, Amara et al. 1994). Juveniles progressively fed during the night on epibenthic and benthic prey (Lagardère 1987, Marchand & Masson 1989, Marchand 1991), which tended to make them settle in shallower and shallower waters (Macquart-Moulin et al. 1991).

The aim of the present study was to examine the role of day/night distribution and feeding pattern in the immigration of common sole larvae and juveniles to coastal nurseries. The specific objectives to achieve this aim were: (1) to develop a sampling strategy for metamorphosing sole, and (2) to assess the impact of metamorphosis on the nutritional status of sole larvae through studies of their diet and feeding rhythm.

Material and methods

Sampling

Stations were located off the Bay of Vilaine (delimited by the 10 m isobath), with station 2 central to the main spawning grounds and the three others shorewards, between Belle-Ile and the mainland (stations 1, 3 and 4) (Figure 1). Station 1 was sampled on 19–20 April 1991, whereas stations 2, 3 and 4 were sampled during two cruises, the first one taking place from 31 March to 8 April 1993, and the second one on 27–28 April 1993 (Table 1). The deepest station (2) was visited twice, at neap- and spring-tides in late March–early April 1993.

At each station, samples were carried out at two-hourly intervals to obtain a 24-hour series of plankton and bottom samples, except at stations 2 and 3, due

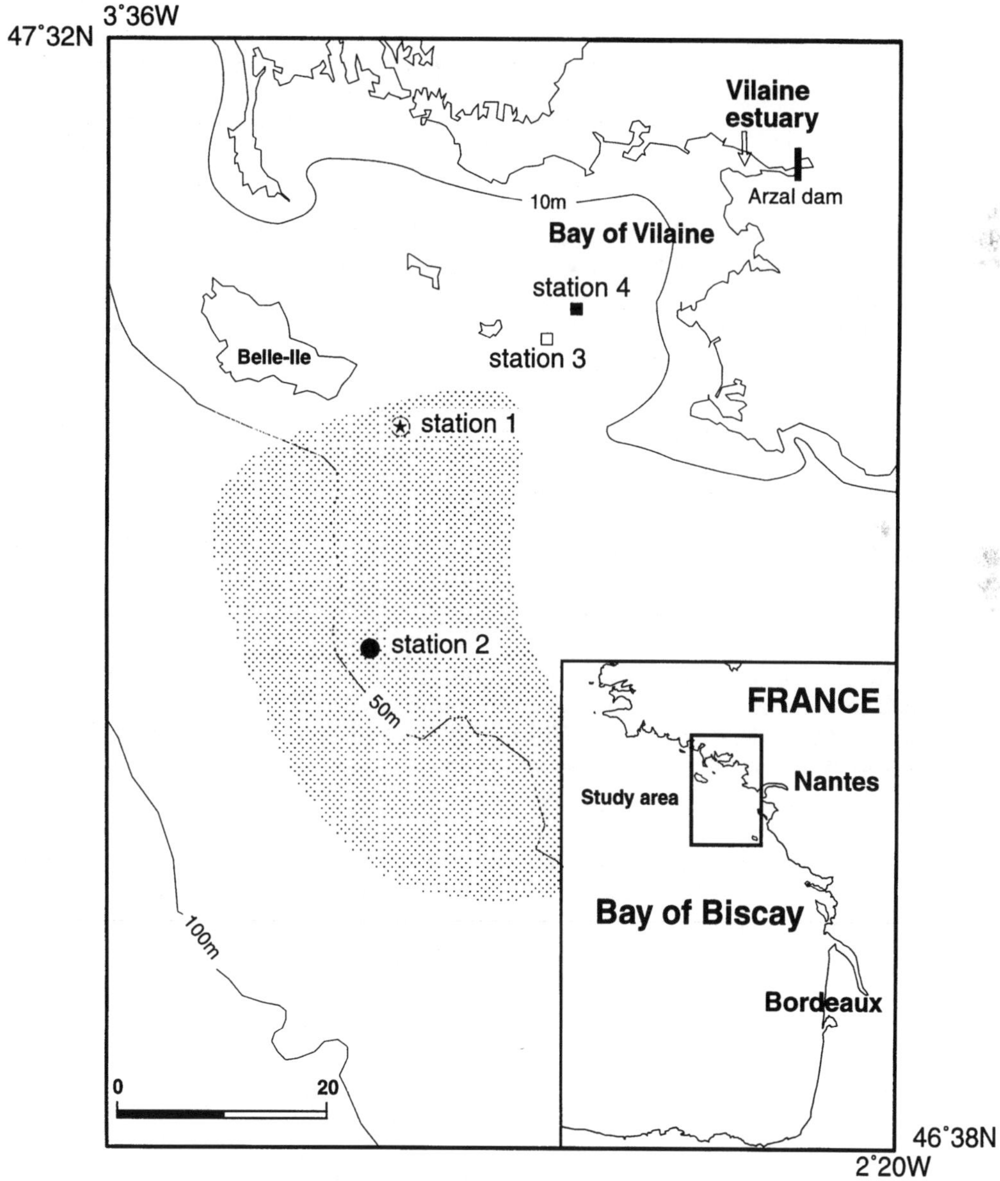

Figure 1. Map of the Bay of Biscay showing the location of the fixed stations sampled in April 1991 (station 1) and in 1993, early April cruise (station 2 and 3), late April cruise (station 4), with regard to the main spawning ground location (dotted area), the Bay of Vilaine and the Vilaine estuary.

Table 1. Dates and sampling conditions in 1991 and 1993 at fixed stations located off the Bay of Vilaine (northern Bay of Biscay).

Date	R/V	Station	Depth (m)	Sampling period (hours)	Tide conditions
1991:					
19–20/4	'Gwen Drez' (24.5 m)	1	35–40	28	spring-tide: 16 March
1993:					
1st cruise					
31/3–2/4	'Thalassa'	2	≈50	18	neap-tide: 31 March
6–8/4	(66.1 m)	2	≈50	44	spring-tide: 7 April
5–6/4		3	30–32	24	
2nd cruise					
27–28/4	'Gwen Drez' (24.5 m)	4	26–28	26	neap-tide: 29 April

to bad weather conditions from late March to early April 1993, which obliged us to shorten the first sampling series and to cancel plankton samples for the second (Table 1). Otherwise, three gears were used for sampling, a suprabenthic sledge and two plankton samplers. The suprabenthic 'Zebulon' sledge had a rectangular aluminium frame (2 m wide by 1 m high) to which two plankton nets (1.3 mm mesh) were fastened; this was used to sample the larvae in the layers 0–0.50 m (low Zebulon) and 0.50–1 m (upper Zebulon) above the sea-bed (see Amara et al. 1994 for gear characteristics and efficiency). Each haul lasted 15 min and covered on average 1.3 km in distance. When suprabenthic hauls were complemented by plankton samples, two different samplers were used. At station 2, the plankton sampler had a 1 m^2 multiple opening and closing net system, using electromagnetic release (MACOF sampler) (P. Bourriau unpublished). This sampler was fitted with 4 nets (0.5 mm mesh) and 'in mouth' flow meters. At stations 1 and 4, the plankton sampler had a 1 m sided frame, fitted with a net (0.5 mm mesh) and 'in mouth' flow meter; it was used for double oblique tows, from the surface to near bottom. All samples taken at station 2 were fixed with buffered (5%) formaldehyde solutions and stored as a whole. For sampling series at stations 1, 3 and 4, larvae were sorted alive on board. They were fixed in 95% alcohol (stations 3 and 4, in part) or randomly selected and isolated in vials for fixation in liquid nitrogen (1993 bottom samples taken at stations 3 and 4, for the other part). These individual samples were stored at around −80°C (1993, early April cruise) or −20°C (1993, late April cruise) until analysed 2–3 months later.

Vertical profiles of water temperature and salinity were recorded during all cruises by CTD (Valeport model 600 MK2), whereas tide level curves for 1993-sampling series at stations 3 and 4 were drawn from SHOM predictions (Service Hydrographique et Océanographique de la Marine, Paris). To obtain inter-cruise differences due to weather conditions during the 1993-sampling series, we used meteorological data for wind speed and direction (recorded at Belle-Ile) and hours of sunshine (pyranometer records at the Arzal dam, 25–30 km from the sampling stations). Wind records indicated that stations 2 and 3 were sampled during southwesterly flows (Figure 2a), with wind speeds $\geq 10\,m\,s^{-1}$, whereas high pressure conditions occurred during the late-April cruise (Figure 2b, wind speeds $\leq 5\,m\,s^{-1}$). Early in April, winter conditions still persisted on the shelf, characterized by a mixed water column, except near the bottom, where temperature and salinity profiles indicated slightly warmer and more saline conditions (Figure 2c). During the entire month of April, virtual non-stop rain affected the rate of flow of the rivers. Vertical profiles of temperature and salinity confirmed an increase in the water-column stratification shorewards: a thermo-halocline started to develop in late April, off the Bay of Vilaine (Figures 2d,e).

Biological analyses

All soleid larvae were identified to species level and their developmental status was determined according to the 5 morphological stages defined for flatfish larvae by

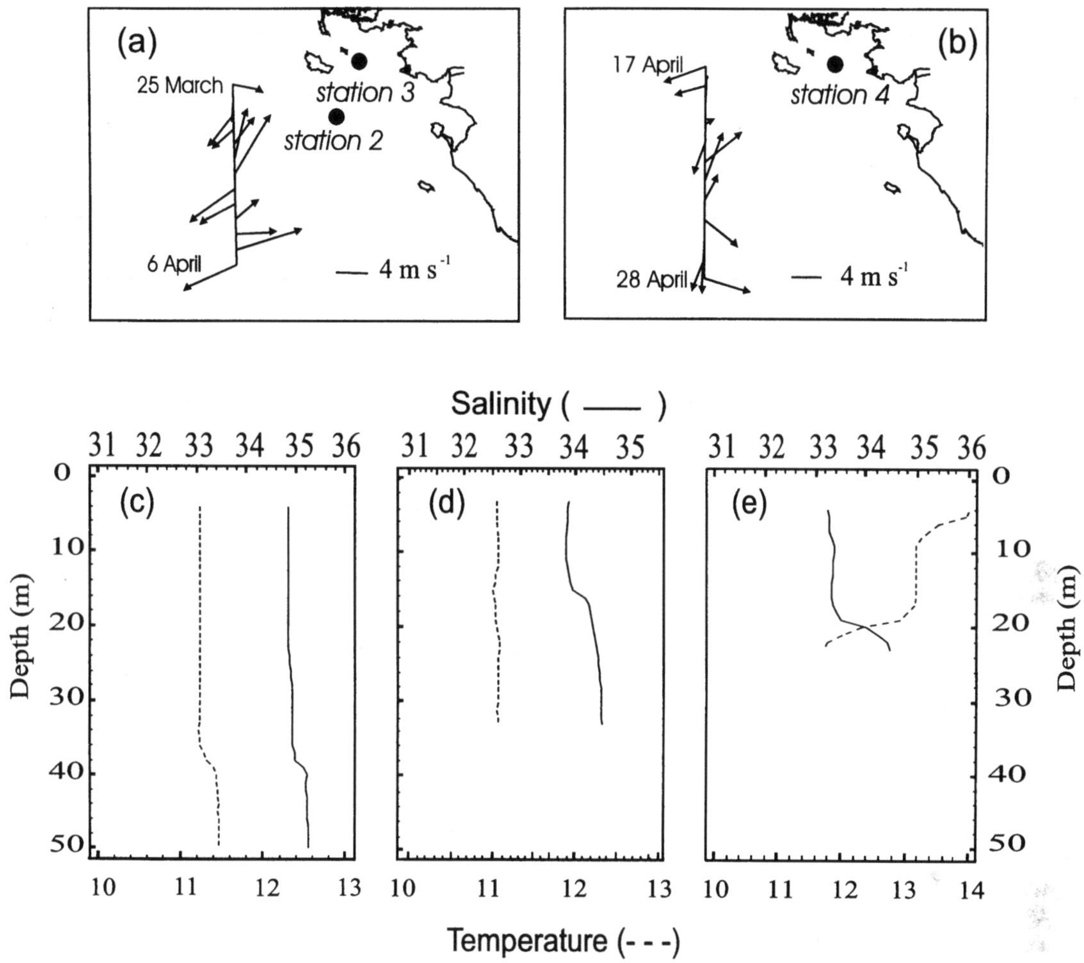

Figure 2. Wind and hydrological conditions of the 1993 cruises in the northern Bay of Biscay. a,b – Wind-vector diagrams (mean direction and speed per day) for 12-days periods corresponding to (a) the early April cruise and (b) the late April cruise. c,d,e – Vertical profiles of temperature and salinity taken at low tide (c) on station 2, (d) station 3 and (e) station 4.

Ryland (1966) and revised by Al-Magazachi & Gibson (1984), which hereafter are referred to as 'steps' following terminology in Balon (1975) (Table 2). Plankton and bottom catches of sole larvae caught during the 1991 cruise were pooled for day and night comparisons of the percentages of larva steps obtained from each sampler. We then concentrated on bottom samples from the 1993 cruises, by subdividing steps 4 and 5 as 'a' (early) or 'b' (late) to get further insights on the abundance and distribution of metamorphosing larvae.

For samples taken during the 1993 cruises, larvae sampled at station 2 were sorted, their developmental status was determined, and the number of larvae per step was summed for each sampling gear to provide the proportion of each larva step per sampling gear and to permit comparison with the shoreward stations 3 and 4. Individual bottom samples of larvae taken at station 3 and 4 ($n = 59$ and 179, respectively) were thawed, identified to species and their developmental status was determined. Owing to the rare occurrence of pre-metamorphic (steps 4a) and late metamorphic sole (steps 5b), these larvae were not included in gut content analyses. Larvae were measured before analysis and they were dissected under a binocular microscope for extraction of the entire gut, from the oesophagus to the anus, which during the larva period includes a small stomach, two intestinal loops and the rectum. Examination was made of the content of each of these gut

Table 2. Step criteria for sole, *Solea solea*, larvae, from hatching to the completion of metamorphosis, adapted from Al-Magazachi & Gibson (1984).

Embryo	1: yolk-sac	Pelagic
Symmetrical larva	2: first-feeding	Pelagic
	3: initiation of fin rays	Pelagic
	4a: urostylar torsion	Pelagic
Symmetrical larva / Metamorphosing larva	4b: initiation of eye migration	Pelagic
Metamorphosing larva	5a: migrating eye on the top of head, from pelagic to epi-benthic	
	5b: late transforming larva	Benthic

Table 3. Summary of regression equations for nitrogen estimates vs. carbon estimates of gut contents and carbon and nitrogen estimates vs. dry weights of metamorphosing sole larvae captured near the bottom during the early and late April 1993 cruises (northern Bay of Biscay).

Cruise date	Sampling size	Regression models	*r*	*F*	p
Station 3:					
5–6 April 1993	58	carbon = 37.879 + 4.452 nitrogen	0.99	4512.7	0.000
		carbon = 48.30 + 390.39 dry weight	0.99	3497.1	0.000
		nitrogen = 3.748 + 86.43 dry weight	0.98	1744.9	0.000
Station 4:					
27–28 April 1993	177	carbon = 87.68 + 3.796 nitrogen	0.95	1701	0.000
		carbon = 59.45 + 395.14 dry weight	0.94	1368.9	0.000
		nitrogen = 1.578 + 96.88 dry weight	0.92	975.7	0.000

segments and a fullness index was used for a rough estimation of feeding activity, from 0 (all segments being empty) to 0.25, 0.50, 0.75 or 1 (all segments being full). Prey were assigned to their main taxonomic groups and counted, with planktonic and suprabenthic items kept separate from each others. Carcasses and gut remains were transferred to pre-tared aluminium capsules, dried at 60°C in an oven, to obtain dry weights to the nearest 0.1 μg.

The number of prey in the guts cannot be used as an accurate measure of actual feeding efficiency nor of the existence of a potential rhythm due to (i) the various sizes of prey (from $<50\,\mu$m to $>500\,\mu$m), (ii) their variable degrees of digestion depending on sample timing and (iii) the presence of exoskeletons. This heterogeneity was smoothed out using carbon and nitrogen analyses of individual gut contents (including whole prey, prey pieces and digestive products). Excess water was removed by filtering these contents on previously calcined GF/F Whatman filters (13 mm in diameter) before further storage at −20°C. Analysis of each gut content was carried out using a Carlo-Erba Elemental Analyser (NA 1500) for equivalent in carbon and nitrogen, estimated to the nearest 0.001 μg, with acetanelide (nitrogen 10.36% – carbon 70.19%) as standard. Owing to some losses during carbon and nitrogen analyses, we present results for 58 larvae from the early April 1993 cruise (station 3) and 177 larvae sampled during late April 1993 (station 4).

Statistical analyses

We could not exclude a loss of dry weight and nitrogen content through freezing storage, although liquid nitrogen was used for fixation (Williams & Robins 1982, Håkanson et al. 1994). We thus studied the relationships between (i) nitrogen and carbon estimates and (ii) these two variables versus the dry weight of larvae with linear regression analyses. Regression analyses showed a strong positive correlation between nitrogen and carbon estimates, and between these two variables and the dry weight of larvae (Table 3). Analysis of covariance (ANCOVA) showed that the slopes of the regression lines fitted between dry weights and nitrogen or carbon contents were not significantly different from one another ($p > 0.05$). We thus assumed that errors, if any, were relative and did not invalidate a discussion of variation patterns in terms of carbon.

From all sampling series, the average densities of larvae per 1000 m^3 and per gear were calculated (Table 4).

Table 4. Date, station and origin of samples taken in 1991 and 1993 of the Bay of Vilaine, using in turn a plankton sampler and a suprabenthic sledge (n = number of samples per station, vol. = mean filtered volume in m^3 per gear and density of larvae = mean density of larvae per 1000 m^3 and per gear, ± standard error).

Date	Station	Plankton samples			Bottom samples		
		n	vol. (m^3)	larva density	n	vol. (m^3)	larva density
1991:							
19–20/4	1	11	990 ± 47.97	36.4 ± 4.52	14	3169 ± 144.4	7.6 ± 2.07
1993:							
31/3–1/4	2	9	4044 ± 290.3	6.5 ± 1.52	9	3700 ± 97.1	1.05 ± 0.39
6–8/4	2	22	3900 ± 164.6	8 ± 0.83	22	3390 ± 187.4	1.46 ± 0.23
5–6/4	3	—	—	—	12	4200 ± 258	2.9 ± 3.3
27–28/4	4	13	360 ± 59.2	2.5 ± 1.1	13	3030 ± 160.4	12 ± 2.3

One-way analysis of variance (ANOVA) was used to test (i) for differences in densities of larvae between plankton and bottom samples and (ii) for differences in feeding activity between stations 3 and 4 and between steps 4b and 5a taken at station 4. The relationships between densities of larvae and carbon estimates with tide level and light intensity (day/night) were analysed using Spearman's rank correlations. Prior to analyses of densities of larvae, the data were $\log(x + 1)$ transformed.

Results

Abundance and vertical distribution

Day and night plankton samples in 1991 provided a higher proportion of symmetrical larvae (mostly steps 2 and 3, Figure 3a) in contrast to bottom samples, where the proportion of metamorphosing larvae (steps 4 and 5) increased, especially during the day (Figure 3b). Differences were found in densities of larvae in both vertical and horizontal dimensions with respect to stations, sampling gears and dates (Table 4). Whatever the station, densities of larvae were significantly lower in bottom samples than in plankton samples (ANOVA station 1 $F = 43.65$, df = 29; station 2 $F = 57.38$, df = 42; $p < 0.05$), except at the inshore station 4, where larvae were significantly more abundant near the bottom than in the water column (ANOVA, $F = 11.74$, df = 24, $p < 0.05$).

Examination of the percentages of larva steps per station showed that sole larvae sampled offshore in late March and early April 1993 were mostly symmetrical larvae, with variations in captures taken in the water column or near the bottom probably due to differences between samplers and cruise conditions (station 2, Figures 4a, b). A lower proportion of symmetrical larvae was sampled shorewards during this cruise (station 3, Figure 4c) and even fewer during the second cruise, despite the plankton sampling series (station 4, Figure 4d). At these shoreward stations, sole larvae were found near the bottom, mostly in steps 4b and 5a. Their density varied according to day/night and tide conditions (Figures 5a–f), but these variations were unclear, which was probably due to insufficient

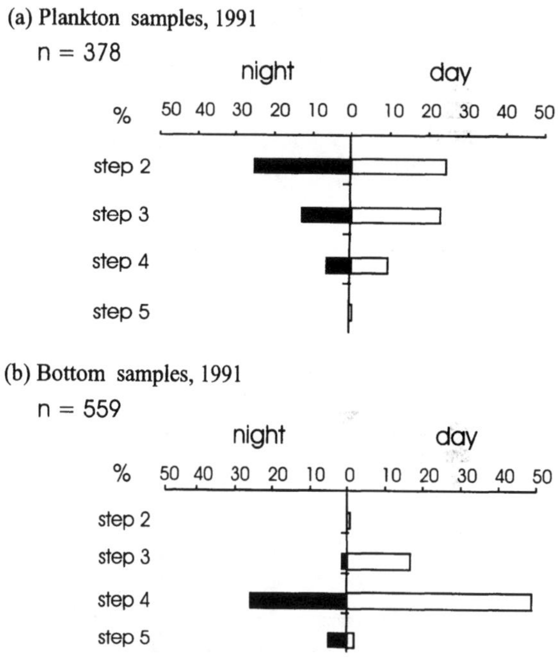

Figure 3. Percentage (combined 24 h catches) of sole larvae of different development status in day/night and plankton/bottom samples at station 1, in 1991.

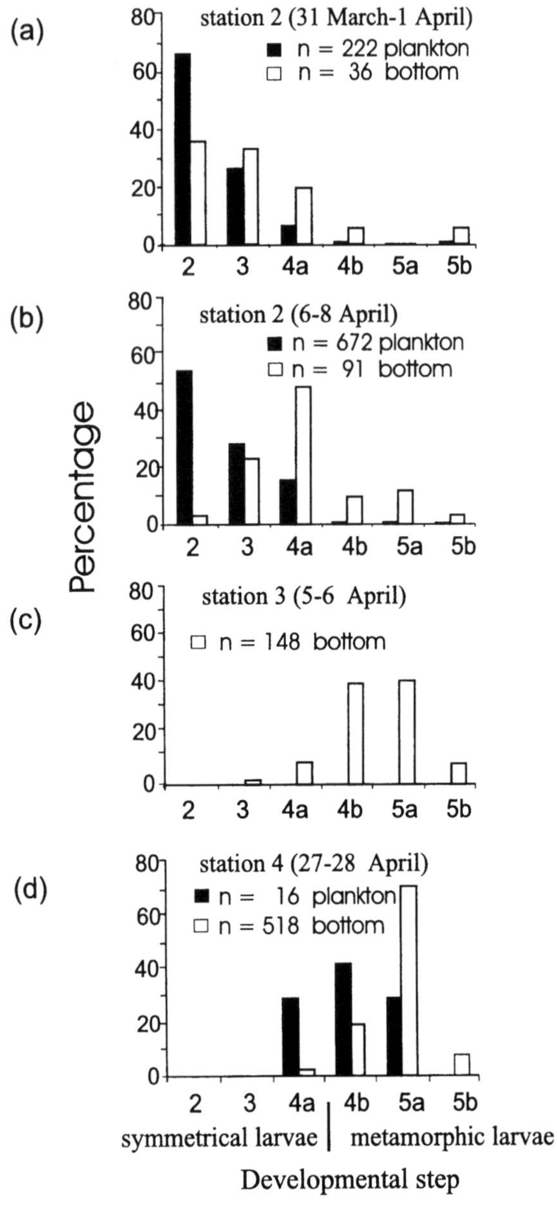

Figure 4. Inter-station comparisons of larva step percentages obtained for each sampling series taken with either a plankton sampler or the suprabenthic Zebulon sledge in 1993.

sampling time. We found no significant correlations between the density of larvae near the bottom, tide level or light intensity, except during the early April cycle, when the densities of larvae in step 5a were positively correlated with light intensity (Table 5).

Feeding behaviour

From gut dissection, the stomach always appeared empty (Figure 6), which was probably due to its very small size and resulted in a maximal fullness index of 0.75. Gut fullness varied between cruises, indicating a reduced feeding activity during the bad weather conditions of early April 1993 (36% of empty guts, compared to only 16% during the second cruise). Lower averages of food items per larva were observed during the first cruise (Figure 7a) compared with the second cruise (Figure 7b) (ANOVA, $F = 4.04$, df $= 186$, $p = 0.0004$). Gut contents mostly consisted of planktonic prey (80–82% of the total number, mainly copepods, cladocerans and nauplii), although some epi-benthic prey (polychaete fragments, harpacticoid copepods, small bivalves and gastropods, and very few amphipods) began to contribute to the diet. Among the miscellaneous prey types were invertebrate eggs, some diatoms, ostracods and unidentified fragments whose number was higher under the reduced feeding activity of the early April 1993 cruise. When active feeding was observed in late April (Figure 7c), larvae in step 4b and 5a fed on the same prey and in the same proportions (ANOVA, $F = 0.117$, df $= 147$, $p = 0.088$).

Concerning feeding rhythm for both cruises, the average estimates in carbon varied according to sampling time, which could depend on light intensity, cruise conditions and larva steps (Figure 8). The highest values ranged on average from 33 to 43 μg and they were observed in the afternoon or at night. The lowest estimates (in the range 4–10 μg) corresponded to gut ranked as 'empty' (fullness index of 0), which was likely due to small remains, unidentifiable as prey, and gut tissues. These low estimates seemed to occur sporadically during the early April cycle (station 3, Figures 8a, b, c), whereas they were mainly observed around sunrise during the late April cycle (station 4, Figures 8c,d,e).

No clear pattern resulted from the low and highly variable carbon estimates obtained at station 3, in contrast to larvae captured at station 4, which seemed to exhibit a diel feeding activity through hourly changes of carbon estimates. These changes were less pronounced for larvae in step 4b (Figures 8d,f) than for larvae in step 5a (Figures 8e,f), possibly due to the smaller sample size of the former. Negative correlations were obtained for both larvae in steps 4b and 5a between the carbon content of guts and irradiance but these correlations were not statistically significant

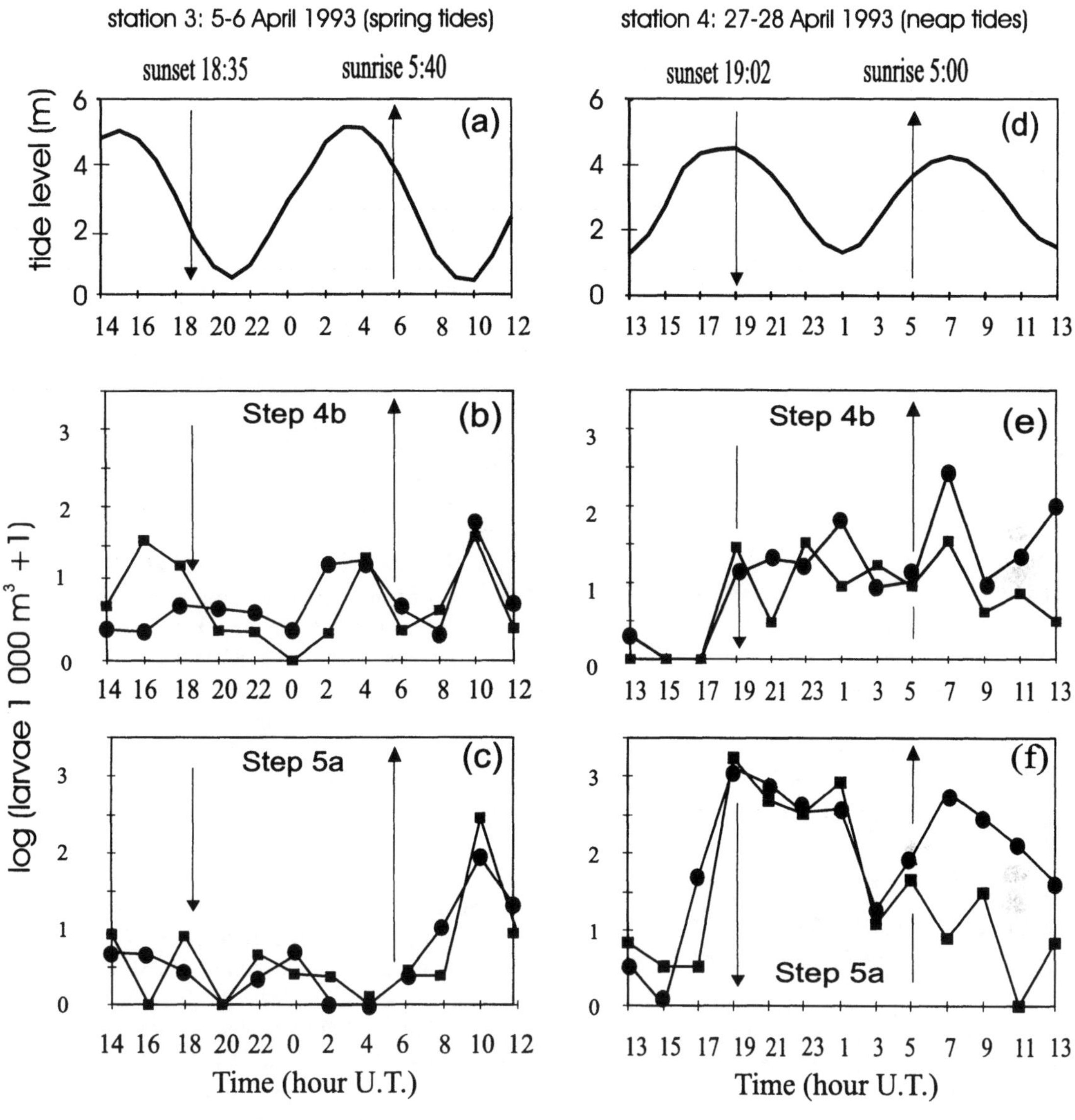

Figure 5. Hourly variations (universal time) in the density, log(larvae·1000 m^3 + 1), of metamorphosing sole caught in bottom samples at stations 3 and 4 in 1993, according to tide conditions, developmental status and sampling levels of the Zebulon sledge. a,d – Tide curves on 5–6 April (spring-tide) and 27–28 April (neap-tide); b,c – early April samples and (e,f) late April samples for larvae in steps 4b and 5a. Squares indicate samples taken with the lower net of the sledge (0–0.50 m above the bottom) and circles, with the upper net (0.50–1 m).

(Table 5). However, for larvae in step 5a, carbon contents started decreasing at sunset and resumed at sunrise, which suggested a feeding rhythm. The occurrence of undigestible remains in the gut (e.g. chitinous crustacean exoskeletons) probably explained the overlap of consecutive estimates and participated to the variability, but digestion during the night very likely explained the lowest values in carbon contents obtained around sunrise.

Discussion

As reported by Amara et al. (1993), metamorphosing sole (steps 4b and 5a) in the northern Bay of Biscay were more abundant in the area 30–40 m deep than previously reported by Koutsikopoulos et al. (1991). Use of a recently developed bottom sampler (see Amara et al. 1994) certainly accounted for these differences (Figure 3), although late metamorphosing sole (step 5b,

Table 5. Spearman's rank correlation coefficients (r_S) of the densities of metamorphosing larvae and gut carbon contents with respect to tide level and irradiance factors, for bottom samples taken at stations 3 and 4 (early and late April 1993 cruises). Correlations significant at the 5% level are marked with *.

	Larva step	Tide level	Irradiance
Density of larvae			
Station 3	4b	0.063	0.371
	5a	−0.07	0.784*
Station 4	4b	−0.174	−0.019
	5a	0.333	−0.333
Carbon estimates of gut contents			
Station 3	4b	−0.257	0.085
	5a	0.257	−0.085
Station 4	4b	−0.217	−0.474
	5a	0.477	−0.419

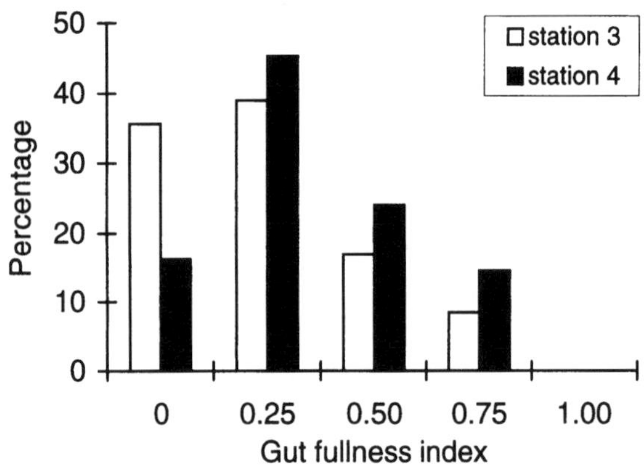

Figure 6. Inter-cruise differences in gut fullness index (in percent) of metamorphosing sole taken in 1993 bottom samples at station 3 (5–6 April) and station 4 (27–28 April).

Figure 4) remained scarce at this depth and even deeper on the shelf (Amara et al. 1993, Amara & Bodin 1995, Amara et al. 1998). In other words, most of the sole began to metamorphose before entering the Bay of Vilaine, and they completed the process by settling as late larvae (step 5b) and early juveniles in the Vilaine estuary (Marchand & Masson 1989, Marchand 1992). The success of transfer to bays and estuaries thus appears to depend not only on the physical context of the area, but also on more complex behavioural performance due to metamorphic changes. By reducing the extent of vertical distribution, they could limit dispersal, whereas survival will depend on the nutritional status of larvae.

Ontogenetic near-bottom shift

How sole larvae undergo eye migration and body rotation had been investigated under laboratory conditions. With food provided ad libitum, metamorphosis lasted from 6–7 days to more than 10–12 days at 19°C and 12°C, respectively (Lagardère 1989), but the switch to benthic life was a sudden event, probably due to rearing in height-limited tanks. A more progressive response, both in time and in vertical distribution, could be expected from the open sea to the estuary, as indicated by the 1991 and 1993 observations (Figures 3, 4). During metamorphosis, larvae completed the organogenesis of mechanosensory receptors, the cephalic neuromasts of the prospective blind side (Harvey et al. 1992), the olfactory and gustatory organs (Appelbaum et al. 1983) and the retinal structure as well, with rod formation (Blaxter & Staines 1970), allowing late larvae to perceive movements at low-light levels (Champalbert et al. 1992). In addition, Boulhic & Gabaudan (1992) showed that the larva swimbladder, differentiated in symmetrical steps, was inflated at metamorphosis and then regressed in early juveniles.

Amara et al. (1993) suggested a shoreward advection of metamorphosing larvae from within-season changes in horizontal distributions. Based on the present study, we hypothesize that metamorphosing sole concentrate near the sea-bed, which probably resulted in the above mentioned coastal accumulation. Irrespective of sampling levels, early April 1993 samples taken in the spawning grounds (Figure 4) provided few metamorphic steps, although spawning peaked in early February (R. Amara unpublished). Most of the larvae were found dispersed in the water column, still at symmetrical steps. Moving shorewards, metamorphic steps reached higher densities during the same early April cruise and even more in late April 1993 (Figures 4, 5). The use of different plankton samplers did not permit inter-station comparisons of larva density in the water column. In bottom samples taken in late April 1993, the density of larvae in step 5a was on average twice the one recorded for larvae in step 4b (8 and 4 larvae 1000 m^{-3}, respectively), irrespective of older ages at step 5a. These estimates were nevertheless far below those of newly settled sole within the Bay of Vilaine or the estuary (up to about 100 larvae 1000 m^{-3}, Marchand & Masson 1989) where the concentration of larvae probably resulted from an increasing 'bottleneck effect'.

In contrast to the complete vertical ascents performed by larvae in steps 4a and b ('stage 4', in

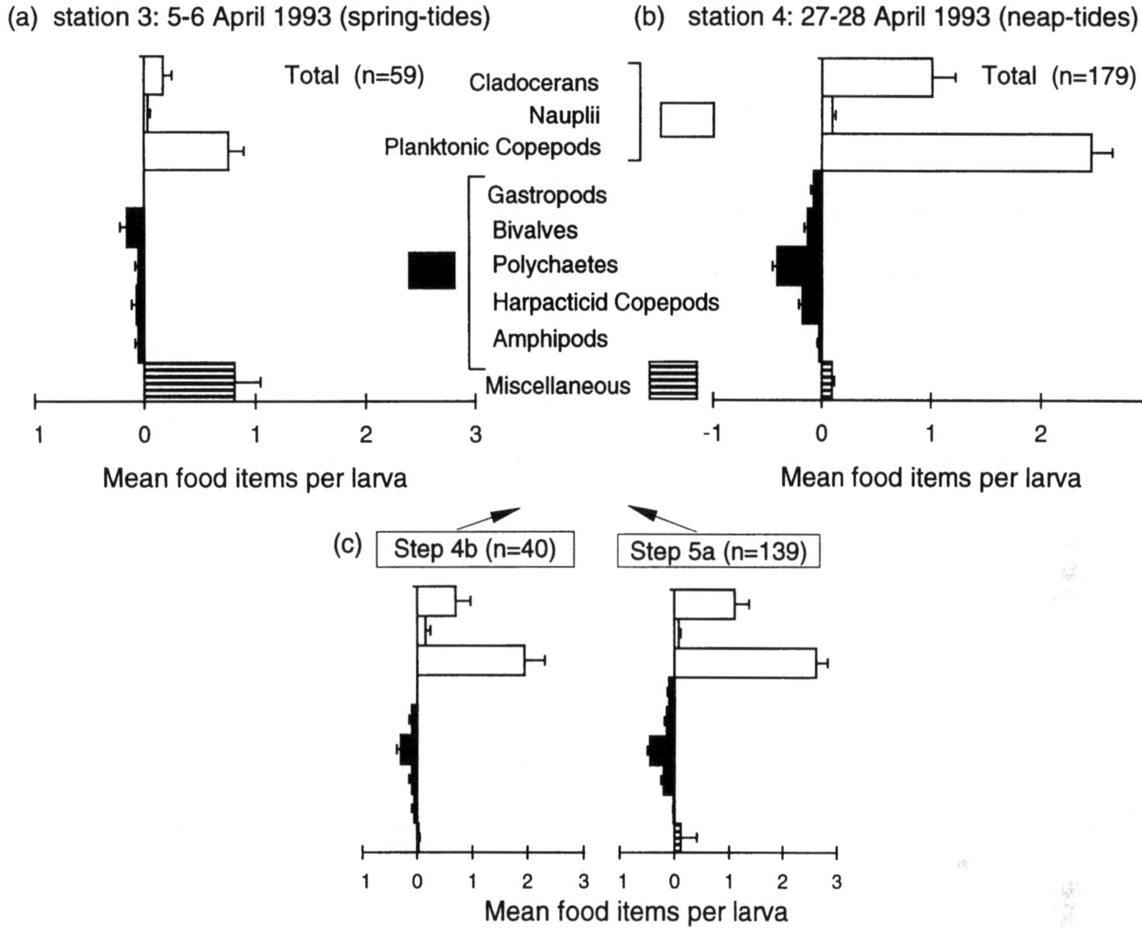

Figure 7. Diet and mean prey numbers in guts of metamorphosing sole larvae captured near the bottom at stations 3 and 4 in 1993; solid bars on the left indicate epi-benthic prey; open and stripped bars on the right indicate planktonic prey and miscellaneous items, respectively. a,b – Within-cycle differences in mean prey numbers for the overall samples (standard error is given by horizontal lines). c – Inter-step resemblance observed from the late-April sampling series.

Koutsikopoulos et al. 1991), the increased captures of metamorphosing larvae obtained from suprabenthic sampling series taken from the spawning grounds to the Bay of Vilaine surroundings (Figure 5) suggest that the extent of vertical distributions was reduced, as sole larvae underwent the eye migration crisis, which mostly affected the larvae in step 5a. This tendency to sink before entering inshore areas could indicate their search for a better-adapted location or their reduced ability to move vertically. A similar pattern of larva distribution, both offshore and close to the sea-bed, has already been described for North Sea plaice (Harding & Talbot 1973) and thickback sole, *Microchirus variegatus* (Fortier & Harris 1989, Amara et al. 1998). This near-bottom location was also inferred from horizontal distribution of co-occurring dab, *Limanda limanda*, and flounder, *Platichthys flesus*, larvae by Campos et al. (1994), who considered this pattern to be step-specific, with responses occurring irrespective of shore distance. Enhanced perceptive abilities and new behaviour patterns of the larvae both fit in well with the step-specific concept of vertical distribution proposed by these authors.

However, we observed movements above the sea-bed in mixed coastal waters due to spring tide conditions, which probably explains the day/night variations in captures near the bottom (Table 5). This assumption is supported by experimental studies of Macquart-Moulin et al. (1989) and Champalbert & Koutsikopoulos (1995). The former authors reported

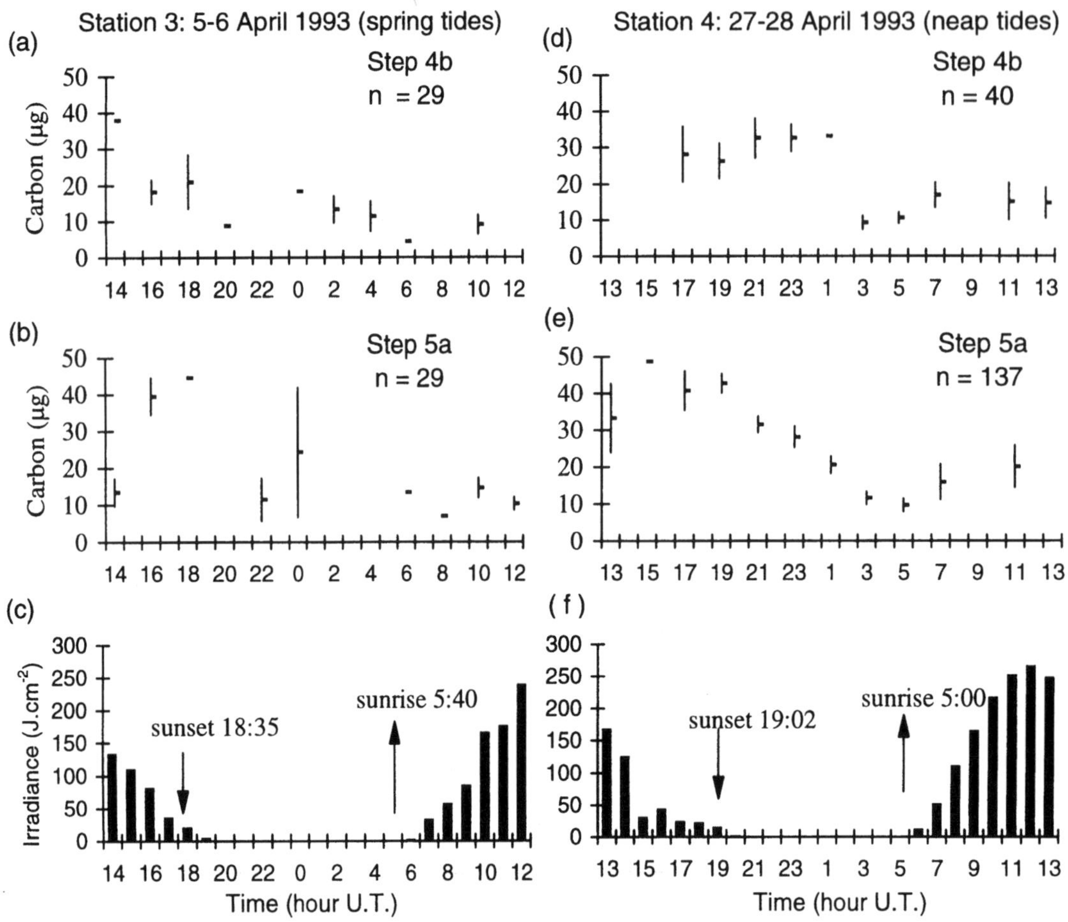

Figure 8. Hourly variations of gut contents (in mean carbon estimates) from metamorphosing sole caught near the bottom in early (a,b) and late (d,e) April 1993, according to daytime illumination (c,f) (vertical bars give the standard error).

swimming activity upwards when pressure is increasing, which led the latter authors to hypothesize that short-period pressure changes, due to tidal cycles, combined with rheopositive behaviour, should generate shoreward displacement. According to Blaxter (1980), swimbladders are not essential to the buoyancy of larvae capable of vertical migration, but they could reduce energy-expenditure by providing extra-buoyancy, and their swimbladder walls are assumed to act as pressure receptors. This function is under reflex control, regardless of the time period (short-term wave effects, semi-diurnal and fornightly tidal cycles, etc). Even though nocturnal ascents were not high enough for larvae to reach the sea surface, they should have at least the same amplitude as in the estuary (2–8 m, depending on the tide). Nevertheless, pressure sensitivity has probably a greater influence to cue the movements of metamorphosing larvae in shallow areas due to the higher relative pressure changes over tidal cycles (Gibson 1997).

Feeding conditions at metamorphosis

Among the factors regulating the feeding activity of larvae are the availability of suitable prey, the larva's capability to detect them and the abiotic and biotic factors prevailing on a small or larger scale. Despite the tendency for metamorphosing sole to sink, gut content analyses revealed that they are plankton feeders, as are symmetrical larvae (Last 1978). Outside the Bay of Vilaine, planktonic prey account for 80–82% of food composition (Figure 7) and in the estuary, for no more than 10% (Marchand & Masson 1989). Inside the estuary, newly-settled sole (presumably larvae in step 5b and early juveniles) shifted to epi-benthic meiofauna (Marchand & Masson 1989). If metamorphosing larvae fed on plankton during immigration towards coastal areas, then it was not for want of more suitable prey. The amount of meiobenthic prey did not limit per se the settlement of soles on the shelf (Amara & Bodin 1995), though planktonic prey were abundant from offshore to coastal areas (Koutsikopoulos et al. 1991).

When they underwent metamorphosis, sole of steps 4b and 5a had the same feeding abilities, irrespective of their difference in size and diel vertical distribution. They both captured the same proportions of planktonic prey (Figure 7) in the lower part of the water column, where diel vertical migrations tended to concentrate larvae in step 4b during the day (Figure 4). Although sole of these steps were not yet able to settle, they learned to feed on small epi-benthic prey, but at a very low rate. Feeding could be triggered by photoperiod, given the enhanced visual performance of these stages (Blaxter 1988, Champalbert et al. 1992). Nocturnal ascents of larvae on the shelf (Koutsikopoulos et al. 1991), and a fortiori of late larvae in the estuary (Marchand & Masson 1989), thus appear independent of feeding activity. Feeding can, however, induce feedback effects on active movements and swimming behaviour and these periodic effects have to be expected if, as suggested by Neilson & Perry (1990), a rhythmic feeding activity resets the internal clock of other circadian rhythms.

Aperiodic events, such as mixing waters due to wind stress, can cause starvation and induce differences between well-fed larvae, which tend to sink towards the bottom, and deprived larvae, which swim for food (Creutzberg et al. 1978, Macquart-Moulin et al. 1991). Low-feeding rates occurred under wind-stress events and spring-tide conditions (Figures 7, 8). According to Sundby (1995), wind speeds above 10 to 15 m s^{-1} yield high feeding rates for cod larvae. Wind stress did not exceed 17 m s^{-1} during the early April 1993 cruise, but tide-induced factors probably contributed to the strong vertical mixing of the water column, which was registered from inshore areas to the spawning grounds. Larvae were probably dispersed and feeding impaired by disruption of the small-scale patchiness of planktonic prey, a situation well-studied at sea. For example, Munk et al. (1989) observed that the size-related vertical distribution of herring, *Clupea harengus*, larvae could be temporarily modified under windy conditions. In contrast, the calmer, neap-tide conditions in late April restarted the larvae's feeding activity. As a result, gut content analyses revealed an underlying periodic feeding activity, with daytime ingestion and night-time digestion until sunrise, which reboosted the diurnal feeding activity.

Similarly, a 30 h experiment carried out with laboratory-reared metamorphosing sole resulted in high levels of feeding activity during the light periods, followed by gut evacuation during the dark period (F. Lagardère unpublished). This diel feeding activity was previously reported for sea-caught larvae of walleye pollock, *Theragra chalcogramma*, by Canino & Bailey (1995), who observed the lowest activity around sunrise due to gut evacuation. These authors emphasized the implications of digestion dynamics in gut content analyses due to the gut residence of prey, estimated up to 8 h from experiments. Gut residence and evacuation probably explained the difference in phase that we observed between the carbon contents of guts and light intensities (Figure 8), resulting in negative but not signicant correlation coefficients (Table 5). Nevertheless, the feeding condition of sole larvae in nature suggests that metamorphosis does not entail starvation. These findings were corroborated by previous analyses of nutritional status: metamorphosing larvae caught offshore and inshore were found in 'good condition' from histological indices and estimation of lipid supplies (Boulhic et al. 1992). This feeding activity can explain that metamorphosing sole did not stop growing during the 10–15 days of their transfer. The average lengths-at-step increased from 9 mm total length (TL) for larva steps 4b (Amara & Lagardère 1995), reaching 10 mm TL when larvae entered the Bay of Vilaine, mostly at steps 5a, and 12 mm TL when they settled in the estuary as step 5b (Marchand 1991).

In conclusion, the timing of coastal immigration of metamorphosing sole to the bays and estuaries of the northern Bay of Biscay appears ontogenetically determined by the initiation of metamorphosis. If cross-shelf dispersion was shown to prevail in early larvae (Koutsikopoulos et al. 1991), then our study reinforces the hypothesis of a coastal transport cued by semi-diurnal and fornightly tidal cycles, modulated by swimming (Le Cann et al. 1992, Champalbert & Koutsikopoulos 1995), and feeding behaviour (this study), and clued by environmental changes encountered as migrants progress shorewards (Gibson 1997). A balance between neap-tide conditions, which result in coastal accumulation of metamorphosing larvae, and spring-tide conditions, which probably override the step-related patterns of vertical distribution, could then orientate immigrating sole towards the surroundings of the Bay of Vilaine and provide the Vilaine estuary with space-time determined 'pulses' (Marchand 1991) of new settlers. Behavioural changes enable them to adjust their vertical position and remain in layers more compatible with their enhanced sensory faculties. Amara et al. (1998) suggested that the swimbladder could help in vertical movement regulation, depending on larva feeding requirements and/or prevailing currents. Metamorphosing sole do not suffer from starvation, neither as immigrating larvae, by maintaining a diurnal planktonic feeding, nor as new settlers, which complete metamorphosis in nearshore nurseries by shifting to epi-benthic prey (Marchand & Masson 1989, Marchand 1992). Both very likely store energetic supplies, since growth neither stops nor decreases.

During metamorphosis, sole larvae thus gain more advantageous than disadvantageous attributes, due to the ontogenetic pathway of body remodeling, physiological preparation and behavioural changes. These feeding and growth performances during metamorphosis improve survival, which probably contributes to the natural selection of features associated with flatfish adaptation to benthic life (Fuiman 1997).

Acknowledgements

This study, as part of the 'Programme National sur le Déterminisme du Recrutement', was supported by PNDR contract (n°93 551 1076) and resulted from combined projects between Ifremer and CREMA-L'Houmeau (CNRS/Ifremer). The authors are grateful to the captains and crews of RV 'Thalassa' and 'Gwen Drez', and to colleagues sharing the tasks before and during the cruises, particularly N. Lacroix, Y. Désaunay and C. Koutsikopoulos. They also thank P. Bourriau for providing them with all the offshore sampling data, V. Loyer for her contribution to data analyses and R. Knutsen for improving the English text.

References cited

Al-Maghazahi, S.J. & R. Gibson. 1984. The developmental stages of turbot, *Scophthalmus maximus*. J. Exp. Mar. Biol. Ecol. 82: 35–51.

Amara, R. & P. Bodin. 1995. L'environnement trophique méiobenthique en dehors des nourriceries côtières est-il favorable à l'installation des jeunes soles? Oceanologica Acta 18: 583–590.

Amara, R. & F. Lagardère. 1995. Taille et âge au début de la métamorphose chez la sole (*Solea solea* (L.)) du Golfe de Gascogne. ICES J. Mar. Sci. 52: 247–256.

Amara, R., F. Lagardère & Y. Désaunay. 1993. Seasonal distribution and duration of the planktonic stage of Dover sole, *Solea solea*, in the Bay of Biscay: an hypothesis. J. Fish Biol. 43 A: 17–30.

Amara, R., Y. Desaunay & F. Lagardère. 1994. Seasonal variation in growth of larval sole, *Solea solea* (L.) and consequences on the success of larval immigration. Neth. J. Sea Res. 32: 287–298.

Amara, R., J.C. Poulard, F. Lagardère & Y. Désaunay. 1998. Comparison between the life cycles of two Soleidae, the common sole, *Solea solea*, and the thickback sole, *Microchirus variegatus*, in the Bay of Biscay (France). Env. Biol. Fish. 53: 193–209.

Appelbaum, S., J.W. Adron, S.G. George, A.M. Mackie & B.J.S. Pirie. 1983. On the development of the olfactory and the gustatory organs of the Dover sole, *Solea solea*, during metamorphosis. J. Mar. Biol. Assoc. U.K. 63: 97–108.

Arbault, S., P. Camus & C. Le Bec. 1986. Estimation du stock de sole (*Solea vulgaris*, Quensel 1806) dans le golfe de Gascogne à partir de la production d'oeufs. J. Appl. Ichthyol. 4: 145–156.

Balon, E.K. 1975. Terminology of intervals in fish development. J. Fish. Res. Board Can. 32: 1663–1670.

Blaxter, J.H.S. 1972. Brightness discrimination in larvae of plaice and sole. J. Exp. Biol. 57: 693–700.

Blaxter, J.H.S. 1980. The effect of hydrostatic pressure on fishes. pp. 369–386. *In*: M.A. Ali (ed.) Environmental Physiology of fishes, Plenum Press, New York.

Blaxter, J.H.S. 1988. Sensory performance, behavior and ecology of fish. pp. 204–232. *In*: J. Atema, R.R. Fay, A.N. Popper & W.W. Tavolga (ed.) Sensory Biology of Aquatic Animals, Springer-Verlag, New-York.

Blaxter, J.H.S. & M. Staines. 1970. Pure-cone retinae and retinomotor responses in larval teleosts. J. Mar. Biol. Ass. U.K. 50: 449–460.

Boulhic, M. & J. Gabaudan. 1992. Histological study of the organogenesis of the digestive system and swim bladder of the Dover sole *Solea solea* (Linnaeus, 1758). Aquaculture 102: 373–396.

Boulhic, M., R. Galois, C. Koutsikopoulos, F. Lagardère & J. Person-Le Ruyet. 1992. Etat nutritionnel, croissance et survie des stades pélagiques de la sole, *Solea solea* (L.) du golfe de Gascogne. Ann. Inst. Océanogr. Paris 68: 117–139.

Campos, W.L., M. Kloppmann & H. von Westernhagen. 1994. Inferences from the horizontal distribution of dab *Limanda limanda* (L.) and flounder *Platichthys flesus* (L.) larvae in the southeastern North Sea. Neth. J. Sea Res. 32: 277–286.

Canino, M.F. & K.M. Bailey. 1995. Gut evacuation of walleye pollock larvae in response to feeding conditions. J. Fish Biol. 46: 389–403.

Champalbert, G. & C. Koutsikopoulos. 1995. Behaviour, transport and recruitment of Bay of Biscay sole (*Solea solea*): laboratory and field studies. J. Mar. Biol. Assoc. U.K. 75: 93–108.

Champalbert, G., C. Macquart-Moulin, G. Patriti & L. Le Direach-Boursier. 1992. Light control of vertical movements of larvae and juvenile sole (*Solea solea* L.). Mar. Behav. Physiol. 19: 263–283.

Champalbert, G., J. Marchand & J. Le Campion. 1994. Rheotaxis in juvenile sole *Solea solea* (L.): influence of salinity and light conditions. Neth. J. Sea Res. 32: 309–319.

Creutzberg, F., A.Th.G.W. Eltink & G.J. van Noort. 1978. The migration of plaice larvae *Pleuronectes platessa* into the western Wadden Sea. pp. 243–251. *In*: D.S. McLusky & A.J. Berry (ed.) Physiology and Behaviour of Marine Organisms, Pergamon Press, Oxford.

Fortier, L. & R.P. Harris. 1989. Optimal foraging and density-dependent competition in marine fish larvae. Mar. Ecol. Prog. Ser. 51: 19–33.

Fuiman, L.A. 1997. What can flatfish ontogenies tell us about pelagic and benthic lifestyles? J. Sea Res. 37: 257–267.

Gibson, R.N. 1997. Behaviour and the distribution of flatfishes. J. Sea Res. 37: 241–256.

Gibson, R.N., J.H.S. Blaxter & S.J. De Groot. 1978. Development changes in the activity rhythms of the plaice (*Pleuronectes platessa* L.). pp. 169–186. *In*: J.E. Thorpe (ed.) Rhythmic Activity of Fishes, Academic Press, New York.

Håkanson, J.L., S.H. Coombs & P. Ré. 1994. Lipid and elemental composition of sprat (*Sprattus sprattus*) larvae at mixed and stratified sites in the German Bight of the North Sea. ICES J. Mar. Sci. 51: 147–154.

Harding, D. & J.W. Talbot. 1973. Recent studies on the eggs and larvae of the plaice (*Pleuronectes platessa*) in the Southern Bight. Rapport et Procès-verbaux des Réunions du Conseil permanent international pour l'Exploration de la Mer 164: 102–103.

Harvey, R., R.H.S. Blaxter & R.D. Hoyt. 1992. Development of superficial and lateral line neuromasts in larvae and juveniles of plaice (*Pleuronectes platessa*) and sole (*Solea solea*). J. Mar. Biol. Assoc. U.K. 72: 651–668.

Koutsikopoulos, C. & N. Lacroix. 1992. Distribution and abundance of sole (*Solea solea* (L.)) eggs and larvae in the Bay of Biscay between 1986 and 1989. Neth. J. Sea Res. 29: 81–91.

Koutsikopoulos, C., L. Fortier & J. Gagné. 1991. Cross-shelf dispersion of Dover sole eggs and larvae (*Solea solea*) in Biscay Bay and recruitment to inshore nurseries. J. Plankt. Res. 13: 923–945.

Koutsikopoulos, C., D. Dorel, Y. Désaunay, B. Le Cann & A. Forest. 1993. Interactions entre processus physiques et comportement individuel: conséquences sur l'organisation et le fonctionnement du stock de sole (*Solea solea* L.) du golfe de Gascogne. pp. 49–74. *In*: D. Gascuel, J.L. Durand & A. Fonteneau (ed.) Premier Forum Halieumétrique, Collection Colloques et Séminaires, ORSTOM Editions, Paris.

Lagardère, F. 1989. Influence of feeding conditions and temperature on the growth rate and otolith-increment deposition of larval Dover sole (*Solea solea* (L.)). Rapport et Procès-verbaux des Réunions du Conseil permanent international pour l'Exploration de la Mer 191: 390–399.

Lagardère, J.P. 1987. Feeding ecology and daily food consumption of common sole, *Solea vulgaris* Quensel, juveniles on the French Atlantic coast. J. Fish Biol. 30: 91–104.

Last, J.M. 1978. The food of four species of pleuronectiform larvae in the eastern English Channel and southern North Sea. Mar. Biol. 45: 359–368.

Le Cann, B., C. Koutsikopoulos, N. Lacroix & G. Champalbert. 1992. Interactions des processus physiques et biologiques dans la dynamiques des stades pélagiques de la sole *Solea solea* (L.), sur le plateau continental du golfe de Gascogne, France. Ann. Inst. Océanogr. Paris 68: 107–115.

MacCall, A.D. 1990. Dynamic geography of marine fish populations. Washington Sea Grant Program, Univ. Washington Press, Seattle. 153 pp.

Macquart-Moulin, C., C. Castelbon, G. Champalbert, D. Chikhi, L. Le Direach-Boursier & G. Patriti. 1989. The role of barosensitivity in the control of migrations of larval and juvenile sole (*Solea solea* L.): influence of pressure variations on swimming activity and orientation. Rapport et Procès-verbaux des Réunions du Conseil permanent international pour l'Exploration de la Mer 191: 400–408.

Macquart-Moulin, C., G. Champalbert, B. Howell, G. Patriti & C. Ranaivoson. 1991. La relation alimentation-fixation benthique chez les jeunes soles *Solea solea* L. métamorphosées. Evidences expérimentales. J. Exp. Mar. Biol. Ecol. 153: 195–205.

Marchand, J. 1991. The influence of environmental conditions on settlement, distribution and growth of 0-group sole (*Solea solea* (L.)) in a macrodital estuary (Vilaine, France). Neth. J. Sea Res. 27: 307–317.

Marchand, J. 1992. Métamorphose et passage pelagos/benthos chez la sole (*Solea solea*): synthèse des données acquises dans le site atelier de la Vilaine (1986–1990). Ann. Inst. Océanogr. Paris 68: 141–150.

Marchand, J. & G. Masson. 1989. Process of estuarine colonization by 0-group sole (*Solea solea*): hydrological conditions, behaviour, and feeding activity in the Vilaine estuary. Rapport et Procès-verbaux des Réunions du Conseil permanent international pour l'Exploration de la Mer 191: 287–295.

Munk, P., T. Kiørboe & V. Christensen. 1989. Vertical migrations of herring, *Clupea harengus*, larvae in relation to light and prey distribution. Env. Biol. Fish. 26: 87–96.

Neilson, J.D. & R.I. Perry. 1990. Diel vertical migrations of fishes: an obligate or facultative process? pp. 115–168. *In*: J.H.S. Blaxter & A.J. Southward (ed.) Advances in Marine Biology, Academic Press, London.

Norcross, B.L. & R.F. Shaw. 1984. Oceanic and estuarine transport of fish eggs and larvae: a review. Trans. Amer. Fish. Soc. 113: 153–165.

Power, J.H. 1984. Advection, diffusion and drift migration of larval fish. pp. 27–37. *In*: J.D. McCleave, G.P. Arnold, J.J. Dodson & W.H. Neill (ed.) Mechanisms of Migration in Fishes, Plenum Press, New York.

Rijnsdorp, A.D., M. van Stralen & H.W. van der Veer. 1985. Selective tidal transport of North Sea plaice larvae *Pleuronectes platessa* in coastal nursery areas. Trans. Amer. Fish. Soc. 114: 461–470.

Ryland, J.S. 1966. Observations on the development of larvae of the plaice, *Pleuronectes platessa* L., in aquaria. J. Cons. Int. Explor. Mer 30: 177–195.

Sundby, S. 1995. On the dome-shaped relationship between wind-generated turbulence and larval feeding. Neth. J. Sea Res. 34: 243–244.

Symonds, D.J. & S. Rogers. 1995. The influence of spawning and nursery grounds on the distribution of sole *Solea solea* (L.) in the Irish Sea, Bristol Channel and adjacent areas. J. Exp. Mar. Biol. Ecol. 190: 243–261.

Veer, H.W. van der. 1986. Immigration, settlement and density-dependent mortality of a larval and early post-larval O-group plaice (*Pleuronectes platessa*) population in the western Wadden Sea. Mar. Ecol. Prog. Ser. 29: 223–236.

Williams, R. & D.B. Robins. 1982. Effects of preservation on wet weight, dry weight, nitrogen and carbon contents of *Calanus helgolandicus* (Crustacea: Copepoda). Mar. Biol. 71: 271–281.

Part 4. Behaviour and ontogeny

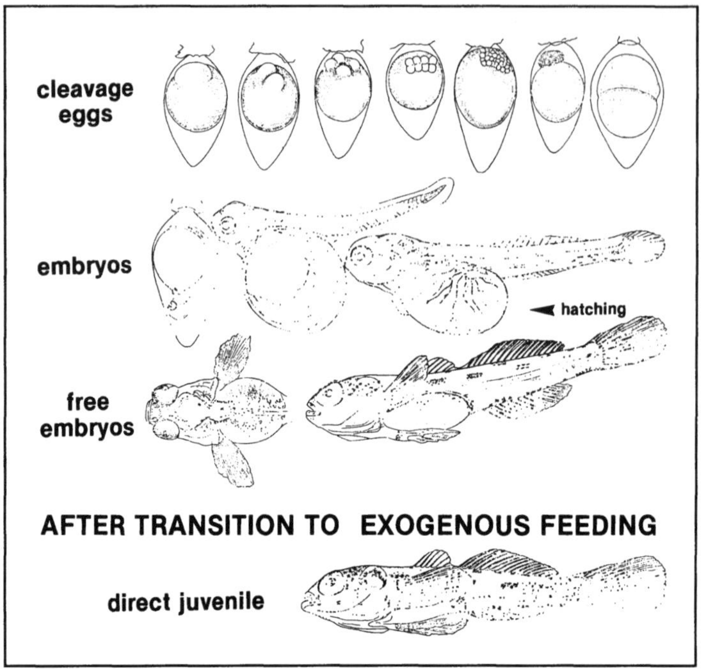

Neogobius melanostomus

Environmental Biology of Fishes **56:** 231–242, 1999.

Ontogeny of aggressive behaviour in schools of yellowtail, *Seriola quinqueradiata*

Yoshitaka Sakakura & Katsumi Tsukamoto
Division of Fisheries Ecology, Ocean Research Institute, The University of Tokyo, Minamidai 1-15-1, Nakano, Tokyo 164, Japan (e-mail: sakakura@ori.u-tokyo.ac.jp)

Received 29 September 1997   Accepted 19 October 1998

Key words: social rank, schooling behaviour, cannibalism, otolith

Synopsis

We studied ontogenetic changes in social interactions, especially in aggressive behaviour of the migratory marine yellowtail, *Seriola quinqueradiata* (Carangidae), and compared these to morphological and physiological changes. No agonistic interactions were observed during the larva period until 10 mm in total length (TL), at approximately 20 days after hatching. Typical shivering behaviour with 'J-posture' was observed during metamorphosis, when fin rays and calcification of vertebra were completed and there was an increase of tissue thyroid hormone. The onset of aggressive behaviour was just after metamorphosis to the juvenile period, and coincided with a significant increase in tissue cortisol levels. The onset of schooling behaviour was at 12 mm TL, slightly after the onset of aggressive behaviour. From observations of individual aggressive behaviour within juvenile schools, we found three categories of social rank: dominants (10–20%), intermediates (10–20%), and subordinates (60–80%). There was an inverse relationship between social rank and cortisol concentration. Otoliths of dominant fish in 8 experimental groups were labeled and the fish were returned to their groups. Six labeled dominants appeared after 1 day and three after 1 week rearing, respectively, indicating that social rank was maintained for at least 1 week (binomial distribution, $p < 0.05$). Dominants were larger than subordinates after 1 day rearing, whereas dominants were smaller after 1 week rearing. From long-term rearing experiments using individual otolith marking, larvae that showed the 'J-posture' more frequently tended to become dominants after metamorphosis, indicating a positive correlation between the 'J-posture' and aggressive behaviour. Synthesizing all results from behavioural experiments, we generated a behavioural model for the triggering mechanism of aggressive behaviour and size selection of school members.

Introduction

Larvae require remodeling – metamorphosis – in order to replace temporary organs by definitive structures (Balon 1984), and habitat and/or niche shifts may occur during this process. Fukuhara (1992) pointed out that behavioural development is the key to understanding the life mode of fish larvae and juveniles in relation to morphological development and organogenesis. Many fish species develop social responses in their early life (Noakes & Godin 1988), and the social behaviour can help define developmental steps. Development of social behaviour is also of practical importance to improve the quality of reared fish for stock enhancement (Olla et al. 1994, Svåsand 1993).

Social behaviour includes all behaviours directly related to actual or potential encounters between individuals within a species (Noakes 1978). This includes agonistic and reproductive behaviour in addition to gregarious and co-operative behaviour. Both agonistic and gregarious behaviour in early life of fishes can be represented by aggressive behaviour with social rank and schooling, respectively. Social rank among fishes is usually related to such important aspects of life history as feeding and breeding (Bond 1979), and is functionally associated with survival in the wild through its influence on competition for food, space (Polis 1981) and mating. Schooling behaviour is closely associated with the enhancement of foraging and predator avoidance (Hamilton 1971, Noakes 1978, Magurran 1990, Pitcher & Parrish 1993). Recently, the presence of individual differences in the behaviour of members of fish schools, such as positioning among individuals, has been reported (Magurran 1993, Pitcher et al. 1982), although schools have long been considered to be egalitarian and leaderless societies (Breder 1954, Shaw 1962, 1978). In fish schools, social rank may also appear as a result of individual differences in behaviours, including aggressive behaviour.

In the early life intervals of the yellowtail *Seriola quinqueradiata* (Temminck & Schlegel, Carangidae), a highly migratory marine species around Japan, seemingly incompatible behaviours such as aggressive and schooling behaviour appear simultaneously (Sakakura & Tsukamoto 1996). After a pelagic larva period (Fukuhara et al. 1986), juvenile yellowtails of uniform body size (Uchida et al. 1958, Anraku & Azeta 1965, Sakakura & Tsukamoto 1996) and age (Safran 1990, Sakakura & Tsukamoto 1997a) aggregate around drifting seaweed. Cannibalism, coinciding with aggressive behaviour, was observed among juvenile schools both in wild (Anraku & Azeta 1965, Sakakura & Tsukamoto 1996) and artificial rearing conditions (Imaizumi 1993, Sakakura & Tsukamoto 1996). The authors of previous studies claimed that there are individual differences in aggressive behaviour termed as aggressive tendencies (Sakakura & Tsukamoto 1996), and there is a social rank in the juvenile schools of this species (Sakakura & Tsukamoto 1997b, 1998a, Sakakura et al. 1998a). Since both aggressive behaviour and cannibalism cause high mortality in the production of young yellowtail (Imaizumi 1993), we need to understand the ontogeny of aggressive behaviour in the early life intervals of this species. There are studies on morphological development of yellowtail because of its importance in fisheries and aquaculture (Uchida et al. 1958, Umeda & Ochiai 1973, Fukuhara et al. 1986, Chantanachookhin et al. 1991). However, little is known of the behavioural development in relation to morphological and physiological development.

In the present study, we examined the ontogeny of aggressive behaviour and compared this to morphological and physiological changes. Next, we studied the influence of school composition, school stability and environmental factors on social rank in juvenile yellowtail. Finally, we examined the transition from the typical behaviour in the larva period (J-posture) to aggressive behaviour (juvenile period), tagging fish individually by otolith marking, which is almost certainly the only marking method applicable in behavioural experiments with fish larvae and juveniles as it is known not to affect behaviour, growth or survival (Tsukamoto 1985).

Material and methods

Rearing conditions

Yellowtail were matured artificially by human chorionic gonadotropin injection (about 10 000 UI kg^{-1} $fish^{-1}$) and allowed to spawn naturally in a 90 m^3 concrete tank at the Goto station of the Japan Sea Farming Association (JASFA) in Nagasaki Prefecture. Fertilized eggs were maintained in a 0.5 m^3 tank, and after 2 days, approximately 900 000 free embryos were obtained (29 April 1993 and 10 May 1995 for the stability of social rank experiment; 24 April 1994 for the experiment on the possible behavioural transition from larva to juvenile). Two days after hatching (day 2), free embryos were transferred to an outdoor concrete rearing pond (90 m^3). The larvae were fed with the rotifer *Brachionus plicatilis* cultivated with *Nannochrolopsis* sp. between days 3 and 20, with newly hatched *Artemia salina* nauplii enriched with feed oil (Riken feed oil, Riken Japan) between days 7 and 24, and with dry pellets (C 400–1000, Kyowa Hakko, Japan) from day 22 until the end of the experiments. They were fed at least three times a day, with food available during the entire day. Water temperature ranged from 22 to 25°C under natural light conditions. We collected samples from the water column from surface to the bottom of the rearing pond using a long plastic pipe (5 cm in diameter) 10 to 15 times in the night (dark condition) and counted the fish density following estimation of live fish numbers in the rearing pond. Dead and suffocating fish were removed from the bottom by siphoning, and numbers were counted every day from day 7 to day 40. 'Suffocation' occurred in some cannibals observed to have a large prey in the mouth; these were used for the index of cannibalism (Sakakura & Tsukamoto 1996).

Ontogeny of aggressive behaviour

Fish used for behavioural observations, and for both of morphological and physiological measurement, were randomly sampled from the rearing pond with 131 buckets at night, at intervals of 3 to 5 days from days 1 to 40. We chose night time for sampling because it was easy to collect a sample while minimizing stress, as juvenile yellowtails cease swimming and drift at the surface in dense patches at night (Sakakura & Tsukamoto 1997a,b).

For the behavioural study, 60 fish were observed at each age. Six groups, each consisting of 10 fish (2 ind. l^{-1}) selected from the bucket sample to minimize size variation, were introduced into experimental tanks (30 cm in diameter, 5 l) in a water bath at 22°C. Fish were kept in the experimental tanks overnight, then fed *Artemia* 4 h before observations. Following Sakakura & Tsukamoto (1996), aggressive behaviours were classified as: (1) **Aim**, a dominant holds position towards a subordinate hovering for a short period (ca. 1 sec); (2) **Chase**, a dominant bursts towards and follows a subordinate for 1–10 sec. These 2 phases were observed in a fixed order. Sometimes a dominant attacks and bites at the tail or body of a subordinate following Chase behaviour. Chase behaviour was observed in every instance of Aim behaviour, and since Chase behaviour was easy to observe and record, its frequency was used as an index of aggressive behaviour. We also observed the frequency of '**J-posture**' during which fish held a position, hovering with pectoral fins and bending the body strongly in a J or reversed J shape for about 10 sec (Sakakura & Tsukamoto 1996). The frequencies of both Chase behaviour and J-posture were recorded for 5 min in each tank. Experiments were conducted during the day (10:00–12:00 h) at intervals of 3 to 5 days from days 7 to 40. Following behavioural observations, all fish were immediately anesthetized and the total length (TL) of each fish was measured with a micrometer or calipers. MS222 (Tricaine, SIGMA) was used as the anesthetics in all experiments.

Fish for physiological measurement were immediately anesthetized and stored frozen at −85°C until analysis. TL and wet body weight (BW mg) of samples at each age were measured individually. Then samples were pooled into three to five groups of ca. 100 mg in wet weight (about 5 to 50 fish for one group) at each age until day 20. From day 20 to day 39, fish were measured individually. Whole-body immunoreactive corticosteroid (cortisol) and thyroid hormone (thyroxine, T4) were measured using the method of Hiroi et al. (1997) and Tagawa & Hirano (1989), respectively.

Fish used for morphological investigation were immediately anesthetized and preserved in 10% buffered seawater formalin solution. TL of 20 to 50 fish at each age was measured under a microscope or by calipers. The morphology of fins were also observed followed by counting of fin rays under a microscope. Fish cleared and stained with alcian blue (Alcian blue 8 GX, Wako Japan) and alizarin red (Alizarin red S, Wako Japan) for determination of cartilage-bone development (Potthoff 1984), were examined for 20 individuals of each age.

Social rank in a school

Preliminary observations to classify the social rank were conducted using 100 fish of comparable size (23.7 ± 2.5 mm TL). Ten fish were transferred to each of 10 white round experimental tanks, under the same experimental conditions as the ontogeny of aggressive behaviour experiment. After 1 h of acclimation, the behaviour of fish in each experimental tank was observed from above for a duration of 10 min, and the first series of aggressive fish was quickly scooped out by a hand net. Four hours after removal of the first group of aggressive fish, a second series of aggressive fish, which appeared after the removal, was also taken out of the tank in the same manner as described above and pooled with the first group of aggressive fish. The behaviour of fish in each experimental tank was observed for an additional hour. All fish were anesthetized and TL was measured by calipers.

Physiological status in social rank was observed using 20 fish (27.0 mm TL). Fish were introduced into an experimental tank (55 cm in diameter containing 100 l seawater at 22°C), and acclimated for 4 h. Fish were classified into three categories based on 10 min of behavioral observation as follows: when fish A was chasing fish B, fish A was classified as 'dominant' and fish B as 'subordinate'; a fish showing no agonistic interactions within 10 min was classified as an 'intermediate'. Four pairs of 'dominants' and 'subordinates' were easily discriminated and scooped out immediately, because these pairs frequently swam away from the school. Following this treatment, 'intermediates' ($n = 12$) consisting of one synchronized school were immediately scooped out. After the above treatments and observations, all fish were immediately scooped out and anesthetized, and then stored at −85°C until used for TL and cortisol measurements.

In order to examine the durability of social rank, 1000 juveniles (ca. 50 mm TL) were transferred from the rearing pond to a round black poly-carbonate tank containing 500 l seawater on 10 July 1995 (day 66). Fish were kept as a stock in running seawater (25°C, flow rate 8 l h^{-1}) and were fed a commercial diet (C-1000, Kyowa Hakko, Japan) at a rate of 3% of the body weight, three times per day. Eight white round tanks (35 cm in diameter) containing 20 l seawater were set up in a water bath at 25°C and aerated. At 17:00 h on 19 July 1995 (1-day experiment) or 11 July 1995 (1-week experiment), a total of 80 fish from the stock tank were introduced into the experimental tanks (10 fish $tank^{-1}$) and were acclimated for 1 h. Based on a 10-min observation of each tank, the most aggressive fish (dominant), with the highest frequency of chase behaviour, was identified and removed by a hand net (first treatment). Using the technique of Tsukamoto (1988), the dominants were labeled individually in 400 ppm of ALC (Alizarin complexone, Wako, Japan) in another group of 8 separate tanks (30 cm diameter, 20 l) for 15 h. The following morning at 9:00 h, the dominants were again returned to the previous tanks from which they originated.

In the 1-day experiment, fish in all the experimental tanks were fed with C-1000 (Kyowa Hakko, Japan) at a rate of 3% of the BW at 10:00 h 1 h after reintroduction of an ALC-labeled individual, and acclimated for 6 h thereafter. We observed fish in each tank for 10 min, and scooped out the dominant fish from each tank with a hand net (second treatment) and anesthetized. The dominant fish and the fish remaining in the experimental tanks were anesthetized and preserved in 90% ethanol in order to prevent degradation of otoliths (Brothers 1984). In the 1-week experiment, all experimental fish were fed the commercial diet C-1000 (Kyowa Hakko, Japan) twice a day (10:00, 16:00 h) at a rate of 3% of the body weight. Half the volume of the seawater was exchanged with fresh seawater after every feeding time. Fish were reared for 1 week; then the dominant fish of each tank were identified and scooped out by a hand net, and all fish were preserved as in the 1-day experiment. TL of the fish of both experiments were measured using calipers and sagittal otoliths were also extracted. ALC labels in the otolith was examined under a UV-light microscope (Tsukamoto et al. 1989).

Possible behavioural transition from larva to juvenile

Fish of 8 mm TL (day 17) were sampled randomly from the rearing pond using a white plastic beaker. The initial size range was from 7.6 to 9.4 mm TL for 30 fish. Based on a 1-min observation, the fish showing J-posture were identified and removed using a large glass pipette and transferred into a tank (30 cm diameter, 20 l, 22°C). Other fish (non-J) were transferred into another tank. This treatment was repeated until the total number of the J-posture fish reached 200. Non-J fish were also collected until 200. Following the procedure of Tsukamoto (1988), the J-posture fish were labeled in 100 ppm of ALC (Alizarin complexone, Wako, Japan) for 15 h. After labeling, J-posture and non-J fish were placed in a 500 l polycarbonate tank and reared until 20 mm TL (18 days) fed with sufficient *Artemia* 3 times a day. Half the volume of seawater (22°C) was exchanged with fresh seawater once a day. We observed fish in the tank, and scooped out the dominant and subordinate fish from the tank with a hand net and anesthetized them. Fish were preserved in a 90% ethanol solution. TL from both experiments was measured using calipers and sagittal otoliths were also extracted. The ALC label in the otolith was examined under a UV-light microscope (Tsukamoto et al. 1989).

Statistical analysis

Between-group comparisons of TL were undertaken using Student's *t*-test for 2 groups, and using Duncan's new multiple range test (Duncan 1955) after a one-way analysis of variance (ANOVA) treatment among 3 or more experimental groups, and after Bartlett's test for comparison of variances ($p < 0.05$, Ichihara 1990).

In the ontogeny of aggressive behaviour experiment, the median frequency of 'chase' among 6 tanks was used as the representative value of an age (size) group, and differences in the frequency of 'chase' among age groups were compared using the Mann-Whitney *U*-test. Variances of cortisol and T4 concentrations were compared using Bartlett's test. ANOVA was applied when there was no significant difference between the variances of the different groups ($p > 0.05$ by Bartlett's test). In cases where significant differences were found among the means by ANOVA ($p < 0.05$), Duncan's new multiple range test was applied for comparison among groups. On the other hand, when significant differences existed among variances ($p < 0.05$ by Bartlett's test), the Kruskal-Wallis test for multiple groups was applied to determine differences among the medians (Ichihara 1990). Significant differences among medians were further examined by the Mann-Whitney *U*-test (Ichihara 1990) against the previous age group.

For the physiological status in social rank experiment, cortisol concentrations were compared using Duncan's new multiple range test after ANOVA treatment among 'dominants', 'subordinates' and 'intermediates', after Bartlett's test for comparison of variances ($p < 0.05$).

In the durability of social rank experiment, a binomial distribution was plotted to determine the theoretical possibility that the dominant fish at the second treatment coincided with the dominant fish with the ALC marking in the first treatment for both the 1-day and the 1-week experiments (Ichihara 1990). G-test (equivalent to Chi-square test) for independence was carried out to determine the correlation between the distribution of J-posture fish and dominant fish (Sokal & Rohlf 1995).

Results

Ontogeny of aggressive behaviour

Total fish survival at day 40 was 71 000, a survival rate of 7.9% (Figure 1a). The highest mortality was observed at day 7 (144 000) and decreased until day 22. After day 22 (10.7 ± 1.1 mm TL), at which suffocation and aggressive behaviour were initially observed in the rearing pond, mortality increased again and both changes in number of dead fish and suffocation showed the same pattern. Fish grew relatively slowly until day 22, but more rapidly thereafter (Figure 1b). The coefficient of variation (CV) of total length increased between day 1 and day 25, and afterwards became stable at about 20%.

The onset of aggressive behaviour occurred after 12 mm TL and increased significantly until 30 mm TL (*U*-test, $p < 0.05$, Figure 2a). No agonistic interactions were observed before this size. J-posture was first recorded at 8 mm TL and showed a peak at 9.8 mm TL and decreased rapidly afterwards. The disappearance of the J-posture coincided exactly with the onset of aggressive behaviour. Polarized synchronized swimming ('schooling' behaviour, Pitcher 1983) was observed from 14 mm TL.

Cortisol concentrations remained at low levels until 9 mm TL, with a slight peak in the concentration observed at about 5 mm TL (*U*-test, $p < 0.05$, Figure 2b). Cortisol concentrations increased significantly at 9 mm TL (Duncan's new multiple range test, $p < 0.05$) and tended to increase further thereafter. T4 concentrations remained at low levels until 7 mm TL,

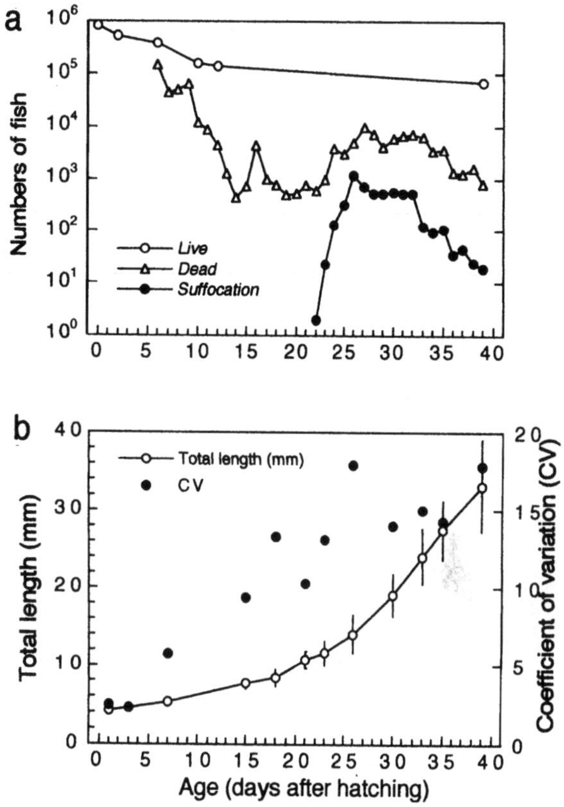

Figure 1. a – Survival curve of yellowtail in the rearing pond. Plots indicate number of fish (open circle = live fish, triangle = dead fish, solid circle = cannibalism determined from suffocation). b – Growth of yellowtail. Open circle indicates mean total length and bars indicate standard deviation. Solid circle represents coefficient of variation in total length.

increased significantly at 8 mm TL (U-test, $p < 0.05$) and increased afterwards.

The process of calcification in pterygiophores started at 9 mm TL and the entire structure retained alizarin red at 12 mm TL (from days 18 to 20, Figure 3). The pectoral fin became round to elongated, and the caudal peduncle was formed at the same body size. Fin-ray completion occurred at 10 mm TL, coincident with the calcification of vertebra.

Social rank in a school

Fish showed typical schooling behaviour as well as chase behaviour in all tanks. The first series of aggressive fish consisted of 1–3 fish (10–30%) from each tank, and the mean number ± SD was 1.4 ± 0.7 fish (14.0 ± 6.6%, $n = 10$). A second series of aggressive fish appeared 10 to 60 min after removing the first

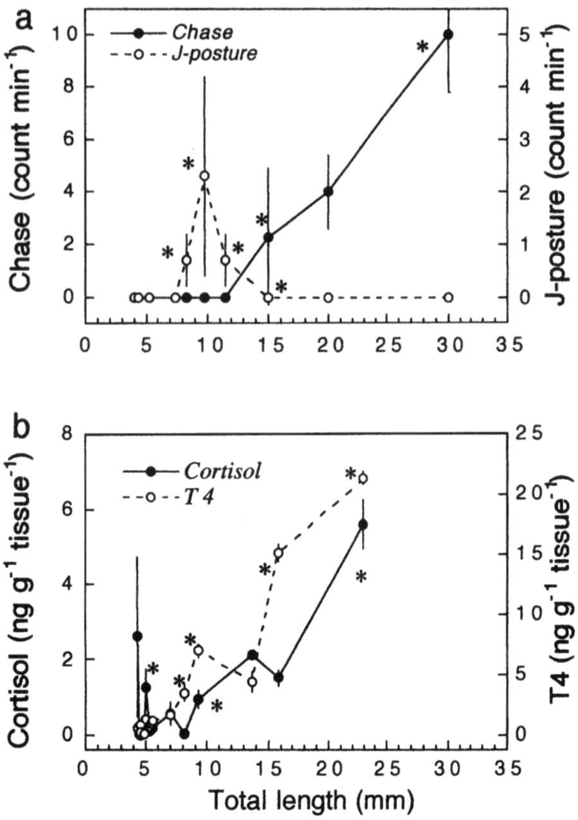

Figure 2. a – Changes in frequency of behavioural traits in experimental tanks. Vertical bars represent quartile deviation of medians ($n = 6$), and an asterisk indicates a significant difference ($p < 0.05$) from the previous point by *U*-test. b – Changes in whole-body thyroxine (T4, open circles) and cortisol concentrations (solid circles) in early life of yellowtail. Vertical bars represent standard errors of the means for both circles ($n = 3 < 7$ mm TL, $n = 5 < 10$ mm TL, $n = 20 \geq 10$ mm TL). Asterisks indicate a significant difference ($p < 0.05$) from the previous point by *U*-test or Duncan's new multiple range test.

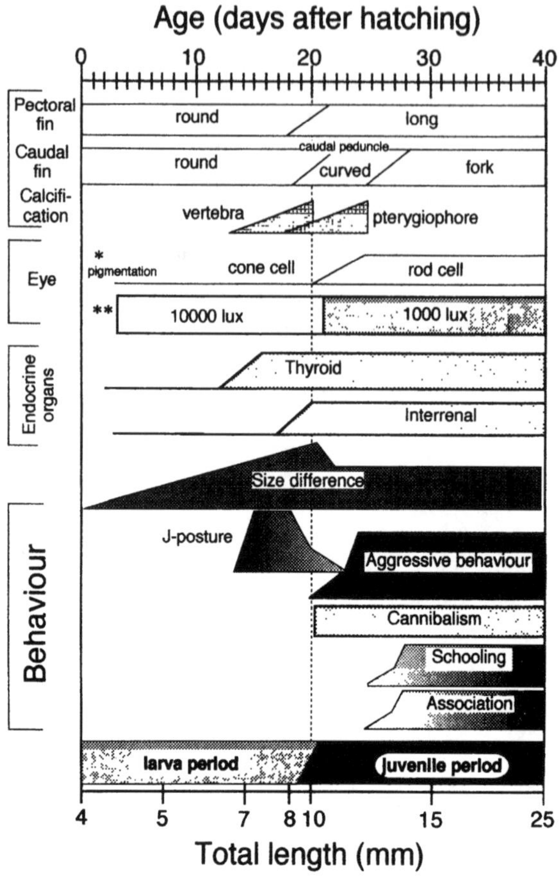

Figure 3. Simplified representation of early morphological and behavioural development in yellowtail. *modified from Ishida (1987 unpublished data), **shows phototaxis modified from Sakakura & Tsukamoto (1997b).

series of aggressive fish. The second series of aggressive fish were found in the same numbers (1 to 3 fish, 11.1 to 42.9%), and mean aggressive fish number ± SD was 1.6 ± 0.8 ($9.1 \pm 10.8\%$). There was no significant difference between aggressive fish occurrence at the first and second observation (*t*-test, $p > 0.1$). After the second observation, no further aggressive fish were observed. TL did not differ between aggressive fish (23.5 ± 2.1 mm, $n = 30$) and non-aggressive fish remaining in the experimental tanks (23.8 ± 2.7 mm, $n = 70$, *t*-test, $p > 0.05$).

The subordinate group (8.6 ± 1.6 ng g^{-1} tissue^{-1}, $n = 4$) showed the highest cortisol level (Duncan's new multiple range test, $p < 0.05$, Figure 4). The dominant group was the lowest (0.6 ± 0.3 ng g^{-1} tissue^{-1}, $n = 4$, Duncan's new multiple range test, $p < 0.05$). The intermediates showed an intermediate level (3.5 ± 1.7 ng g^{-1} tissue^{-1}, $n = 12$). The total lengths of the groups did not differ (dominant, 27.3 ± 1.8 mm; subordinate, 25.2 ± 3.0 mm; intermediate, 27.7 ± 3.1 mm: ANOVA, $p > 0.1$).

In the 1-day experiment, six labeled fish were found from the 8 dominant fish (75%) of second treatment (Table 1), with dominant fish at first treatment coinciding with the dominant fish at second treatment significantly ($p < 0.01$). Mean TL of dominants was significantly larger than that of subordinates (*t*-test, $p < 0.05$). In the 1-week experiment, three of 8 dominant fish (37%) had ALC label, with dominant fish at first treatment coinciding significantly with the dominant fish at second treatment ($p < 0.05$). TL of dominants were significantly smaller than those of subordinates (t-test, $p < 0.05$).

Possible behavioural transition from larva to juvenile

The survival rate was very high in this experiment (over 80%). Comparing the distribution to a Chi-square table (Table 2), dominant fish coincided significantly with the J-posture fish, indicating the possibility that fish formerly showing strong J-posture become aggressive fish as they grow (G-test, $p < 0.05$). Dominant fish were significantly larger than subordinate fish. On the other hand, there was no significant difference in size between J-posture fish and non-J fish. When 30 fish were sampled randomly, the numbers of J-posture fish were 14 and non-J fish were 16, respectively. There was no significant difference between the TL of J-posture fish (21.4 ± 2.7 mm) and non-J fish (21.6 ± 2.1 mm, t- test, $p > 0.05$).

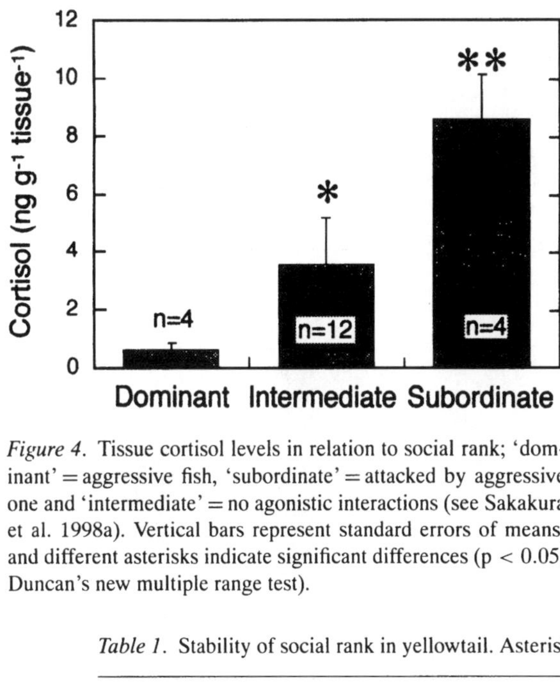

Figure 4. Tissue cortisol levels in relation to social rank; 'dominant' = aggressive fish, 'subordinate' = attacked by aggressive one and 'intermediate' = no agonistic interactions (see Sakakura et al. 1998a). Vertical bars represent standard errors of means, and different asterisks indicate significant differences ($p < 0.05$; Duncan's new multiple range test).

Discussion

Larva and juvenile period of yellowtail

Organs for swimming (skeleton, fin) and visual acuity were almost completely differentiated at 10 mm TL (Figure 3), the size at which the yellowtail is defined morphologically as a 'juvenile' according to the taxonomy of Kendall et al. (1984).

A small temporary peak of cortisol occurred on day 7 (about 5 mm TL, Figure 2b), when the interrenal first appears histologically (Figure 3, Chantanachookhin et al. 1991) and high mortality occurred in the rearing pond as a result of failure to ingest food properly (Figure 1a). We assume that this so-called critical period (point of no return, May 1974) for starvation in the larva yellowtails is the consequence of the occurrence of gluconeogenesis from muscle, resulting from the failure in exogenous feeding. Significant cortisol secretion and/or accumulation corresponded to the transition period from larva to juvenile in yellowtail at 9 mm TL, whereas thyroid hormones increased just before the beginning of transition (Figure 2b). Since thyroid hormone is involved in metamorphosis in fishes (Inui & Miwa 1985), the drastic changes in thyroid hormone and in calcification (Figure 3) of yellowtail at 8 mm TL can be explained as the onset of 'metamorphosis'.

In the metamorphosis interval of the Japanese flounder, *Paralichthys olivaceus*, increases in tissue cortisol level herald the increase of thyroid hormones (Jesus et al. 1991). Cortisol accelerates dorsal fin-ray resorption

Table 1. Stability of social rank in yellowtail. Asterisks indicate significant difference.

Rearing period	Dominant: Coincidence with firstly observed dominant (8 tanks)	N	Theoretical possibility	Total length (mm) (mean ± SD)		Subordinate N	Total length (mm) (mean ± SD)	
1 day	Label	6	0.000023*	65.1 ± 6.7	63.1 ± 6.8	2	67.5 ± 9.9	58.4 ± 5.9**
	Non-label	2		57.2 ± 0.6		72	58.1 ± 5.7	
1 week	Label	3	0.038*	55.6 ± 1.1	54.4 ± 7.1	5	58.4 ± 3.7	59.4 ± 5.9**
	Non-label	5		53.8 ± 6.0		67	59.4 ± 6.3	

*is $p < 0.05$, binomial distribution and **is $p < 0.05$, by Student's t-test, respectively.

Table 2. Chi-square table to determine the correlation between aggressive dominant (juvenile period) and prior tendency of J-posture (larva period). Asterisks indicate significant difference.

Category	J-posture	Non-J	Sum
Dominant	6*	2	8 (21.2 ± 3.0 mm TL **)
Subordinate	7	14	21 (16.3 ± 1.8 mm TL)
Sum	13 (18.9 ± 3.7 mm TL)	16 (16.6 ± 2.1 mm TL)	29

* = G-test $p < 0.05$, ** = Student's t-test $p < 0.01$, respectively.

in combination with thyroid hormones in vitro (Jesus et al. 1990). However, there are differences in cortisol concentration patterns during metamorphosis between these two species (this study, Tanaka et al. 1995). Subordinate yellowtail in a school showed significantly higher cortisol levels than dominant fish (Figure 4). We propose that subordinate yellowtail juveniles are physiologically stressed by their social rank in a school in experimental tanks, similar to the effects of social rank in rainbow trout (Noakes & Leatherland 1977) and in coho salmon (Ejike & Schreck 1980). Therefore, the cortisol increase from 9 mm TL (Figure 2b) indicates the onset of stress response and aggressive behaviour. Thus, cortisol in yellowtail is presumed to be concerned with the stress response and social rank rather than metamorphosis (Sakakura et al. 1998a).

'J-posture' appeared at the same time as the increase of T4 and disappeared with the onset of aggressive (social) behaviour. Therefore, in the point of behavioural development in yellowtail, metamorphosis is defined as the period of 'J-posture' and juvenile period is from the onset of social behaviour (Figure 3). However, at 10 mm TL, aggressive behaviour had not begun, indicating a slight delay thereof as well as in morphological development. Fish size at the onset of aggressive behaviours can vary (12 mm TL in the present study, 10 mm TL in Sakakura & Tsukamoto 1996). In the striped jack, *Pseudocaranx dentex*, which is in the same family as yellowtail, a similar time lag between the morphological complement as the juvenile (10 mm TL) and the onset of schooling (social) behaviour (over 12 mm TL) has been reported (Masuda & Tsukamoto 1998). The onset of social behaviours, both aggressive and schooling, can even be delayed in yellowtail considered to be morphologically juvenile, if they have received insufficient nutrients, such as docosahexaenoic acid (Masuda & Tsukamoto 1999 this volume) or vitamin C (Sakakura et al. 1998b). Since these nutrients are known to be essential for development of brain and nerve systems in fishes (Watanabe & Kiron 1994, Koshio et al. 1997), the development of central nervous system can be the most important factor for the social behaviour as well as morphological and physiological development.

Possible mechanism of aggressive behaviour

Formerly, fish schools were thought to be egalitarian and leaderless societies (Breder 1954, Shaw 1962), and there are few studies on the stability and durability of social rank in fishes (Elwood & Rainey 1983, Oliveira & Almada 1996). We are the first to demonstrate individual differences in aggressive behaviour (Figure 5) and the stability of social rank

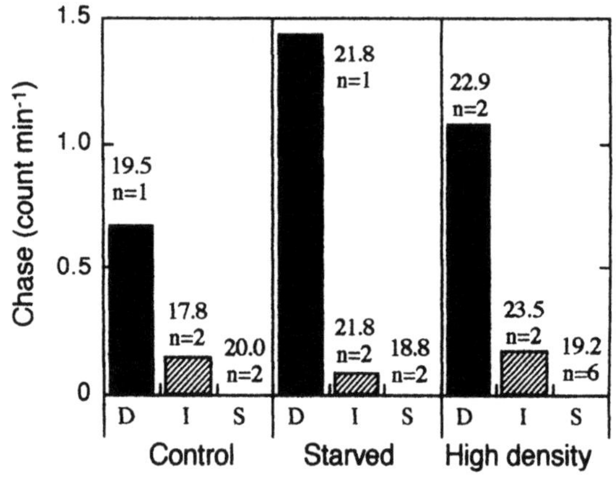

Figure 5. Individual aggressive behaviour in relation to social rank (D = dominant, I = intermediate, S = subordinate) in different environmental conditions (modified from Sakakura & Tsukamoto 1998b). Individual aggressive behaviour were counted using video-image analysis system in 60 min, and mean frequency are shown according to their social rank (control = 5 fish, starved = 5 fish starved 12 h, high density = 10 fish in 5 l experimental tank). Numerals on the bars indicate the total length (mm) of fish belonging to the each social rank.

(Table 1) in schools of small juvenile marine fish (under 50 mm TL).

In all experiments, aggression was directed only toward fish of lower rank. The results of preliminary observations to classify the social rank reflected the three categories of social rank in yellowtail: dominants (the most aggressive fish, which showed chase behaviour over 5 times min^{-1}), intermediates (the secondary aggressive fish, which appeared after the removal of dominants), and subordinates (non-aggressive fish). Therefore, the social rank of this species resembles the 'pecking order' of chickens (Schjelderup-Ebbe 1935), and the social rank of rainbow trout, *Oncorhynchus mykiss* (Abbot et al. 1985, Yamagishi 1962).

Since no agonistic interactions were observed in dark conditions (Sakakura & Tsukamoto 1997b), aggressive behaviour in yellowtail can be regarded as a response to visual stimuli (counterpart). Aggressive individuals may not be able to distinguish which fish is a specific subordinate, suggesting the stimuli may be always at the same level in an experimental tank. However, biotic factors such as starvation and high-density accelerated the aggression of dominant fish, but did not affect that of intermediate or subordinate fish (Figure 5, Sakakura & Tsukamoto 1998a,b). Moreover, size advantages were not observed between dominants and subordinates (1 week experiment of Table 1).

To explain these difference in aggressiveness of the dominant fish, we assume the presence of 'aggressive drive', or motivation for an aggressive behaviour, similar to the migratory drive for jumping behaviour in the amphidromous ayu, *Plecoglossus altivelis* (Tsukamoto 1988, Tsukamoto & Uchida 1992). Aggressive drive is put as an intermediate function between a stimulus (counterpart) and a response (aggressive behaviour, Figure 6). In this model, the aggressive drive of dominant fish is enhanced by environmental factors, i.e. biotic factors such as high density, starvation (Figure 5), size-difference (Figure 1a, Sakakura & Tsukamoto 1998b), and abiotic factors such as high water-temperature (25 to 30°C), and low light intensity (10^2 to 10^3 lx, Sakakura & Tsukamoto 1997b). Thus, increased 'aggressive drive' leads to an increase in response (frequency of aggressive behaviour) despite a fixed stimulus (visual stimuli of counterparts in a school). On the other hand, a subordinate fish is always physiologically stressed by dominants (Figure 4) and so its aggressive drive is assumed to be extremely suppressed.

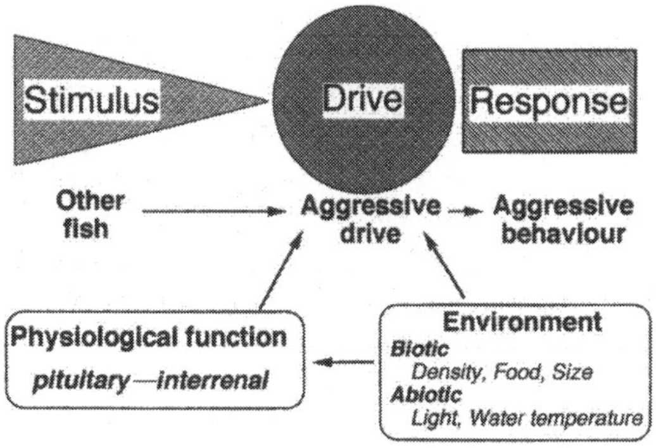

Figure 6. Behavioural model for aggressive behaviour in the yellowtail. An endogenous drive of aggression is located as an intermediate function between stimulus and response.

J-posture, which was observed during the metamorphosis (Figure 3), had a positive correlation with aggressive behaviour (Table 2). Thus, yellowtail undergoing metamorphosis are assumed to develop a similar motivation as the aggressive drive. However, during the metamorphosis, yellowtail may be unable to show aggressive behaviour toward counterparts because they are morphologically and physiologically not sufficiently well developed. Therefore, J-posture is determined as a precursor to aggressive behaviour (Sakakura & Tsukamoto 1996) and is assumed to be some kind of threat behaviour toward counterparts or predators to show the body larger in width as a result of flexion.

Possible function of aggressive behaviour

In 1-day experiment, almost all the dominants were previous dominant fish and were larger in size, whereas the hierarchy changed after 1 week, and so that TL of both previous and new dominants became smaller (Table 1). This result indicates that social rank of this species is not strictly stabilized and has some flexibility to experience a rank reversal between dominant and intermediate fish over a long-term period. The reversal of social rank will be caused by changes in relative physical strength of fish, in which dominants tend to lose much energy over the long-term by chasing other fish resulting in depression of growth rate, and intermediates have a chance to overtake the previous dominant. Thus, social rank in schools of this species may not bring the benefit of growth to the dominant fish as reported in some salmonid species (e.g. Ruzzante 1994).

In the rearing ponds, cannibalism among juvenile yellowtail of one batch causes high mortality

(Figure 1a). Chase behaviour is observed not only in every sequence of aggressive behaviour but also in cannibalism. Cannibalism is interpreted as the final phase of aggressive behaviour (Sakakura & Tsukamoto 1996). Size variations in TL in a rearing pond showed constant levels from the onset of aggressive behaviour and cannibalism (CV < 20%, Figure 1b). Larger fish (dominants), which can cannibalize smaller fish in a school with less than 50% of body size, are assumed to lose energy by chasing other subordinates or prey, and thus to lose benefits for growth (Table 1). Thus, reversals of social rank and body size of dominants can also occur in rearing ponds, and social rank may act to minimize the size variation of school members.

In wild conditions, juvenile yellowtail schools are formed by the same batch as estimated from age composition (Sakakura & Tsukamoto 1997a). They aggregate to drifting seaweed in current rips, and forage on pelagic copepods and fish larvae that are not associated with the seaweed (Anraku & Azeta 1965, Sakakura & Tsukamoto 1996). Therefore, schools of juveniles in the wild are assumed to be in a similar condition as those of rearing ponds, and do not compete for the food resources of drifting seaweed. Field observations also revealed that the body sizes of school members are uniform (CV = 15%), and that cannibalism occurs among school members (Sakakura & Tsukamoto 1996). Therefore, aggressive behaviour and social rank in schools of yellowtail are assumed to make the body size of school members uniform both in wild and artificial rearing conditions.

The uniformity in body size among fish schools is considered to minimize the individual predation risk by predator confusion (Pitcher & Parrish 1993). Social rank of the yellowtail seems to be inconsistent with schooling behaviour at a glance, but the balance between schooling and social rank derived from aggression regulates the size of school members as uniform, and it is interpreted as a behaviour which increase predator avoidance in early life.

Acknowledgements

The authors are grateful to S. Shiozawa, K. Maruyama, Y. Mizuta and T. Furusawa of JASFA for support and kind help, without which the study could not have been made. The authors express great thanks to David L.G. Noakes for his critical reading of this manuscript. We are also grateful to two anonymous reviewers for their helpful comments to improve this manuscript. Thanks are also due to Eugene K. Balon, and all members who attended the FONE workshop for their helpful comments and continuous encouragement. Y.S. was supported by the Postdoctoral Fellowships for Research Abroad of the Japan Society for the Promotion of Science.

References cited

Abbott, J.C., R.L. Dunbrack & C.D. Orr. 1985. The interaction of size and experience in dominance relationships of juvenile steelhead trout (*Salmo gairdneri*). Behaviour 92: 241–253.

Anraku, M. & M. Azeta. 1965. The feeding habits of larval and juveniles of yellowtail, *Seriola quinqueradiata* Temminck et Schlegel, associated with floating seaweeds. Bull. Seikai Reg. Fish. Res. Lab. 33: 14–15.

Balon, E.K. 1984. Reflections on some decisive events in the early life of fishes. Trans. Amer. Fish. Soc. 113: 178–185.

Bond, C.E. 1979. Behavior. pp. 425–453. *In:* C.E. Bond (ed.) Biology of Fishes, Saunders College Publishing, Philadelphia.

Breder, C.M. 1954. Equations descriptive of fish schools and other animal aggregations. Ecology 35: 361–370.

Brothers, E.B. 1984. Otolith studies. pp. 50–57. *In:* H.G. Moser, W.J. Richards, D.M. Cohen, M.P. Fahay, A.W.J. Kendall & S.L. Richardson (ed.) Ontogeny and Systematics of Fishes, The American Society of Ichthyologists and Herpetologists Spec. Publ. 1, Lawrence.

Chantanachookhin, C., T. Seikai & M. Tanaka. 1991. Comparative study of the ontogeny of the lymphoid organs in three species of marine fish. Aquaculture 99: 143–155.

Duncan, D.B. 1955. Multiple range and multiple F tests. Biometrics 11: 1–42.

Ejike, C. & C.B. Schreck. 1980. Stress and social hierarchy rank in coho salmon. Trans. Amer. Fish. Soc. 109: 423–426.

Elwood, R.W. & C.J. Rainey. 1983. Social organization and aggression within small groups of female Siamese fighting fish, *Betta splendens*. Aggressive Behav. 9: 303–308.

Fukuhara, O. 1992. Study on the development of functional morphology and behaviour of the larvae of eight commercially valuable teleost fishes. Contributions to the Fisheries Researches in the Japan Sea Block 25: 1–122.

Fukuhara, O., T. Nakagawa & T. Fukunaga. 1986. Larval and juvenile development of yellowtail reared in the laboratory. Bull. Japan. Soc. Sci. Fish. 52: 2091–2098.

Hamilton, W.D. 1971. Geometry for the selfish herd. J. Theor. Biol. 31: 295–311.

Hiroi, J., Y. Sakakura, M. Tagawa, T. Seikai & M. Tanaka. 1997. Developmental changes in low-salinity tolerance and responses of prolactin, cortisol and thyroid hormones to low-salinity environment in larvae and juveniles of Japanese flounder, *Paralichthys olivaceus*. Zoological Science 14: 987–992.

Ichihara, K. 1990. Statistics for bioscience. Nankoudo, Tokyo. 378 pp. (in Japanese).

Imaizumi, K. 1993. Recent progress of Japan Sea-Farming Association in marine finfish hatchery activities. pp. 173–177. *In:* C.-S. Lee, M.-S. Su & I.C. Liao (ed.) Finfish Hatchery in Asia, Proceedings of Finfish Hatchery in Asia '91, 3, Tungkang Marine Labolatory TFRI and The Oceanic Institute, Tungkang.

Inui, Y. & S. Miwa. 1985. Thyroid hormone induces metamorphosis of flounder larvae. Gen. Comp. Endocrinol. 60: 450–454.

Jesus, E.G.D., T. Hirano & Y. Inui. 1991. Changes in cortisol and thyroid hormone concentrations during early development and metamorphosis in the Japanese flounder, *Paralichthys olivaceus*. Gen. Comp. Endocrinol. 82: 369–376.

Jesus, E.G.D., Y. Inui & T. Hirano. 1990. Cortisol enhances the stimulating action of thyroid hormones on dorsal fin-ray resorption of flounder larvae in vitro. Gen. Comp. Endocrinol. 79: 167–173.

Kendall, A.W.J., E.H. Ahlström & H.G. Moser. 1984. Early life history stages of fishes and their characters. pp. 11–22. *In:* H.G. Moser, W.J. Richards, D.M. Cohen, M.P. Fahay, A.W.J. Kendall & S.L. Richardson (ed.) Ontogeny and Systematics of Fishes, The American Society of Ichthyologists and Herpetologists Spec. Publ. 1, Lawrence.

Koshio, S., Y. Sakakura, Y. Iida, K. Tsukamoto, T. Kida & K. Dabrowski. 1997. The effect of vitamin C intake on schooling behavior of amphidromous fish, ayu *Plecoglossus altivelis*. Fish. Sci. 63: 619–624.

Magurran, A.E. 1990. The adaptive significance of schooling as an anti-predator defense in fish. Ann. Zool. Fennici 27: 51–66.

Magurran, A.E. 1993. Individual differences and alternative behaviours. pp. 441–477. *In:* T.J. Pitcher (ed.) Behaviour of Teleost Fishes, Chapman & Hall, London.

Masuda, R. & K. Tsukamoto. 1998. The ontogeny of schooling behaviour in the striped jack. J. Fish Biol. 52: 483–493.

Masuda, R. & K. Tsukamoto. 1999. School formation and concurrent developmental changes in carangid fish with reference to dietary conditions. Env. Biol. Fish. 56: 243– 252 (this volume).

May, R.C. 1974. Larval mortality in marine fishes and the critical period concept. pp. 3–19. *In:* J.H.S. Blaxter (ed.) The Early Life History of Fish, Springer-Verlag, Berlin.

Noakes, D.L.G. 1978. Social behaviour as it influences fish production. pp. 360–382. *In:* S.D. Gerking (ed.) Ecology of Freshwater Fish Production, Blackwell, Oxford.

Noakes, D.L.G. & J.-G.J. Godin. 1988. Ontogeny of behavior and concurrent developmental changes in sensory systems in teleost fishes. pp. 345–395. *In:* W.S. Hoar & D.J. Randall (ed.) Fish Physiology, Volume 6, Academic Press, San Diego.

Noakes, D.L.G. & J.F. Leatherland. 1977. Social dominance and interrenal cell activity in rainbow trout, *Salmo gairdneri* (Pisces, Salmonidae). Env. Biol. Fish. 2: 131–136.

Oliveira, R.F. & V.C. Almada. 1996. On the (in)stability of dominance hierarchies in the cichlid fish *Oreochromis mossambicus*. Aggressive Behav. 22: 37–45.

Olla, B.L., M.W. Davis & C.H. Ryer. 1994. Behavioural defects in hatchery-reared fish: potential effects on survival following release. Aquacult. Fish. Manag. 25: 19–34.

Pitcher, T.J. 1983. Heuristic definitions of shoaling behaviour. Animal Behav. 31: 611–613.

Pitcher, T.J. & J.K. Parrish. 1993. Functions of shoaling behaviour in teleosts. pp. 363–439. *In:* T.J. Pitcher (ed.) Behaviour of Teleost Fishes, Chapman & Hall, London.

Pitcher, T.J., C.J. Wyche & A.E. Magurran. 1982. Evidence for position preferences in schooling mackerel. Anim. Behav. 30: 932–934.

Polis, G.A. 1981. The evolution and dynamics of intraspecific predation. Ann. Rev. Ecol. Syst. 12: 225–251.

Potthoff, T. 1984. Clearing and staining techniques. pp. 35–37. *In:* H.G. Moser, W.J. Richards, D.M. Cohen, M.P. Fahay, A.W.J. Kendall & S.L. Richardson (ed.) Ontogeny and Systematics of Fishes, The American Society of Ichthyologists and Herpetologists Spec. Publ. 1, Lawrence.

Ruzzante, D.E. 1994. Domestication effects on aggressive and schooling behavior in fish. Aquaculture 120: 1–24.

Safran, P. 1990. Drifting seaweed and associated ichthyofauna: floating nursery in the Tohoku waters. La mer 28: 225–239.

Sakakura, Y., M. Tagawa & K. Tsukamoto 1998a. Whole-body cortisol concentrations and ontogeny of aggressive behavior in yellowtail (*Seriola quinqueradiata* Temminck & Schlegel; Carangidae). Gen. Comp. Endocrinol. 109: 286–292.

Sakakura, Y., S. Koshio, Y. Iida, K. Tsukamoto, T. Kida & J.H. Blom. 1998b. Dietary vitamin C improves the quality of yellowtail, *Seriola quinqueradiata*, seedlings. Aquaculture 161: 427–436.

Sakakura, Y. & K. Tsukamoto. 1996. Onset and development of cannibalistic behaviour in early life stage of yellowtail. J. Fish Biol. 48: 16–29.

Sakakura, Y. & K. Tsukamoto. 1997a. Age composition in the schools of juvenile yellowtail, *Seriola quinqueradiata*, associated with drifting seaweeds in the East China Sea. Fish. Sci. 63: 37–41.

Sakakura, Y. & K. Tsukamoto. 1997b. Effects of water temperature and light intensity on the aggressive behavior in the juvenile yellowtails. Fish. Sci. 63: 42–45.

Sakakura, Y. & K. Tsukamoto. 1998a. Social ranks in schools of juvenile yellowtail, *Seriola quinqueradiata*. J. Appl. Ichthyol. 14: 69–73.

Sakakura, Y. & K. Tsukamoto. 1998b. Effects of density, starvation and size difference on aggressive behaviour in juvenile yellowtails (*Seriola quinqueradiata*). J. Appl. Ichthyol. 14: 9–13.

Schjelderup-Ebbe, T. 1935. Social behavior of birds. pp. 947–972. *In:* T. Murchison (ed.) Handbook of Social Psychology, Clark University Press, Worcester.

Shaw, E. 1962. The schooling of fishes. Scient. Amer. 206: 128–138.

Shaw, E. 1978. Schooling fishes. Amer. Scient. 66: 166–175.

Sokal, R.R. & F.J. Rohlf. 1995. Analysis of frequencies. pp. 685–793. *In:* R.R. Sokal & F.J. Rohlf (ed.) Biometry, W.H. Freeman and Company, New York.

Svåsand, T. 1993. Are reared juveniles suited for release? ICES Council Meeting F34: 1–13 (mimeogr.).

Tagawa, M. & T. Hirano. 1989. Changes in tissue and blood concentrations of thyroid hormones in developing chum salmon. Gen. Comp. Endocrinol. 76: 437–443.

Tanaka, M., J.B. Tanagonan, M. Tagawa, E.G.D. Jesus, H. Nishida, M. Isaka, R. Kimura & T. Hirano. 1995. Development of the pituitary, thyroid and interrenal glands and applications of endocrinology to the improved rearing of marine fish larvae. Aquaculture 135: 111–126.

Tsukamoto, K. 1985. Mass-marking of ayu eggs and larvae by tetracycline-tagging of otoliths. Bull. Japan. Soc. Sci. Fish. 51: 903–911.

Tsukamoto, K. 1988. Migratory mechanisms and behavioral characteristics in ayu. pp. 100–133. *In:* T. Ueno & M. Okiyama (ed.) Ichthyology Currents, Asakura Shoten, Tokyo (in Japanese).

Tsukamoto, K., Y. Seki, T. Oba, M. Oya & M. Iwahashi. 1989. Application of otolith to migration study of salmonids. Physiol. Ecol. Japan. 1: 119–140.

Tsukamoto, K. & K. Uchida. 1992. Migration mechanism of the ayu. pp. 145–172. *In:* V.I. Ilyichev & V.V. Anikiev (ed.) Oceanic and Anthropogenic Controls of Life in the Pacific Ocean, Kluwer Academic Publishers, Dordrect.

Uchida, K., Y. Dotu, S. Mito & K. Nakahara. 1958. The spawning and early life history of 'buri', Japanese yellow tail, *Seriola quinqueradiata* Temminck et Schlegel. Sci. Bull. Fac. Agr., Kyushu Univ. 16: 329–342.

Umeda, S. & A. Ochiai. 1973. On the development of the structure and function of the alimentary tract of the yellowtail from the larval to the juvenile stage. Bull. Japan. Soc. Sci. Fish. 39: 923–930.

Watanabe, T. & V. Kiron. 1994. Prospects in larval fish dietetics. Aquaculture 124: 223–251.

Yamagishi, H. 1962. Growth relation in some small experimental populations of rainbow trout with special reference to social relations among individuals. Japan. J. Ecol. 12: 13–53.

Environmental Biology of Fishes **56:** 243–252, 1999.

School formation and concurrent developmental changes in carangid fish with reference to dietary conditions

Reiji Masuda[a] & Katsumi Tsukamoto[b]
[a]*The Oceanic Institute, Makapuu Pt., Waimanalo, HI 96795-1820, U.S.A. (e-mail: reiji@compuserve.com)*
[b]*Ocean Research Institute, University of Tokyo, Nakano, Tokyo 164, Japan*

Received 29 September 1997 Accepted 14 September 1998

Key words: schooling behaviour, docosahexaenoic acid, behavioural ontogeny, juveniles, *Pseudocaranx dentex*, *Seriola quinqueradiata*

Synopsis

The ontogeny of schooling behaviour was studied in comparison to the development of sensory and swimming organs and taxis in carangid fish. Striped jack, *Pseudocaranx dentex*, larvae showed strong phototaxis at 3 days after hatching (3.5 mm in TL) when they developed pigmentation in the retina. Rheotaxis and optokinetic responses were apparent at 4.0–6.0 mm TL as larvae completed development of the basic structure of their eyes. A major inflection of allometric growth occurred at 9 mm, and fin ray formation was completed at these stages. Schooling behaviour, represented by one TL of inter-individual distance and parallel orientation, only appeared at 16 mm TL, and just prior to this behaviour, fish showed mutual attraction through vision at 12 mm TL. Canalization of buccal lateral lines was complete at 18 mm TL, whereas that of trunk lateral lines started at 23 mm TL and was complete at 30 mm TL. With these results, we assumed that critical factors of the ontogeny of schooling behaviour in carangid fish include not only the development of sensory or swimming organs, but also other factors such as development of the central nervous system. To show this, we reared another carangid species, the yellowtail *Seriola quinqueradiata*, with dietary depletion of DHA (docosahexaenoic acid), which is indispensable for the development of the central nervous system. Although DHA-free fish showed optokinetic response, they did not show schooling behaviour when they attained their schooling size. Tracer experiments using radioisotope labelled DHA showed that DHA is incorporated into the brain, spinal cord, and retina of juvenile fish. Under natural conditions, carangid fish larvae should intake enough DHA through diet to develop schooling behaviour; the fluctuation of dietary quality in zooplankton might therefore influence the development of indispensable antipredatory behaviour. Morphological changes of striped jack occurred in two steps; first at 9–12 mm (fin formation and inflection of allometric growth) and then second at 20–30 mm (scale and lateral line formation), and these changes corresponded with the development of schooling and recruitment to coastal waters, respectively. Since the onset of schooling is the first step of active antipredatory behaviour, we considered that 12 mm TL is the size at which they attain the juvenile period.

Introduction

Schools of fish have attracted the attention of many scientists from ancient ages. As early as the 4th century BC, Aristotle, in his work Historia Animalium, defined the 'shoal-fishes' as 'those caught by netting' (Peck 1970). During the subsequent 2000 years, 'shoal-fishes' have been heavily exploited because they form schools, which should be per se an ecologically antipredatory behaviour. This overfishing has brought about the collapse of some commercially important fish stocks (Pitcher & Hart 1982). Only in the past 100 years has the interest of scientists shifted from how to catch these fishes to how to understand and protect them (reviewed by Chambers & Trippel 1997).

Apart from fishing pressures, natural stocks are known to fluctuate greatly from year to year. This fluctuation is attributed to either food availability and/or predation at some period of the early life history, the so called 'critical period' (Cushing 1974, 1990, May 1974). As schooling behaviour is an indispensable antipredatory behaviour, knowledge of developmental changes in this behaviour is crucial to understand fluctuations in fish stocks. Despite much work on the schooling behaviour of juvenile and adult fish (reviewed by Pitcher & Parrish 1993), there are only a few studies focussing on developmental aspects in larvae and juveniles.

In this paper, we review and synthesize our laboratory experiments on the ontogeny of schooling behaviour in the carangid fish, striped jack *Pseudocaranx dentex*. Generally, behavioural modifications are considered to be concomitant with morphological and physiological changes (Youson 1988). We therefore investigated morphological changes in sensory and swimming organs, which are indispensable for school formation. These changes corresponded well with the onset of phototaxis and optokinetic responses rather than schooling behaviour. We then focused on the development of the central nervous system and tried to control its development by manipulating the amount of DHA (docosahexaenoic acid) in the yellowtail *Seriola quinqueradiata*, another carangid fish. We found that DHA-free fish could not develop schooling behaviour when they had attained their schooling size. Development of schooling behaviour in natural waters is discussed with reference to the environmental factors that may be involved.

Morphological changes of sensory and swimming organs

Morphological changes in sensory and swimming organs were investigated in artificially reared embryos, larvae and juveniles (Figure 1). Conventional histological work (staining paraffin sections by Haematoxylin and Eosin stains) showed that pigmentation of the retina in embryos at 3.4 mm in total length (TL) (2 days after hatching), was only slight, although the lens, ganglion cell layer, inner plexiform layer, inner nuclear layer and outer nuclear layer had developed. At 3.5 mm TL (3 days after hatching), the embryos developed a well

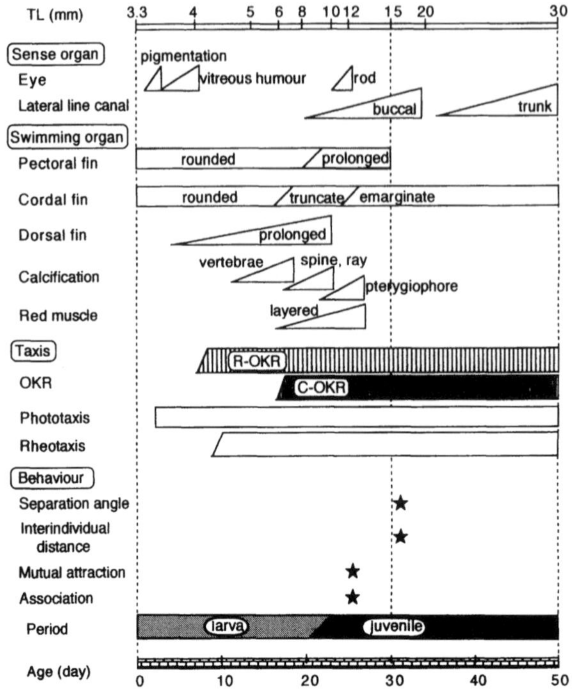

Figure 1. Development of sense organs, swimming organs, taxis and behaviour of striped jack. Note that onset of phototaxis (3.5 mm TL) and rotatory optokinetic response (4 mm TL) corresponded with the eye pigmentation and the formation of vitreous humour, whereas behavioural characteristics related to schooling only appeared at 12 mm TL or later.

pigmented epithelium (Masuda & Tsukamoto 1996). At 4.3 mm TL (10 days), vitreous humour developed, which should improve far-seeing ability in larvae. From 8 mm (20 days) to 12 mm TL (25 days), the density of nuclei in the outer nuclear layer increased, suggesting the formation of rod cells (Sandy & Blaxter 1980).

In most teleosts studied so far, free neuromasts are present at hatching (Blaxter & Fuiman 1989). Scanning electron microscopic analysis showed that free neuromasts were present in striped jack from at least 3.5 mm TL (3 days). Cephalic canal formation began at 8 mm, and was completed at 18 mm TL. Trunk lateral line formation occurred at a much larger size, starting at 23 mm and complete at 30 mm TL.

Allometric growth was observed in the pectoral fins, dorsal fin, caudal peduncle height, as well as anal and pectoral body height. A major inflection of allometric growth occurred at 9 mm TL and a minor inflection at 20 mm in all the parameters (Masuda & Tsukamoto 1996). Calcification of vertebrae started at 5 mm and was complete at 8 mm TL, and the development of fin spines and rays was completed at 9 mm TL. Scale formation started at 20 mm TL and was complete at 30 mm TL. In lateral muscles, superficial red muscle layers increased from five to six layers at 8 mm TL (20 days) to more than 10 layers at 12 mm TL (25 days), which might correspond to the onset of cruise swimming ability.

Development of phototaxis, rheotaxis and optokinetic response

We considered that the earliest form of aggregation should be caused by phototaxis. To examine the ontogeny of phototaxis, 20–80 individual striped jack were released into a long rectangular aquarium (100 × 10 × 7 cm) with even illumination (1000 lux). After 5–10 min acclimation, light intensity was set to provide four areas of 100, 1000, 10 000, and 100 000 lux within the aquarium (Masuda & Tsukamoto 1996). The number of fish in each section was counted 5–6 times in a 5–10 min interval, depending on fish size. The number of fish appearing in each block was compared using the G-test. As a control, a similar aquarium was prepared, but illumination was maintained at 1000 lux throughout the experiment. Fish were introduced, counted as above. As a result at 3.4 mm TL (2 day old) they showed no phototaxis and dispersed equally in all areas. At 3.5 mm TL (3 day old), they showed strong phototaxis and most chose the brightest area (100 000 lux). A similar tendency was observed at 4.4 mm (10 day), 8.7 mm (20 day), and 10 mm TL (23 day old) stages. Fish at 12 mm TL (25 day old) tended to appear in the second brightest area (10 000 lux), as did the 16 mm (30 day) and 24 mm (35 day) individuals.

Optokinetic response (OKR) was measured by putting fish into a transparent aquarium around which a screen with black and white pattern (6 mm and 8 mm wide stripes, respectively) was rotated at 10 revolutions per min. The reaction of fish that faced the screen and maintained their position was defined as rotatory OKR (R-OKR). R-OKR was apparent in fish larger than 4 mm TL. The reaction of fish which followed the pattern was defined as circular OKR (C-OKR) and appeared in 6 mm TL larvae (Masuda & Tsukamoto 1998). Development of rheotaxis was observed in 38 fish from 4.3 to 11.0 mm TL (9 to 29 day old) using a circular aquarium with constant water current. Positive rheotaxis was apparent in 4.5 mm TL individuals (Masuda & Tsukamoto 1996).

Development of schooling and association behaviour

To analyse development of schooling behaviour, separation angle and inter-individual distance were defined and measured (Masuda & Tsukamoto 1998). Twenty fish were put in an aquarium (30 × 20 cm, 7 cm depth), and their movements were video-recorded from above. Angle to the nearest neighbour was measured for all individuals in the video frame (Figure 2a) and the mean of 20 angles was calculated. Ten frames were sampled in a 10 sec interval. Separation angle (SA) is expected to be close to 90° when fish are randomly located, and to decrease when parallel orientation develops. The distance to the nearest neighbour was also measured for each individual and was defined as inter-individual distance (ID). Development of mutual attraction was investigated by the following method. Three transparent tanks were arranged in parallel in a water bath and 20 fish were put in the central tank. After 20 mm, another group of 20 fish were put in one of the neighbouring tanks (Figure 2b). The behaviour of fish in the central tank was video recorded from above and the number of fish attracted to each neighbouring tank was counted. As a control, a video recording with no fish in the neighbouring tank was analysed as above.

Some pelagic fishes, especially carangids, are known to show an association with floating objects (flotsam) (reviewed by Kingsford 1993). Many of the flotsam-associated fishes are schooling species (Hunter & Mitchell 1967) and since the habits of schooling and of association may be related, we investigated the ontogeny of association behaviour in the striped jack (Masuda 1995). Four different flotsam conditions were tested; a grey vinyl object (grey), a transparent acrylic object (transparent), a shadow of a grey vinyl object (shadow), and a control without any flotsam. Twelve 30 l tanks were put in one water bath, each with one of the above flotsam conditions suspended above the tank (three replicates of each), and 10 fish were released in each tank. After 24 h, the distribution of fish in each tank with flotsam was compared to that of the control (Figure 2c).

Both SA and ID decreased in fish from 12 mm to 16 mm TL, showing that schooling behaviour appeared between these lengths (Figure 3a, b). Average SA (±SD) in 10 mm and 12 mm TL fish were 78 ± 13.5° and 82 ± 9.5°, respectively, which were not significantly different from 90° (Student's *t*-test). In juveniles of 16 mm TL, SA was 57 ± 13.1°, which was significantly smaller than 90°. ID also decreased with

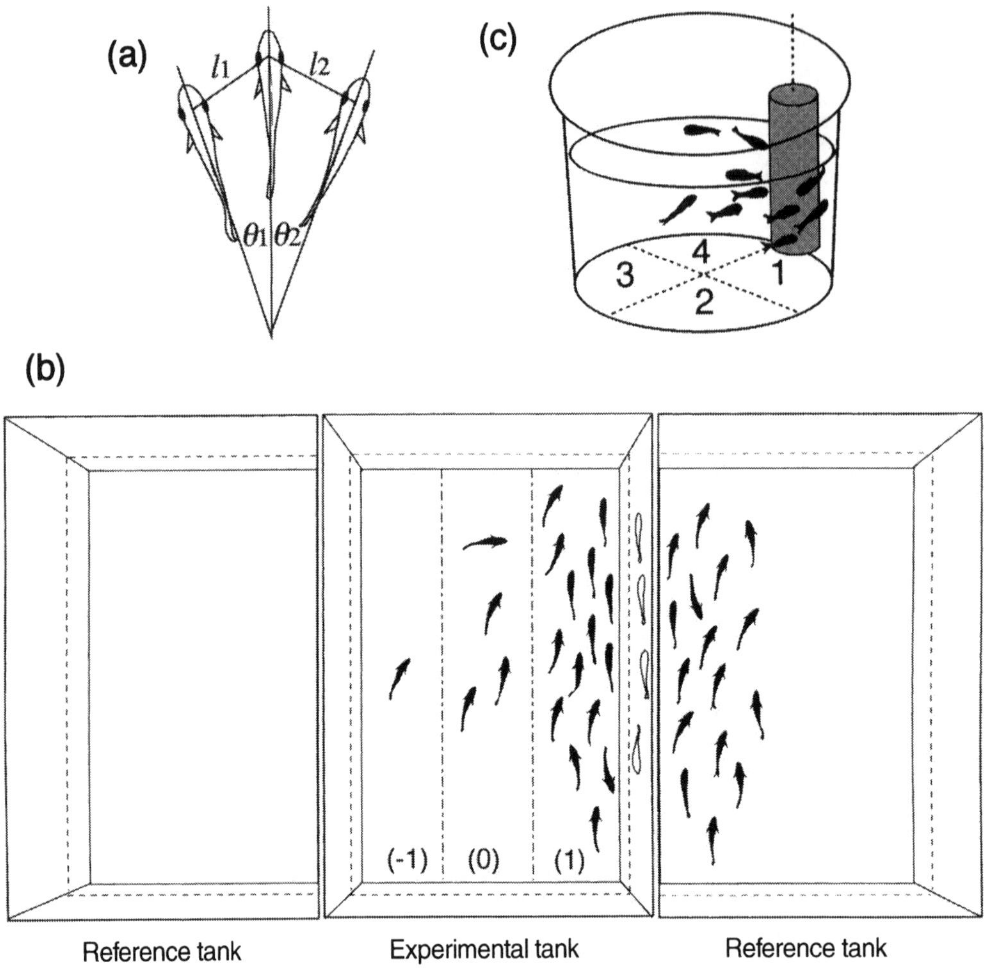

Figure 2. Measurement of behavioural characters: a – Separation angle ($\theta 1, \theta 2$) and inter-individual distance ($l1, l2$) were measured for each individual. b – For the measurement of mutual attraction, the experimental tank was divided into three sections, and each section given a score of $+1, 0, -1$. The number of fish in each section was multiplied by the score, summed and then divided by 20 to give the mutual attraction index. c – For association behaviour, the experimental tank was divided into four, and the number of fish appearing in each section was counted five times at 5 min interval. The average number of fish in the section with flotsam was calculated and then divided by 10 to give the association rate.

the growth of fish, and showed a significant decrease between 12 and 16 mm TL (Student's *t*-test). The ID/TL ratio decreased from 12 mm until 30 mm TL; the ID/TL ratio of 30 mm fish was 0.79 ± 0.15. For mutual attraction index, there was no difference between test and control at 10 mm TL. At 12 mm TL, they showed slight but significant mutual attraction, and this value increased with growth (Figure 3c).

The first association behaviour was observed at 12 mm TL to both transparent and grey flotsam (Figure 3d). Fish larger than 20 mm TL showed very strong association to the grey object, and a weaker association to the transparent one. Association to the shadow was never observed, thus association behaviour of striped jack early juveniles seems to be evoked by mechanical stimuli and is also supported by visual stimuli.

Development of sensory and swimming organs and ontogeny of taxis and behaviour are summarized as follows (Figure 1): Basic sensory organs are formed as early as 4 mm and phototaxis, rheotaxis, and optokinetic response appeared correspondingly. Pectoral, caudal and dorsal fins changed their form from rounded shape at < 8 mm TL to elongated or indented shape in > 10 mm TL. The climax of these changes are the

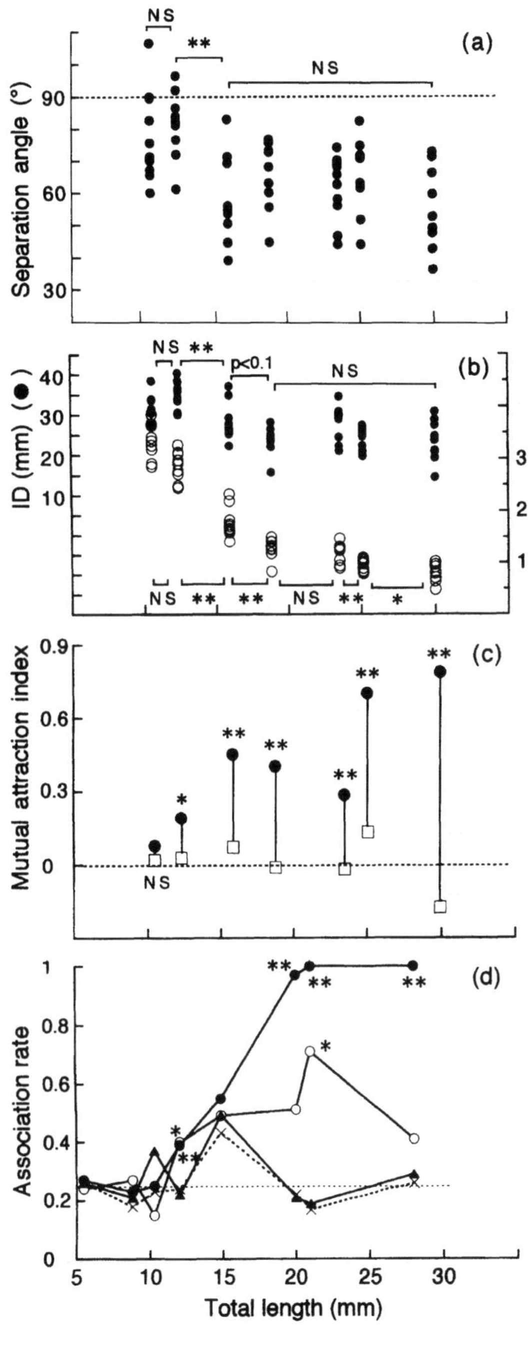

Figure 3. Development of (a) separation angle, (b) inter-individual distance (●), inter-individual-TL ratio (b; ○), (c) mutual attraction index, and (d) association rate to grey flotsam (●), transparent flotsam (○), shadow flotsam (▲), and control (×). Dotted lines show expected value of random direction (a) or random distribution (c, d). For separation angle and inter-individual distance, average value of different intervals were compared by t-test ($n = 20$), and for mutual attraction, tests and control were compared by t-test ($n = 60$). In association rate each flotsam condition was compared to the control by t-test ($n = 6$).

fin ray complement and major inflection of allometric growth, which both occurred at 9 mm TL. Calcification of the pterygiophore was complete at 12 mm TL, and the red muscle layer increased from 8 mm to 12 mm TL, a change that should enhance maneuvering and cruising abilities. Schooling behaviour only appeared at 16 mm TL, and just prior to this event, mutual attraction and association behaviour appeared at 12 mm TL.

There have been several studies to identify the onset of schooling behaviour in fish. Gallego & Heath (1994) showed that herring start to form schools at 35 mm TL. Silversides and anchovy start to school at 12 mm SL (Shaw 1960, Hunter & Coyne 1982), and guppy start at birth (Magurran & Seghers 1990). Hunter & Coyne (1982) attributed the onset of schooling to the development of sensory organs and changes in respiratory and locomotor systems. School formation by newborn guppies is considered to be the consequence of viviparity, which results in the production of relatively developed newborn fish (Magurran & Seghers 1990). These authors also pointed out that newborn guppies might be subject to considerable levels of predatory or cannibalistic attacks immediately after birth.

There is a time lag from the appearance of taxis to the development of schooling behaviour (Figure 4). This time lag cannot be fully explained by a lack of development of sensory or swimming organs, as this is complete at an earlier state, just before taxis appears. The degree of development of the central nervous system may be responsible, as has been pointed out by several authors (Neave 1984, Noakes & Godin 1988, Browman 1989).

Development of behaviour is probably programmed genetically and some factors such as learning or physical and chemical environmental factors may have influence on it. However, juvenile silversides reared without contact with mates showed almost immediate school formation after being exposed to mates (Shaw 1961). Therefore, early stages of learning do not seem to be critical for school formation.

Involvement of dietary DHA in the development of schooling behaviour

To control the development of the central nervous system, we focused on dietary condition and manipulated the amount of DHA, or docosahexaenoic acid. Yellowtail form schools at about 13 mm TL and show a similar developmental repertoire to striped jack but grow faster (Sakakura & Tsukamoto 1996). We reared yellowtail

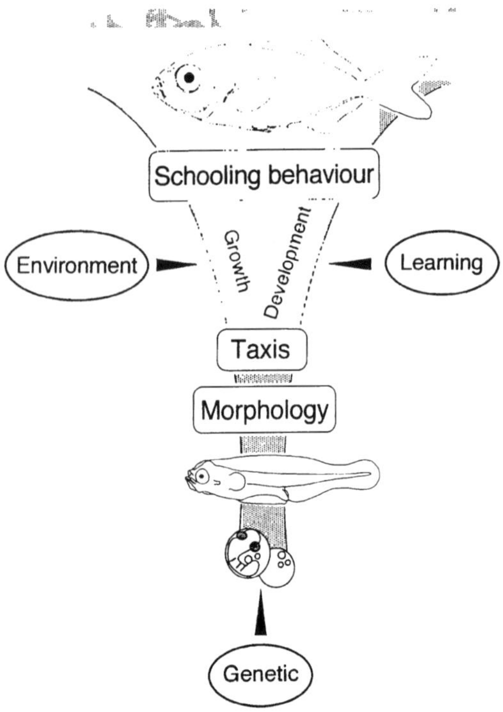

Figure 4. Developmental time lag from the onset of taxis to the school formation can be explained by environmental factors, learning process, and the development of the central nervous system.

Table 1. Effect of dietary DHA content on the survival, growth, activity, optokinetic response (OKR), and mutual attraction index (MAI) in the yellowtail *Seriola quinqueradiata* larvae.

Experiment code	OA	1/2 EPA	1/2DHA	DHA
Survival rate (%)	44	90	94	86
Growth rate (mm day^{-1})	0.41	0.62	0.65	0.77
Activity test (%) 30 sec	40	66	48	92
60 sec	14	64	60	76
OKR*	7/7	4/5	3/5	5/5
MAI** (9th day)	–	–	++	+
(11th day)	–	–	++	++
Diet EPA dry %	0.2	3.5	0.8	1.5
DHA dry %	nd	nd	1.5	2.4
Fish body EPA dry %	0.5	1.2	0.4	0.4
DHA dry %	0.2	0.1	1.1	1.5

*Expressed as the number of positive (= more than 31 points) individuals per tested individuals.
** + and ++ show significant difference between test and control conditions at $p < 0.05$ and 0.01, respectively ($n = 60$, t-test).

embryos and larvae under different dietary conditions by manipulating the amount of DHA, EPA, and OA (Masuda et al. 1998). DHA is a highly unsaturated fatty acid that is the main component of brain tissue in vertebrates. EPA, eicosapentaenoic acid, is also a highly unsaturated fatty acid, but marine fish cannot convert EPA to DHA. Oleic acid (OA) is another fatty acid that is not highly unsaturated. Fish were reared in four tanks receiving *Artemia* of different dietary enrichment conditions: two levels of DHA (1 g or 0.5 g DHA enrichment in 10 l of incubated *Artemia* with a density of 120 individuals per 1 ml, defined as DHA and 1/2DHA), 1/2EPA, and OA. Growth, survival, activity test (survival subsequent to air exposure), OKR, and mutual attraction were investigated (Table 1). Fatty acid components of the diet and fish body were also analysed. On the 12th day of rearing, total survival rate of OA group was 44%, whereas the 1/2DHA group showed the highest survival rate (94%), followed by 1/2EPA (90%), and DHA (86%) groups (Table 1). The DHA group showed highest growth rates, followed by the 1/2DHA and 1/2EPA groups, whereas the OA group had lowest growth rate. In activity tests, the DHA group showed greatest survival both in the 30 sec and the 60 sec air exposure, and the OA group had least. The inferiority of OA group was obvious, while difference among two groups of DHA and EPA group was only slight.

In the OKR test, most fish from each rearing condition showed a significant response, and there was no significant difference among the average of the tests in all groups (one-way ANOVA). Mutual attraction appared on the 9th day of the experiment in the 1/2DHA and DHA groups, when average total lengths were 11.3 mm and 12.1 mm, respectively. On the 11th day, fish of these groups showed strong mutual attraction. In the 1/2EPA and OA groups, however, mutual attraction had not appeared by the 11th day of experiment, when the fish were 11.1 mm and 13.4 mm TL, respectively.

Artemia enriched with OA contained little (0.2%) EPA and DHA was not detected. *Artemia* enriched with 1/2EPA contained EPA (3.5%), but no DHA was detected. In 1/2DHA or DHA enriched *Artemia*, DHA was detected (1.5 and 2.4%). The fish bodies reflected the fatty acid composition of these diet The fish bodies of the 1/2DHA and DHA groups contained high percentages of DHA (1.1 and 1.5%, respectively), whereas those of the OA and 1/2EPA group contained only small amounts (0.2 and 0.1%, respectively). These results show that DHA deficiency directly affects fish body composition, and that DHA deficient fish attain size of school formation (13 mm TL) and show normal OKR, but cannot form schools at the same size as normal fish.

Incorporation of DHA into the central nervous system

To show that DHA is incorporated into the central nervous system, we carried out a tracer experiment using radioisotope labelled DHA (Masuda 1995, Masuda et al. unpublished). Three rearing conditions were set up for 10 days. In aquarium 1, carbon 14 labelled ('hot') *Artemia* nauplii were fed for 10 days. In aquarium 2, hot *Artemia* were fed for 8 days and then unlabelled, or 'cold', *Artemia* were fed for 2 days. In aquarium 3, only cold *Artemia* were fed for 10 days. After 10 days, 3–4 individuals from each aquarium were dissected and radioactivity in the eyes, brain, gill takers, liver, gut, and other bones and muscles, was measured by liquid scintillation counter. In fish from aquarium 1, differences among organs were not obvious, however, the brain of fish from aquarium 2 showed significantly stronger radioactivity compared to other body parts (one-way ANOVA with Scheffé tests). Fish from aquarium 3 did not show any radioactivity.

Whole body autoradiography on the 4th and 11th day samples was conducted by preparing 10 µm thick frozen sections. They were dried and then exposed to imaging plates (Fuji Film Co.) and the radioactivity measured by a bioimaging analyser (BAS 1000 Mac, Fuji Film Co.). The conventional method of autoradiography with X-ray sensitive film was also conducted. Using both methods, the entire body was already radioactive in the sample from the 4th day, and the brain and gut showed strong radioactivity. In the 11th day, the distribution of radioactivity in fish from aquarium 1 was basically the same as on the 4th day. In the 11th day sample from aquarium 2 (fed hot *Artemia* for 8 days and cold for 2 days), however, only brain and nerve tissue showed strong radioactivity. This shows that DHA used in brain and nerve tissue is retained for 2 days or longer, whereas DHA used in other parts of the body has been converted. These results suggest that the fish used DHA to form brain and nerve tissue.

Ecological speculations on migratory behaviour

The spawning grounds of striped jack are unknown. Mature female striped jack are often caught off the Ryukyu Islands (Kanashiro & Ebisawa 1993) and Yakushima Island (Masuda 1995), both southern islands of Japan. Harada et al. (1984) also showed that there are no reports of matured striped jack around the Honshu mainland of Japan. The spawning ground of this species is therefore speculated to be offshore from the southern islands of Japan.

Embryos, larvae and juveniles are only rarely seen and we have essentially no information about the ecology of embryos, larvae and early juveniles. Our laboratory experiments showed that they are strongly phototactic at 3.5 mm TL (3 days). In large scale rearing tanks in hatcheries (25 m^3), they tend to appear at the water surface at 3 days and are patchy from 7 days (Masuda 1995). Under natural conditions, they probably remain in oceanic surface waters and drift with the current.

Climax of morphological changes in the swimming organs was at 9 mm TL, when the fin ray counts were complete and relative growth changed from strong allometry to weak allometry or almost isometry. These changes should be the preparation for following events, such as association to flotsams and schooling.

At 12 mm TL, light intensity preference changed, and fish began choosing areas of lower illumination. At the same time, association with floating objects and mutual attraction using vision appeared. At this size, they might aggregate with flotsam offshore, as reported in other carangid fish such as *Caranx caballus* (Hunter & Mitchell 1967) and yellowtail (Sakakura & Tsukamoto 1997). The smallest size of fish associating with floating objects in both *Caranx caballus* and yellowtail was reported to be 12 mm TL. At this length, they show mutual attraction and soon form schools around flotsam (Figure 5). If fish cannot find flotsam when they attain their association size, then they might alternatively form schools and then associate with objects. The latter possibility is supported by Druce & Kingsford (1995) who reported that *Trachurus* associate with algae in groups rather than singly.

Our research, including sample collections from setnets and scuba diving observation, showed that striped jack recruit to coastal areas at 40 mm TL (Masuda et al. 1993, 1995). They remain there until reaching 150–200 mm TL, then migrate to sandy areas. After growing to 200 mm TL or larger, striped jack migrate to offshore reefs in deeper areas (Masuda et al. 1993).

School formation in natural waters and implications for fisheries

The developmental time lag between the appearance of taxis and that of schooling behaviour (Figures 1, 4) can be explained as follows. Both taxis and schooling involve the movement of swimming organs responding to the information from visual and mechanical stimuli. But, taxis seems to operate through the peripheral nervous system, whereas schooling behaviour operates through the central nervous system. And in the development of the central nervous system, the amount of dietary DHA can be the critical factor.

As schooling is an important antipredatory behaviour, the developmental timing of this behaviour should be ecologically important to survival in natural waters. Conspicuousness of fish larvae increases rapidly as they grow because of the opacity of the internal body structure and proliferation of pigment cells (Langsdale 1993). Therefore, if they attain the size and colour to form schools but fail to do so, then they may be more vulnerable to predators.

As we mentioned earlier, fluctuation of fish stocks has been attributed to two factors: starvation and predation. Most research linking plankton prey and fishes have focused on the quantity rather than quality of plankton in the diet (reviewed by May 1974). DHA content of plankton in natural waters, however, differs greatly depending on time and place (Davis & Olla 1992). Therefore, if larvae feed on DHA deficient plankton, they might grow to juveniles without developing proper schooling behaviour, and such juveniles would be extremely vulnerable to predation. Further multidisciplinary studies are needed in this respect.

When do striped jack become juveniles?

Kendall et al. (1984) defined the juvenile period as starting with the completion of fin ray counts and the beginning of squamation, a definition generally accepted among marine fishery biologists. The timing of the squamation, however, differs depending on species (Copp & Kováč 1996). Therefore, fin ray counts are more often used as a criteria for the start of the juvenile period. This definition is very convenient especially when the samples of the species in natural waters are limited. As all the teleosts, by definition, have calcified bone, the stages of bone formation is a useful criterion to compare different species of fishes.

Balon (1990) stated that a larva is 'the transitory vegetative form, often inhabiting an entirely different niche than the definitive form, and being equipped with numerous temporary organs and different body shape'. Copp & Kováč (1996) stressed the importance of the

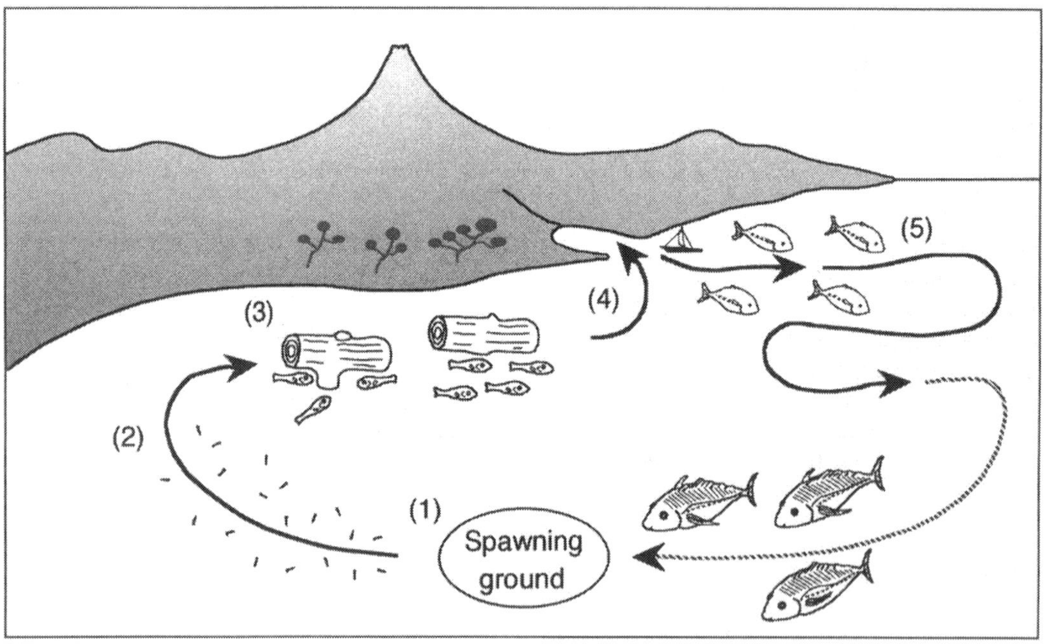

Figure 5. Schematic drawing of the life history of striped jack: (1) They are spawned and hatched in waters around southern Japan, (2) drift by current, (3) show association to flotsams at 12 mm TL and form schools, (4) recruit to coastal areas at 40 mm TL, and (5) migrate offshore at 200 mm TL.

stabilization of relative growth, as this corresponds well with a habitat shift in roach. Relative growth is an applicable comparison to show timing of metamorphosis not only among Pisces but also other classes, such as amphibians, or even invertebrates. This criteria, however, requires measurement of a considerable number of specimens and careful sampling. Two methods are commonly used: a plot of original measurement of a body part against body length (Huxley 1924, Gould 1966) or the plot of the ratio of a body part against body length (Copp & Kováč 1996). Comparing these two methods, Marr (1955) concluded that original measurements are easier to interpret and less likely to lead to erroneous conclusions. This latter approach is adopted by Kováč et al. (1999 this volume).

Our study revealed that in the striped jack a major morphological changes such as fin ray formation and inflection of allometric growth occurred at 9 mm TL, just prior to the onset of association behaviour and visual mutual attraction; these behavioural changes probably correspond with the association to flotsams and the start of aggregation in natural waters (from 2 to 3 in Figure 5). Therefore in case of striped jack fin ray formation worked as a good criterion of the first step of metamorphosis. Morphological changes, however, continued after this event, and so the development of behaviour. There was a minor inflection point of the allometric growth at 20 mm TL, and the scale formation started at 20 mm TL and was completed at 30 mm TL. Schooling behaviour with parallel orientation and one body length of inter-individual distance is completed in this interval (Figure 3). These morphological and behavioural changes are probably the preparation for the recruitment to coastal waters (which occurs at 40 mm TL or larger; from 3 to 4 in Figure 5). Development of association to flotsams and schooling are the first step of their active antipredatory behaviour and are essential to develop at their proper size, whereas the size of recruitment may differ depending on the availability of coastal area. Therefore the former change seems more important in their life history and thus 12 mm TL striped jack can be defined as juvenile.

Acknowledgement

We are grateful to M. Okiyama who provided helpful suggestions and constructive criticism throughout. Comments from two anonymous referees and guest editors improved the manuscript. Thanks are also due to L. Nickell for revising our English and providing precious comments. R. M. was supported by a postdoctoral fellowship from the Japan Society for the Promotion of Science.

References cited

Balon, E.K. 1990. Epigenesis of an epigeneticist: the development of some alternative concepts on the early ontogeny and evolution of fishes. Guelph Ichthyol. Rev. 1: 1–42.

Blaxter, J.H.S. & L.A. Fuiman. 1989. Function of the free neuromasts of marine teleost larvae. pp. 481–499. *In*: S. Coombs, P. Görner & H. Münz (ed.) The Mechanosensory Lateral Line: Neurobiology and Evolution, Springer-Verlag, New York.

Browman, H.I. 1989. Embryology, ethology and ecology of ontogenetic critical periods in fish. Brain Behav. Ecol. 34: 5–12.

Chambers, R.C. & E.A. Trippel. 1997. Early life history and recruitment in fish populations. Chapman & Hall, London. 596 pp.

Copp, G.H. & V. Kováč. 1996. When do fish with indirect development become juveniles? Can. J. Fish. Aquat. Sci. 53: 746–752.

Cushing, D.H. 1974. The possible density-dependence of larval mortality and adult mortality in fishes. pp. 103-111. *In*: J.H.S. Blaxter (ed.) The Early Life History of Fishes, Springer-Verlag, Berlin.

Cushing, D.H. 1990. Plankton production and year-class strength in fish populations: an update of the match/mismatch hypothesis. Adv. Mar. Biol. 26: 249–293.

Davis, M.W. & B.L. Olla. 1992. Comparison of growth, behavior and lipid concentrations of walleye pollock *Theragra chalcogramma* larvae fed lipid-enriched, lipid deficient and field-collected prey. Mar. Ecol. Prog. Ser. 90: 23–30.

Druce, B.E. & M. Kingsford. 1995. An experimental investigation on the fishes associated with drifting objects in coastal waters of temperate Australia. Bull. Mar. Sci. 57: 378–392.

Gallego, A. & M.R. Heath. 1994. The development of schooling behaviour in Atlantic herring *Clupea harengus*. J. Fish Biol. 45: 569–588.

Gould, S.J. 1966. Allometry and size in ontogeny and phylogeny. Biol. Rev. 41: 587–640.

Harada, T., O. Murata & S. Miyashita. 1984. Maturation and egg yield of reared striped jack, *Caranx delicatissimus*. Bull. Fish. Lab. Kinki Univ. 2: 143–149 (in Japanese).

Hunter, J.R. & K.M. Coyne. 1982. The onset of schooling in northern anchovy larvae, *Engraulis mordax*. CalCOFI Rep. 23: 246–251.

Hunter, J.R. & C.T. Mitchell. 1967. Association of fishes with flotsam in the offshore waters of central America. U.S. Fish. Bull. 66: 13–29.

Huxley, J. 1924. Constant differential growth-ratios and their significance. Nature 114: 895–896.

Kanashiro, K. & A. Ebisawa. 1993. Some ecological aspects of striped jack, *Pseudocaranx dentex* around the Okinawa Islands from the viewpoint of catching data. Suisanzoshoku 41: 105–112 (in Japanese).

Kendall, A.W., Jr., E.H. Ahlstrom & H.G. Moser. 1984. Early life history stages of fishes and their characters. pp. 11–22. *In*: H.G. Moser (ed.) Ontogeny and Systematics of Fishes, Allen Press, Lawrence.

Kingsford, M.J. 1993. Biotic and abiotic structure in the pelagic environment: importance to small fishes. Bull. Mar. Sci. 53: 393–415.

Kováč, V., G.H. Copp & M. Francis. 1999. Morphometry of the stone loach, *Barbatula barbatula*: do mensural characters reflect the species' life history thresholds? Env. Biol. Fish. 56: 105–115 (this volume).

Langsdale, J.R.M. 1993. Developmental changes in the opacity of larval herring, *Clupea harengus*, and their implications for vulnerability to predation. J. Mar. Biol. Ass. U.K. 73: 225–232.

Magurran, A.E. & B.H. Seghers. 1990. Population differences in the schooling behaviour of newborn guppies, *Poecilia reticulata*. Ethology 84: 334–342.

Marr, J.C. 1955. The use of morphometric data in systematic, racial and relative growth studies in fishes. Copeia 1955: 23–31.

Masuda, R. 1995. The ontogeny of schooling behaviour in the striped jack *Pseudocaranx dentex*. D. Phil. Thesis, University of Tokyo, Tokyo. 199 pp. (in Japanese).

Masuda, R., K. Tsukamoto, S. Shiozawa & K. Imaizumi. 1993. Some ecological aspects of striped jack *Pseudocaranx dentex* in Kyushu and the Ogasawa islands, Japan. Saibai Giken 22: 55–65 (in Japanese).

Masuda, R., T. Kamaishi, T. Kobayashi, K. Tsukamoto & K. Numachi. 1995. Mitochondrial DNA differentiation between two sympatric morphs of striped jack near Japan. J. Fish Biol. 46: 1003–1010.

Masuda, R. & K. Tsukamoto. 1996. Morphological development in relation to phototaxis and rheotaxis in the striped jack, *Pseudocaranx dentex*. Mar. Fresh. Behav. Physiol. 28: 75–90.

Masuda, R. & K. Tsukamoto. 1998. The ontogeny of schooling behaviour in the striped jack. J. Fish Biol. 52: 485–493.

Masuda, R., T. Takeuchi, K. Tsukamoto, Y. Ishizaki, M. Kanematsu & K. Imaizumi. 1998. Critical involvement of dietary docosahexaenoic acid in the ontogeny of schooling behaviour in the yellowtail. J. Fish Biol. 52: 471–484.

May, R.C. 1974. Larval mortality in marine fishes and the critical period concept. pp. 3–19. *In*: J.H.S. Blaxter (ed.) The Early Life History of Fishes, Springer-Verlag, New York.

Neave, D.A. 1984. The development of visual acuity in larval plaice (*Pleuronectes platessa* L.) and turbot (*Scophthalmus maximus* L.). J. Exp. Mar. Biol. Ecol. 78: 167–175.

Noakes, D.L.G. & J.-G.J. Godin. 1988. Ontogeny of behavior and concurrent developmental changes in sensory systems in teleost fishes. pp. 345–395. *In*: W.S. Hoar & D.J. Randall (ed.) Fish Physiology, Volume 11B, Academic Press, New York.

Peck, A.L. 1970. Aristotle: history of animals, Books IV–VI. Harvard University Press, Cambridge. 409 pp.

Pitcher, T.J. & P.J.B. Hart. 1982. Fisheries ecology. Chapman & Hall, London. 414 pp.

Pitcher, T.J. & J.K. Parrish. 1993. Functions of shoaling behaviour in teleosts. pp. 363–439. *In*: T.J. Pitcher (ed.) Behaviour of Teleost Fish, Chapman & Hall, London.

Sakakura, Y. & K. Tsukamoto. 1996. Onset and development of cannibalistic behaviour in early life stages of yellowtail. J. Fish Biol. 48: 16–29.

Sakakura, Y. & K. Tsukamoto. 1997. Age composition in the schools of juvenile yellowtail *Seriola quinqueradiata* associated with drifting seaweeds in the East China Sea. Fish. Sci. 63: 37–41.

Sandy, J.M. & J.H.S. Blaxter. 1980. A study of retinal development in larval herring and sole. J. Mar. Biol. Ass. U.K. 60: 59–71.

Shaw, E. 1960. The development of schooling behavior in fishes. Physiol. Zool. 33: 79–86.

Shaw, E. 1961. The development of schooling in fishes, II. Physiol. Zool. 34: 263–272.

Youson, J.H. 1988. First metamorphosis. pp. 135–196. *In*: W.S. Hoar & D.J. Randall (ed.) Fish Physiology, Volume 11B, Academic Press, New York.

Environmental Biology of Fishes **56:** 253–262, 1999.

Ontogeny of diel pattern of stream-margin habitat use by emerging brown trout, *Salmo trutta*, in experimental channels: influence of food and predator presence

Jean-Marc Roussel & Agnès Bardonnet
INRA, Laboratoire d'Ecologie Aquatique, 65 rue de Saint-Brieuc, 35042 Rennes cedex, France
(e-mail: roussel@roazhon.inra.fr)

Received 25 September 1997 Accepted 16 July 1998

Key words: salmonid fish, early life, habitat shift, predation risk, foraging behavior

Synopsis

Age-0 brown trout, *Salmo trutta*, inhabit shallow and slow-flowing habitats where they can easily maintain stationary swimming positions. However, recent results have shown that they use deeper and faster habitats during daylight than at night, suggesting the occurrence of a nocturnal movement toward stream-margin habitats. Experiments were conducted to describe precisely when this diel pattern of habitat use appears during ontogeny. In two indoor channels, free-embryo brown trout were deposited under the gravel. When emerging, alevins were free to choose between margin (2 cm deep, 0–2 cm s^{-1}) or deep habitat (12 cm, 2–4 cm s^{-1}), or to leave the channel (upstream or downstream). During the week of emergence, upstream and downstream catches, fish habitat use (deep habitat or margin), and fish behavior (resting or swimming) were measured by direct observations and trap counts. Three treatments were performed: (1) fish artificially fed on drifting invertebrates, (2) fish exposed to predators (bullhead, *Cottus gobio*), and (3) control channels (no food, no predator). In control and food channels, a diel pattern of habitat use was observed 1–2 days after the emergence started. Most fish rested in the margin at night, whereas they moved towards the deep habitat during daylight to hold stationary swimming positions. In the presence of bullhead, most trout were cryptic, and visible fish stood in the margin during both daylight and at night. The importance of predation risk and foraging behavior on the ontogeny of the diel pattern of habitat use is discussed. Results support the direct development without larva from free-embryo via alevin in brown trout.

Introduction

Salmonid early ontogeny is characterized by an under-gravel life (lasting from 1 to 6 months depending on species and temperature), which concerns the embryos and the free-embryos. According to Balon (1975), salmonid development is distinguished by the absence of a larva period, passing directly from the embryonic to the juvenile period, which begins with the alevin phase. This phase starts with the depletion of the yolksac and the emergence from the gravel (physiological and behavioral cues), and ends with the complete scalation (morphological cue). However, Coregoninae or Thymallinae (Salmonidae) exhibit morphological changes after the emergence (for Thymallinae) or mixed feeding (for Coregoninae), which are important enough to recognize them having a larva period (Peňáz 1975, Zil'ukas et al. 1983). In Salmoninae, emergent fish exhibit few morphological changes during their early growth in comparison with Thymallinae and Coregoninae, suggesting that the juvenile period starts with emergence from the gravel. However, in addition to morphology, behavioral cues could be helpful to describe the transition to juvenile period.

The ontogenesis of behavior in young trout has been studied as it relates to swimming, aggressiveness and territoriality, but little work has been conducted on habitat use. From observations in the wild, Hubert et al. (1994) reported that age-0 brown trout, *Salmo trutta*, use stream-margins at night and move off-shore during daylight. This stimulated research into the relationship between the ontogenesis of the alevin and the development of the diel pattern of habitat use. For drift-feeders like trout, habitat use is often regarded as a trade-off between potential energy intake (Fausch 1984) and exposure to potential predators (Greenberg 1992). The tendency of young fish to move in-shore at night seems to be widespread among several species (Copp & Jurajda 1993, Sempeski & Gaudin 1995), with the motivation typically involving predator avoidance and foraging. To investigate this phenomenon further, brown trout habitat choice and behavior were monitored under experimental conditions at the time of emergence both by day and by night. The experiment was designed to answer to the following questions: (1) At what point does the diel pattern of stream-margin habitat use appear during ontogeny? (2) What is the role of predation on its development? (3) What is the role of foraging on its development?

Material and methods

Experiments were conducted in two indoor channels (4 m long, 0.48 m wide, 0.25 m high) supplied with re-circulated water at constant temperature (12 ± 1°C). The channels were enclosed by black plastic walls with small windows (5 × 20 cm) to allow observations. Light came from neon tubes (Mazdafluor, Daylight 365), beginning at 7:30 for 11.5 h (150 lux). Twice a day, 7:00 to 7:30 and 19:00 to 19:30, 30 min of gradated illumination simulated dawn and dusk.

The channels were filled with a 4 cm gravel bed (10–30 mm diameter) set above the bottom at different heights across the channel to create three habitats in terms of water depth and water velocity: a margin (2 cm deep, 0–2 cm s^{-1}), a slope (40 degrees), and a deep habitat (12 cm deep, 2–4 cm s^{-1}). Each of these three habitats covered a third of the area. Mean water velocities were less than the critical velocity for the emerging trout (Heggenes & Traaen 1988). Upstream and downstream ends of the channels were closed by a plate with three holes (64 mm diameter), two in the deep part (one at the bottom and one at the water surface) and one in the shallow part. Each hole was connected to a trap to catch alevins moving upstream or downstream. A screen across the holes (1 cm mesh) prevented predators from leaving the channel.

Sixty trout free-embryos were placed 60 cm apart in groups of ten under the gravel in the deep habitat of the channel. In two 'control' channels (1 and 2), neither predators nor food were added. In two 'food' channels (3 and 4), food (15–25 frozen daphnia and copepodia per min, delivered for 1.5 h) was added at the upstream part of the deep habitat twice during daylight and twice at night. Finally, in two 'predator' channels (5 and 6), two male bullheads, *Cottus gobio*, caught in the wild (River Oir, Normandy, France) were added the day before the expected date of emergence. A specific tank of re-circulating water was used for the predator channels, as young salmonids may modify their behavior in response to predator odour (Keefe 1991). Experiments were conducted over two consecutive springs with two different batches of eggs (Table 1). In both years, eggs resulted from the mixture of gametes of several males and one female caught in the wild (River Oir, Normandy, France). The temperature at which the eggs were incubated allowed the dates of emergence to be manipulated (Table 1).

Traps were checked twice a day (at the beginning of dusk and after dawn). Throughout the experimental period, the number of fish seen in the channels and their exact location (margin, slope or deep habitat) and behavior were reported during daylight and at night. Three postures were identified: (1) fish lying motionless on the bottom (**rest**), (2) fish alternating between stationary swimming and bottom contact (intermittent swimming: **i.s.**), and (3) fish stationary swimming within the lower third of the water column (**swim** 1), in the middle third (**swim** 2), or in the upper third (**swim** 3).

Three observations were made during daylight and at night for batch b and two for batch a (see Table 1). In the food channels, day and night behavior and position of fish were monitored during food distribution. During night samples, fish were observed using a narrow beam torch (20 cm ring) with a low brightness light (20 watt) equipped with a red filter. Ten seconds of light were necessary to scan 40 cm of the channel length. Then, the observer moved along to the upstream window and shone the torch to sample the following 40 cm. The alevins did not seem to react to the light. and alevin flights were rarely recorded during the experiment (<5%). At the end of the experiment, each channel was emptied to collect alevins and to assess fish mortality.

Emergence and first downstream displacement of trout take place during the first hours of darkness

Table 1. Summary of the experiment schedule, including type of brown trout material.

Treatment	Control		Food		Predator	
Channel	1	2	3	4	5	6
Time of emergence	May 1996	February 1997	March 1997	April 1997	May 1996	May 1997
Batch of eggs	a	b	b	b	a	b

(Bardonnet et al. 1993). As behavioral sampling performed at night took place at least 2.5 h after complete darkness, the dynamics of a group of fish observed at night was assumed to correspond to those of the following daylight observations. Consequently, a 'Day' of observation is represented by 'night + following daylight'. The reference dates chosen corresponded to Day-1 (D1), Day-50% (50%) and Median-Day (M). Day-1 is the first day fish were caught in the following conditions: at least three fish were caught (cumulative catches, 5% of the stock) and there was no day with zero catches after Day-1, unless the cumulative catches amounted to six fish (10% of the stock). Day-50% corresponded to the day when half the stock had left the channel. Considering the period that gathers 80% of the fish together at the centre of the catches distribution, the median corresponded to Day-M. When the number of days necessary for 80% catches was even, the median was systematically chosen as the following day. The percentage of fish visible each day was calculated from the number of fish still present in the channel at the time of observation (by subtracting the catches on day n from the number of fish present on day $n-1$). When survival was affected, a constant rate of mortality was applied by day, from the first day of capture, up to the end of the experiment.

In order to compare day and night proportions of visible fish, the Wilcoxon signed-rank test was used, with equivalent day and night samples (same day) as matched pairs. Mann-Whitney tests were used to compare day and night differences in the mean percentage of fish observed in each habitat type (margin, slope or deep habitat) for the entire experimental period. Day-by-day tests were also conducted to compare day and night proportions of fish present in the margin and in the deep habitat. The one-tailed Fisher exact probability test for a 2 × 2 table was used to determine on which day(s) the proportion of fish in the margin at night exceeded the proportion of fish in the margin during daylight (alternative hypothesis H1a), and on which day(s) the proportion of fish in the deep habitat during daylight exceeded the proportion of fish in the deep habitat at night (H1b).

Post-emergence distribution of fish within the channels was followed day after day by calculating an index of habitat use for each day:

$$\text{Q ratio} = \frac{(\text{deep habitat})_n - (\text{margin})_n}{(\text{deep habitat})_n + (\text{margin})_n + (\text{slope})_n},$$

where $(\text{deep habitat})_n$, $(\text{margin})_n$ and $(\text{slope})_n$ are respective numbers of fish in the three habitat types for the day of observation n. Two different Q ratios were calculated for daylight and night-time observations. When the Q ratio value is around zero, fish are evenly distributed between the margin and the deep habitat. The further the Q ratio diverges from zero, the more fish use the margin (Q ratio < 0) or the deep habitat (Q ratio > 0).

Results

Survival and pattern of trapping

No mortalities occurred in the control and food channels (Table 2). For this reason, we assigned mortality in channels 5 and 6 to predation. The number of trout staying in the control or food channels after the end of the experiment was quite similar ($\approx$6 ind. m^{-2}), but differed from the predator channels ($\approx$2 ind. m^{-2}). Total number of catches recorded in the six channels was similar (48 fish on average). The proportions of fish leaving the channel exceeded 90% in predator channels, but averaged 80% in the control and food channels (Table 2). The proportion of fish caught during daylight was always less than 5%. Downstream catches were the most numerous, but upstream ones were not negligible and involved up to 32% of the fish (control 1).

Visible fish

Percentages of visible fish tended to increase with time to reach 60–100% at the end of the experiment

Table 2. Percentage of migrant brown trout and predation rate for each channel at the end of the experiment.

Treatment	Batch of eggs	Number of fish trapped (% trapped upstream)	Number of fish in the channel at the end of the experiment	Number of fish eaten	Survival (%)	Migrants (%)
control (1)	a	50 (32)	10	–	100	83.3
control (2)	b	47 (15.2)	13	–	100	78.3
food (3)	b	47 (23.4)	13	–	100	78.3
food (4)	b	48 (25)	12	–	100	80
predator (5)	a	50 (12.5)	3	7	88.3	94
predator (6)	b	44 (4.5)	4	12	80	91.5

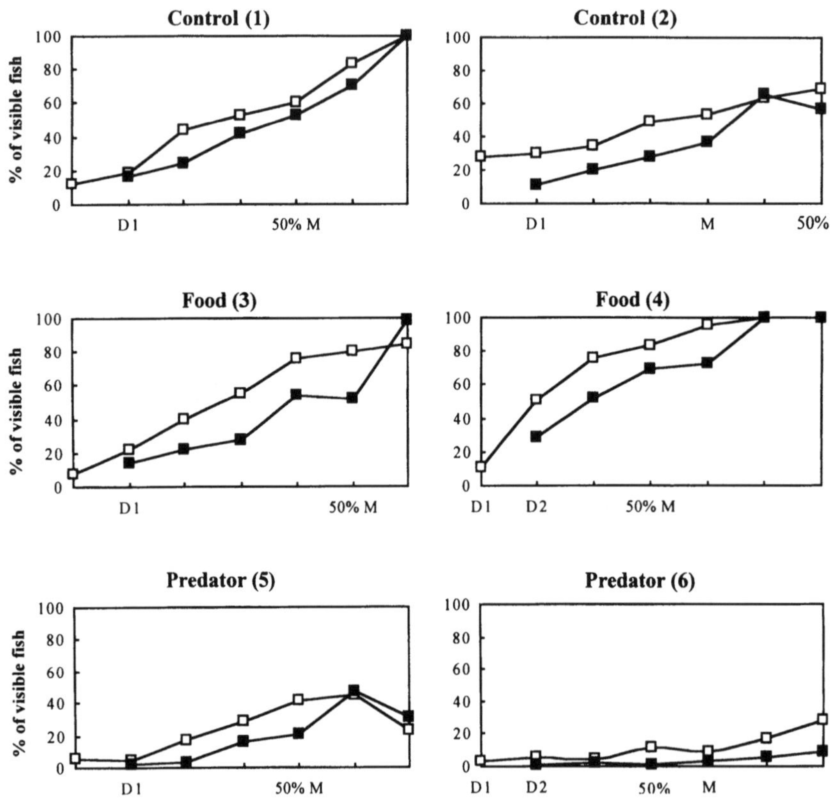

Figure 1. Day-by-day percentage of fish visible at night (solid squares) and during daylight (open squares) for each channel (D1 = first day at least 3 fish caught, 5% of the stock with no subsequent zero catches; 50% = day when half the stock had left the channel; M = the median of the catches distribution).

(Figure 1), except in the channels with predators (10 and 30%). On average, more fish were visible during the day than at night, and this difference was significant in the control (Wilcoxon, $p < 0.05$) and food ($p < 0.10$) channels. In the predator channels, a much smaller number of fish was seen, especially in channel 6. The percentages of fish visible between night and day were not significantly different in channel 5 ($p > 0.10$), but it was in channel 6 ($p < 0.05$).

Habitat preferences and time dynamics of habitat use

Whatever the treatment, the slope was rarely used by emerging trout (less than 10% on average) (Figure 2).

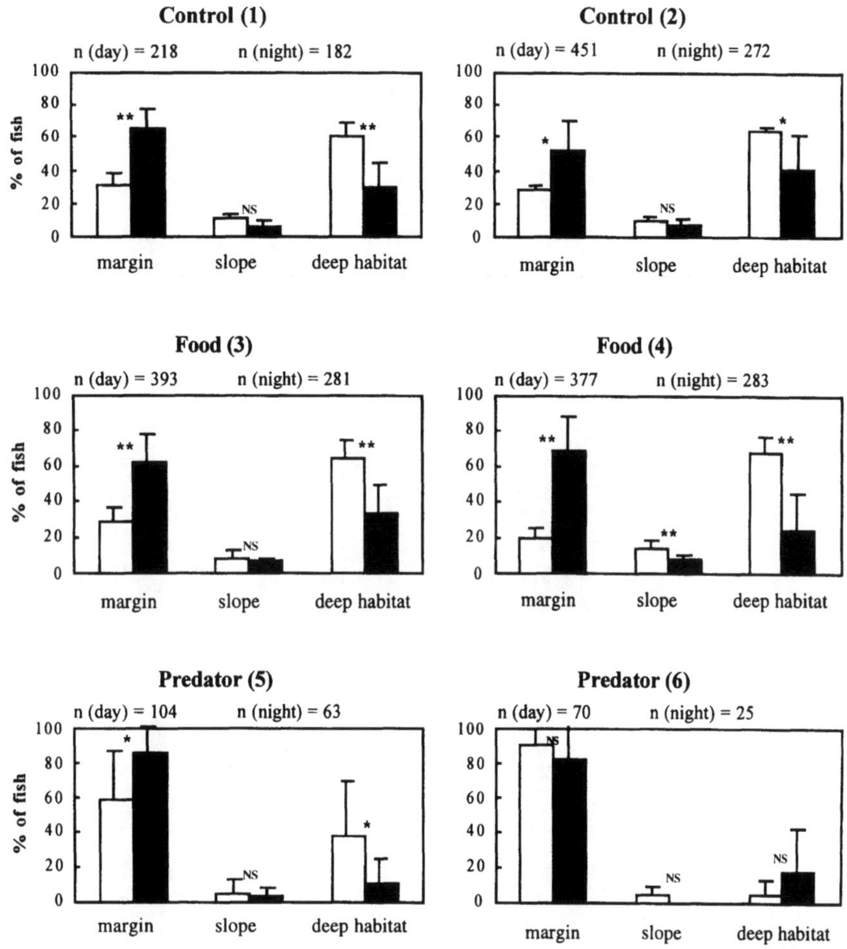

Figure 2. Mean percentages of fish present in the margin, the slope and the deep habitat during daylight (open bars) and at night (filled bars), at the end of each treatment (n = the total number of fish observations. Vertical lines represent standard deviations, SD, with significant differences, Mann-Whitney tests given as: NS = not significant; * = $p < 0.05$; ** = $p < 0.01$).

A diel pattern of habitat use was observed for the control and food channels. Most fish (64% on average) used the deep habitat during daylight and the margin during the night (62% on average). Differences in the percentages of fish observed during daylight and at night in these two habitat types were significant (Mann-Whitney; $p < 0.01$ for control 1, food 3 and food 4; $p < 0.05$ for control 2). In predator channels, the margin was preferred both during the day and at night. However, 37% of fish on average used the deep habitat during daylight in predator 5.

All fish were initially settled under the gravel of the deep habitat. Since fish were seen in the margin from Day-1, they were capable to get to the margin immediately after emergence. The day-by-day results on fish position confirmed that fish were not randomly distributed among the three habitat types, but differences were not significant for the 6 days of each experiment (Figure 3). In most cases, for the control and food channels, the proportions of fish present at night in the margin and during daylight in the deep habitat respectively, exceeded the proportions of fish present during daylight in the margin and at night in the deep habitat (Fisher exact test, $p < 0.05$). No significant differences were found for Day-1 in channels 2, 3 and 4, and for Day-1 and Day-2 in channel 1 ($p > 0.10$). In the case of predator channels, no significant differences were found ($p > 0.10$), except on the last day in channel 5 ($p < 0.05$).

The temporal changes in fish distribution within the channel showed a similar tendency for control and food experiments (Figure 4). Preference for the deep

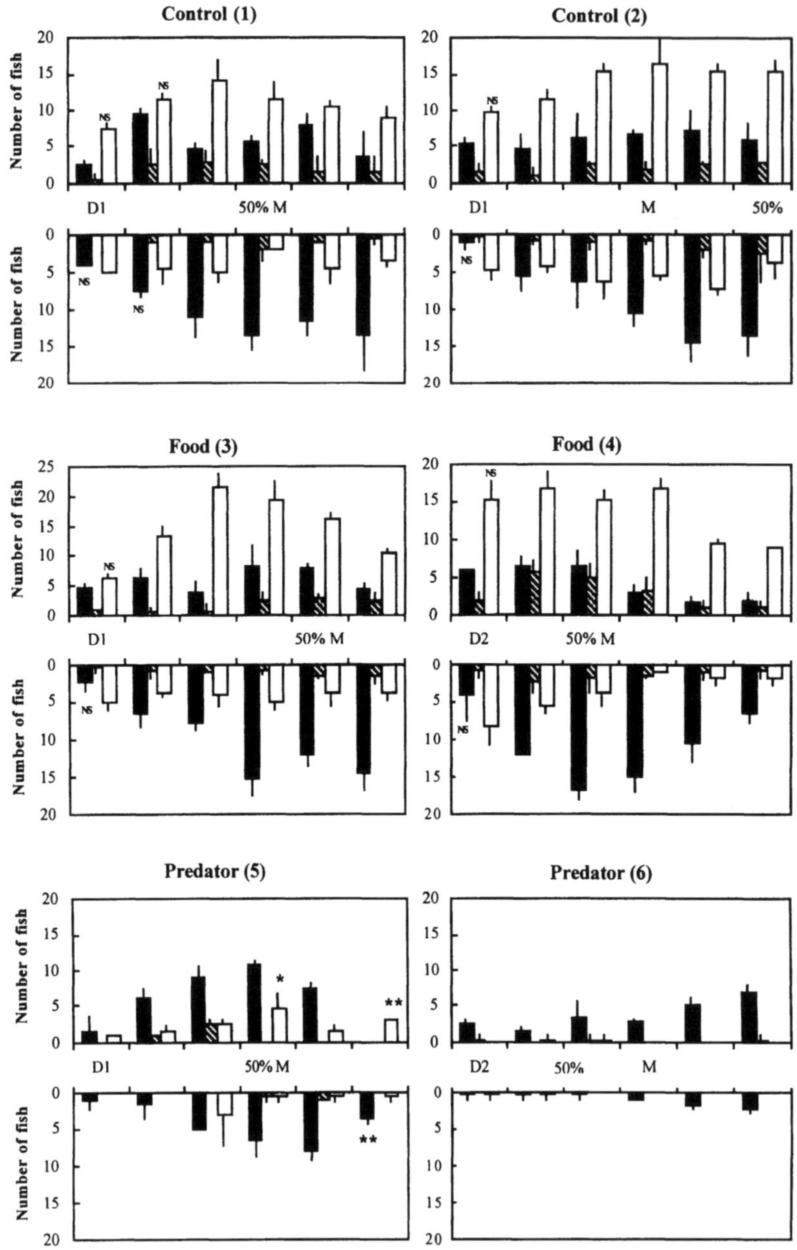

Figure 3. Number of fish present in the margin (filled bars), the slope (hatched bars) and the deep habitat (open bars) during daylight (upper histogram) and at night (lower histogram), for each treatment. For control and food channels, all are significantly different (Fisher exact method, $p < 0.05$) except where indicated (NS: $p > 0.1$). For predator channels, only significant differences are indicated (* = $p < 0.1$; ** = $p < 0.05$) (D1 = first day at least 3 fish caught, 5% of the stock, with no subsequent zero catches; 50% = day when half the stock had left the channel; M = the median of the catches distribution).

habitat during daylight was virtually constant (with slight exceptions on Day-2 and Day-5 in control 1). On the contrary, observations at night revealed a change soon after the beginning of emergence (between Day-1 and Day-3 depending on the channel), which indicated an increased use of the margin at night for most fish of the group. When predators were present, this diel displacement was almost never observed. Whatever

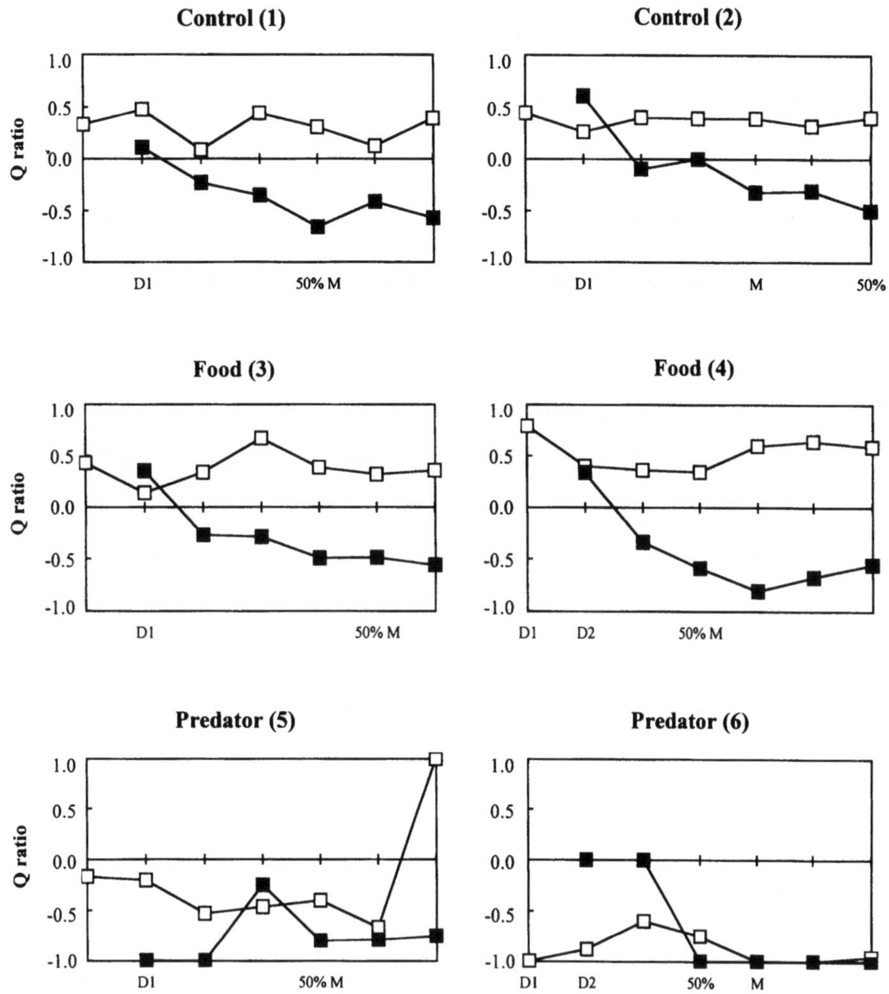

Figure 4. Progressive establishment of the habitat use patterns during daylight (open squares) and at night (filled squares) through ontogeny and for each treatment. When Q ratio values were around zero, fish were evenly distributed between the margin and the deep habitat. The more Q ratio values diverged from zero, the more fish occupied the deep habitat (Q ratio $>$ 0) or the margin (Q ratio $<$ 0) (D1 = first day at least 3 fish caught, 5% of the population, with no subsequent zero catches; 50% = day when half the population had left the channel; M = the median of the catches distribution).

the time of observation, most fish used the margin from the beginning to the end of the emergence period (except Day-2 and Day-3 in channel 6, when the few fish observed were evenly distributed between the margin and the deep habitat at night). However in channel 5, the diel pattern of transversal movement existed the last day of the experiment.

Use of the water column

In control channels (Figure 5), most fish were seen lying on the bottom (**rest**) at night, the others (about 15%) either swam intermittently (**i.s.**) or close to the bottom (**swim** 1). During daylight, these three postures were recorded for almost all the fish observed, but patterns differed with those observed at night in two ways. First, there were fewer fish in **rest** (less than 40% on average), and fish in **swim** 1 averaged more than 50%. Second, there was an apparent time tendency: **rest** fish percentages decreased with time, and correspondingly **swim** 1 fish increased. In food channels, the night pattern was similar to control channels. During daylight, the tendencies were also similar to control channels, except that **swim** 2 and **swim** 3 proportions were higher (20–60%). Fish in channel 4 tended to swim higher in the water column than fish in channel 3, but in both

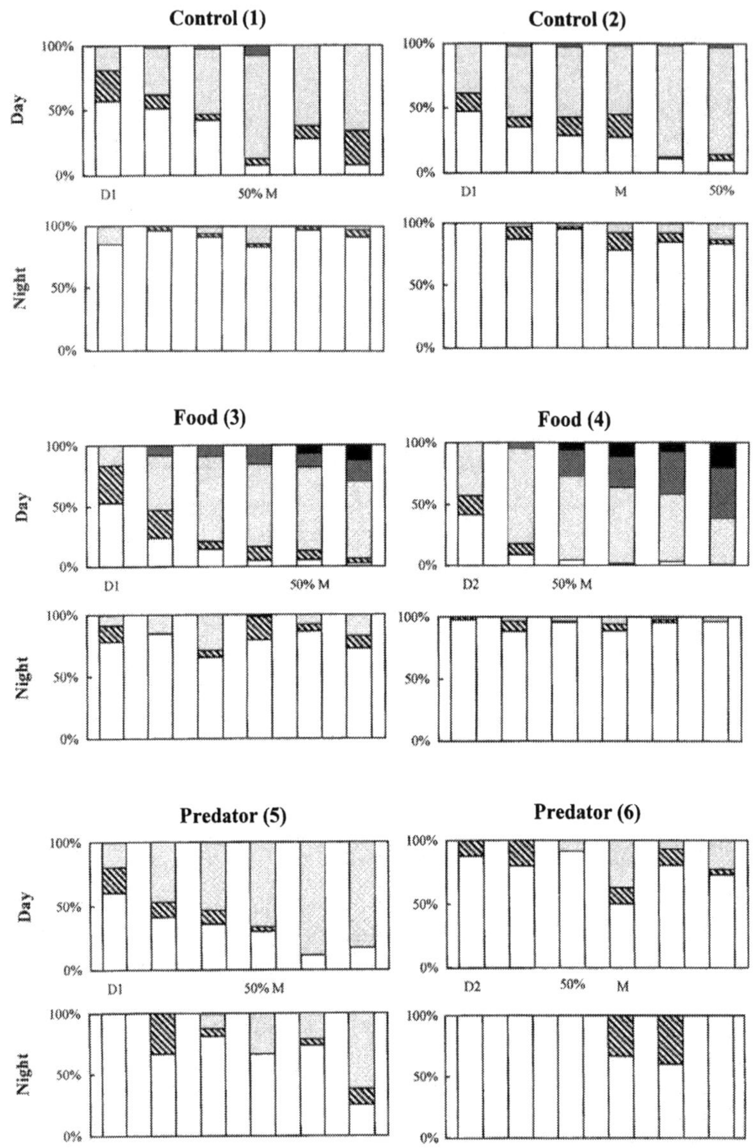

Figure 5. Percentages of fish exhibiting different postures during daylight and at night for each treatment. Blank bars, fish lying motionless on the bottom (**rest**); hatched bars, fish alterning between stationary swimming and bottom contact (**i.s.**); light-grey bars, dark-grey bars and black bars, fish stationary swimming within the lower (**swim** 1), the middle (**swim** 2) and the upper third (**swim** 3) of the water column, respectively.

cases use of the water column increased with time. In predator channels, there were differences between the two replicates. In channel 6, fish were associated with substrate both during the day and at night. In channel 5, fish exhibited a pattern similar to control channels by daylight, i.e. an increased tendency to use the water column with time. However, the night pattern was particular in that quite a large proportion of fish adopted the **i.s.** and **swim** 1 postures.

Discussion

Emergence from the gravel is a specific behavior distinguished by differences between the characteristics of the undergravel habitat and those of the free-water habitat (especially concerning water velocity, the presence of predators and the necessity of external feeding). Amongst emerging brown trout, some fish will establish territories close to the redd while others will move downstream and establish territories later (Héland 1980). The carrying capacity of the stream will be determined by environmental factors of which physical habitat suitability and food availability are preponderant. The number of fish establishing territories also depends on biotic interactions between individuals, since older and bigger juveniles tend to defend larger territories (Grant & Kramer 1990). Despite the addition of food to channels 3 and 4, a similar proportion of fish left the channels, indicating that the carrying capacity was not modified by food delivery. The overall consequences of food addition were subtle and concerned only the use of the water column during daylight. In contrast, the predator presence modified almost all the measured variables. The diel pattern of habitat use was altered, as fish mainly used the margin, regardless of time of the day. A tendency to shift towards shallower habitats in the presence of large piscivorous fish has been reported for older juvenile brown trout (Greenberg 1992, Greenberg et al. 1997). Despite shallowness of the margin, bullhead were sometimes seen there and the safety of this habitat may be relative. Concealment under the gravel was probably more effective. Most fish adopted this style and so results on habitat use only concern the part of the stock that did not become cryptic by burying itself in the substrate. The decrease in percentage of visible fish noticed here was not due to the absence of feeding, as it has also been observed in emerging brown trout when fed with live food (Bardonnet & Héland 1994). Within these general tendencies in the predator channels, differences were noticed between the two replicates.

Loss by predation was heavier in channel 6 (12 fish were eaten) than in channel 5 (7 fish eaten). This difference can be explained by differences in the size of the predators. Bullheads were bigger in channel 6 (5.6 and 6.1 g) than in channel 5 (3.6 and 3.8 g), so emerging trout probably underwent greater predation pressure in channel 6. This difference could represent the uncontrolled factor responsible for the variation between the two replicates. Fish have been found to respond to a predation risk in a graded manner, which reflects the degree of threat posed by the predator (Helfman 1989). A graded response may have acted here, the response of trout to predators in channel 5 being weak (in terms of percentage of visible fish, posture, habitat use) compared with the channel 6, which underwent the strongest predation pressure. Additional variation may have introduced bias as a result of the treatments being initiated on several dates with two different batches of eggs. However, the experiments were performed under a strict environmental regime (temperature and light) and the brood stock came from the same river (Oir). Thus, we believe the possible bias to be low and do not interfere with the among treatment analysis.

The shift from the stream-margin habitat at night to the off-shore habitat during daylight in age-0 salmonids has been described in the field by Sempeski & Gaudin (1995) in grayling, *Thymallus thymallus*, by Harris et al. (1992) in brown trout and by Campbell & Neuner (1985) in rainbow trout, *Oncorhynchus mykiss*. A diel habitat shift has also been observed for one-year-old brown trout under natural (Roussel & Bardonnet 1996, 1997) and experimental (Roussel & Bardonnet 1995) conditions, with fish moving from fast flowing habitats during daylight towards deep and slow flowing habitats at night. The present study is the first to address the nocturnal pattern of stream-margin habitat use at the exact time of emergence. Under natural conditions, stream-margin habitats require little energy expenditure by fish to maintain stationary swimming positions but provides low invertebrate drift. As distance from the edge increases, water velocity usually increases, which means more energy expenditure for fish to maintain position, but it also means more drifting invertebrates. In this experiment, the use of the deep habitat during daylight without food distribution meant no profitability in terms of feeding opportunities and an increase of energy expenditure. Consequently, the diel pattern of stream-margin habitat use is likely to be an unlearned behavior established through natural selection.

Two environmental pressures may have acted on the development of the observed diel pattern. As feeding becomes impossible with a decrease in light intensity, young brown trout leave the feeding habitats for resting areas where energy expenditure is lower. This first explanation is in accordance with optimal foraging theory, which predicts how animals can achieve a maximum net energy gain with the objective of maximizing lifetime reproducing success (Gerking 1994). The other environmental pressure that could have been acting here was the need for a safe habitat against predation. Indeed, in the presence of bullheads, visible trout almost exclusively used the margin, which was a relatively safe habitat against predation. This behavior is in accordance with the risk-balancing hypothesis, which refers to the trade-off between foraging and predator avoidance (Metcalfe et al. 1987). Therefore, as light intensity decreases at night, it could be hypothesized that alevins leave the feeding habitat because catching prey becomes unprofitable for them and/or because predation risk increases. Alternatively, moving towards very shallow habitats at night could also represent a shift to an area with a reduced risk of accidental downstream movement.

In summary, neither predator avoidance nor feeding motivation were necessary for emerging brown trout to exhibit a diel shift in habitat use, and this behavior is likely to be present immediately after emergence. Reaching the stream-margin would be the immediate goal of the fish emerging at night, followed by a movement towards slightly deeper and faster water during daylight. However, this movement could be altered in the case of strong predation risk. That young trout exhibited the diel habitat shift from the time of emergence, lends support to the theory of direct development in brown trout, as no transitional period corresponding to the larva period was observed in the present study.

Acknowledgements

We would like to thank D. Azam for his help in setting the experiment, Y. Lequennec, P.-M. Lucas and F. Marchand for technical assistance and R. Delanoé for providing biological materials. This work was partly financed by the French Conseil Supérieur de la Pêche.

References cited

Balon, E.K. 1975. Terminology of intervals in fish development. J. Fish. Res. Board Can. 32: 1663–1670.

Bardonnet, A., P. Gaudin & J.E. Thorpe. 1993. Diel rhythm of emergence and of first displacement downstream in trout (*Salmo trutta*), Atlantic salmon (*S. salar*) and grayling (*Thymallus thymallus*). J. Fish Biol. 43: 755–762.

Bardonnet, A. & M. Héland. 1994. The influence of potential predators on the habitat preferenda of emerging brown trout. J. Fish Biol. 45(Suppl. A): 131–142.

Campbell, R.F. & J.H. Neuner. 1985. Seasonal and diurnal shifts in habitat utilized by resident rainbow trout (*Salmo gairdneri*) observed in western Washington Cascade Mountain streams. pp. 38–49. *In*: Olson & White (ed.) Proceeding of Symposium on Small Hydropower and Fisheries, American Fishery Society, Denver.

Copp, G.H. & P. Jurajda. 1993. Do small riverine fish move inshore at night? J. Fish Biol. 43(suppl. A): 229–241.

Fausch, K.D. 1984. Profitable stream position for salmonids: relating growth rate to net energy gain. Can. J. Zool. 62: 441–451.

Gerking, S.D. 1994. Feeding ecology of fish. Academic Press, San Diego. 416 pp.

Grant, J.W.A. & D.L. Kramer. 1990. Territory size as a predictor of the upper limit to population density of juvenile salmonids in streams. Can. J. Fish. Aquat. Sci. 47: 1724–1737.

Greenberg, L.A. 1992. The effect of discharge and predation on habitat use by wild and hatchery brown trout (*Salmo trutta*). Regul. Riv. Res. Manag. 7: 205–212.

Greenberg, L.A., E. Bergman & A.G. Eklöv. 1997. Effects of predation and intraspecific interactions on habitat use and foraging by brown trout in artificial streams. Ecol. Freshwat. Fish 6: 16–26.

Harris, D.D., W.A. Hubert & T.A. Wesche. 1992. Habitat use by young-of-year brown trout and effects on Weighted Usable Area. Rivers 3: 99–105.

Heggenes, J. & T. Traaen. 1988. Downstream migration and critical water velocities in stream channels for fry of four salmonid species. J. Fish Biol. 32: 717–727.

Héland, M. 1980. La dévalaison des alevins de truite commune, *Salmo trutta* L. II – Activité des alevins dévalants comparés aux sédentaires. Annales de Limnologie 16: 257–254.

Helfman, G.S. 1989. Threat-sensitive predator avoidance in damelfish–trumpetfish interactions. Behav. Ecol. Sociobiol. 24: 47–58.

Hubert, W.A., D.D. Harris & T.A. Wesche. 1994. Diurnal shifts in use of summer habitat by age-0 brown trout in a regulated mountain stream. Hydrobiologia 284: 147–156.

Keefe, M. 1991. Chemically mediated avoidance behavior in wild brook trout, *Salvelinus fontinalis*: the response to familiar and unfamiliar predaceous fishes and the influence of fish diet. Can. J. Zool. 70: 288– 292.

Metcalfe, N.B., F.A. Huntinford & J.E. Thorpe. 1987. The influence of predation risk on the feeding motivation and foraging strategy of juvenile Atlantic salmon. Anim. Behav. 35: 901–911.

Peňáz, M. 1975. Early development of the grayling *Thymallus thymallus* (Linnaeus, 1758). Acta Scientiarium Naturalium Academiae Scientiarum Bohemoslovacae Brno 9: 1–35.

Roussel, J.-M. & A. Bardonnet. 1995. Activité nycthémérale et utilisation de la séquence radier-profond par les truitelles d'un an (*Salmo trutta* L.). Bull. Fr. Pêche Piscic. 337/338/339: 221–230.

Roussel, J.-M. & A. Bardonnet. 1996. Changements d'habitat de la truite (*Salmo trutta*) et du chabot (*Cottus gobio*) au cours du nycthémère. Approches multivariées à différentes échelles spatiales. Cybium 20: 43–53.

Roussel, J.-M. & A. Bardonnet. 1997. Diel and seasonal patterns of habitat use by fish in a natural salmonid brook: an approach to the functional role of the riffle-pool sequence. Bull. Fr. Pêche Piscic. 346: 573–588.

Sempeski, P. & P. Gaudin. 1995. Size-related changes in diel distribution of young grayling (*Thymallus thymallus*). Can. J. Fish. Aquat. Sci. 52: 1842–1848.

Zil'ukas, V.J., M. Peňáz & M. Prokeš. 1983. The post-hatching steps in the early ontogeny of *Coregonus peled*. Folia Zool. 32: 85–93.

Environmental Biology of Fishes **56:** 263–275, 1999.

Size-based variation in somatic energy reserves and parental expenditure by male smallmouth bass, *Micropterus dolomieu*

Robert W. Mackereth[a,c], David L.G. Noakes[a] & Mark S. Ridgway[b]
[a]*Department of Zoology, University of Guelph, Guelph, Ontario N1G 2W1, Canada*
[b]*Harkness Laboratory of Fisheries Research, Aquatic Ecosystem Science Section, Ontario Ministry of Natural Resources, 3rd Floor North, 300 Water St., Peterborough, Ontario K9J 8M5, Canada*
[c]*Present address: Centre for Northern Forest Ecosystem Research, Ontario Ministry of Natural Resources, Lakehead University, 955 Oliver Rd., Thunder Bay, Ontario P7B 5E1, Canada (e-mail: rmackere@sky.lakeheadu.ca)*

Received 2 July 1998 Accepted 12 January 1999

Key words: parental care, Centrarchidae, energetics, bioenergetics, allometry, reproductive success

Synopsis

Male smallmouth bass show size-based variation in both probability and timing of reproduction. The objective of this research was to determine seasonal and size-based patterns of depletion of energy reserves and determine if parental defense is related to males' energy reserves. We sampled male smallmouth bass in the spring, during the parental care period and in the fall to measure energy reserves (lipid stores in muscle and viscera tissue) over a two year period. Energy stores, which were not built up before nesting, declined to a minimum level by the end of the parental care period. Small males had consistently lower energy reserves than larger males and did not utilize these reserves at the same rate during the parental care period. All parental males complimented endogenous energy reserves by feeding during parental care, however, small males appear to rely proportionately more on exogenous energy intake than do larger males. Parental defense by all sizes of males declined over the parental care period, the decline being the most obvious by small males. Small males' lower energy budget may make them less effective parents and decrease their probability of survival over the following winter relative to larger males.

Introduction

Several studies have demonstrated that fishes with parental care lose weight or deplete stored energy during the parental care period. Male threespine sticklebacks, *Gasterosteus aculatus*, showed large declines in tissue dry weights as well as in lipid and glycogen levels in their livers, gonads and carcasses (Chellappa et al. 1989, FitzGerald et al. 1989). This weight loss and reduction in stored energy was correlated with a higher rate of mortality for breeding males compared to non-reproductive males (Chellappa et al. 1989, Dufrense et al. 1990). For European river bullheads, *Cottus gobio*, there was significant weight loss and increased mortality for males during the parental care period (Marconato et al. 1993). Somatic growth was reduced for parental male longear sunfish, *Lepomis megalotis*, in response to reproductive investment (Jennings & Philipp 1992). The body weight of parental male bluegill sunfish, *L. macrochirus*, declined significantly during the parental care period due in part to declines in stored lipids (Coleman & Fisher 1991). Body weight declined during the parental care period for male rock bass, *Ambloplites rupestris*, with increased weight loss corresponding to a reduced probability of recapture in subsequent years (Sabat 1994).

The observed decline in the physiological state of parental male fish is constrained by allometric relationships between body size and metabolic rate (negatively allometric), body size and cost of locomotion (negatively allometric), and body size and energy reserves (positively allometric) (Schmidt-Nielsen 1972, Brett & Groves 1979, Robinson et al. 1983, Shuter & Post 1990). These allometric relationships suggest that small males may require relatively more stored energy and utilize these reserves at a faster rate than larger males. As a consequence of these constraints large males may be able to acquire the energy reserves required to provide parental care prior to small males within a season, as hypothesized for smallmouth bass (Ridgway et al. 1991a), and large males would loose relatively less weight than small parental males (*A. rupestris*: Sabat 1994).

Male smallmouth bass, *M. dolomieu*, provide solitary parental care of their offspring for a number of weeks (Ridgway & Friesen 1992), and sustain a high level of fanning and guarding 24 hours per day through most of this period (Hinch & Collins 1991). Foraging opportunities for nesting males are limited because their nest range is substantially smaller than the summer home ranges they use for foraging (Scott et al. 1997, Ridgway & Shuter 1996). In a number of smallmouth bass populations, large males precede small males in initiating parental care (Ridgway et al. 1991a, Wiegmann et al. 1992), an observation consistent with the reproductive constraints hypothesis based on allometric relationships between body size and various energetic parameters (Ridgway et al. 1991a, Schultz et al. 1991).

The first objective of this study is to determine the seasonal and size-based patterns of energy depletion for male smallmouth bass. Smallmouth bass should be subject to two periods of energy depletion in each year, although these have not been previously measured. One period should occur during the quiescent period of late fall and winter months, when smallmouth bass gather in large groups and remain inactive until the following spring (Webster 1954, Kolok 1991). The other period should occur in the reproductive season when parental males are restricted to a habitat patch defined by their nest site (Scott et al. 1997). We will examine the hypothesis that large males should have larger energy reserves relative to smaller males before, during and after each depletion period.

The second objective of this study is to determine if parental defense of offspring reflects the size-based patterns of energy depletion among males of different size. Because aggressive defense of offspring has an energetic cost that may be higher than other activity costs (Chellappa & Huntingford 1989), changes in parental investment that occur as a result of offspring development and number (Ridgway 1988, 1989) may be modified by these size-based patterns of energy depletion.

Materials and methods

Data were collected from the population of smallmouth bass in Lake Opeongo, in the south eastern region of Algonquin Park, Ontario. Lake Opeongo (45°42′N, 78°22′W) is a large oligotrophic lake with a total surface area of 58.6 km^2, mean and maximum depths of 14.8 m and 52 m respectively and secchi disc readings of 6 m (Martin & Fry 1972). Additional physical and biological descriptions are provided in Martin & Fry (1972) and Ridgway et al. (1991).

Male bass were sampled in the spring, during the nesting season and in the fall of 1991 and 1992. In order to minimize the sample size required to address our question we treated size as a discrete, rather than continuous, variable by dividing the size range of male bass into three categories: 21–25 cm, 28–32 cm and 35–39 cm fork length which we refer to as small, medium and large, respectively. The categories were chosen to be representative of the full range of sizes of mature male bass in the population (Ridgway et al. 1991a, Ridgway & Friesen 1992) and to insure that fish in each category were as different in size as was reasonably possible to test our size-based hypotheses.

Spring and fall samples were taken using trap nets set at several locations throughout the lake in the littoral zone and checked daily. Spring sampling began after the ice cover completely melted and smallmouth bass began moving into the littoral zone (between 7–19 May 1991 and between 19–25 May 1992). Fall sampling occurred between 21–23 September 1991 and between 25 September and 1 October 1992. Nesting males were collected at three times during the nesting period. Timing of the sample was determined by the developmental period of young in the nest. Parental males were sampled early in the embryo period (embryo), generally when the egg envelope was still intact, at the time of transformation from embryo to larva period (larva) and at the time of transformation from larva to juvenile (juvenile). Nesting males were mainly angled from the nest by a diver who could insure that the appropriate fish was caught.

Energetic analyses

When males of appropriate size were sampled they were killed by a blow to the head and returned to the laboratory for dissection. Dissection involved measuring and weighing the fish (to the nearest 0.01 g), separating the testes, liver and the remaining visceral tissue and emptying all gut contents. Gut contents were dried and weighed (to the nearest 0.01 g). Testes were weighed (to the nearest 0.01 g) and this weight was then divided by the total body weight to calculate a gonadosomatic index (GSI). Muscle tissue was removed from the fish and skin was removed from the muscle. The muscle tissue was then homogenized in a Waring blender. All tissue samples were weighed and then frozen at −25°C until drying.

Liver, testes, viscera and a subsample of homogenized muscle tissue were freeze-dried (VirTis RePP sublimator) to a constant weight. Lipid content of the dried muscle and viscera samples were determined using a chloroform-methanol extraction technique adapted from the Herbes & Allen (1983) modification of the method of Bligh & Dyer (1959). Two replicates of each sample were analyzed. The total lipid content of each tissue type was calculated by multiplying the dry tissue weight by the proportion of lipid in the sample. Energy stored as lipid in the tissue was determined by multiplying total lipid content by the constant of 9.3 Kcal g^{-1} (Voet & Voet 1990). An index of stored energy (energy index) was calculated to compare the relative amount of energy, stored as lipid, in the tissue of different sized fish. An energy index was calculated by dividing total energy stores (Kcal) by fork length (cm) cubed (FL^3).

Parental defense behaviour

During the 1992 season the nest defense behaviour of males was recorded prior to sampling. Conspecific models were used to elicit male nest defense behaviour. Models were prepared following Ridgway's (1988) modification of Helfman's (1983) procedure of preparing resin-coated fishes. Briefly, smallmouth bass were captured, fixed in strong formalin solution (>20%), coated in fiber glass resin, painted with acrylic paint to restore natural colours and mounted on a metal fork. Model sizes were 23 cm, 30 cm, and 37 cm fork length. These model sizes ensured that nesting males were presented models within 2 cm of their own fork length in an attempt to standardize the threat posed by the model.

During behaviour recording, the model was presented to a nesting male by inserting the end of the metal fork into the end of a 1.2 m aluminum pole. A swimmer could then use the pole to position the model. Models were positioned at the edge of the nest for the embryo and larva period recordings and in the centre of the school during the juvenile recording. These positions have been shown to be effective in eliciting parental defense behaviour (Ridgway 1988).

Behaviour recording followed a standard procedure and occurred between 10:00 h and 16:00 h. Two swimmers (with mask and snorkel) would approach the nest and position themselves side by side approximately 2 m from the nest and the presenter would extend the pole to the appropriate position. After a 2 min period for the nesting male to become accustomed to the observers the model was placed in position and behaviour was recorded for 5 min. The observer used a stopwatch and a plastic slate to record behaviour. The behaviours recorded (jaw display, lateral display, opercular display, approach model, bite and tail beat) are described by Ridgway (1988). The amount of time the parental male spent within approximately 2 body lengths of its brood (time with brood) was recorded.

Data analysis

All energetic data were analyzed using factorial analysis of variance (ANOVA). All interaction terms were included in the ANOVA models, however, interaction terms are referred to only if they were significant. All percentage and proportion data were arcsine square root transformed to normalize variance (Zar 1984). Independent variables used in the analyses were size (small, medium and large) and sample period (spring, embryo period, larva period, juvenile period and fall). In preliminary analyses, year (1991 and 1992) was also included as an independent variable, however, in no case did it explain a significant proportion of the variation in the dependent variable or contribute to a significant interaction term. Data from both years were subsequently combined to reduce the imbalance of sample sizes within cells, particularly for the juvenile period. In addition, lipid analysis was done on a subsample of 7 fish of each size taken from the spring sample in 1991 to balance the sample sizes.

Behaviour data were analyzed as total parental defense behaviour per minute (all parental defense behaviours performed by the nesting male divided by the time with brood (min), hereafter called 'total defense') to control for variation in time with brood. Total defense behaviours per min were also divided into 2 categories: contact and non-contact defense. Contact parental defense behaviours (bite and tailbeat) were assumed to be the most aggressive behaviours because the male made physical contact with the model. Non-contact defense (jaw display, lateral display, opercular spread and approach model) were assumed to be less aggressive than contact defense. Data were analyzed using factorial ANOVA with sample period and male size as independent variables and an interaction term. Behaviour data were square root transformed ($(x + 0.5)^{1/2}$) to normalize the variation in the data (Zar 1984). Brood number and time with brood were log transformed ($\ln(x + 1)$) for analyses.

In all analyses a significance level of 5% (alpha = 0.05) was chosen as the critical level of difference among groups. Interpretation of analyses was limited to describing the patterns of variation that contributed significant effects in 2-way ANOVA models.

Results

Seasonal energetics

In total 174 male bass were sampled over the 2 years. The sample size is broken down by year, size and sampling period in Table 1. In 1991 it was not possible to reach the target sample size of 7 fish in each size category at each sampling period. Many nests failed before the young reached the juvenile period, possibly due to high winds and rough water late in the parental care period, and as a result the 1991 juvenile period sample was incomplete. Preliminary analyses of 1991 data showed that sample size could be reduced to 5 fish in each category without a large increase in variance. After subsampling fish captured in spring 1991 the remaining sample size used for analyses of tissue lipid content was 153 fish.

The muscle lipid content of male bass tissue varied significantly among the sample periods over the year (sample period: $F_{4,152} = 58.774, p < 0.001$). Muscle lipid level remained relatively constant between spring and embryo samples in medium and large males and declined slightly in small males (Figure 1). During the parental care period muscle lipid levels declined from the embryo to the larva sample and again from the larva to the juvenile sample. Fish sampled in the fall had the highest muscle lipid levels, approximately 1.5 times higher than in the spring. Although the pattern of variation was consistent for males in the three size classes there were significant differences among the size classes in muscle lipid levels (size: $F_{2,152} = 26.55, p < 0.001$). Large males had higher levels of muscle lipid than medium sized males which had higher levels than small males.

The seasonal variation in visceral lipid content showed the same general pattern as muscle lipid variation (Figure 2). Visceral lipid levels varied significantly among the samples (sample period: $F_{4,152} = 66.719, p < 0.001$), however, the magnitude of the variation was higher than in muscle lipids. This was particularly obvious in the fall sample when visceral lipid levels were more than double levels in the spring. Size differences in visceral lipid levels were significant (size: $F_{2,152} = 19.166, p < 0.001$) with lipid levels positively related to male size. One difference

Table 1. Numbers of male smallmouth bass sampled during 5 sampling periods in 1991 and 1992 (sample date denotes beginning and end of sample period). Total sample size = 174.

Year	Fork length (cm)	Sampling period				
		Spring	Embryo	Larva	Juvenile	Fall
1991	21–25	14	7	4	0	5
	28–32	14	7	5	3	5
	35–39	14	7	5	5	5
	Sample date	7–19 May	27 May to 5 Jun	4–19 Jun	18–24 Jun	21–23 Sep
1992	21–25	5	5	5	5	5
	28–32	5	5	5	5	5
	35–39	5	5	5	5	4
	Sample date	19–25 May	5–15 Jun	18 Jun to 2 Jul	6–16 Jul	24 Sep to 1 Oct

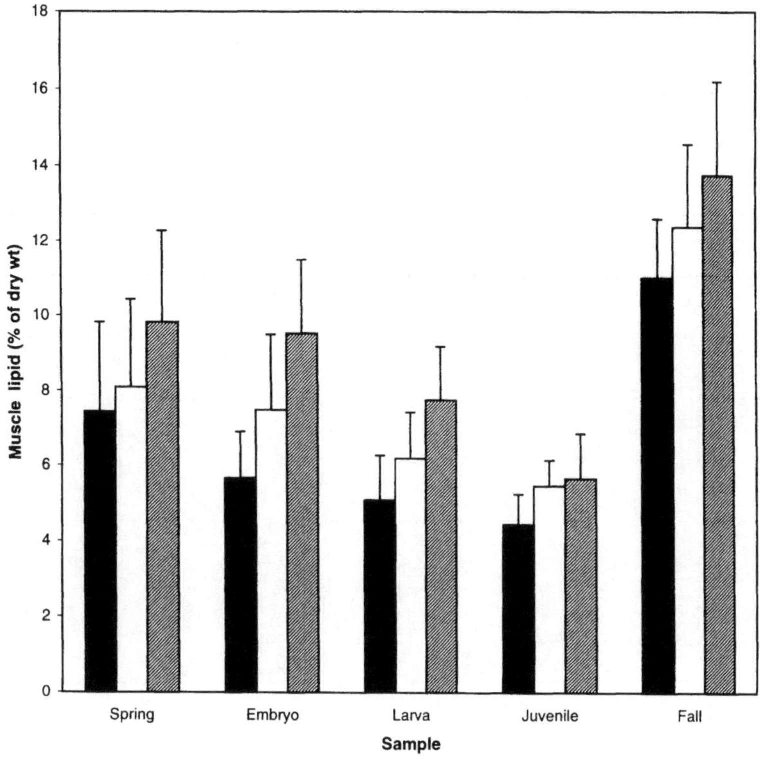

Figure 1. The mean (+ S.D.) amount of lipid in muscle tissue, as a percentage of total muscle dry weight, of small (fork length 21–25 cm: solid bars), medium (28–32 cm: open bars) and large (35–39 cm: hatched bars) male smallmouth bass sampled in the spring, when the males' young were at the embryo, larva and juvenile period, and in the fall. Sample sizes are presented in Table 1 (total n = 153).

between muscle and visceral lipid levels was seen in small males whose visceral lipid levels remained at a constant, low level throughout the parental care period, although this difference did not result in a significant interaction between size and sample period.

There were significant differences in the energy indices of males among sample periods (sample period: $F_{4,152} = 115.72$, $p < 0.001$, Figure 3). The energy index did not change between spring and embryo samples for medium and large males and declined in small males. Energy index declined over the parental care period for all males and increased to a maximum in the fall. There was also a significant positive relationship between energy index and size (size: $F_{2,152} = 33.09$, $p < 0.001$). The decline in energy index from the embryo to juvenile period was 45.5% in large males, 30.4% in medium males and 17.0% in small males, however, this trend did not result in a significant size by sample period interaction.

Although the proportion of lipid in muscle tissue was not high compared to viscera (Figures 1, 2) the majority of total energy stored as lipid was in the muscle. From the spring to the end of the parental care period between 70% and 75% of lipid energy was stored in the muscle. In the fall, muscle lipid increased but the proportion of total lipid energy stored in the muscle declined significantly to approximately 50% (sample period: $F_{4,152} = 43.361$, $p < 0.001$). This is because bass sampled in the fall had very large quantities of lipid in the visceral tissue. There were no significant differences among the different sized males in the proportions of total lipid energy stored in muscle.

GSI increased to a maximum from spring to the embryo period and then declined during the parental care period, particularly in large males (Figure 4). In the fall, GSI increased in larger males to levels similar to the embryo period. This seasonal variation, along with the greater variation in GSI for larger males, contributed to a significant interaction between sample period and size ($F_{8,152} = 4.431$, $p < 0.001$).

During the parental care period the majority of males had some gut contents, however, the proportion of males with gut contents was independent of size and sample period ($X^2_{0.05,4} = 0.108$, $p = 0.99$). For the

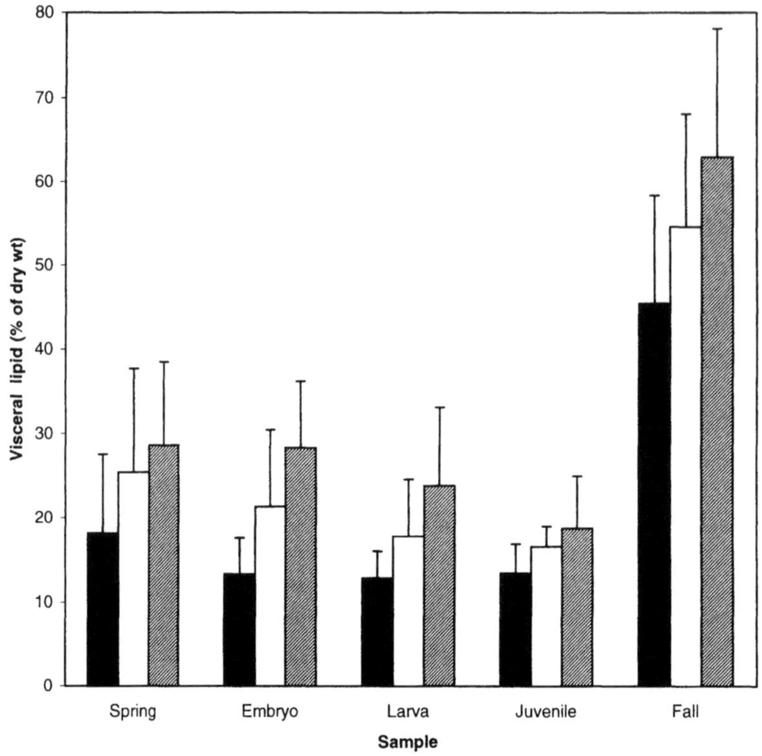

Figure 2. The mean (+ S.D.) amount of lipid in visceral tissue, as a percentage of total visceral dry weight, of small (solid bars), medium (open bars) and large (hatched bars) male smallmouth bass sampled in the spring, during parental care (embryo, larva and juvenile period) and in the fall (total n = 153).

males with gut contents there were no significant differences in the dry weight of gut contents among different sized males ($F_{2,58} = 0.67$, $p = 0.548$) or among the three sample periods ($F_{2,58} = 1.201$, $p = 0.308$: Table 2). Gut contents were not identified but generally consisted of crayfish, aquatic insects and fish. Because males caught in trap nets were in the nets for a variable length of time and often regurgitated when they were removed from the net, gut contents data from these males were not included in Table 2.

Parental defense behaviour

In total 45 males were sampled during the 1992 nesting season (sample dates in Table 1). Behaviours were recorded for 43 males prior to sampling; 5 males of each size during the embryo and larva periods, and for 5 small, 4 medium and 4 large males at the juvenile period.

Brood number varied significantly among sampling periods (sample period: $F_{2,42} = 62.693$, $p < 0.001$: Figure 5) and was largest at the embryo period. During the larva period, brood number of small and large males tended to be slightly less than during the embryo period, whereas brood number was about equal between the 2 periods for medium males. Brood number at the juvenile period was an order of magnitude lower than at the embryo or larva periods. Brood number did not differ significantly among the three size classes of males at each period although large males had slightly larger broods than small males at all periods.

Time with brood differed significantly among the three sample periods (sample period: $F_{2,42} = 15.365$, $p < 0.001$: Figure 6). Males' time with brood was highest at the embryo period, slightly lower at the larva period and by the juvenile period males spent only about 10% of the 5 min observation period with their brood. Time with brood did not differ among different sized males.

Total defense behaviour (per min) did not differ significantly among the three sample periods (Figure 7a). There were significant differences among the three size classes (size: $F_{2,42} = 3.346$, $p = 0.046$). Large males showed a consistent level of defense over the three

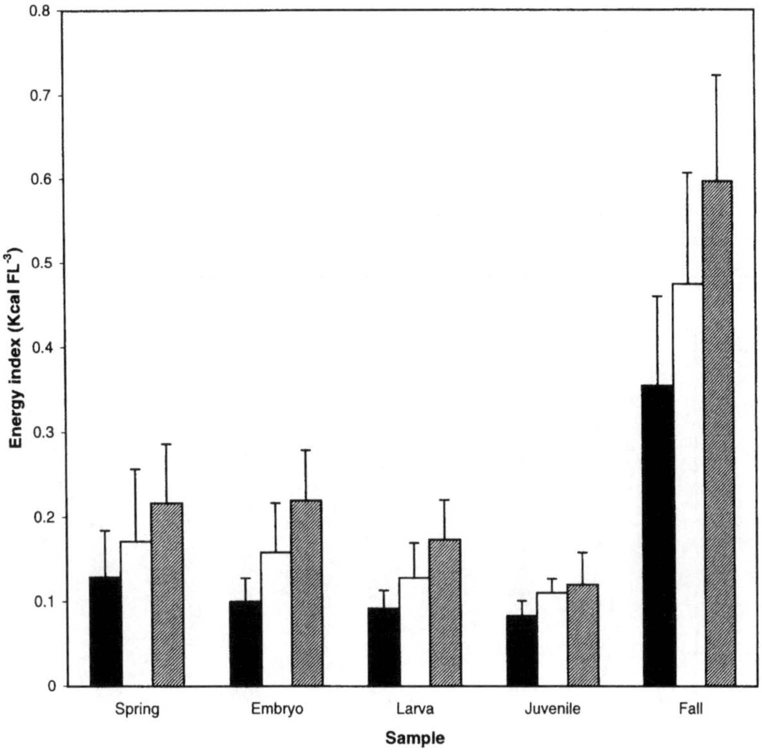

Figure 3. Mean (+ S.D.) energy index (calculated as total lipid energy (Kcal) divided by FL^3) of small (solid bars), medium (open bars) and large (hatched bars) male smallmouth bass sampled in the spring, during parental care (embryo, larva and juvenile period) and in the fall (total n = 153).

sample periods which was higher than the level of defense shown by medium males. Small males showed parental defense during the embryo and larva period, but none of the small males sampled at the juvenile period showed any defense.

Total defense (per min) was divided into contact and non-contact defense. Contact defense was infrequent and occurred mainly at the embryo period (Figure 7b). In the larva period contact defense was performed mainly by large males and by the juvenile period no males performed any contact defense. The reduction in contact defense over the sampling period was significant (sample period: $F_{2,42} = 4.415$, $p = 0.019$), but there were no significant differences due to male size.

Non-contact defense made up the majority of total defense (Figure 7c) and, therefore, showed the same pattern as total defense. The level of non-contact defense differed slightly among sample periods (sample period: $F_{2,42} = 2.597$, $p = 0.088$). There were no significant differences in non-contact defense due to male size.

Discussion

The results of this study show that male smallmouth bass are subject to two periods of energy depletion through the year. Energy depletion occurred during the winter, as shown by the large drop in lipid levels from the fall to the spring samples. Energy depletion also occurred during the parental care period, as shown by a decline in lipid levels from embryo to larva period and again from larva to juvenile period. The reduction in lipid levels during both periods was consistent between muscle and viscera. While the majority of total lipid energy was stored in the muscle, as has been reported for other fishes (Brett & Groves 1979), visceral lipid levels were more dynamic, particularly in the fall when they comprise almost half of the fish's total lipid energy stores.

One potentially confounding factor in this study was that different males were sampled at each time during the parental care period. Use of stored energy during nesting would be ideally measured by repeatedly

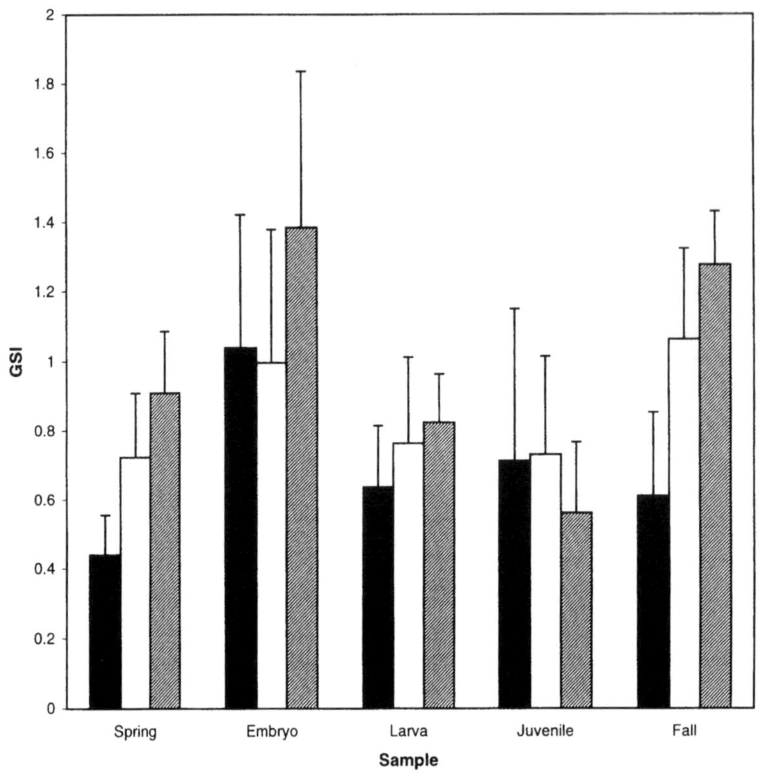

Figure 4. The mean (+ S.D.) gonadosomatic index (GSI) of small (solid bars), medium (open bars) and large (hatched bars) male smallmouth bass sampled in the spring, during parental care (embryo, larva and juvenile period) and in the fall (total n = 153).

Table 2. The percentage of male smallmouth bass in each size category with gut contents and the mean dry weight in grams (± S.D.) of gut contents (see Table 1 for sample sizes, total n = 88).

Fork length (cm)	Sampling period		
	Embryo	Larva	Juvenile
21–25	67% 0.47 ± 0.54 g	78% 0.48 ± 0.39 g	100% 1.21 ± 1.54 g
28–32	83% 0.73 ± 0.86 g	80% 0.91 ± 0.76 g	75% 1.12 ± 1.07 g
35–39	83% 0.83 ± 0.94 g	70% 1.33 ± 1.20 g	60% 0.83 ± 0.92 g

measuring the energy stores in the same individual during nesting. This type of repeated sampling was not technically possible. The potential problem of destructive sampling is that by the juvenile period only a fraction of the original nesting stock remains (34% in 1991 and 52% in 1992) and these may be the males with the largest energy reserves. Those males that have abandoned their nests may have done so because their energy reserves were too low to continue parental care. Therefore, the decline in energy reserves we have demonstrated for successful males may be an underestimate of the overall rate of decline of stored energy for all nesting males.

Although males build large lipid stores in the fall, which are depleted during the winter, we did not find a similar increase in lipid levels between the spring

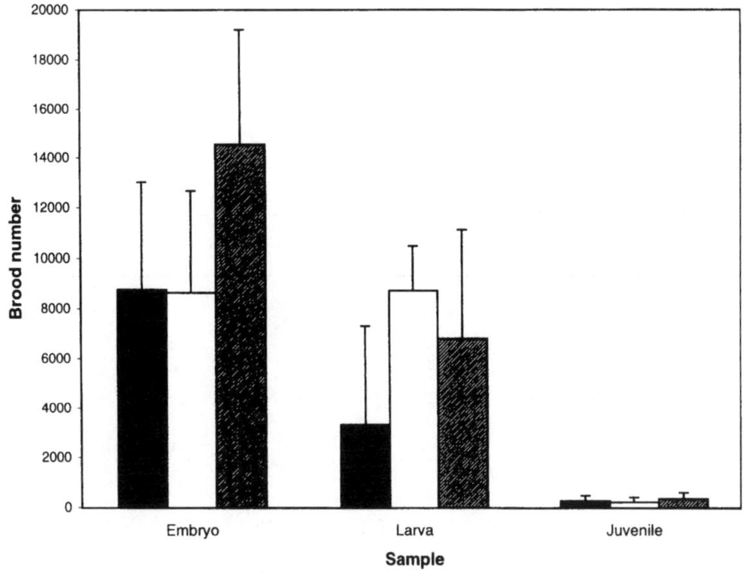

Figure 5. The mean (+ S.D.) brood number of small (solid bars), medium (open bars) and large (hatched bars) male smallmouth bass with young at the embryo, larva and juvenile period of development (total n = 43).

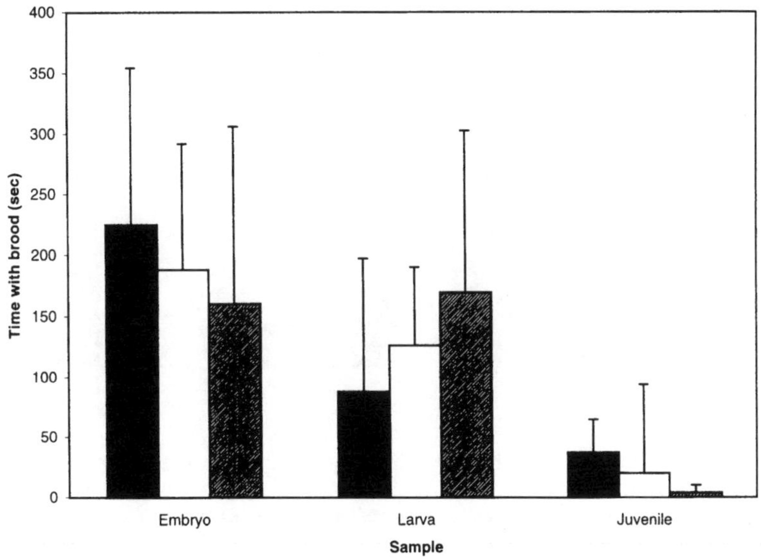

Figure 6. The mean (+ S.D.) time (seconds), per 300 sec model presentation, within 2 body lengths of their brood by small (solid bars), medium (open bars) and large (hatched bars) male smallmouth bass with young at the embryo, larva and juvenile period of development (total n = 43).

and embryo samples. One contributing factor could be that energy is directed towards gonad development during the spring rather than to somatic lipid stores. The large increase in GSI between the spring and embryo samples requires an energetic investment (Wootton 1985). While the GSI of smallmouth bass is smaller than in species such as longear sunfish (Jennings & Philipp 1992), three-spined sticklebacks (Chellappa et al. 1989), and northern pike (*Esox lucius*: Medford & Mackay 1978), there are two lines of evidence that

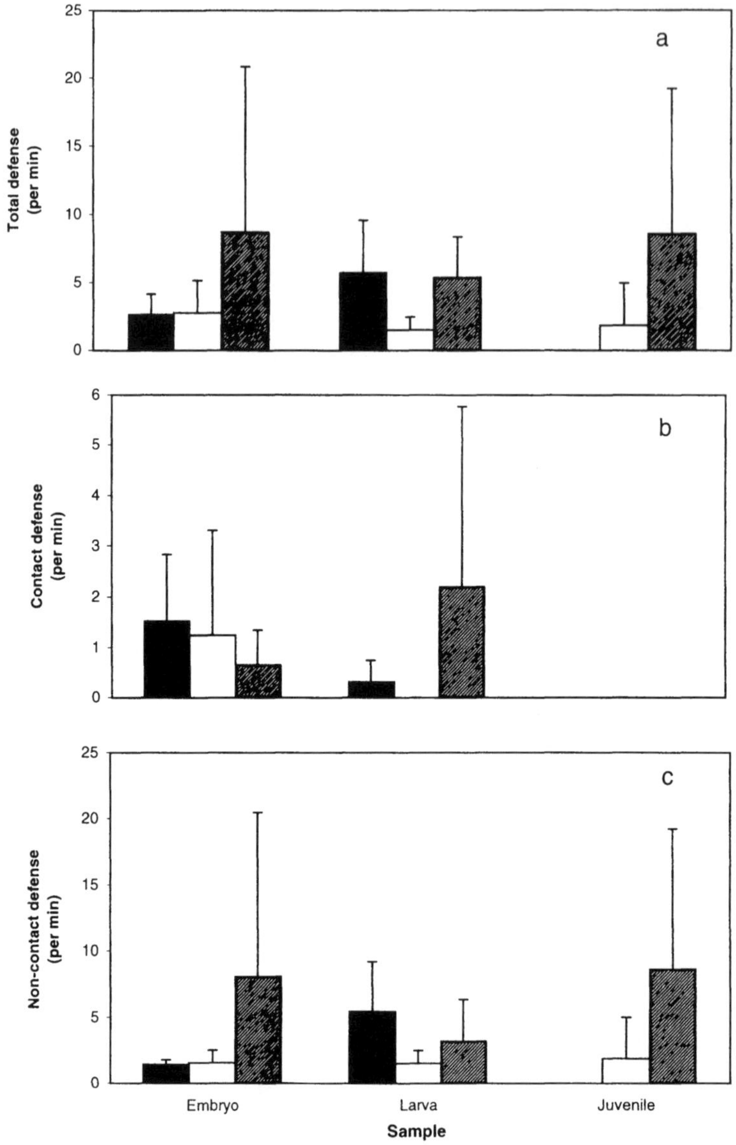

Figure 7. Mean frequency (+ S.D.) of defense behaviour, per minute with brood, of small (solid bars), medium (open bars) and large (hatched bars) male smallmouth bass with young at the embryo, larva and juvenile period of development (total n = 43 for each). a – total defense behaviour, b – contact defense behaviour and c – non-contact defense behaviour.

gonad development is energetically costly for male bass. First, males do not maintain GSI over the parental care period possibly because energy stored in the gonads is utilized. Second, males invest energy in gonad development in the fall which could reduce the time and energy required for gonad development in the spring.

It is important to consider that our expectation that males would build energy stores prior to the nesting period was based on the assumption that successful nesting is constrained by the availability of stored energy. A previous explanation for larger males nesting earlier in the season is that they are in relatively better condition in the spring and require less time to build reserves (Ridgway et al. 1991). An alternative, though not necessarily mutually exclusive hypothesis, is that nesting males rely on a combination of endogenous (somatic lipids) and exogenous (available food) reserves during the nesting period. This explanation is consistent with the reserve complementation model of Shultz et al. 1991. Further evidence for reliance on exogenous reserves comes from the fact that the majority of nesting males sampled had food in their stomachs. While foraging is likely opportunistic, because males remain in their territory (Hinch & Collins 1991, Scott et al. 1997), this energy may be an important supplement to endogenous reserves.

We also predicted that small males would deplete relatively more stored energy than large males. The results show that there was size-based variation in how parental males used somatic energy reserves during the parental care period. However, although small males had consistently lower energy indices than larger males, they did not deplete energy reserves at a greater rate than larger males as predicted. Indeed, during the parental care period, when energy stores declined significantly for all males, the percentage decline in energy stores tended to be less for small males (energy index decline of 17%) than for medium (30.4% decline) and large (45.5% decline) males.

Size-based variation in the level of stored energy may occur because differences exist in the relative reliance on endogenous and exogenous energy reserves among different sized males. Large males, with larger somatic reserves, may be able to nest earlier in the season and utilize somatic energy reserves during parental care. The greater energetic cost of reproduction, which large males incur by nesting earlier and relying more heavily on endogenous reserves, may give their young the advantage of a longer growing season. The young can grow to a larger size before the winter starvation period and thus have a greater chance of survival (Shuter et al. 1980). Small parental males may nest later in the season, when food availability may be greater, and rely mainly on exogenous energy reserves to supplement lower endogenous reserves (Schultz 1991). Although nesting later in the season may put the young of small males at a growth disadvantage, small males use relatively less endogenous energy which may increase their probability of survival during the winter starvation period. These results support the use of the reserve complementation model to explain size based variation in nesting patterns, which appear more complex than can be explained by the allometrically based reproductive constraints hypothesis (Ridgway et al. 1991).

Prior to sampling parental males, we recorded parental defense behaviour to determine if parental defense levels were related to endogenous energy stores. Parental defense behaviour directed towards a model predator has been used by other investigators as a measure of parental expenditure and the level of risk the parent takes during parental care (Pressley 1981, Sargent & Gross 1986). The results are consistent with the prediction that parental expenditure by smaller males would be less than that of larger males. Parental defense declined over the sampling periods for all males but the decline was greatest for small males who ceased parental defense behaviour by the juvenile period. Total defense was highly variable for all sizes of males at all sample periods but, on average, did not differ among sample periods within size classes (Figure 7a). Larger males performed more parental defense than medium sized males at all sample periods while small males only performed parental defense behaviour during the embryo and larva periods.

Although total defense did not differ among sample periods, the amount of contact defense performed by all males was significantly lower during the juvenile period. Because contact defense is more likely to result in injury to the nesting male (Pressley 1981, Ridgway 1988), and may require more energy than less active behaviour (Chellappa & Huntingford 1989), it may represent a greater risk of mortality to the parent and higher parental expenditure than non-contact defense. Therefore, although males continue to defend their brood, they may reduce their parental expenditure and potential risk of injury by switching to less risky defense behaviour.

Another indicator of parental expenditure is the amount of time that the parent spends with the young (Clutton-Brock & Godfray 1991). Time with the brood was highest during the embryo period, lower at the larva period and very low at the juvenile period (Figure 6). The low time with brood at the juvenile period is complicated by the fact that the young are free swimming and both the area occupied by the young and the male's territory are larger than at earlier periods (Ridgway 1988, Scott 1993). The larger territory of the male may partially explain the decline of time with the brood, but the model, in the centre of the group of young, still represents a risk to the male's brood. Both the decline in time with the brood and total parental defense behaviour directed towards the model suggests that parental males reduce parental expenditure during the juvenile period.

The decline in parental defense can be explained by the decline in the value of the current brood as it develops which is predicted to reduce the optimal level of parental effort (Sargent & Gross 1986). This pattern of decline in parental defense has been previously observed in male smallmouth bass (Ridgway 1988). While variation in the value of the current brood may partly explain the decline in male parental defense over sample periods, it does not explain the size-based variation in parental defense. Brood number, which does decline as the young develop, did not differ among different sized males. The probability of the young surviving without parental care is likely also independent of male size. Therefore, the value of the current brood is probably similar for all sizes of males.

Our results suggest that size-based differences in available energy may influence parental defense. While nesting males utilize both endogenous and exogenous energy, their ability to utilize endogenous reserves may be constrained by their energetic requirements over the winter. Because males rely on endogenous reserves to survive the winter, depletion of reserves during parental care may increase the probability of overwinter mortality for parental males. Increased probability of mortality may represent a greater cost to small males, in terms of reduced future reproduction, because they have more potential breeding seasons in the future than large males. Small males may rely on exogenous energy to a much greater extent than larger males and this relatively lower energy budget, along with relatively higher metabolic demands, means they are not able to expend energy on parental defense as long as larger males can. Further study, including manipulation of energy budgets, is required to clarify size-based energy budget differences.

The present study demonstrates that male smallmouth bass utilize stored energy during both the winter and the parental care period. Energy stores, which were not built up before nesting, declined to a minimum level by the end of the parental care period. Small males had consistently lower energy reserves than larger males and did not utilize these reserves at the same rate during the parental care period. All parental males complimented endogenous energy reserves by feeding during parental care, however, small males appear to rely proportionately more on exogenous energy intake than do larger males. Parental defense by all sizes of males declined over the parental care period, the decline being the most obvious in parental defense by small males. Small males' lower energy budget may make them less effective parents and decrease their probability of survival over the following winter relative to larger males.

Acknowledgements

We thank T. Friesen, R. Scott and the staff and summer students at Harkness Laboratory for their assistance in the study. J. Leatherland and W. Beamish provided advise and laboratory facilities for lipid analyses. The research was supported by a Ontario Renewable Resources Research Grant to D. Noakes and an Electric Power Research Institute Fellowship to R. Mackereth.

References cited

Adams, S.M., R.B. McLean & J.A. Parrotta. 1982. Energy partitioning in largemouth bass under conditions of seasonally fluctuating prey availability. Trans. Amer. Fish. Soc. 111: 549–558.

Bligh, E.G. & W.J. Dyer. 1959. A rapid method of total lipid extraction and purification. Can. J. Biochem. Physiol. 37: 911–917.

Blumer, L.S. 1979. Male parental care in the bony fishes. Quart. Rev. Biol. 54: 149–161.

Brett, J.R. 1965. The relation of size to rate of oxygen consumption and sustained swimming speed of sockeye salmon (*Oncorhynchus nerka*). J. Fish. Res. Board Can. 22: 1491–1501.

Brett, J.R. & T.D.D. Groves. 1979. Physiological energetics. pp. 279–352. *In*: W.S. Hoar, D.J. Randall & J.R. Brett (ed.) Fish Physiology, Vol. 8, Academic Press, New York.

Chellappa, S. & F.A. Huntingford. 1989. Depletion of energy reserves during reproductive aggression in male three-spined stickleback, *Gasterosteus aculeatus* L. J. Fish Biol. 35: 315–316.

Chellappa, S., F.A. Huntingford, R.H.C. Strang & R.Y. Thomson. 1989. Annual variation in the energy reserves in male three-spined stickleback, *Gasterosteus aculeatus* L. (Pisces, Gasterosteidae). J. Fish Biol. 35: 275–286.

Clutton-Brock, T.H. & C. Godfray. 1991. Parental investment. pp. 234–262. *In*: J.R. Krebs & N.B. Davies (ed.) Behavioural Ecology: An Evolutionary Approach. 3rd ed., Blackwell Scientific Publications, London.

Coleman, R.M., M.R. Gross & R.C. Sargent. 1985. Parental investment decision rules: a test in the bluegill sunfish. Behav. Ecol. Sociobiol. 18: 59–66.

Coleman, R.M. & R.U. Fischer. 1991. Brood size, male fanning effort and the energetics of a non-shareable parental investment in bluegill sunfish, *Lepomis macrochirus* (Teleostei: Centrarchidae). Ethology 87: 177–188.

Dufresne, F., G.J. FitzGerald & S. Lachance. 1990. Age and size-related differences in reproductive success and reproductive costs in threespine sticklebacks (*Gasterosteus aculeatus*). Behav. Ecol. 1: 140–147.

FitzGerald, G.J., H. Guderley & P. Picard. 1989. Hidden reproductive costs in the threespine stickleback (*Gasterosteus aculeatus*). Exper. Biol. 48: 295–300.

Gross, M.R. & R.C. Sargent. 1985. The evolution of male and female parental care in fishes. Amer. Zool. 25: 807–822.

Helfman, G.S. 1983. Resin-coated fishes: a simple model technique for in situ studies of fish behaviour. Copeia 1983: 547–549.

Herbes, S.E. & C.P. Allen. 1983. Lipid quantification of freshwater invertebrates: method modification for microquantitation. Can. J. Fish. Aquat. Sci. 40: 1315–1317.

Hinch, S.G. & N.C. Collins. 1991. Importance of diurnal and nocturnal nest defense in the energy budget of male smallmouth bass: insights from direct video observations. Trans. Amer. Fish. Soc. 120: 657– 663.

Hirshfield, M.F. 1980. An experimental analysis of reproductive effort and cost in the Japanese medaka, *Oryzias latipes*. Ecology 61: 282–292.

Jennings, M.J. & D.P. Philipp. 1992. Reproductive investment and somatic growth rates in longear sunfish. Env. Biol. Fish. 35: 257–271.

Keast, A. 1968. Feeding of some Great Lakes fishes at low temperatures. J. Fish. Res. Board Can. 25: 1199–1218.

Kolok, A.S. 1991. Temperature compensation in two centrachid fishes: do winter-quiescent fish undergo cellular temperature compensation? Trans. Amer. Fish. Soc. 120: 52–57.

Mackereth, R.W. 1995. Size-based variation in the allocation of energy to parental care by male smallmouth bass (*Micropterus dolomieu*). Ph.D. Dissertation, University of Guelph, Guelph. 119 pp.

Marconato, A., A. Bisazza & M. Fabris. 1993. The cost of parental care and egg cannibalism in the river bullhead, *Cottus gobio* L. (Pisces, Cottidae). Behav. Ecol. Sociobiol. 32: 229–237.

Martin, N.V. & F.E.J. Fry. 1972. Lake Opeongo: effects of exploitation and introductions on the salmonid community. J. Fish. Res. Board Can. 29: 795–805.

Medford, B.A. & W.C. McKay. 1978. Protein and lipid contents of gonads, liver, and muscle of northern pike (*Esox lucius*) in relation to gonad growth. J. Fish. Res. Board Can. 35: 213–219.

Nelson, J.A. & J.J. Magnuson. 1992. Metabolic stores of yellow perch (*Perca flavescens*): comparison of populations from an acidic, dystrophic lake and circumneutral, mesotrophic lakes. Can. J. Fish. Aquat. Sci. 49: 2474–2482.

Pressley, P.H. 1981. Parental effort and the evolution of nest-guarding tactics in the threespine stickleback, *Gasterosteus aculeatus* L. Evolution 35: 282–295.

Ridgway, M.S. 1988. Developmental stage of offspring and brood defense in smallmouth bass (*Micropterus dolomieui*). Can. J. Zool. 66: 1722–1728.

Ridgway, M.S. 1989. The parental response to brood size manipulation in smallmouth bass (*Micropterus dolomieui*). Ethology 80: 47–54.

Ridgway, M.S. & T.G. Friesen. 1992. Annual variation in parental care in smallmouth bass, *Micropterus dolomieu*. Env. Biol. Fish. 35: 243–255.

Ridgway, M.S. & B.J. Shuter. 1996. Effects of displacement on the seasonal movements and home range characteristics of smallmouth bass in Lake Opeongo. North Amer. J. Fish. Manage. 16: 371–377.

Ridgway, M.S., B.J. Shuter & E.E. Post. 1991a. The relative influence of body size and territorial behaviour on nesting asynchrony in male smallmouth bass, *Micropterus dolomieui* (Pisces: Centrarchidae). J. Anim. Ecol. 60: 665–681.

Ridgway, M.S., J.A. MacLean & J.C. MacLeod. 1991b. Nest-site fidelity in a centrarchid fish, the smallmouth bass (*Micropterus dolomieui*). Can. J. Zool. 69: 3103–3105.

Robinson, W.R., R.H. Peters & J. Zimmerman. 1983. The effects of body size and temperature on metabolic rate of organisms. Can. J. Zool. 61: 281–288.

Sabat, A.M. 1994. Costs and benefits of parental effort in a brood-guarding fish (*Ambloplites rupestris*, Centrarchidae). Behav. Ecol. 5: 195–201.

Sargent, R.C. & M.R. Gross. 1985. Parental investment decision rules and the Concorde fallacy. Behav. Ecol. Sociobiol. 17: 43–45.

Sargent, R.C. & M.R. Gross. 1986. Williams' principle: an explanation of parental care in teleost fishes. pp. 275–293. *In*: T.J. Pitcher (ed.) The Behaviour of Teleost Fishes, Croom Helm, London.

Schmidt-Nielsen, K. 1972. Locomotion: energy cost of swimming, flying, and running. Science 117: 222– 228.

Schultz, E.T. 1991. The effect of energy reserves on breeding schedule: is there a saturation point? Func. Ecol. 5: 819–824.

Schultz, E.T., L.M. Clifton & R.R. Warner. 1991. Energetic constraints and size-based tactics: the adaptive significance of breeding-schedule variation in a marine fish (Embiotocidae: *Micrometrus minimus*). Amer. Nat. 138: 1408–1430.

Scott, R.J., M.S. Ridgway & D.L.G. Noakes. 1997. The nest range of male smallmouth bass: parental care after swim-up. Can. J. Zool. 75: 2058–2062.

Scott, W.B. & E.J. Crossman. 1973. Freshwater fishes of Canada. Bull. Fish. Res. Board Can. 184. 966 pp.

Shuter, B.J., J.A. MacLean, F.E.J. Fry & H.A. Regier. 1980. Stochastic simulation of temperature effects on first-year survival of smallmouth bass. Trans. Amer. Fish. Soc. 109: 1–34.

Shuter, B.J. & J.R. Post. 1990. Climate, population viability, and the zoogeography of temperate fishes. Trans. Amer. Fish. Soc. 119: 314–336.

Townshend, T.J. & R.J. Wootton. 1985. Adjusting parental investment to changing environmental conditions: the effect of food ration on parental behaviour of the convict cichlid, *Cichlasoma nigrofasciatum*. Anim. Behav. 33: 494–501.

Trivers, R.L. 1972. Parental investment and sexual selection. pp. 136–179. *In*: B. Campbell (ed.) Sexual Selection and the Descent of Man 1871–1971, Aldine Publishing Company, Chicago.

Voet, D. & J.G. Voet. 1990. Biochemistry. John Wiley and Sons, New York. 1223 pp.

Webster, D.A. 1954. Smallmouth bass, *Micropterus dolomiuei*, in Cayuga Lake. Part 1. Life history and environment. Memoir 327, Agriculture Experimental Station, Cornell University, Ithaca. 39 pp.

Weigmann, D.D., J.R. Baylis & M.H. Hoff. 1992. Sexual selection and fitness variation in a population of smallmouth bass, *Micropterus dolomieui* (Pisces: Centrarchidae). Evolution 46: 1740–1753.

Williams, G.C. 1966. Adaptation and natural selection. Princeton University Press, Princeton. 307 pp.

Wootton, R.J. 1985. Energetics of reproduction. pp. 231–254. *In*: P. Tyler & P. Calow (ed.) Fish Energetics, Croom Helm, London.

Zar, J.H. 1984. Biostatistical analysis, 2nd ed.. Prentice-Hall, Englewood Cliffs. 718 pp.

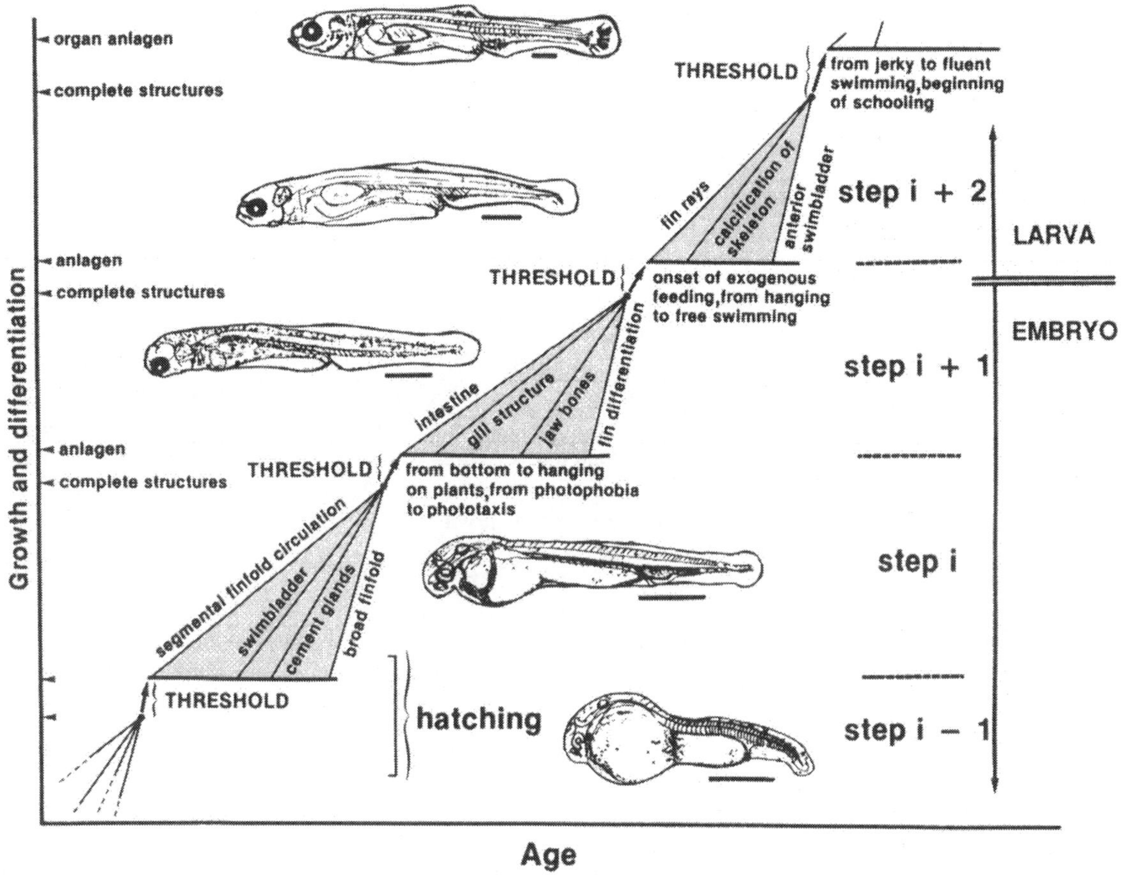

Figure 1. Diagrammatic representation of a few steps in the development of *Abramis ballerus* (from Balon 1959) constructed by Balon (1986) to illustrate his 'theory of saltatory ontogeny'. While not an entirely satisfactory illustration of this theory, it was later used by Greenwood (1989, figure 1) and Bruton (1994, figure 2, reproduced here from the cover of this journal).

embryo leaves the egg during the process of hatching, 'is not a matter of terminology but a matter of understanding what we are talking about'. How true!

Yet, hatching may not be considered a synonym of birth. It may seem strange, but I have met with this problem even in the manual on developmental biology published lately which, right after the first image illustrating the development of a frog, have the terms 'hatching (birth)' printed explicitly. Obviously, neither release, nor egg deposition in oviparous fishes is parturition, as Balon (1999, p. 20) writes in this volume adding: 'Birth, however, is in a broad sense a synonym of parturition . . . '. In this, I would beg to differ from him, because the Latin noun 'parturitio' is derived from the verb 'parturire' meaning 'to prepare for labour', 'to work towards giving birth', 'to feel the throes of childbirth', etc. Therefore, while 'parturitio' is the process of being born (leading to birth), 'birth' is a state resulting from being born. I only fear that some might understand my remark in terms of the Latin adage: 'parturiunt montes, nascetur ridiculus mus'.

Wald (1981, p. 1) wrote: 'With metamorphosis it is easy to know where to start, but hard to know where to stop'. During metamorphosis there is, on the one hand, regression of larval structures and functions, on the other, the formation of some new structures and functions that are essential to adults. 'The larva frequently possesses rudiments of adult structures, the adult vestiges of larval structures (. . .) the larva and adult developing not so much in succession as side by side', reminds us Wald (op. cit., p. 5). The consequences of metamorphosis are changes of anatomy,

Environmental Biology of Fishes **56:** 277–280, 1999.

To be a juvenile and not to be a larva: an attempt to synthesize

Karol Hensel
Department of Zoology, Comenius University, Mlynská Dolina B-2, 842 15 Bratislava, Slovakia
(e-mail: hensel@fns.uniba.sk)

The question – when does a fish becomes a juvenile – might seem strange to some, and even totally worthless, mainly to those fish and fishery biologists who designate all the small ontogenetic stages of fishes by the banal term 'fry' (for a commentary on this unfortunate term, see Balon 1990). Despite this, a group of predominantly younger scientists met at a workshop in Bratislava to exchange views on this topic. Ontogeny is a process during which one event is related to another and everything is related to everything else. Therefore, besides the title subject, the participants of the workshop discussed also such topics as fish metamorphosis, whether the larva period begins with hatching or with the onset of exogenous feeding, and eventually, whether fish ontogeny is saltatory or otherwise.

A variety of views emerged on ontogeny being either a gradual process (insisting on this view were some scientists engaged in the domain of fishery biology, or behavioral ecology), or a saltatory one. It also became obvious that certain specialists looked at ontogeny from the narrow angle of that group of fishes with which they were familiar, failing to consider other fish groups (not to mention other animals). Some rooted opinions and strongly held views were defended by 'practicality'.

Ontogeny is a formative process taking place in every species at a specific rate and rhythm, in which the so-called heterochrony, i.e. change in rates and timing of developmental events, becomes manifest. Hence, ontogeny cannot be understood as a gradual process during which small and inconspicuous changes in form and function accumulates continuously, but neither as a discontinuous process (as some erroneously understood 'the theory of saltatory ontogeny'). Ontogeny is a continuous sequence of longer stabilized states (steps) alternating with shorter less stable intervals (thresholds, Figure 1), in short, some sort of 'a stepwise progression' (Balon 1986). At the same time, changes in anatomy, physiology, behavior and ecology may take place at different rhythms and at varying rates. Consequently, it is at times quite difficult to pinpoint the onset, but especially the completion, of the various steps and therefore to compare individual ontogenies of different species of fishes. This also causes the existing confusions, leaving several questions open.

One of these questions is whether insemination, activation, or fusion of male and female pronuclei may be considered as the beginning of ontogeny, or only the first division of a zygote. Balon (1985, 1990), because of the existence of gynogenesis and parthenogenesis, considers activation the only precise beginning of ontogeny for bisexual organisms.

A further question was whether the embryo period of fishes ends with hatching, or with the onset of exogenous feeding. Balon (1985, 1999 this volume) has presented arguments showing that the start of external feeding is characteristic of larvae or juveniles. Therefore, the term 'yolksack larva' is inappropriate[1] and we should speak of a free embryo or eleutheroembryo. Hatching is thus a process in which the embryo leaves the egg envelope and this occurs at different steps during the embryo period; in certain animals this already takes place at the blastula stages (hatched blastula of sea urchins, or hatching blastocyst of mammals). According to Balon (1999 this volume), the popular lay usage of 'egg hatched' or of 'freshly hatched larva', when the

[1] This similarly applies to the term 'prelarva'. Its illogical use was pointed out by Kryzhanovsky (1956). Nonetheless, Makeyeva (1988) gave priority to this term and objected to that of 'free embryo' as not corresponding to the state of organism freed from an egg envelope, because, allegedly, the Greek word embryo designates 'in envelope'. That, however, is not true. The term embryo has its root in the Greek expression 'em bryein' – to swell inside.

physiology, behavior and ecology (Balon 1985, 1999 this volume). The outcome of metamorphosis is that the fish turns into a juvenile.

Evidence to the fact that it is possible to learn 'when fish become juvenile' has been provided by several participants of the workshop. They determined the onset of the juvenile period precisely by comparing morphological and ecological changes. Thus, for instance, Kováč et al. (1999 this volume) found that the coincidence in shifts in morphometric (mensural) values with those in microhabitat use suggests that thresholds do occur during this interval of stone loach, *Barbatula barbatula*, life history, and that the larva period ends with the completion of this shift in relative growth. Similarly, Simonović et al. (1999 this volume) found significant shifts in relative growth of six mensural characters of the European minnow, *Phoxinus phoxinus*, in concurrence with significant changes in microhabitat use, which they consider as the threshold between larva and juvenile development. Gozlan et al. (1999a this volume) have shown that the onset of the juvenile period in sofie, *Chondrostoma toxostoma*, takes place at 50 mm SL, coinciding with a change in mouth position to inferior. Gozlan et al. (1999b this volume) interpret the shift of relative growth in sofie (at 23 mm SL) to be rather a threshold that initiates the last interval of the remodeling process and designate it as 'pre-juvenile' interval that is principally metamorphic and completes the 'end-of-larva-period' metabiosis. That is why, in their view, we should consider the last metamorphic (transitional) step of sofie (at 23–50 mm SL) as either the last step of their larva period (metamorphic larvae) or as a separate period (metamorphosis period). However, it seems appropriate here to mention Balon's statement (1999, p. 23 this volume) that 'some decisive events like ... metamorphosis ... may not be a part of a saltatory threshold, but an interval undergoing heterochronous shifts according to environmental stimuli'. Metamorphosis may therefore not be a part of a saltatory threshold nor of a specific step.

Is it then possible to determine when a fish becomes juvenile? The results of the above authors have shown that it is possible, though other authors think it is not always possible to do so with precision. Arguing that in certain cases there exists a transitional state (neither larva, nor juvenile), Pavlov (1999 this volume) has attempted to use the term 'state' which has no saltatory significance. To determine the transition from the larva to the juvenile and to determine the onset of the 'juvenile state', Pavlov (op. cit.) made use of exclusively morphological criteria. He advocates the formation of main definitive organs and skeletal elements and the disappearance of larval characters. The terms 'state' then differs from that of 'step' (sensu Balon) which designates the main natural (homeorhetic) intervals of ontogeny within the saltatory process. According to Pavlov (op. cit.), even despite different types of early ontogeny (indirect, transitory and direct), the beginning of the juvenile state that he studied in sea fishes (herring, wolffish and eelpout) occurred at similar total lengths (approx. 35 mm). On the other hand, the results of a composite study of morphology, physiology, behavior and ecology of fishes during ontogenetic development, presented by Sakakura & Tsukamoto (1999 this volume) and Masuda & Tsukamoto (1999 this volume) solve the problem well. Vagelli (1999 this volume) then proved beyond any reasonable doubt that the direct development without a larva exists also in purely marine fish.

The participants in the Bratislava workshop agreed that the question of early ontogeny and particularly metamorphosis of fishes requires more attention. They parted with the conviction that the presentation of their results and mutual exchange of views was of great value and should be continued in the future. Who knows, maybe the organizers of some future workshop will focus attention not solely on 'early', but also 'late' ontogeny and on the interval 'when fish become adults'. Perhaps then P.J. O'Rourke's sentence will prove appropriate: 'You realize that your children have come of age when they stop asking you where they came from and refuse to tell you where they are going.'

References cited

Balon, E.K. 1959. Die embryonale und larvale Entwicklung der Donauzope (*Abramis ballerus* subsp.). Biologické práce 5: 1–87.

Balon, E.K. (ed.) 1985. Early life histories of fishes: new developmental, ecological and evolutionary perspectives. Developments in Environmental Biology of Fishes 5, Dr W. Junk Publishers, Dordrecht. 280 pp.

Balon, E.K. 1986. Saltatory ontogeny and evolution. Rivista di Biologia/Biology Forum 79: 151–190 (in English and Italian).

Balon, E.K. 1990. Epigenesis of an epigeneticist: the development of some alternative concepts on the early ontogeny and evolution of fishes. Guelph Ichthyol. Rev. 1: 1–42.

Balon, E.K. 1999. Alternative ways to become a juvenile or a definitive phenotype (and on some persisting linguistic offenses). Env. Biol. Fish. 56: 17–38 (this volume).

Bruton, M.N. 1994. The epigenesis of an epigeneticist: an interview with Eugene Balon. South African Journal of Science 90: 270–275.

Gozlan, R.E., G.H. Copp & J.-N. Tourenq. 1999a. Early development of the sofie, *Chondrostoma toxostoma*. Env. Biol. Fish. 56: 67–77 (this volume).

Gozlan, R.E., G.H. Copp & J.-N. Tourenq. 1999b. Comparison of growth plasticity in the laboratory and field, and implications for the onset of juvenile development in sofie, *Chondrostoma toxostoma*. Env. Biol. Fish. 56: 153–165 (this volume).

Greenwood, P.H. 1989. Ontogeny and evolution: saltatory or otherwise? pp. 245–259. *In*: M.N. Bruton (ed.) Alternative Life-History Styles of Animals, Perspectives in Vertebrate Science 6, Kluwer Academic Publishers, Dordrecht.

Kováč, V., G.H. Copp & M.P. Francis. 1999. Morphometry of the stone loach, *Barbatula barbatula*: do mensural characters reflect the species' life-history thresholds? Env. Biol. Fish. 56: 105–115 (this volume).

Kryzhanovsky, S.G. 1956. Materials on development of clupeid fishes. Trudy Inst. Morf. Zhiv. A. N. Severtsova 17: 1–256 (in Russian).

Makeyeva, A.P. 1988. Review of 'Early life histories of fishes: new developmental, ecological and evolutionary perspectives (ed. by E.K. Balon)'. Voprosy ichtiologii 28: 697–700 (in Russian).

Masuda, R. & K. Tsukamoto. 1999. School formation and concurrent developmental changes in carangid fish with reference to dietary conditions. Env. Biol. Fish. 56: 243–252 (this volume).

Pavlov, D.A. 1999. Features of transition from larva to juvenile in fishes with different types of early ontogeny. Env. Biol. Fish. 56: 41–52 (this volume).

Sakakura, Y. & K. Tsukamoto. 1999. Ontogeny of aggressive behaviour in schools of yellowtail, *Seriola quinqueradiata*. Env. Biol. Fish. 56: 231–242 (this volume).

Simonović, P.D., P. Garner, E.A. Eastwood, V. Kováč & G.H. Copp. 1999. Correspondence between ontogenetic shifts in morphology and habitat use in minnow *Phoxinus phoxinus*. Env. Biol. Fish. 56: 117–128 (this volume).

Vagelli, A. 1999. The reproductive biology and early ontogeny of the mouthbrooding banggai cardinalfish, *Pterapogon kaudermi* (Perciformes, Apogonidae). Env. Biol. Fish. 56: 79–92 (this volume).

Wald G. 1981. Metamorphosis: an overview. pp. 1–39. *In*: L.I. Gilbert & E. Frieden (ed.) Metamorphosis, a Problem in Developmental Biology, Plenum Press, New York.

Environmental Biology of Fishes **56:** 281–289, 1999.

Species and subject index

Acanthochromis polyacanthus 21
Acantocyclops vernalis 170, 173
Activation 25, 28, 32, 50, 68, 69, 84, 85
Activity rhythm 194
Adenomera 18
Aggressive
  drive 239
  defence 82, 83
  harassment 84
Albula vulpes 27
Alderfen Broad 172
Allometry 17, 105, 106, 114, 249 (see also Growth)
  gradual 105, 107
Allopatric 32
Alprehost 17, 23, 29, 93
Altricial 17, 25, 26, 29, 30, 32, 91, 100
  phenotype 23
  form 10, 17, 99, 154, 163
Ambassis agassizii 137
Ambicolouration 53
Ambloplites rupestris 170
Ametamorphic 93
Amphidromous
  ayu 239
  goby 25
Amphidromy, freshwater 28
Amphipoda 210, 220
Anabranch 132, 140, 142
Anarhichadidae 42
Anarhichas lupus 41, 42
Anchovy 247
Anguilla anguilla 93
Aplodinotus grunniens 176
Apogon 81, 89
  affinis 89, 90
  erythrinus 90
  imberbis 90
  lineatus 90
  maculatus 90
  maculiferus 90
  menesemus 90
  niger 83, 90
  notatus 90, 91
  rueppelii 89, 90
  semilineatus 89, 90
Apogonichthys 89
  waikiki 90
Apogonidae 79, 88, 89
Archamia 89
Artemia 54, 88, 146, 233, 248, 249
  salina 81, 232
Astrapogon 89
Auditory capsule 43

Bacillariaphycea 194
Barbatula barbatula 105, 106, 118
Barbel 184, 199, 201, 204–210
  zone 184
Barbus barbus 184, 199
Bass, sea 55
Bautzen Reservoir 174
Bay of Biscay 213, 215
Bearer 26, 28, 49
  live 19, 48
Behaviour
  adaptation 157
  aggressive 68, 231–240, 253
  aim 233
  antipredatory 243, 244, 250, 251
  association 244, 246, 247, 251
  avoidance 185
  change 112
  chase 233–235, 240
  co-operative 231
  dominant 233–240
  feeding 63, 93, 154, 220, 225
  foraging 253
  gregarious 231
  intermediate 233–239
  mating 79, 80
  migratory 249
  model 239
  optimal foraging 183
  post-spawning 83
  pre-spawning 81, 82
  reproductive 91, 231
  rheopositive 224
  schooling 45, 231, 232, 235, 240, 243-246, 249–251
  shivering 231
  social 75, 99, 231
  subordinate 233–240
  swimming (see Swimming)
  territorial 139, 253
  transition 232, 234, 237
Benthivory 96, 97
Bidyanus bidyanus 137
Billabong 140–142, 145
Bifurcation 119
Biodiversity 18
Bioenergetics 169
  model 173–175, 177

Bivalvia 210, 220
Blackfish
  freshwater 146
  river 133, 139
Blastoderm 69
Blastodisc 22, 85
Bleak 184
Bluegill 172
Bonefish 27
Bosmina
  longirostris 170–172
  longispina maritima 171
  sp. 170, 175
Bottleneck effect 222
Brachionichthys hirsutus 21
Brachionus plicatilis 232
Bream
  sea 55
  zone 184
Breeding 142, 232
Buccal
  cavity 17, 32
  pouch 33, 42, 51, 79, 83
Bullhead 118, 253, 254, 261
Burbot 176
Bythotrepes 176

Calcification 33, 88, 231, 235–237, 244, 247
Cannibalism 68, 99, 231, 232, 235, 236, 240
Capsule 19, 43
Carangid 244, 245, 249
Caranx caballus 249
Cardinal fish, Banggai 21, 79, 89
Carnivore, opportunistic 141
Carp 93–100, 131
  gudgeon 133, 134, 140, 146
  shoot 97
Carrying capacity 261
Caterpillar 18, 20
Catfish 146
  freshwater 133, 137, 139, 140, 146
  marine ariid 32
Cell
  epithelial 53, 55, 56, 58, 60, 62
  eucaryotic 19
  somatic 19
Cephalization 69
Ceriodaphnia quadrangula 170
Circadian rhythm 225
Cladoceran 146
Chaetognatha 23
Chaoborus 176
Cheilodipterus 81, 89
  lineatus 90
Chironomidae 145, 146, 192, 199, 205, 210
Chlorophycea 194
Chlorophyll a 135
Chondrification 86–88
Chondrostoma 163
  nasus 67, 154, 184
  toxostoma 10, 67–73, 76, 153–161
Chorion 19–22, 84
  filament 79, 80, 84, 91
Chromatophores 59, 86
Chub 74, 183–194
Chydorid 135
Cichlid 32
Circulation 87
  blood 70, 85
  system 67, 75, 76
Cladoceran 98, 135, 146, 169, 170, 172, 175, 176, 178.
    210, 220
Cleavage 22, 69, 80, 84, 85
Clupea harengus 170, 173, 225
  pallasi marisalbi 41, 42
Clupeonella delicatula 28
Cod 225
  Murray 130, 133, 136, 137, 140, 142, 145, 146
Coelacanth 21, 33
Coleopteran 210
Colloid 53, 55, 56, 58
Colonisation 192, 194
Competition
  interspecific 209
  intraspecific 98
Conservation genetics 154
Consumption/biomass ratio 173
Copepoda 135, 146, 169, 170, 172, 173, 175–177, 210,
    220, 254
  calanoid 146, 173, 175
  cyclopoid 170, 172
Coral reef 24
Coregonidae 253
Cortisol 53–55, 63, 233, 235
  accumulation 237
  concentration 231, 235, 236, 238
  level 231, 236, 237
  secretion 237
Cottus gobio 118, 253, 254
Courtship 80, 89
  display 81, 91
Cover
  riparian 209
  instream 209
Crappie, black 176
Craterocephalus fluviatilis 137
Crayfish, Murray 140
Critical
  period 237, 244
  swimming speed 192
Crustacean 173, 221
Cyclops sp. 173
  strenuus abyssorum 170
  vernalis 171, 175
Cyphotilapia frontosa 21, 25, 32, 33, 42, 49, 51
Cyprinidae 74, 98, 118, 153, 177, 183, 184, 192, 200, 209

Cyprinus carpio 93–95, 131
haematopterus 95
Cytoplasm 22, 23

Dab 223
Dace 75, 183–194, 209
Damselfish 21
Daphnia 146, 173–177, 254
cuculatta 171, 174
galeata 171, 173, 174
galeata mendotae 171, 173, 175
hyalina 170–172
longispina 173
pulex 171, 173
retrocurva 171, 173
sp. 170
Day/night 219
variation 223
Definitive form 114
Demersal 21, 133, 213
Development
continuous 8
direct 17, 24, 29, 32, 33, 51, 79, 89, 93, 262
indirect 17, 26, 30, 51, 118
non-gradual 7
precocial 33
saltatory 24
threshold 9
Diademia sp. 90
Diadromous 134
Diatoms 220
Dicentrarchus labrax 55
Dichotomy 18
Diel 68, 183
activity 213
cycle 183
displacement 258
dynamics 183, 184, 186, 187, 191–195
feeding 220, 225
migration 213–225
pattern 187, 194, 253, 254, 257, 259, 261
variation 136, 163
vertical migration (see Migration)
Dietary
overlap 201
shift (see Shift)
Differentiation 25, 154, 178
bipolar 22
fin 186
Diffusion 214
Digestibility 178
Digestive system 210
Diptera 192, 205, 209
Discharge 144
Dispersal 17, 24
night 194
Dispersion index 204
Displacement, downstream 254

Dixidae 210
Dorosoma cepedianum 170, 172
Drag reduction 163
Drift 138, 139, 205, 233, 250, 261
density 210
drifting seaweed 232, 240
drum, freshwater 176
dwarf form 10, 163

Ecological
species 183
transition 25
Ecosystem 18
integrity 183
Eel 93
catadromous 25
leptocephalus 24
Eelpout 41–43, 46, 49, 50, 51
Egg 254
demersal 132
deposition 49, 50
diameter 68–70
envelope 9, 19, 20, 22, 42, 91
planktonic 22
secondary envelope 19, 20
size 21–26, 49
transfer 80–83, 91
Electivity
index 204
relativized 201
Eleotridae 130, 146
Elodea sp. 108, 111
Elopomorph 24
Emergence 146, 253, 254, 257, 258, 260, 262
Emigration 140, 183, 192
Endocrinological cascade 63
Endogenous 25, 28, 29, 97
diel rhythm 214
Energetic
cost 75
gain 261
Environmental perturbation 118
Eleutherodactylus 18
Ephemeroptera 199, 205, 210
Epiboly 22, 69
Epigenetic process 23, 26
Etap 18, 94
Etymology 21
Euastacus armatus 141
Euphasia pacifica 80
superba 80
Eurytemora affinis 170, 171, 173, 175
Evolution 19, 32
divergence 23
hypermorphic 29
paedomorphic 29
Extinction 32, 143

Eye
  lens 85
  migration 53, 54, 56, 58, 62, 63, 218, 222, 223
  pigmentation 244

Fecundity 17, 26, 29, 49, 50, 80, 91
Feeding
  ability 225
  activity 213, 219, 220, 224, 225
  behaviour (see Behaviour)
  exogenous 7, 9, 17, 19, 25, 26, 28, 32, 42, 50, 51, 68, 70, 72, 81, 119, 120, 127, 159, 161, 237
  efficiency 218
  endogenous 144, 146
  exogenous feeding 144, 146
  first oral 23, 25, 50, 218
  function 127
  mixed 25, 33
  pattern 214
  rhythm 213, 214, 221
  selectivity 174
Feral form 99
Fertility 83
Fertilization 25
Fetal tissue 33
Finfold 2l, 30, 42, 43, 46, 73, 85, 119, 153, 161, 204
  differentiation 72, 74
  resorption 94
Fishway 138
Flatfish 53, 58, 59, 61, 63, 213, 214
Flood
  pulse concept 132, 136, 141, 142
  recruitment model 131, 132, 138
Flooding 129, 131–133, 136, 138–140, 142, 143, 147, 199
Flounder 93, 223
  Japanese 54, 60, 62, 237
  starry 54
  winter 59
Flow 129, 133–136, 138, 139, 141–143, 145, 147, 185, 200
Foa 89
  brachygramma 90
  madagascariensis 90
Follicle 53, 55
  cell 19, 20
Food web 140
Foraging
  activity 163
  behaviour (see Behaviour)
  optimal 176
  pattern 163
Frog 18, 29
Fullness index 213, 218
Functional describer 183

Gadopsis marmoratus 139
Galaxias 134
Galaxias olidus 133
  rostratus 133, 137
Galeichthys feliceps 32
Gamete 30, 68, 80, 84
Gametogenesis 136, 138
Gape 133–135, 145
  limited 177, 199
  size 133, 134, 146
  width 176
Gas bladder 71
Gasterosteus aculeatus 171, 175
Gastric
  evacuation rate 173
  gland 58
Gastropod 220
Generalist 163
Genetic heterogeneity 99
Germ ring 69
Gerontomorphosis 29
Gill chamber 17
Glosamia 89
Gluconeogenesis 237
Gobio gobio 184
Goby 28
  round 42
Gonad 30, 127, 132, 138, 163
Gonadosomatic index 80
Gradualist 8, 9
Grayling 117, 261
Growth
  allometric 25, 93–96, 105–107, 117, 126, 153, 243, 244, 247, 251
  hormone 164
  isometric 95, 105, 107, 110, 112
  opportunity 164
  proportional 119
  plasticity 153
  relative 10, 68, 74, 76, 96, 105, 106, 112–114, 118, 122, 126, 127, 153, 156, 158, 163, 249
Guarder 26–29
Gudgeon 130, 133, 137, 146, 184
  flathead 13, 134, 137, 140, 145
  western carp 137, 146
Guppy 247
Gut
  capacity 201
  evacuation 225
  fullness 201, 205, 209, 220, 222
Gymnocephalus cernuus 177
  baloni 16

Habitat
  calm 194
  diel shift (see Shift)
  partitioning 183
  suitability 126, 261
  transition 24
  use 7
Halibut 55–57, 59, 61–63
  Atlantic 11, 53, 63

Handfish, spotted 21
Hardyhead, Murray 137
Harpacticoid 220
Hatching 9, 20, 23, 25, 28–30, 48–50, 54, 68, 71, 81, 88, 89, 93, 133, 244
Hemoglobin 30
Herring 41–46, 49, 50, 173, 175, 225, 247
  bony 137, 138, 140
Heterochrony 23, 41, 42
Hippoglossus hippoglossus 11, 53
Hirudinea 210
Homeorhesis 29, 42, 95
Hormone cascade 53
  exogenous 58, 60
Hydracarina 205, 210
Hypseleotris
  klunzingeri 137
  spp. 134

Ichthyoplankton 19
Immigration 140, 183, 213, 214, 225
Inflexion point 105
Insemination 42, 49, 50, 67, 68, 91
  artificial 43
Inshore
  -offshore movement 183, 194
  refuge (see Refuge)
  transfer 213
Instantaneous state 23
Interaction
  agonistic 231, 233, 235, 239
  cell-to-cell 9
  organ-to-organ interaction 8, 9, 42, 114
  organism-to-environment interaction 8, 9, 42, 114, 154, 163
  social 93, 97, 231
  tissue-to-tissue 9
  trophic 169
Inter-individual distance 245–247
Intestinal loop 186, 217
Isochrysis galbana 54
Isometry 95, 96, 105, 106, 249 (see also Growth)
Isopoda 210

J-posture 231–238
Jolytail, Murray 137
Juvenile (see Transition)
Juvenilization 29

k-selection 26
killifish, rainwater 94
krill 80

Labeotropheus 49
  fuelleborni 28
  trewavasae 32
Lake
  Hume 139
  Mulwala 139
  Opinicon 172
  Shelbyville 172
  Tanganyika 32, 42
Lamprey 10
Larva (see Transition)
Lateral line 243
Latimeria chalumnae 25, 28, 33, 49
Lecithotrophic 18
Leiopotherapon unicolor 137, 138
Lepomis
  gibbosus 170
  macrochirus 170, 172
Leptodora 176
Leuciscus
  cephalus 74, 183, 184, 186–193
  leuciscus 75, 183, 184, 186–193, 209
Life cycle 129, 132, 141, 143
Life history 19, 21, 28, 131
  interval 205
  mode 147
  model 10, 17, 19, 23, 24, 28, 29, 91, 93, 94, 111
  style 26, 144
  theory 20
  threshold 127
  transitory 23
Limanda limanda 223
Lipid
  content 28
  droplet 85
  globule 84
Littoral 135
Loach 113
  stone 105–108, 111, 114, 118
Locomotion 96, 99
Longevity 144
Lota lota 176
Low flow 194
  recruitment hypothesis 129, 131, 132, 134, 136, 144, 146, 147
Lucania goodei 30
  parva 30, 94
Lumen 53, 55, 56, 62
Luvarus imperialis 26

Maccullochella macquariensis 137, 139
  peelii peelii 130, 137
Macquaria ambigua 130, 137
  australasica 130, 137
Margin 192, 205
Match/mismatch hypothesis 144
Mating 89
  process 91
Maturation 25, 30, 131, 136, 138
  ovary 23, 30
Maturity 26
Meckel's cartilage 86
Melanophore 43, 46, 71–75, 85–87, 118, 120

Melanotaenia fluviatilis 134, 137
Membrane 19
Mensural character 7, 105–107, 109–114, 117–119, 121–124, 126, 127, 155, 160, 161
Mesocosm 169, 172–175
Metabiosis 10, 11, 25, 153, 154, 163, 164
Metabolic
 cost 178
 rate 95, 99
 requirement 177
Metamorphic 10, 164
Metamorphosis 7, 10, 11, 17, 20, 23–25, 27, 28, 30, 32, 42, 49, 53, 54, 56, 57–63, 67, 74, 91, 93–96, 106, 118, 119, 132, 153, 154, 163, 164, 213, 214, 218, 222–226, 231, 237–239, 251
 cataclysmic 24, 25, 30
 second 10, 11, 25
Microchirus variegatus 223
Microcrustacean 135, 145
Migration 48, 74, 99, 131, 138, 139, 143, 214, 231, 232, 249
 drive 239
 nocturnal 194
 vertical 213, 214 (see also Diel)
Minnow 10, 117, 118, 120–127, 163, 184, 209
Mogurnda adspersa 137
Morone americana 171, 174
Morphotype 155, 156, 163
Morula 22, 69
Mouth gape (see Gape)
Mouthbrooder 32, 89, 90
Mucus 10, 161, 163
Mussel 80
Mutual attraction 245–249, 251
Myotome 54–59, 69, 71, 72
Myriophyllum spp. 108, 111
Mysidacea 176

Nannochrolopsis sp. 232
Nannoperca australis 137
Nase 67, 74, 75, 154, 184
Nematalosa erebi 137, 138
Nematoda 210
Neogobius melanostomus 28, 42
Nervous system 250
 central 247, 249, 250
Nest
 building 144
 guarding 32
Niche 10, 157, 250
 breadth 95
 niche shift 205
Nocturnal pattern 261 (see also Migration)
Nonequilibrium thermodynamics 8, 18
Non-guarder 26, 28, 46, 74
Non-migratory 139
Notochord 68, 69, 72, 73, 85, 87, 154, 161
Nutrition
 endogenous 19, 20, 21, 96
 exogenous 96
 value 178

Ocellate spot 118
Oil
 droplet 46
 globule 21, 81, 84
Olfactory pit 10
Oligochaeta 210
Olive perchlet 137
Oncorhynchus mykiss 131, 261
Ontogenic
 interval 119, 169, 178, 187, 193
 process 157
 rate 10, 153, 155, 158, 163, 164
 scale 118
 trajectory 95–99, 119, 126
Ontogeny 9, 22, 23, 41
 direct 22–25, 41, 50, 91
 gradual 9, 94
 indirect 21–25, 41, 49, 50
 intermediate 24
 non-gradual 41
 non-saltatory 7
 saltatory 7–9, 18, 23, 41, 51, 67, 94, 105, 111, 127, 153
 transitory 41
Oocyte 20, 80, 81, 84
Optokinetic response 243–248
Oral cavity 42, 79, 87, 89, 91
Oreochromis niloticus 32
Organogenesis 222, 231
Osmerus eperlanus 171, 174
Ossification 43, 46, 49, 59, 88
 skeletal 48
Ostracoda 135, 210, 220
Ovary 81
Overwinter period 191, 194
Oviparity 20, 41
Ovulation 49, 136, 138
Ovum 9, 24, 67, 68, 81, 91
 envelope 69
Oxbow lake 132

Paedomorphosis 24, 29, 33
Paralichthys olivaceus 54, 237
Parental care 29, 133, 134, 144
 protection 91
Parturition 20, 25, 33, 49, 50, 51
Pecking order 239
Pelagic phase 49, 214
Perca flavescens 170, 172
Perca fluviatilis 131, 170, 173, 194
Perch 177, 194
 Eurasian 173–175, 177
 golden 130, 131, 133, 136–140, 142, 145, 146
 Macquarie 130, 137

Perch – *Continued*
  redfin 131
  silver 133, 137, 139, 145, 146
  southern pygmy 133, 137
  spangled 137, 138
  white 174
  yellow 172, 173, 176, 177
Percina caprodes 170
Performance
  metabolic 170, 172, 176
  morphological 169
Perivitelline fluid 42
  space 21, 22, 68, 70, 81, 84, 85
Perna canaliculus 80
Phaeoptyx 89
  conklini 90
Phenotype 7, 17, 20, 154
  definitive 17, 24–32, 53, 61, 93, 94, 163
  dispersal 23
  final 61
  plasticity 164
  precocial 23
Philypnodon grandiceps 134, 137
Photophobia 70, 75
Photoperiod 136–138
Photoreaction 46
Phototaxis 236, 243–246, 249
Phoxinus phoxinus 10, 117, 118, 184, 209
Phylogenetic position 75
Phytolithophilous 184
Pigmentation 46, 53, 56, 58, 59, 61–63, 69, 70–76, 86–88, 95, 118, 120, 134, 243
  retina 244
  visual 30
Piscivory
  facultative 177
Placenta 17
  analogue 23, 25
Plaice 54
  North Sea 214, 223
Planktonic
  dispersal 21, 22, 91
  larvae 21
  phase 89
  soup 24
  transport 21
Platichthys flesus 223
  stellatus 54
Plecoglossus altivelis 239
Plecoptera 210
Pleuronectes americanus 59
  platessa 54, 214
Point abundance sampling 118, 154, 200
Polychaete 220
Pomoxis nigromaculatus 170, 176
Precocial 17, 25, 26, 29, 32, 79, 164
  form 10, 99, 154, 163, 164
Predation
  avoidance 157, 183, 209, 240, 254, 261
  particulate 176
  piscivorous 177
  pressure 261
  risk 253, 262
  selective predation 176
Prey preference 201
Primary male 82
Propulsion 163
Protein synthesis 178
Proteolytic
  activity 178
  enzyme 19
Protoplasm 20
Pseudocaranx dentex 10, 238, 243
Pseudopleuronectes herzensteini 93
Pterapogon 89
  kauderni 21, 79, 81, 82, 87, 89–91
Pterygiophore 59, 86, 88
  migration 58

r-selection 26
Rainbowfish, crimson-spotted 133, 134, 137, 140, 145
Rank reversal 239
Ranunculus fluitans 122, 126
Ranunculus sp. 118
Recolonization 138
Recruitment 75, 129, 131, 132, 136, 143, 147, 183, 213, 243
Refuge 135, 194
  nearshore shallow 194
  predation 192
Remodelling 27, 42, 67, 74, 75, 93, 112, 126, 127, 153, 163, 226, 231
Reproductive
  behaviour (see Behaviour)
  capacity 175
  guild 26, 184
Resource
  availability 176
  partitioning 183
Respiration
  cutaneous 177
  function 127
  system 247
Retropinna semoni 134, 137
Rhabdamia 89
Rheotaxis 243, 245, 246
River
  Amazon 141
  Amur 99
  Danube 67, 99, 209
  Darling 144
  Frome 117, 118
  Garonne 74, 153–155
  Great Ouse 106, 108
  Lachian 136
  Lee 117, 118, 192

River – *Continued*
  Mekong 141
  Meuse 184
  Mississippi 98
  Murray 98, 100, 139, 144
  Murrumbidgee 98, 100, 142
  Ourthe 183, 185
  regulation 146–148, 200
  Rhine 200
  Sieg 199, 200
  Thames 135
  Volga 99
  Zaire 141
Roach 10, 94, 106, 117, 164, 172, 175, 178, 184
Rotifera 135, 145, 146, 169, 176, 192, 232
Ruffe 177
Rutilus rutilus 10, 94, 106, 117, 164, 169, 172, 184
Ryukyu Islands 249

Salmo trutta 131, 194, 253
Salmonidae 51, 164, 183, 239, 253, 254
  anadromous 24
  coho 238
Saltatory 7–10, 18, 23, 41, 51, 94, 105, 106, 154 (see also Ontogeny)
Salvelinus namaycush 28
Sanddiver 21
Scale 30, 49, 74, 163, 185, 204, 243
  cover 10, 11, 118, 119, 153, 157, 161, 186
  formation 5, 117, 245, 250, 251
School formation 243, 244, 247–249, 249 (see also Behaviour)
Scottish loch 173
Sea
  Baltic 173, 175
  bass 55
  bream 55
  urchin 90
  White 42, 43, 46, 48, 49
Secondary male 82
Sensory ability 177
Separation angle 245–247
Septum 10, 161
  nasal 95, 119, 153, 157
Seriola quinqueradiata 8, 231, 232, 243, 244, 248
Sessile
  habit 21
  state 91
Settlement 57, 63, 89, 213, 214, 225, 226, 257
Shad, gizzard 172, 173, 176
Shelter 185
  predator 209
  habitat 191, 194
  riparian 183, 185, 187, 189, 191
  vegetation 187, 191
Shift
  diel 261, 262
  dietary 8, 178
  displacement avoidance 262
  habitat 10, 93, 106, 111, 127, 135, 163, 178, 200, 209, 213, 222, 231, 251, 253
  heterochronous 23
  morphological 11, 105, 112
  niche 199, 205, 231
  ontogenetic 117, 199
  resource use 126
Shooting 97–99, 100
Sicyopterus lagocephalus 28
Silverside 247
Simulidae 210
Siphamia 89
  corallicola 90
Size
  structure 184, 187, 192
  range (see Transition)
Skeleton
  axial 33, 43, 46, 49
  calcification 30, 32
  cartilaginous axial 32
Smelt 174
  Australian 133, 134, 137, 140, 145
Smolting plasticity 164
Social rank 231, 233, 235, 237–240
Sofie 10, 11, 67–76, 153–164
Sole 216–219, 221, 223–225
  common 213
  thickback 223
Solea solea 213, 218
Sparus aurata 55
Spawning 74, 75, 84, 98, 131–142, 144, 147, 214, 249
  mode 142
  protracted 134
  serial 134
Sperm 9, 84, 91, 154
Spharaemia 89
Spharaemia orbicularis 89, 90
Split linear regression 107–110
Sprat 175
Sprattus sprattus 171, 175
Squamation (see Scale formation)
Stabilised state 8, 9, 41, 114, 127
Starvation 213, 214, 226, 237, 239
Stizostedion
  lucioperca 171
  vitreum 28, 171, 173
Stress response 163, 238
Stickleback, threespine 175
Striped jack 10, 11, 238, 243, 244, 249–251
Suffocation 8, 235
Suitability score 120
Survival 145, 226, 250, 255
Swimbladder 43, 46, 72–75, 86, 95, 119, 126, 213, 224, 226
  differentiated 222

Swimming
  ability 10, 75, 105, 106, 135, 204, 209, 245
  activity 214, 224
  behaviour 214, 225
  capacity 75, 112, 154, 163, 192, 209
  costs 209
  sustainable speed 199
Sympatric 32
Synchronisation 163

Tadpole 18, 20, 21, 29
Tandanus tandanus 137, 139
Taoism 18
  dichotomy 41
  dualism 8
Tench 131
Tetraselmis sp. 54
Theragra chalcogramma 225
Threshold 8, 9, 23, 41, 51
  ontogenetic 106, 158, 163
  plasticity 154, 158
  saltatory 8, 9, 23
  stabilized 17
Thymallus thymallus 117, 261
Thyroid 53, 61, 236
  follicle 56, 58, 60, 61, 63
  hormone 54, 59, 233, 237
Thyroxine 53–55, 58, 62, 63, 233
  exogenous 53, 54, 56, 57, 59, 62
Tilapia rendalli 32
Tinca tinca 131
Tobi koi 97
Top-down force 176
Trachipterus trachypterus 31
Trachurus 249
Transfer process 214
Transition 10, 41
  interval 144
  juvenile 172
  larva-to-juvenile 7, 10, 93, 99, 105, 159, 164, 169, 172
  morphological 25, 41
  period 262
  size range 186–189, 193
Transport, longitudinal 142
Transversal movement 259
Trichonotus 21
Trichoptera 192, 210
Triggering mechanism 231
Trout 194, 253, 254
  brown 131, 255, 256, 261
  rainbow 131, 238, 239, 261
Trout cod 133, 137, 139

Urostylar torsion 218
Urostyle 68, 70, 74, 75, 154
Uterine cannibalism 25
  histotroph 17
  milk 25

Vascular system 71
Vicentia 89
  conspersa 90
Vitelline
  circulation 85
  membrane 20
  sac 86
Vitellogenesis 22, 24
Vitreous humour 244
Viviparity 17, 247
  facultative 41, 48
  obligate 41

Walleye 173, 177
  pollock 225
Wentworth scale 118, 185
Window of opportunity 11, 53, 61–63
Wolffish 41–43, 46–51

Yakushima Islands 249
Yellow tail 8, 231–249
Yolk 17, 20, 25, 30, 32, 84, 85, 87, 88
  density 21, 24, 25, 26, 32, 91
  diameter 85, 86
  size 22, 28
  volume 23, 29
Yolksac 42, 46, 49, 67, 69–72, 75, 79, 89, 146, 218, 253

Zander 174
Zoarces viviparus 41, 42
Zoarcidae 42
Zoarcoidei 42
Zona radiata 20, 22, 50
Zooplanktivory 96, 97
Zygote 9, 25, 28

GPSR Compliance

The European Union's (EU) General Product Safety Regulation (GPSR) is a set of rules that requires consumer products to be safe and our obligations to ensure this.

If you have any concerns about our products, you can contact us on ProductSafety@springernature.com

In case Publisher is established outside the EU, the EU authorized representative is:

Springer Nature Customer Service Center GmbH
Europaplatz 3
69115 Heidelberg, Germany

FSC
www.fsc.org
MIX
Papier | Fördert gute Waldnutzung
FSC® C083411

Kolibri 360 GmbH
Ferdinand-Jühlke-Straße 7
99095 Erfurt, Deutschland
produktsicherheit@kolibri360.de